DEVELOPMENTS IN SEDIMENTOLOGY 13

SILURIAN REEFS OF GOTLAND

TYPOLOGY, PALAEOECOLOGY AND STRATIGRAPHICAL IMPLICATIONS

FURTHER TITLES IN THIS SERIES

1. *L. M. J. U. VAN STRAATEN, Editor*
DELTAIC AND SHALLOW MARINE DEPOSITS

2. *G. C. AMSTUTZ, Editor*
SEDIMENTOLOGY AND ORE GENESIS

3. *A. H. BOUMA and A. BROUWER, Editors*
TURBIDITES

4. *F. G. TICKELL*
THE TECHNIQUES OF SEDIMENTARY MINERALOGY

5. *J. C. INGLE Jr.*
THE MOVEMENT OF BEACH SAND

6. *L. VAN DER PLAS Jr.*
THE IDENTIFICATION OF DETRITAL FELDSPARS

7. *S. DŻUŁYŃSKI and E. K. WALTON*
SEDIMENTARY FEATURES OF FLYSCH AND GREYWACKES

8. *G. LARSEN and G. V. CHILINGAR, Editors*
DIAGENESIS IN SEDIMENTS

9. *G. V. CHILINGAR, H. J. BISSELL and R. W. FAIRBRIDGE, Editors*
CARBONATE ROCKS

10. *P. McL. D. DUFF, A. HALLAM and E. K. WALTON*
CYCLIC SEDIMENTATION

11. *C. C. REEVES Jr.*
INTRODUCTION TO PALEOLIMNOLOGY

12. *R. G. C. BATHURST*
CARBONATE SEDIMENTS AND THEIR DIAGENESIS

14. *K. W. GLENNIE*
DESERT SEDIMENTARY ENVIRONMENTS

DEVELOPMENTS IN SEDIMENTOLOGY 13

SILURIAN REEFS OF GOTLAND

BY

A.A. MANTEN

Utrecht (The Netherlands)

ELSEVIER PUBLISHING COMPANY Amsterdam London New York 1971

ELSEVIER PUBLISHING COMPANY
335 JAN VAN GALENSTRAAT, P.O. BOX 211, AMSTERDAM, THE NETHERLANDS

ELSEVIER PUBLISHING CO. LTD.
BARKING, ESSEX, ENGLAND

AMERICAN ELSEVIER PUBLISHING COMPANY, INC.
52 VANDERBILT AVENUE, NEW YORK, NEW YORK 10017

LIBRARY OF CONGRESS CARD NUMBER: 68-12477
ISBN 0-444-40706-5

WITH 230 ILLUSTRATIONS, 24 TABLES AND 2 MAP ENCLOSURES

PRINTED IN THE NETHERLANDS

PREFACE

It was the late Prof. Dr. M.G. Rutten who focussed my attention on the intriguing geological problems which are envoked by the Silurian reefs of the island of Gotland. For several years he was an amiable advisor, supervisor, critic, and much more, of my Gotland work. I very much regret that he has not lived to see the appearance of this book. I remember him with great gratitude.

The interest shown in my work by several Swedish geologists over the years is gratefully acknowledged. The inhabitants of Gotland were always so hospitable and helpful that I count the Gotland years among the happiest of my life. By mentioning in particular the Rev. and Mrs. Joh. Siltberg of Öja I wish to thank them all.

A .A.M.

CONTENTS

Chapter I

INTRODUCTION

THE SCOPE OF THIS BOOK

The Swedish island of Gotland, in the Baltic, has long attracted the attention of many geologists. Of these, several have visited the island, and more than a few have contributed to the extensive literature concerning its Middle Palaeozoic sediments and fossils. A large number of museums and institutions contain beautiful petrifactions derived from Gotland.

Nevertheless it took a long time before a more or less definite stratigraphy of this island could be established. Although the stratigraphical subdivision, proposed by Hede, has been rather generally accepted since 1921, a few dissentient views were still subsequently recorded (Wedekind and Tripp, 1930; Jux, 1957).

These stratigraphical difficulties stem from the variable genesis of the sediments involved. The stratigraphical column of the Middle Palaeozoic of Gotland shows an alternation of sediments, typical of deposition in a shallow sea, bordered by a base-levelled continent. Small-sized epeirogenetic movements of the sea floor resulted in very marked differences in the facies of the sediments (marlstone, limestone, oolite, sandstone). Moreover, in several localities in Gotland, the succession of sediments which is normal for an epicontinental basin, is overshadowed by the occurrence of reefs. These reefs formed protruding elevations on the sea bottom and influenced the process of sedimentation. This caused a number of local differences in the facies of the deposits, even within a certain bathymetrical zone. Besides, each of the several facies may contain a fauna largely typical of the special environment, more so than of the time of deposition, making stratigraphical work even more difficult.

It is clear from these short remarks, that for a good understanding of the sedimentary succession of Gotland, it is highly important that the distribution of all the different kinds of sediments be carefully mapped and their faunal contents intensively studied. This has been done especially by Hede. But this alone is not sufficient. An attempt should also be made to deduce the mode of formation of the various sediments, both stratified and unstratified, and to study their interrelationships.

In 1955 it appeared to Professor M.G. Rutten of the State University in Utrecht (The Netherlands), that in this respect much work still needed to be done. Until then, the Middle Palaeozoic reefs especially had not received the attention they deserved. Upon his advice five of his students, the present author included, spent a summer of field work in Gotland during 1956. Their results have been briefly reported by Rutten (1958). From 1957 onwards, the present author has continued studies on the reef formations in Gotland.

Attention has been paid to the typology and constitution of the reefs, and more especially to their genesis and their relation to the stratified Middle Palaeozoic deposits.

This book is a survey of the observations and ideas gathered during these years, combined with a review of the more important data contained in earlier literature about the stratigraphy and reefs of Gotland. After a few introductory chapters about the Baltic area in general, the development of the present stratigraphical subdivision of the Middle Palaeozoic of Gotland, and some tectonic and pseudo-tectonic phenomena, six chapters will be devoted to the organic reefs found on the island. This will be followed by an extensive chapter dealing with the stratigraphy of the island, giving full attention not only to the stratified but also to the unstratified sediments. A subsequent chapter will deal with a number of palaeoecological observations on Gotlandian fossils. Some comparisons of the reefs of Gotland with reefs in a few other areas, and a short retrospect conclude the book.

GOTLAND

Gotland, off the east coast of southern Sweden, is situated geographically between 18° and 19° east and 57° and 58° north. The total surface amounts to 3160 km^2, including a number of minor islands surrounding the main island. The latter has an elongated, rhomboidal shape, with a length of about 117 km and a breadth of about 45 km.

Gotland is built up by a Middle Palaeozoic table-land, with an average height of 20–30 m, but with some higher hills (up to 68 m) in the inland and with steep cliffs along part of the coast (up to 77 m high). Large parts of the land are covered with Pleistocene boulder clay, minor parts with Holocene sand and gravel, and some lower parts with peat.

The Middle Palaeozoic strata consist predominantly of limestones and marlstones. As a result Gotland is devoid of rivers. There are a few brooks that usually carry water only part of the year.

The very permeable limestone areas of Gotland often have a scattered vegetation of pines and junipers, or are more or less naked. These areas are called "alvar". They have a rather unique flora and vegetation. The slightly lower parts are inundated in winter and give rise to lake vegetation, which at its periphery, gradually passes into xerophilous vegetation. In summer the whole alvar can be completely dry. This alternation in environmental conditions causes great annual differences in the occurrence of many species. Frequent alternation of temperatures below and above zero, in autumn and spring, causes frost upheaval which can eradicate even deep-rooted perennials. The introduction by man of certain factors, in particular the grazing of sheep, has also affected the flora and vegetation of the limestone areas. During the last few decades, however, this has diminished or ceased. As a result, vegetation is again spreading over those parts which are covered by moraine or gravel, or are criss-crossed by fissures. Herb-rich pine vegetation is already found over several lower limestone areas, and narrow strips of vegetation indicate the pattern of fissures in the higher areas. On the marlstones there were originally forests of deciduous trees (birch, ash, alder, lime, hazel, oak), but these areas have been extensively cleared for arable and pasture lands.

Rainfall is slight, and the growing of grapes, mulberries, peaches and apricots points to a climate sunnier than that of other parts of Sweden. The occurrence of several orchids, some of which have mainly a south European distribution (Rosvall and Pettersson, 1951), is also a climatic indicator. In general, however, the flora of Gotland is poorer in southern species than that of the island of Öland, southwest of Gotland and much closer to the Swedish mainland. This may have been caused by the fact that an immigration of the Late Glacial flora of southern Sweden was easier in Öland, after the ice cover and Baltic water had withdrawn from there. Gotland was at all times separated from the mainland by a rather broad area of water.

The capital of Gotland is Visby, an old and charming little town with its medieval town wall still almost complete. The record of the habitation of the island however, goes back to the later Stone Age. A large number of prehistoric finds have made it an "El Dorado" for archaeologists.

During Antiquity and the Middle Ages, Gotland was an important commercial centre. Many finds of Roman, Byzantian and Arabic coins have been made. Gotland's wealth in medieval times led to the flowering of architecture and local art. About a hundred highly interesting churches dating from this time, still remain. The cruel plundering of the island by the Danish king Valdemar, in 1361, ended its prosperity. Gotland became a dilapidated island, and was alternatedly occupied by the Danes and the Swedes, until, in 1645, it became definitely Swedish.

At the end of the last century Gotland was still very backward. Since then, however, the discovery of the island by an increasing number of tourists has led to new prosperity.

For more information about the interesting prehistoric age and the old and medieval history of Gotland, see Noreen et al. (1959) and further: Lindblom and Svahnström (1959), Lundquist (1940), Manten (1959, 1960d, 1961b), Roosval (1950, 1952) and Söderberg (1948).

In addition to its growing significance as a tourist attraction, Gotland has also regained its military importance in recent times. Military establishments are to be found all over the island. For this reason the northeasternmost part of the island, east of the line Kappelshamnviken–Hydeviken[1], is forbidden to non-Swedes. Notwithstanding all the efforts that were made and supported by several Swedish authorities, it was impossible for the present author to receive permission to include this part of the island in his studies also.

Because Sweden has a state church, the civil and religious units into which the country is subdivided are generally identical. This implies that the word *parish* is used in Sweden in a civil and geographical sense also. Outside its capital, Visby, Gotland is traditionally divided into 92 parishes. At present, most of these are combined to form larger units, but the ancient parish boundaries are still recognized in several respects. In this book, the approximate position of various localities will, therefore, often be indicated by stating the parish in which they are situated.

---

[1]A short note should be inserted about the use of the article in the Swedish language. The indefinite article precedes the noun; thus: en vik = a bay, and et berg = a mountain. The definite article, however, is generally attached at the end of the noun; thus: viken = the bay, and berget = the mountain. Because of this, many Swedish geographical terms often include the definite article (e.g., Kappelshamnviken, Galgberget). In such cases, the English article will be omitted in this book. The reader should also note that Västerberg and Västerberget, for example, indicate the same locality.

THE MORPHOLOGY OF GOTLAND

Although an extensive discussion of the morphology of Gotland is outside the scope of this book, a few general remarks are in order.

The relief of Gotland is almost completely determined by the geological constitution of the island. The limestone complexes are always topographically higher than the marlstone areas, due to more resistance against erosion. A glance at the geological map (Fig.11) shows that in this way Gotland is divided into three higher parts, separated by two lower marlstone areas. This morphology is developed by the action of both glacial erosion and abrasion by the post-Glacial sea.

Within the limestones, the reef limestones are again more resistant than the stratified ones. Most inland hills consist mainly of fossil bioherms. In part, on the side of the elevation that has not been attacked by the post-Glacial sea, they may contain some stratified sediment. This greater

Fig.1. Coastal cliff about 0.5 km northwest of Axelsro, seen from the southwest. The cliff is built up by the Visby Beds. During on-shore winds, waves attack the lower part, causing undercutting. Protruding is an Upper Visby reef.

Fig.2. Detail of the uppermost part of the northeastern face of the Högklint, south of Visby, with a protruding nose of reef limestone. Between Högklint and Axelsro no large reef-limestone masses occur. This corresponds with a landward curve of the shore line and an absence of the main Högklint cliff. Along the sea shore, however, a low cliff, only a couple of metres high, of Visby rocks is exposed.

resistance of the reef limestones is also apparent in the present coast. A good example is presented by the southern peninsula of Gotland where the highest part of the west coast is formed by the reef-limestone masses of Hoburgen and Hallbjäns. Since at Hoburgen the bioherms were not continuous, but were separated by stratified parts, the hill fell apart into four hillocks upon the erosion of most of the stratified sediments. Also along the northwest coast of Gotland it can be observed that the cliff, built up by the Högklint Beds, bends inwards towards the land where no bioherms are present.

The resistance of the bioherms against marine erosion is also clearly demonstrated by the raukar fields at some places along the east coast. For instance, the raukar areas at Holmhällar and Hammarshagahällar in the south, project seawards, while between the two there is a bay, called Sqvalpvik.

In the remarks about the tectonics (Chapter IV) it will be pointed out that the occurrence of cliffs along the northwest coast is mainly caused by the southeastward dip of the strata. The height of these cliffs has been influenced by both the character of the exposed sediments and the post-Glacial rise of the land (cf. Fig.1,2). Steep cliffs are found there especially, where the lowermost parts are built up by marlstone. Marine, and in some cases atmospheric erosion causes undercutting and fall of the higher parts.

The dip of the strata not only explains the rhomboidal shape of the island as a whole, but also the orientation of several bays, as Burgsviken, Gansviken (Grötlingbo Parish) and Lauviken (see further Chapter IV).

EXPOSURES

Exposures of Middle Palaeozoic rocks are found mainly in ancient and recent cliffs. Of these the ancient ones have been formed by the *Ancylus* lake and the *Littorina* sea. After the retreat of the sea, weathering attacked the sediments and crustaceous lichens now cover several of the cliffs. Nevertheless, many of these cliffs have provided very useful information. Occasional cliff falls and quarries in the walls have led locally to fresh exposures.

Examples of *Ancylus* cliffs are, among others, Galgberget (Visby), Klinteberget, Fröjelklint, Torsburgen (the highest inland hill of the island), Gannberget (Östergarn), Klinteklint (Gammelgarn), Petsarveklint and Kaupungsklint (Ardre), Lindeklint, Klev (Sundre). *Littorina* cliffs are, among others, part of Högklint, Korpklint (Snäckgärdsbaden), Brissund klintar, Lännaberget and Solklint (Slite), Bogeklint, Grogarnsberget (Östergarn), Hoburgen.

The present coastal cliffs also contain by far the best exposures, many of which are of great importance. This is especially true of the northwest coast, from Nyrevsudde to Kappelshamnsviken. Except for the three Upper Visby reefs at Snäckgärdsbaden Hotel, almost all the reefs that can be studied in this stratigraphical unit occur in these cliff walls. The majority of Högklint reefs occur as well in the present coastal cliff along the northwest coast. Recent cliffs also expose many important details of the reefs of Karlsöarna.

The bizarre raukar fields on the southeastern coast at, e.g., Fågelhammar, Ljugarn and Holmhällar have also been formed by the action of the present sea. Their origin will be more fully discussed in Chapter VIII.

In connection with lime burning in earlier years, several quarries were opened in the unstratified limestones. At present most of these have disappeared again. Nowadays quarries are not of much use for the study of the Gotland reefs proper. Stratified sediments can be observed in several quarries made for other purposes. These expose limestones (large quarries, e.g., in the north of the Graunsklint, Lärbro; in File Haidar; in the Gannberg, Östergarn; and in the Klinteberg), marlstones (the huge quarries of the Slite cement factory) and sandstones (grindstone industry at Burgsvik). Small, often private quarries, used only occasionally when a farmer is in need of building stones, are moreover encountered in many places in Gotland.

Chapter II

PALAEOZOIC GEOLOGY OF THE BALTIC AREA OUTSIDE GOTLAND

THE PALAEOZOIC BALTIC BASIN

Murchison (1846) was the first to state that the Silurian strata of Gotland form part of one large Baltic basin, the oldest deposits of which occur on the mainlands of Sweden and Russia, whereas the Upper Silurian sediments are found closer to the centre of the basin on the main land and on the islands of Gotland and Saaremaa (Oesel). This view, also adopted by Schmidt (1890, 1891) and Dames (1890), has since been proved entirely valid (cf. Fig.3,4).

This Silurian transgression over the Baltic shield was preceded by a number of earlier ones. The first of these took place during the Early Cambrian. During that time the weathered surface of the Fennoscandian crystalline basement was inundated by a shallow sea, in which sand and gravel were sedimentated close to the shore. Part of these near-shore deposits can still be studied on the Swedish mainland, in a narrow belt (about 15 km wide) of sandstones and sandstone conglomerates, along the east coast between the island of Runnö, at the height of Paskallavik in the north, and close to Karlskrona in the south. Much of it, however, is covered with Quaternary deposits. To the west of this sandstone belt the crystalline basement is exposed.

Towards the east the Lower Cambrian deposits can be pursued. They form the floor of Kalmarsont, the strait between the mainland and the island of Öland. The layers here show a very slight dip eastwards (varying between hardly recognizable and about 2°). Cambrian sandstones in westernmost Öland are younger than those at Kalmar, on the mainland. Crossing Öland from west to east, Middle and Upper Cambrian and Ordovician deposits are encountered. As will be discussed in the next section of this chapter, Cambrian and Ordovician sedimentation in the area of Öland was not continuous. Still further eastwards, in Gotland, deposits of Silurian (Gotlandian) age are found. Here the layers dip very gently (on the average about 0°30') towards the southeast. They comprise the Llandoverian, Wenlockian, Ludlowian, and perhaps also part of the Downtonian, of the English standard profile (see also Chapter III, pp.42–44).

On the islands of Hiiumaa (Dagö) and Saaremaa (Oesel) and in the east Baltic mainland, the layers generally dip southwards. From north to south the same succession as is found in Sweden is encountered, from crystalline basement via Cambrian and Ordovician to Silurian and south of this also Devonian (Fig.3). Of the Silurian, Llandoverian and Wenlockian are encountered in southern Hiiumaa, northern Saaremaa and the west coast of Esthonia (between Hapsal and Pernau) and the Ludlowian is found in southern Saaremaa. On the floor of the Baltic the stratigraphical boundaries turn from

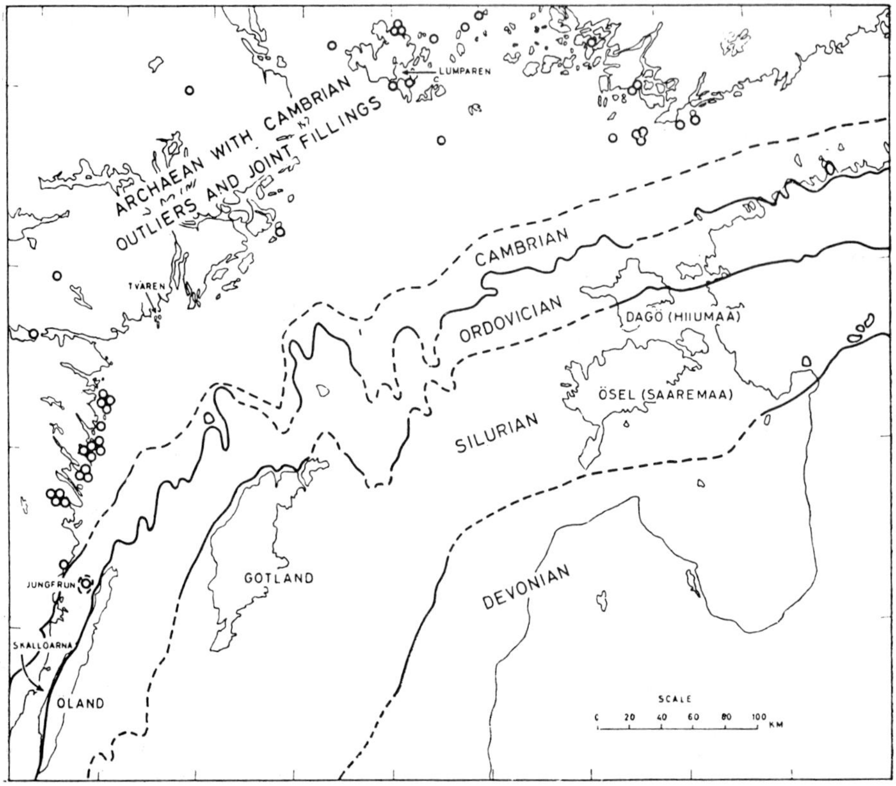

Fig.3. Outline map of the Baltic area, showing the submarine distribution of Palaeozoic rocks at the floor of the Baltic Sea. The boundaries are based on the extension of submarine escarpments (clints [1]) and their connections with the supramarine clints of Sweden and Esthonia. Dashed lines indicate those parts of the boundaries without morphological features, and they are, therefore, less well established. The rings indicate fissure fillings ("clastic dikes") in the strongly-jointed Archaean rocks. (After Martinsson, 1958; reproduced with kind permission of the author).

---

an about south–north direction (Öland) to an about west–east direction (Esthonia); this can be well observed when looking at the coastal and submarine clint[1] complexes of the Palaeozoic (Martinsson, 1958).

---

[1]According to Martinsson (1958, p.15) *clint* is originally a Danish and Swedish word, used to indicate an escarpment, particularly in sedimentary rocks, without pointing out its morphogenesis. Usually it comprises marine abrasion cliffs, and fluviatile erosion scarps in Palaeozoic and Cretaceous rocks. The word was already adopted in medieval English. Unfortunately in the English language "clint" is also used synonymously with *grike*, signifying a type of hollow formed by karst weathering (Martinsson, 1958). In this book the word "clint" will be used in its original Scandinavian sense.

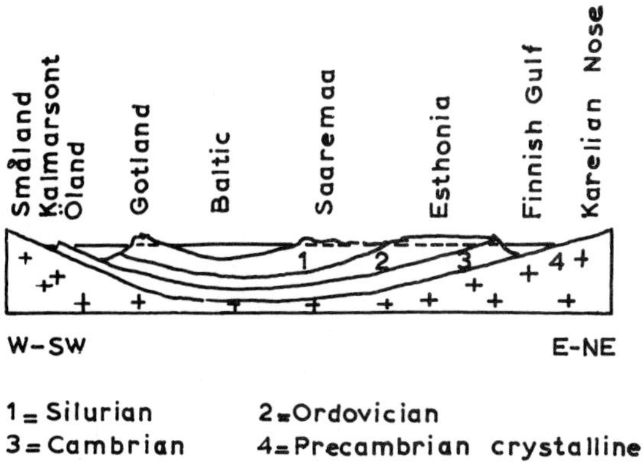

Fig.4. Schematized section through the Baltic, vertical scale about 100 × exaggerated.

From a schematized section through the Baltic area from Kalmar to the Karelian Nose, it becomes clear that the Baltic is enclosed as a saucer-like depression in the Fennoscandian crystalline basement. The successive systems overlie each other as a pile of plates (Fig.4).

THE STRATIGRAPHY OF ÖLAND

Before starting with a discussion on the Middle Palaeozoic sediments of Gotland, it may be valuable to briefly describe the older deposits first, in order to present some idea of the previous history of sedimentation in the Baltic basin. In the first place, these older deposits are represented by the Cambrian and Ordovician strata of the neighbouring island, Öland (Table I, Fig.5).

Sandstones of the Lower Cambrian, or *Holmia* Series, have been observed in only one outcrop on the island of Öland, but have also been encountered in a few borings. These sandstones indicated that the Lower Cambrian sea had transgressed over an uneven morphology. Quartzite monadnocks gave rise to small islands that drowned much later in the Lower Cambrian sea and consequently are covered with a much thinner sedimentary complex than the surrounding areas. Thus of two borings, a little less than 10 km apart, in one were found 78 m of Lower Cambrian sandstones, in the other only 2 m (Westergård, 1936). This has been interpreted as evidence that the Lower Cambrian sea presented a picture comparable with the Recent skerry coast off Stockholm. According to the present author this conclusion is not yet warranted. Much more data would be necessary before it could be decided that no other morphological features might have caused these sedimentary differences in thickness.

The Middle Cambrian strata are known as the *Paradoxides* Series. They are developed as conglomerates, sandstones, shales and limestones. The Lower *Paradoxides* Series (*P. oelandicus* Stage) starts with a thin conglomerate, covered by grey and greenish shales with lenses of impure limestone; the total thickness near Borgholm was about 57 m (of which the uppermost

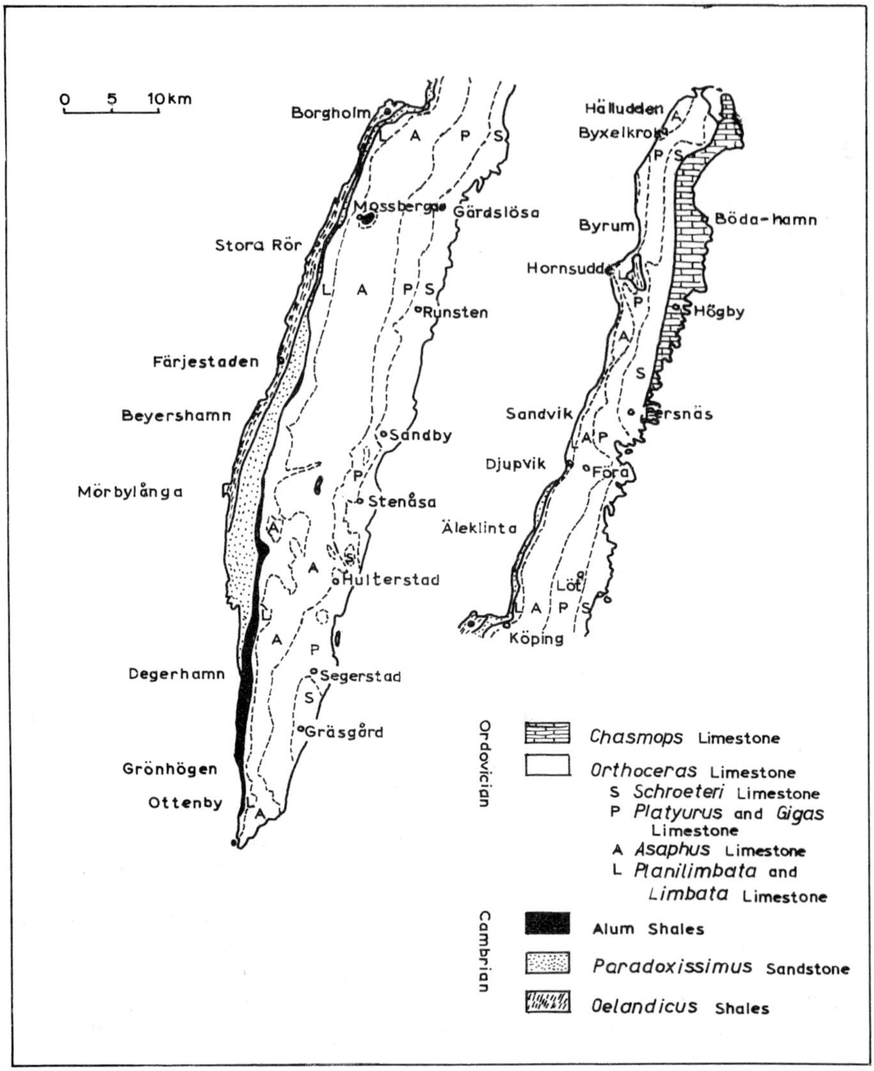

Fig.5. Simplified geological map of Öland, showing the distribution of the Palaeozoic sediments.

---

ca. 18 m have now been removed by erosion), decreasing towards the north and south to less than 25 m (Regnéll, 1948; Westergård, 1936). The middle part of the series (*P. paradoxissimus* Stage) also has a thin conglomerate at its base. It is 10–15 cm thick and contains pebbles derived from the *P. oelandicus* Stage. It is overlaid by a 7–30 m thick complex of thin-bedded

TABLE I

Subdivision of the Palaeozoic of Öland and its correlation with Västergötland and Scania

Stratigraphy		Öland	Västergötland	Scania
Upper Ordovician	*Dalmanitina* Series	not observed	*Dalmanitina* Limestone	*Dalmanitina* Shales and Limestone
	Tretaspis Series	Östersjö Limestone Masur Limestone *Paleoporella* Limestone (all three known from loose slabs only)	Sandstone *Tretaspis* Mudstone Masur Limestone *Tretaspis* Shales	*Tretaspis* Shale and Limestone
Middle Ordovician	*Chasmops* Series	(*Chasmops*) *macrourus* Limestone (loose slabs only)	Shale *Chasmops* Limestone with metabentonite Shale	Zone with *Dicranograptus clingani* Zone with *Amplexograptus vasae* Zone with *Diplograptus molestus*
		Ludibundus Limestone (*Echinosphaerites* Limestone)	*Ludibundus* Limestone	Zone with *Nemagraptus gracilis*
		Crassicauda Limestone	*Crassicauda* Limestone	Zone with *Climacograptus haddingi* (Bronni Limestone)
		Schroeteri Limestone	*Schroeteri* Limestone and Shale	Zone with *Glossograptus hincksi*
Lower Ordovician	*Asaphus* Series (*Orthoceras* Series)	*Platyurus* Limestone	Upper Red Limestone	Upper *Didymograptus* Shale
		(*Megistaspis*) *gigas* Limestone		
		Asaphus raniceps Limestone	Lower Grey Limestone	
		Asaphus expansus Limestone		
		Asaphus lepidurus Limestone		
		(*Megistaspis*) *limbata* Limestone	Lower Red Limestone	*Orthoceras* Limestone
		Upper (*Megistaspis*) *planilimbata* Limestone		Lower *Didymograptus* Shale
	Ceratopyge Series	Lower (*Megistaspis*) *planilimbata* Limestone	*Ceratopyge* Limestone	*Ceratopyge* Limestone
		Ceratopyge Shales and Limestone		
		Dictyonema Shales		
Upper Cambrian	*Olenus* Series			
Middle Cambrian	*Paradoxides* Series	*Paradoxides forchhammeri* Stage		
		Paradoxides paradoxissimus Stage		
		Paradoxides oelandicus Stage		
Lower Cambrian	*Holmia* Series	*Holmia kjerulfi* Stage		
		Schmidtiellus torelli Stage		

sandstones and sandy limestones with thin beds of shale. The upper stage (*P. forchhammeri* Stage) is, on the whole, developed in southern Öland only. There it consists in the extreme south of a thin conglomerate with *Oligomys exporrecta*, overlaid by a bed of alum shale and bituminous limestone, with a thickness of 0.5 m. Elsewhere, it is exclusively represented by the *Oligomys exporrecta* conglomerate that thins out towards the north and can be traced in the northern part of the island in scattered localities only. There it contains an association of fossils that is characteristic partly of the upper *P. forchhammeri* Stage and partly of the Upper Cambrian and even the *Dictyonema* Shales of the lowermost Ordovician. This indicates that conglomerate formation continued into the basal part of the Lower Ordovician (Westergård, 1946; Thorslund, 1960).

The Upper Cambrian or *Olenus* Series, is only developed in its normal succession of alum shales with some limestones in the southern part of Öland; it is incomplete even there. The maximum thickness, established in a boring in the south of the island, is 13.2 m, built up by alum shales, that contain at various levels nodules, balls, or beds of bituminous limestone (anthraconite or stinkstone) (Hadding, 1958c). Towards the north it is to a greater or lesser extent replaced by conglomerate horizons, finally wedging out completely (Westergård, 1922, 1944, 1947).

The Lower Ordovician is divided into two series, the *Ceratopyge* Series and the *Asaphus* (or *Orthoceras*) Series. The lower series, resting upon the Upper Cambrian conglomerate mentioned above, consists of shales with a maximum thickness of 78 m in the south of Öland, overlaid by about 3 m of limestones, of which the lower 3/4 part in the south is replaced by shales. The *Asaphus* Series is a complex of limestones, coloured red or grey, and about 32 m thick.

Limestones also build up the Middle Ordovician *Chasmops* Series. These limestones are the youngest Palaeozoic rocks exposed as solid rocks on the island. The top of this series and the *Tretaspis* Series (lower Upper Ordovician) only occur as loose fragments, whereas all younger Palaeozoic sediments are lacking completely.

For further information about the Lower Palaeozoic of Öland the reader is referred to reviews by Regnéll (1948) and Manten (1960a,b) and to the more specialized publications cited in these papers or given in the list of references at the end of this book (e.g., Bohlin, 1949, 1955; Hessland, 1953; Jaanusson, 1955, 1957; Jaanusson and Strachan, 1954; M. Lindström, 1963; Regnéll, 1940, 1942; Waern, 1949; Westergård, 1929).

THE LOWER PALAEOZOIC OF GOTSKA SANDÖN

North of Gotland a deep boring was carried out on the island of Gotska Sandön, 40 km north of Fårö (Thorslund, 1958). In addition to that which has been said earlier in this chapter about the Lower Palaeozoic south of Gotland, and in comparison to that which borings have revealed about the basement of

---

Fig.6. Diagram showing the lithology of the Cambrian, Ordovician and Lower Silurian underneath Gotska Sandön, a small island, about 40 km north of Gotland. (Drawn after data from Thorslund, 1958.)

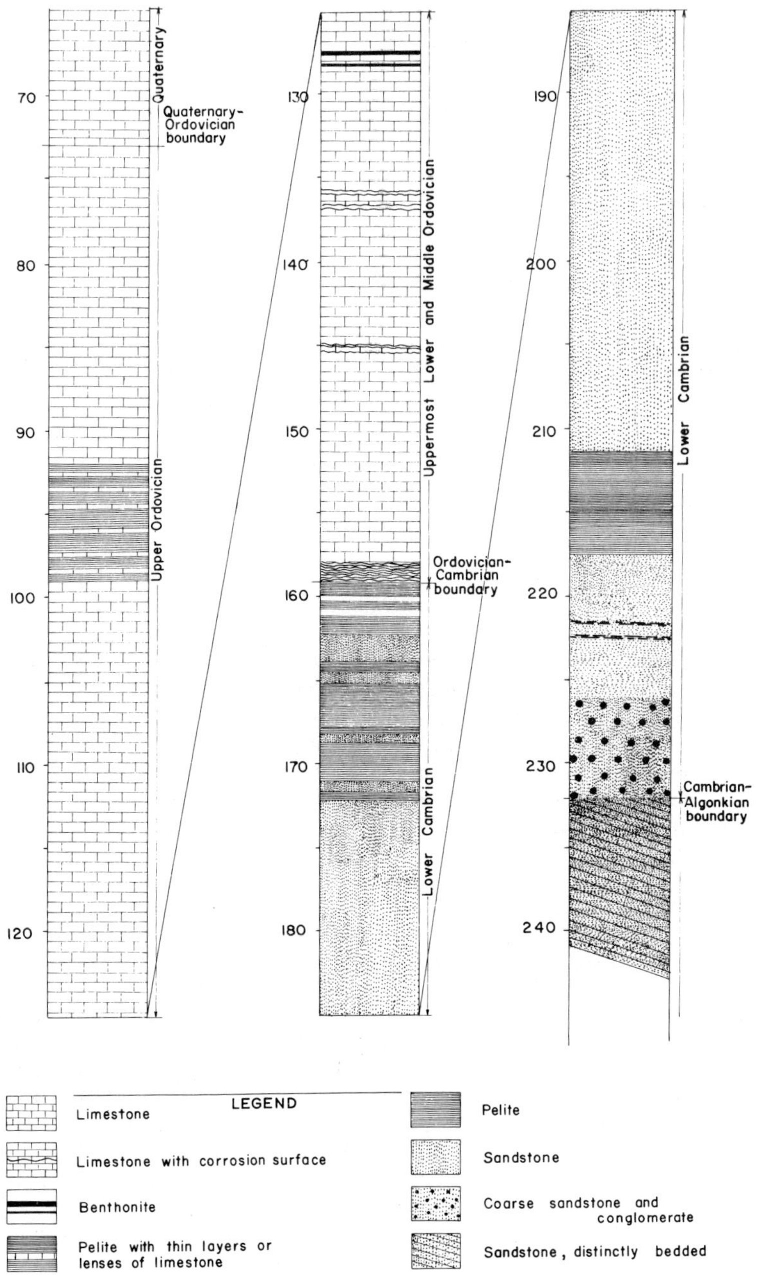
70
80
90
100
110
120
Quaternary
Quaternary-Ordovician boundary
Upper Ordovician
130
140
150
160
170
180
Uppermost Lower and Middle Ordovician
Ordovician-Cambrian boundary
Lower Cambrian
190
200
210
220
230
240
Lower Cambrian
Cambrian-Algonkian boundary
LEGEND
Limestone
Limestone with corrosion surface
Benthonite
Pelite with thin layers or lenses of limestone
Pelite
Sandstone
Coarse sandstone and conglomerate
Sandstone, distinctly bedded

Gotland (Chapter III, pp.29–32), some data about the correlating deposits of Gotska Sandön will be useful (see Fig.6).

The boring was made near Hamnudden, in the southwest of Gotska Sandön. It reached a depth of 241 m below the surface, or 239 m below sea level.

The lowermost 9 m of the core consist of a generally coarse-grained quartzitic sandstone, distinctly bedded, with a dip of 10–15°. Its age is presumably Jotnian, perhaps younger (Thorslund, 1958, p.193). This sandstone has been described in detail by Gorbatschev (1962). The underlying crystalline was not reached.

The core section between the depth of 232 and 159.5 m is referred to the Cambrian. *Skolithos* Sandstone, common in the borings in Gotland, was not found in Gotska Sandön. At the base of the Cambrian is a loose, coarse-grained sandstone, with conglomerate horizons, containing small-sized gravel. The bulk of the Cambrian is formed by fine-grained sandstone, with intercalated pelite layers, especially in its lower and upper parts. The whole section is correlated with the Lower Cambrian, the thickness of which (72.5 m) is thus comparable with that at Böda in Öland (78 m), but smaller than that under Gotland (130 m at File Haidar, 106 m at Visby). It may be mentioned here that at Tallinn (Esthonia) the Lower Cambrian measures 130 m and at Leningrad 170 m in thickness.

Both the Middle and Upper Cambrian are missing in the Gotska Sandön core.

The Ordovician section, 86.5 m thick, between 159.5 and 73 m in depth, consists almost completely of limestone. Only between 99 and 92 m do pelite layers with thin layers or lenses of impure limestone occur. The oldest Ordovician layers may possibly represent the *Megistaspis gigas* Limestone. Both the Lower and Middle Ordovician layers contain several surfaces of discontinuity. The highest Ordovician strata which were found belong to the *Paleoporella* Limestone. It is overlaid by Quaternary deposits, which attain considerable thickness (73 m).

THE WEST BALTIC AREA DURING POST-GOTLANDIAN TIME

The Silurian stratigraphy of Gotland will be discussed in Chapters III and XI. For some time it was thought that the youngest Palaeozoic sediments of this island belong to the basal Devonian (cf. pp.43–44). From almost the whole of the west Baltic area, no sediments are known to have been deposited during the long stretch of time between the Early Devonian and the Late Quaternary. In the east Baltic area however, sedimentation continued during the Devonian (see the next section of this chapter). It is not known whether the connection with the western sea (Fig.7–9) was still in existence. The possible Devonian sediments that have been described from the Oslo area are of continental character. However, the connection between the Baltic area, east of Oslo, and the western sea may have been situated further south of Oslo, as in fact it was during, e.g., the Early Cambrian. Stronger negative evidence is that no Devonian and younger Palaeozoic sediments are found in Scania or anywhere else in Sweden.

It is difficult to ascertain whether the west Baltic area was above sea level during the entire interval between the Early Devonian and the Quaternary or only during part of it. The latter possibility would imply that younger deposits had been present, but were again eroded. Two facts might point to this possibility. One is the established deposition of Devonian sediments further to the east but still within the limits of the Baltic area. The second is the preservation of uppermost Silurian (probably Downtonian) sediments in Gotland. One would expect that these Middle Palaeozoic deposits would have been removed by subaerial processes, had they been exposed to the atmosphere for the entirety of this long time interval.

Quaternary land ice left great amounts of morainic material. During post-Glacial time much of the land in the Baltic area was inundated. Many fossil beach ridges and marine cliffs are proof of the former lower position of the land relative to sea level, which caused a greater extension of the sea and lakes. Examples of such cliffs in Gotland have already been mentioned in Chapter I.

THE EAST BALTIC PALAEOZOIC

In the east Baltic, sediments comparable to those in Öland and Gotland may be found in Esthonia, where the basement shows a general fall in the southern direction (3–4 m/km on the average). This is similar to the dip of the overlying Lower Palaeozoic deposits. The basement rises again only in the southeastern part of Esthonia. This basement is built up of crystalline rocks of varying composition (gneiss, granite, pegmatite, amphibolite). Its surface is uneven, with an amplitude of some tens of metres. Moreover, this surface is strongly weathered in its uppermost few metres; in many places, kaolin clay is found.

This basement is overlaid by a 150–700 m thick complex of sedimentary rocks belonging to the Cambrian, Ordovician, Silurian and Devonian. Its stratigraphical subdivision is shown in Table III. The Lower Palaeozoic is in its turn covered by Quaternary deposits almost everywhere.

The Cambrian in Esthonia begins with a complex of coarse-grained sediments called the Gdovi Stage or, in older literature, the "Lower Sandstone and Conglomerate". Conglomerates and gravelites directly overlie the basement in some places. Coarse-grained sandstones are dominant in the Lower Gdovi and fine-grained sandstones or aleuroliths in the Upper Gdovi. The overlying Kotlini Stage (Laminarite Clay) is built up of greenish-grey or brownish clays, in which thin layers of sandstone are interbedded. The thin bedding is characteristic; it is further emphasized by the presence of sapropel films on the bedding planes. The Kotlini is well developed in east and southeast Esthonia, but is absent or thin in the west and centre. Gdovi and Kotlini are united to the Valdaic Series or Waldai Series.

Next is the Baltic Series, which opens with the Lomonossovi Stage, or Upper Laminarite Sandstones, built up mainly of coarse- and fine-grained sandstones; these are overlaid by clays of the Lontova Stage (Lower Cambrian Blue Clay, in the strict sense). At the top, the fine-grained sandstone of the Pirita Stage (*Eophyton* Sandstone) is found.

The stratigraphical position of the Tiskre Stage is questionable. During recent years, many geologists have classified it as Middle Cambrian, whereas others favour the upper Lower Cambrian. The Tiskre consists of fine-grained sandstone with generally thin interbedded layers of sandy-clayish material.

TABLE II

Earlier and present subdivisions of the Ordovician in Esthonia

F.B. Schmidt (1881)		K. Orviku, A. Öpik (1927–1940)			Present stratigraphy			
Borkholm-schichten	F_2							
Lyckholm-schichten	F_1	Harju	Saaremõisa	F_I	Upper Ordovician	Kohila	Pirgu	F_{1c}
							Vormsi	F_{1b}
						Oandu	Nabala (Saunja)	F_{1a}
Wesenberg-schichten	E		Rakvere	E			Rakvere	E
Jõhvischichten	D	Viru	Keila–Vasalemma	D_2–D_3	Middle Ordovician	Kurma (Saue)	Vasalemma	D_{III}
							Keila	D_{II}
			Jõhvi	D_1			Jõhvi	D_I
Itfersche Schichten	C_3		Kukruse–Idavere	C_2–C_3			Idavere	C_{III}
Kukersche Schichten	C_2					Purtse	Kukruse	C_{II}
Echinosphaeriten-kalkstein	C_1		Uhaku	$C_{1\delta}$			Uhaku	C_{Ic}
		Tallinn	Lasnamägi	$C_{1\gamma}$			Lasnamägi	C_{Ib}
			Aseri	$C_{1\beta}$			Aseri	C_{Ia}
Vaginaten-kalkstein	B_3		Kunda	B_{III}	Lower Ordovician	Ontika	Kunda	B_{III}
Glauconit-kalkstein	B_2		*Megalaspis* Limestone	B_{II}			Toila (Volchov)	B_{II}
Glauconitsand	B_1	Iru	Glauconite Sandstone	B_I			Leetse	B_I
			Dictyonema Shales	A_3		Iru	Pakerordi	A_{2-3}
			Obolus Sandstone	A_2				

No further Middle Cambrian is present in Esthonia and the Upper Cambrian is completely missing. The thickness of the Esthonian Cambrian is 100–250 m.

The Ordovician in Esthonia is mainly built up of limestones. Only in the lower part do sandstones and slates occur. Table II shows the recent stratigraphy of the Esthonian Ordovician, which is based mainly upon publications of the last two decades, and predominantly upon those published in Russian or Esthonian. It is compared to the well-known earlier stratigraphy of Öpik and others and to the first stratigraphy of that area, which was given by Schmidt. A few other authors have published more or less deviating stratigraphical subdivisions, but these need not be reviewed in this section.

There has been some difference of opinion about the lower boundary of the Ordovician. Schmidt (1858, 1881) posited the base of the Glauconite Sandstone for this. During the first half of this century, almost all stratigraphers also included the *Dictyonema* Shales in the Ordovician. In recent years, there

TABLE III

Stratigraphy of the Palaeozoic in Esthonia

	Series				Local stage names	Index	Thickness (m)
Devonian	Upper			Frasnian	Svinordi	D_3^{so}	1.5
					Šeloni	$D_3^{\check{s}l}$	8.8
					Čudovo (Tsuudovo)	$D_3^{\check{c}d}$	6–12
					Pskovi	D_3^{ps}	12.5–17
					Snetogori	D_3^{sn}	5–55
					Amata	D_3^{am}	10–45
					Gauja	D_3^{gj}	55–65
	Middle			Givetian	Burtnicki	D_2^{br}	82–105
					Aruküla	D_2^{ar}	70–100
					Narva	D_2^{nr}	25–100
					Pärnu	D_2^{pr}	7–72.5
Silurian	Upper			Ludlowian	Ohesaar	K_4	approx. 10
					Kaugatuma	K_3	approx. 15
					Paadla	K_2	approx. 15
					Kaarma	K_1	35–65
	Lower			Wenlockian	Jaagarahu	J_2	approx. 25
					Jaani	J_1	30.5–56.5
				Llandoverian	Adavere	H	27.3–35.4
					Raikküla	G_{III}	30–52
					Tamsalu	G_{II}	6.8–14.4
					Juuru	G_I	8.4–22
					Porkuni	F_2	3–16
Ordovician	Upper	Harjuan	Kohila		Pirgu	F_{1c}	25–57
					Vormsi	F_{1b}	6–20
			Oandu		Nabala	F_{1a}	22–42
					Rakvere	E	8–25
	Middle	Viruan	Kurna		Vasalemma	D_{III}	0.5–10
					Keila	D_{II}	1.5–26.8
					Jõhvi	D_I	5–12.9
					Idavere	C_{III}	0.4–11
			Purtse		Kukruse	C_{II}	4–14
					Uhaku	C_{Ic}	4–16
					Lasnamägi	C_{Ib}	5–13
					Aseri	C_{Ia}	0.1–7
	Lower	Ölandian	Ontika		Kunda	B_{III}	0.1–14
					Toila	B_{II}	0–14
					Leetse	B_I	0.3–4
			Iru		Pakerordi	A_{2-3}	2–21
Cambrian	Middle (?)				Tiskre	A_{1d}	10–30
	Lower	Baltic			Pirita	A_{1c}	10–20
					Lontova	A_{1b2}^{2}	25–85
					Lomonossovi	A_{1b2}^{1}	25–30
		Valdaic			Kotlini	A_{1b1}	2–47
					Gdovi	A_{1a}	15–65

has been a growing tendency to return to Schmidt's idea.

Similar difficulties exist for the upper boundary. In 1944, Jaanusson published the opinion that this boundary belonged between the Vormsi and Pirgu Stages, since several characteristically Silurian genera occur already in the Pirgu. Most Esthonian stratigraphers, however, still do not seem to share his opinion but stick to the boundary as defined earlier by Öpik (1934). This also seems to be true for Jaanusson's (1945) lower and upper boundaries of the Middle Ordovician, which he drew between the Aseri and Lasnamägi and between the Keila and Vasalemma, and even more for Bekker's (1922) earlier definition of the lower boundary of the Middle Ordovician at the base of the Kukruse Stage.

The Lower Ordovician is called Ölandian in Esthonia; it is subdivided into two subseries, the Iru and Ontika.

The Iru comprises one stage, the Pakerordi. This begins with fine-grained sandstone (Ülgase member; Sandstone with *Acrotreta*). Next is the Maardu member which begins with a basal conglomerate that is not developed everywhere. It is overlaid by one or two layers of the *Obolus* Conglomerate, consisting of quartz-sandstone lenses very rich in shells of *Obolus* and their fragments. Over this conglomerate, sandstone with interbedded *Dictyonema* Shale is found. Locally, the Maardu member ends with the so-called Pyrite layer, a nearly 10 cm thick layer of quartz sandstone with *Obolus* fragments cemented with pyrite. Underlying this layer is the well-known Detritus layer, of a few centimetres to 1–5 m thick, consisting of sandstone rich in *Obolus* fragments. The Türissalu member (*Dictyonema* Shales) is only developed in northern Estonia.

The Ontika Subseries has three stages. The oldest is the Leetse Stage or Glauconite Sandstone. Next is the Toila (or Volchow), formerly known as Glauconite Limestone or *Megalaspis* Limestone. The youngest is the Kunda, which is bounded at both its base and top by discontinuities. In the east, the succession consists of oolitic limestone, dolomitic limestone, limestone and oolitic limestone. In the west, a phosphoritic conglomerate, limy sandstone and limestone are present; the latter farther westward passes into sandy limestone.

The Middle Ordovician or Viruan Series is comprised of not less than eight stages. The oldest is the Aseri Stage, which consists mainly of limestones, generally containing ferruginous ooids. The Lasnamägi Stage was originally known as Building Limestone. The Uhaku Stage is represented by thin-bedded marly limestones. Next is the Kukruse Stage, with marly limestones with interbedded bituminous limestone and kukersite (oil shale). The Idavere Stage is built up of limestones with a large number of locally interbedded marlstone layers. About the same sediments are found in the Jõhvi Stage. The Keila Stage shows marly limestone and harder, purer limestone with uneven bedding planes. The youngest is the Vasalemma Stage. It is comprised of three members, which are not represented everywhere. The Vasalemma member consists of coarse-grained cystoid limestone and unstratified, dense and marly reef limestone. The Saku member is of an extremely varied lithological composition with, from the base upwards, coarse-crystalline dolomitic limestone, marly dolomite and dolomitic limestone with interbedded layers of clayey marlstone, and marly limestone. The Oandu member contains marly limestone, marlstone and claystone.

The Upper Ordovician is known in Esthonia as the Harjuan Series. This begins with the characteristic limestone of the Rakvere Stage. This limestone is relatively homogeneous, yellowish in colour, microcrystalline, and shows shelly fracturing; it reminds one of lithographic limestone. The next or Nabala Stage contains similar limestone and dolomitized variations of it. The Vormsi Stage consists of nodular limestone with varying marl content. There is a discontinuity between this and the preceding stage. At the top, the Vormsi limestone passes gradually into the marly nodular limestones of the overlying Pirgu Stage, which often contain interbedded marlstone layers. In the Pirgu Stage, some reefs are also present (cf. Rõõmusoks, 1960, p.67).

The Silurian is exposed in central Esthonia and on the islands. It is mainly represented by limestones, dolomites and marlstones. Most geologists draw the lower boundary at the base of the Porkuni Stage; however, some include the upper part of the Porkuni, which is devoid of fossils, in the next or Juuru Stage, and then draw the boundary below the thus-defined Juuru Stage. The top of the Silurian in Esthonia is the Ohesaar Stage. Younger Silurian sediments are not found there. The discontinuity continues up to the Middle Devonian. In Latvia and Lithuania, younger Silurian deposits are found by drilling.

The Silurian sediments of Esthonia were deposited in a shallow sea during a period of regression. The direction in which the sea withdrew can be established from the outcrops of the various stages of the Esthonian Silurian. Whereas the oldest deposits of the Llandoverian extend up to the town of Mustvee at Lake Čuds, the Upper Silurian is only exposed in the island of Saaremaa and smaller islands in the environment of that island.

The lower three quarters of the Porkuni Stage consist of pure, marly and bituminous limestones, which locally are very fossiliferous. Small reefs occur, built up by corals and Algae. The Upper Porkuni consists of quartz sandstone or sandy limestone overlaid by limestone, and is devoid of fossils.

The Juuru is unquestionably of Llandoverian age. It is represented by marly limestones with many interbedded marlstone layers.

The Tamsalu Stage occurs in a belt from the west coast of Hiiumaa to the vicinity of the town of Mustvee, and is composed mainly of limestones and marly limestones. Five members are distinguishable. Of these, the Ridala member composes the lower beds of the stage, to a thickness of 3.3 m. It is represented by three lithological complexes which, from the base upwards, are (1) *Pentamerus* limestone, containing shelly detritus and limestone pebbles (0.5 m); (2) limestone consisting of shelly detritus and partly of rounded psammitic limestone grains (1.5 m); and (3) thin bedded marly limestone rich in *Aulopora* sp. (1 m). The Hilliste member is 7.7 m thick and also comprises three lithological complexes, which are (4) medium-grained limestone with limestone pebbles and coarse shelly detritus (1.7 m); (5) fine-grained marly limestone with very small shell fragments (1 m); and (6) crinoid limestone, coarse-crystalline or with varying granulation and containing coarse shell detritus (3–5 m). A characteristic of the Hilliste member is the occurrence of relatively small reefs, mainly built up by corals. The Purga member (1.5–10.15 m) is mainly composed of thin-bedded marly limestones which often contain redeposited carbonate particles. The Tammiku member (1.85–10 m) is mainly represented by *Pentamerus* limestone and is very rich in shells and shell fragments of *Pentamerus borealis* Eichwald, together with corals and stromatoporoids. The Kose member consists of marlstones and marly limestones which, except for small shell fragments, are very poor in fossils.

South of the Tamsalu Stage, the Raikküla Stage is exposed. In western Esthonia it is represented by grey limestone, limestone rich in corals and stromatoporoids and thin-bedded limestone. These are overlaid by limestone with interbedded conglomerate layers, coral limestone, light-grey limestone and cavernous dolomite. In eastern Esthonia, irregularly-dolomitized limestone and marly dolomite occur, which are overlaid by cavernous dolomite with corals and stromatoporoids, marlstone, and thin-bedded fossiliferous limestone.

The Adavere Stage represents the uppermost Llandoverian. In the west of the Esthonian mainland, there is a succession comprised of dolomitized marly limestone with *Pentamerus oblongus* Sowerby, marly limestone, marlstone and claystone with *Catazyga furcata* (Sowerby), dolomitized limestone with *Pentamerus oblongus* Sowerby and dolomitized marly limestone with *Pentamerus estonus* Eichwald. Towards the west and south, there is an increase in terrigenous components. In drillings in Saaremaa, the Adavere Stage was found to consist mainly of claystones and marlstones. In the east, there is a succession of light grey dolomite with pentamerids, yellowish coarse-crystalline dolomite and bluish-grey coarse-crystalline dolomite with pentamerids.

The Wenlockian begins with the Jaani Stage, which is exposed as a narrow belt along the northern shore of the islands of Saaremaa and Muhu; in the Esthonian mainland, this belt is broader (up to 15 km), but does not extend farther to the east than the village of Vändra. From here it is covered by the Middle Devonian. The lower part of the stage consists of dolomitized marlstone and clayish marlstone. The content of terrigenous components increases in the westward direction. The upper part shows fossiliferous marly limestone and marlstone, which are dolomitized over large areas.

The Jaagarahu Stage is exposed in the islands of Vaika and Vilsandi, the northern part of the island of Saaremaa, the island of Muhu and the southwest of the Esthonian mainland. The stage consists of limestones, part of which are dolomitized. Three members are distinguishable. In Saaremaa, west of the Triigi Peninsula, the Pangamägi member is present, with light-coloured limestone, crinoid limestone and, to a lesser extent, marly limestone. Reefs and biostromes of corals and stromatoporoids are characteristic. East of the Triigi Peninsula, the Kesselaid member can be found with dolomitized limestone and relatively many reefs, mainly built up by calcareous Algae and bryozoans. In Muhu and in the northeast of Saaremaa, this member is overlaid by the Maasi member, which consists of yellowish, dolomitized limestone locally rich in spongiostromids.

The Ludlowian of Esthonia is only exposed in the island of Saaremaa. Four stages are distinguishable. The oldest of these is the Kaarma Stage. In the west of the island, it consists of thin-bedded limestone which rhythmically alternates with dolomite containing *Eurypterus*, and marl. These are overlaid by light-coloured dolomite. In central and east Saaremaa, the stage is mainly built up of light-yellowish dolomite which is thick-bedded and partly cavernous.

The Paadla Stage is found from the southwest coast, over Kogula, to the southeast coast. In the west of this stage the Atla member is found, which consists of stratified limestone and reefs built up by corals and stromatoporoids. In the central and eastern part of the area where the Paadla is exposed, the Irase member crops out, which consists of stratified limestone. In the environs of Lümanda and Paadla, these members are overlaid by the Kogula member, with thick-bedded, fine-grained limestone containing *Solenopora* sp. and *Ilionia prisca* (Hisinger).

The Kaugatuma Stage is exposed in the south of Saaremaa, including the north of the Sõrve Peninsula. In the lower part of this stage, grey marly limestone and marlstone occur, and in the upper part, grey crinoid limestone and fossiliferous marly limestone are present.

The Ohesaar Stage is only exposed in the southern part of the Sõrve Peninsula and in the skerry Ohesaar and its surroundings. It comprises limestone, marly limestone and marlstone. In the top of the Ohesaar section, sandy limestone is also found.

During the end of the Silurian, the Early Devonian and part of the Middle Devonian, continental conditions prevailed in Esthonia. Denudation occurred with the result that deposits from these periods are missing. On the washed-out surface of the Silurian, Ordovician and even Cambrian and Precambrian (in the drilling at Mõniste), Middle Devonian (Givetian) sediments of continental origin can be found, which were deposited on the extensive alluvial coastal plains (sandstones, claystones, marlstones). These deposits are followed by Upper Devonian sediments. There is no distinct lithological boundary and the sediments of the lower three stages of the Frasnian consist also of material of continental origin. They are overlaid by marine sediments (dolomites, limestones).

The deposits of the Middle Devonian and the lower part of the Upper Devonian have a cyclic character. Individual cycles consist mainly of sands and sandstones in the lower part and mainly of marlstones and claystones, alternating with aleuroliths and sandstones, in the upper part. The fossil content is comprised of primitive plants and armoured fishes, but only a few invertebrates. The upper part of the Upper Devonian, on the other hand, is again rich in invertebrate fossils.

CAMBRIAN AND EARLY ORDOVICIAN PALAEOGEOGRAPHY

Since the areas with Cambro-Silurian sediments in Sweden are relatively small and in most cases situated at considerable distances from each other, it is a difficult task to draw palaeogeographical maps for various times during the Early Palaeozoic. However, such a wealth of data has been collected on the Cambrian and Early Ordovician that such maps can be made for certain stages of these systems (Fig.7–9). They present a general picture of how the Early Cambrian sea reached the Baltic shield through an arm covering northernmost Jutland, Kattegat and southwestern Scania. The southwestern boundary of this arm is uncertain. The Early Cambrian sea inundated vast parts of the Baltic area. The maps also show the culmination of this transgression in the middle of the Early Cambrian. During the Middle Cambrian the sea again withdrew from most of the Baltic area. It still covered vast areas of Norway however, and from there intermittent, smaller transgressions in an eastern direction took place during both the Middle and Late Cambrian. During these, Öland was usually inundated in contrast to

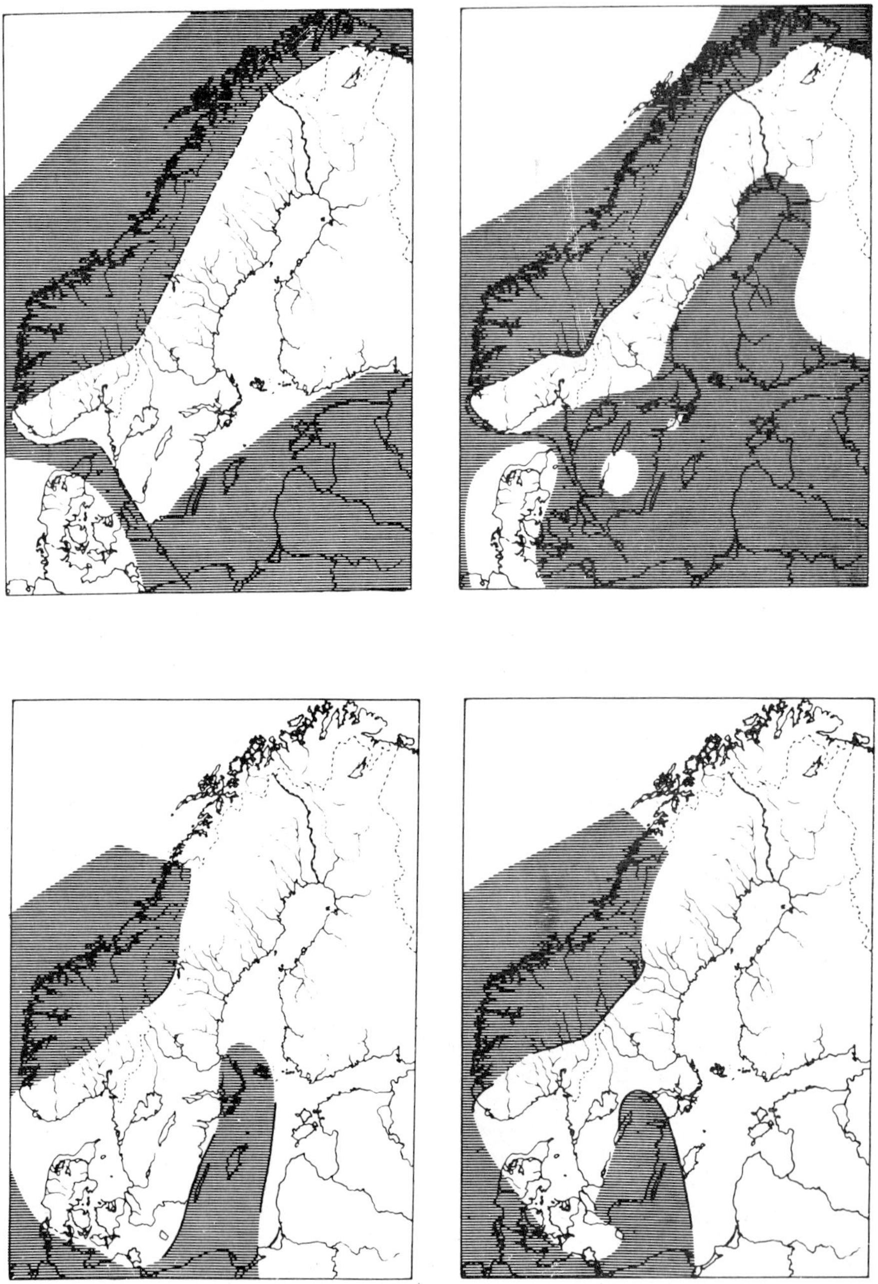

Fig.7. Palaeogeographical map of two phases during the Early Cambrian (above) and two phases during the early Middle Cambrian (below). (After Thorslund, 1960; reproduced with the kind permission of the author and Sveriges Geologiska Undersökning, Stockholm).

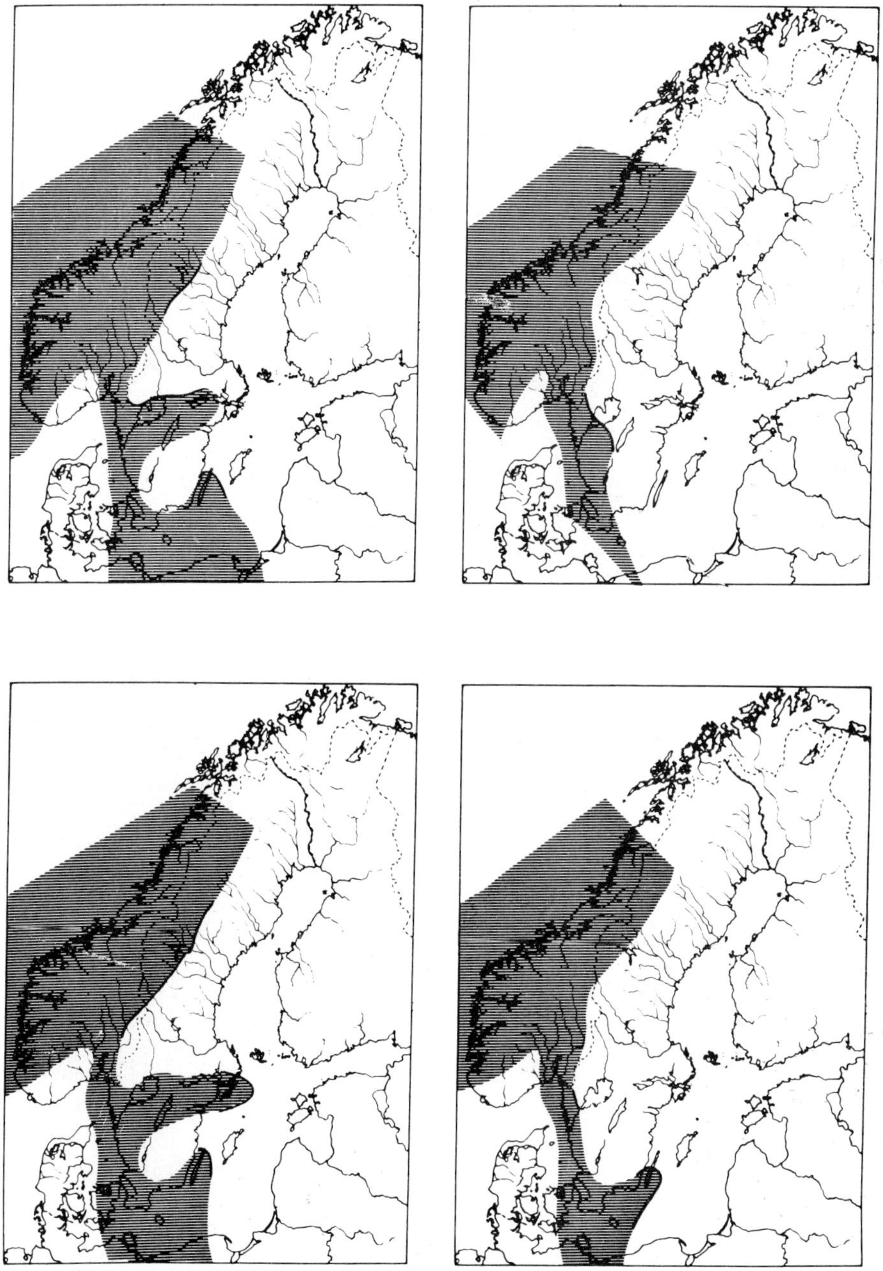

Fig.8. Palaeogeographical maps of an early and a late phase during the middle Middle Cambrian (above) and of two phases during the Late Cambrian (below). (After Thorslund, 1960; reproduced with the kind permission of the author and Sveriges Geologiska Undersökning, Stockholm).

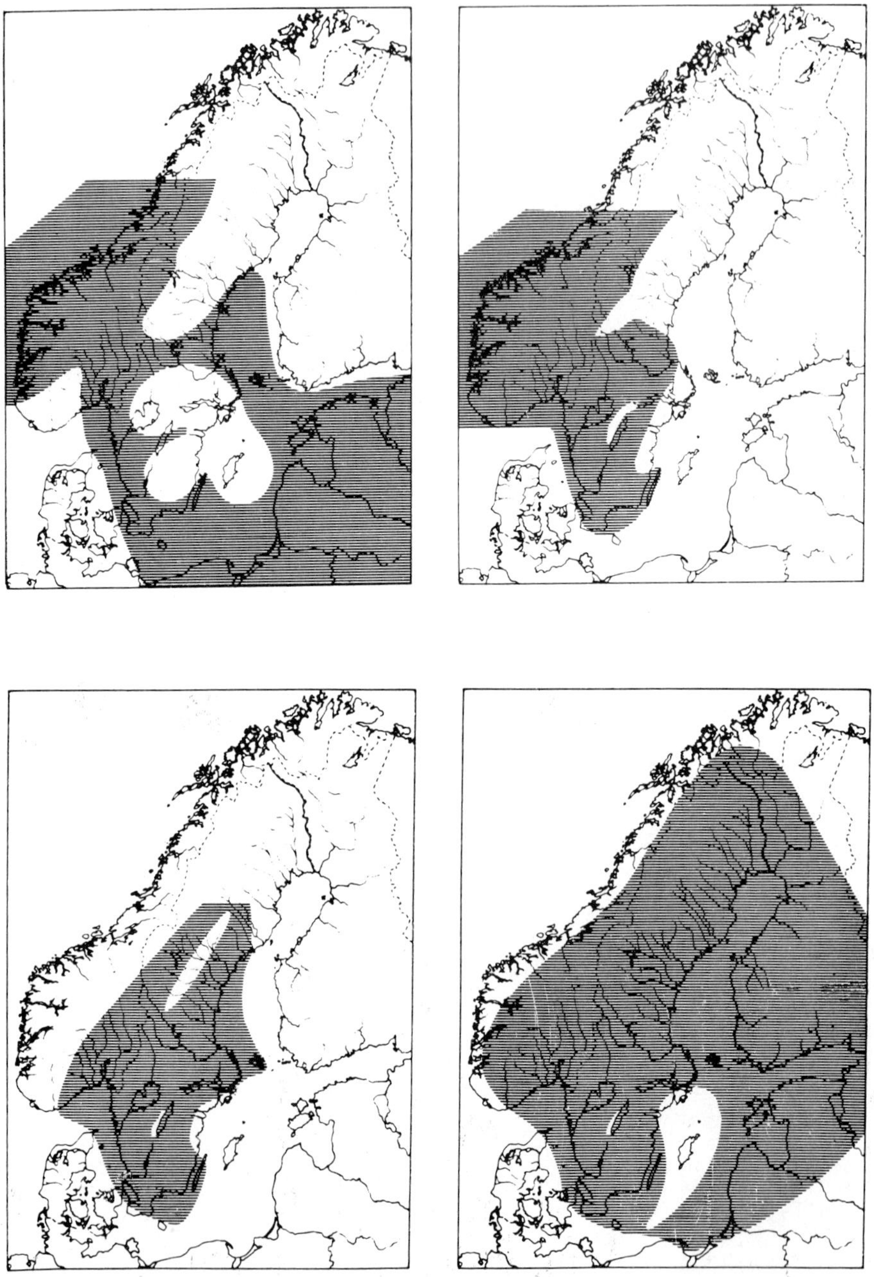

Fig. 9. Palaeogeographical maps of four phases during the Lower Ordovician Epoch: Tremadocian (upper left), Early Arenigian (Hunneberg Age in the Swedish geochronology; upper right, below left) and Middle Arenigian (Billingen Age; below right). (After Thorslund, 1960; reproduced with the kind permission of the author and Sveriges Geologiska Undersökning, Stockholm).

Gotland which generally remained dry. A new large transgression began in the Early Ordovician, beginning with and including the Early Arenigian (Early *Asaphus* Epoch), when the sea expanded more generally over the area of resistance (Rõõmusoks, 1960, p.66; Thorslund, 1960, p.75)[1].

Unfortunately no palaeogeographical maps of the Middle and Late Ordovician and the Silurian have been published up till now. There seems to have been a reduction of the area of the sea covering the Baltic area during the Middle and Late Ordovician Period. Middle Ordovician sediments have been recorded from several regions. Those of Jämtland exhibit a general change of facies, expressed by a diminishing content of limestones and an increasing wealth of shales and sandstones. In Dalecarlia and in the south of the Gulf of Bothnia the so-called Kullsberg Limestone is found, which includes reef-like formations.

Of the Upper Ordovician the *Tretaspis* Series has a larger extension than the overlying *Dalmanitina* Series. The latter is missing in the Gulf of Bothnia area, in Esthonia and in Gotland (Chapter III, p.32). In Dalecarlia there is a second sequence with reef limestone (Boda Limestone) in the Upper Ashgillian. The lithological development of the *Dalmanitina* Series in Västergötland points to deposition in shallow water (Thorslund, 1960, p.83). In Östergötland the *Tretaspis* shales and finely nodular limestones are overlaid by a limestone conglomerate belonging to the *Dalmanitina* Series. There this series otherwise consists of argillaceous limestone with layers of calcareous shale.

SILURIAN PALAEOGEOGRAPHY

For a better understanding of the Silurian of Gotland a few lines about the palaeogeography during that period are still necessary. An attempt will be made here to give a generalized picture of the distribution of the Silurian sediments, as has been recorded until now.

It has been mentioned in one of the previous sections of this chapter that Silurian sediments were deposited in the east of the Baltic, on the islands of Hiiumaa and Saaremaa and in the east Baltic main land.

Silurian sediments are also well-known in southwestern Scania. They start with the black and dark-grey *Rastrites* Shale (*Rastrites* Series). The boundary with the underlying Ordovician is indistinct and has been a matter of discussion (Troedsson, 1936; Waern, 1948). The series is correlated with the English Llandoverian.

The *Rastrites* Series is overlaid by the *Cyrtograptus* Series that is correlated with the two uppermost zones of the Llandoverian, characterized by *Monograptus spiralis*, and *Cyrtograptus lapworthi*, respectively, and with

---

[1]In a personal letter professor Thorslund stressed once more that the maps, reproduced here in Fig.7–9 are only tentative drawings, as all palaeogeographic maps must be. He kindly informed me that the Middle Baltic Island during the Billingen Age very likely extended further to the east and southeast than drawn in Fig.9. The cover of the Early Palaeozoic sea over Denmark is unknown but for the island of Bornholm. However, according to Holmsen (1958) the boring at Ringe (Fyn, Denmark) very likely reached Permian or Triassic arkose which was previously dated as Precambrian. Consequently, it is possible that there are Early Palaeozoic strata below as found in the Oslo region (Thorslund, personal information).

the whole of the Wenlockian. It consists mainly of grey to black shales. Lenses or thin beds of grey or yellowish, dense and hard limestone are common. In its lower parts thin layers of metabentonite are present.

Presumably following upon the *Cyrtograptus* Series is the *Colonus* Shale. It is remarkable however, that in no locality has the *Colonus* Series been seen in direct contact either with underlying or with overlying strata (Regnéll, 1960). The series comprises mainly light-grey, or greenish, bluish, or reddish micaceous shales, which are generally somewhat calcareous. Lenses or intercalated thin beds of grey limestone occur frequently. The *Colonus* Series is correlated with the Lower and Middle Ludlowian.

The youngest beds of Palaeozoic age in Scania belong to the Öved-Ramsåsa Series, of Upper Ludlowian age. It is comprised of a composite group of limestones, marlstones, shales and sandstones.

From a stratigraphical point of view, the Silurian is far more completely represented in Scania than in any other part of Sweden, Gotland not excepted. With regard to facies there is a great difference between these two areas, with shales predominating in Scania, and limestones, marlstones and sandstones in Gotland. Not only is the Silurian of Scania very complete, it is also enormously thick in comparison with sediments of similar age in other parts of Sweden. No definite figure can be given, but the entire thickness of the Silurian of Scania may be estimated to be in the order of 1000–1500 m (*Rastrites* Series approx. 40–100 m; *Cyrtograptus* Series approx. 100 m?; *Colonus* Series approx. 600 m; Öved-Ramsåsa Series approx. 300 m?) (Regnéll, 1960; Thorslund, 1960).

In Östergötland and Västergötland no Silurian stages higher than Upper Llandoverian are present. In Jämtland, along the border of the Scandinavian mountain range, the Llandoverian and lowermost Wenlockian are represented. No Silurian sediments have been reported from Närke, in the area of the great lakes. In Dalecarlia there are shales belonging to the *Rastrites* Series and shales and limestones that are correlated with the uppermost Llandoverian and lowermost Wenlockian. Most of the Wenlockian and all of the Ludlowian are missing. The youngest Palaeozoic beds are those of the Orsa Sandstone. The basal beds of this rock unit are coarse-grained, in part conglomeratic and cross-bedded. The main mass of the unit consists of a fine-grained, soft, sometimes calcareous sandstone. Its colour varies from almost white to red, its clayey, shaly bedding planes occasionally show beautiful mud cracks (Thorslund, 1960). The Orsa Sandstone is probably of uppermost Ludlowian age and may have been deposited during an eastward transgression of the western sea.

From this survey it may be concluded that during the Llandoverian the sea covering the Baltic area again had a larger extension than during the uppermost Ordovician. The northwestern shore line of the sea, however, showed a general trend to withdraw southeastward. During the time the Silurian sediments of Gotland were deposited, Gotland on the average was situated not far away from the northwestern shore line of the sea which covered part of the present Baltic area. Between this sea and the western sea the connection over Scania and the Kattegat still existed. In its main points the situation may have been more or less comparable with that of the Early Cambrian.

A different interpretation of the Silurian situation around Gotland is given by Von Bubnoff (1952). He assumes that the facies with graptolite shales in central southern Sweden has been deposited in deeper water than the limestone in the areas east (Gotland, Esthonia) and west (Oslo) of it. Consequently, he assumes that a large sea covered the whole area between Oslo and Esthonia. But, whereas the limestones in the marginal areas show a thickness of a few hundred metres, the shales in central southern Sweden, according to him, do not measure more than about 50 m. This is interpreted by Von Bubnoff (1952, p.76) as being due to a stronger subsidence of the sea bottom in the central area, which has not been compensated by sedimentation to the same extent as the marginal areas. He, therefore, regards the limestones in Gotland and Esthonia as directly related to those in the Oslo area.

Several arguments can be advanced to dispute this view of Von Bubnoff:

(1) The Silurian in central southern Sweden is indeed relatively thin. But this is due to the fact that no sediments of Wenlockian and Ludlowian age are known from Östergötland, Västergötland and Närke. There are no indications whatsoever of an inundation of these areas of Sweden during the middle and upper stages of the Silurian.

(2) On the Swedish mainland the Silurian is developed most completely in Scania (see above). However, the thickness of the Silurian there should be estimated as in the order of 1000–1500 m, not about 50 m. It is true that rather considerable vertical displacements must have taken place in connection with the deposition of the *Colonus* Series, but these can adequately and in fact more satisfactorily be explained on the basis of block faulting (Regnéll, 1960).

(3) The bathymetric chart and the sections published by Fromm (1943) for the northwestern Baltic show a northwestern boundary of the Silurian basin that is in accordance with the picture presented in the first part of this section, and not with that of Von Bubnoff. Both the chart and the sections show that the crystalline basement takes up an increasingly lower position towards the east and southeast. Also, the sequence of the Cambro-Silurian increases in thickness with increasing distance from the eastern shore of Sweden.

(4) Epeirogenetic movements in Gotland during the Silurian took place around a dominantly northeast striking hinge line (Chapter IV, pp.46–48). Occasionally the strike of the hinge line was about east or north. Von Bubnoff's interpretation is not supported by any of the directions of dip.

(5) The distribution of both the bioherms and the stratified sediments in Gotland indicates formation not far from the northwestern shore line of the Baltic (see, e.g., Chapter XI).

In conclusion, the validity of Von Bubnoff's view of a large Silurian sea must be denied. A palaeogeographical situation with a smaller sea covering part of the present Baltic area is more likely. The area now constituting Gotland formed part of this sea, but was relatively close to the northwestern shore. Further south this sea had a connection with an other, western sea. This connection occupied the area of the present Scania and Kattegat.

Chapter III

THE PALAEOZOIC DEPOSITS OF GOTLAND

THE BASEMENT OF GOTLAND

Knowledge of the basement of the Silurian strata exposed in Gotland has been obtained from the cores of two deep borings, one near Visby (Hedström, 1923d), the other in the area of File Haidar, between Tingstäde Träsk and Slite (Thorslund and Westergård, 1938).

Both borings were brought down into the Precambrian basement, which was reached at a depth of 382.3 m near Visby and of 500 m in File Haidar (Fig.10). It consists of grey-veined gneiss, with red pegmatite. These rocks are considered Archaean of age and were found to be rather strongly weathered.

The Cambrian overlies the basement with a basal conglomerate, 35 cm thick, which is very loosely cemented and crowded with small pebbles which do not exceed 15 mm in diameter and which are, as a rule, well worn. The bulk of them are made up of light-grey quartz[1]; to a minor extent the pebbles consist of light quartzites.

The Cambrian deposits, mainly consisting of sandstones, reach a thickness of 140.9 m under Visby and of 157.3 m under File Haidar. They represent the Lower Cambrian (*Holmia* Series) and the lowest stage of the Middle Cambrian (*Paradoxides oelandicus* Stage) (cf. Table I). The lower as well as the upper boundaries of the Cambrian in the borings are very sharp. The bedding planes are, broadly speaking, horizontal. However, at some levels they show a dip which is usually slight but which may reach a maximum of 13° (Thorslund and Westergård, 1938).

The Lower Cambrian sandstones, in part with seams of shale, are very poor in fossils. Burrows and trails (*Skolithos linearis* Haldeman, and *Diplocraterion parallelum* Thorell), on the other hand, are common in nearly the whole sequence. The *Paradoxides oelandicus* Stage in Öland and other Scandinavian regions consists predominantly of shales. This sediment is also the predominating lithological constituent of the *P. oelandicus* section in the core of the Visby drilling (Hedström, 1923d). In the File Haidar core, however, sandstone prevails. The fauna of the Middle Cambrian too, is

[1]Thorslund and Westergard (1938) consider it evident that the quartz of the pebbles originated from pegmatite dikes piercing the subjacent gneiss. This conclusion is not considered obvious by the present author. Between the Archaeic and the Cambrian a long time has elapsed during which several peneplanations must have occurred, which may well have left behind quartz of an other origin. This is also suggested by the presence of quartzites, which are not directly derived from the Archaeic basement of gneiss with pegmatite dikes.

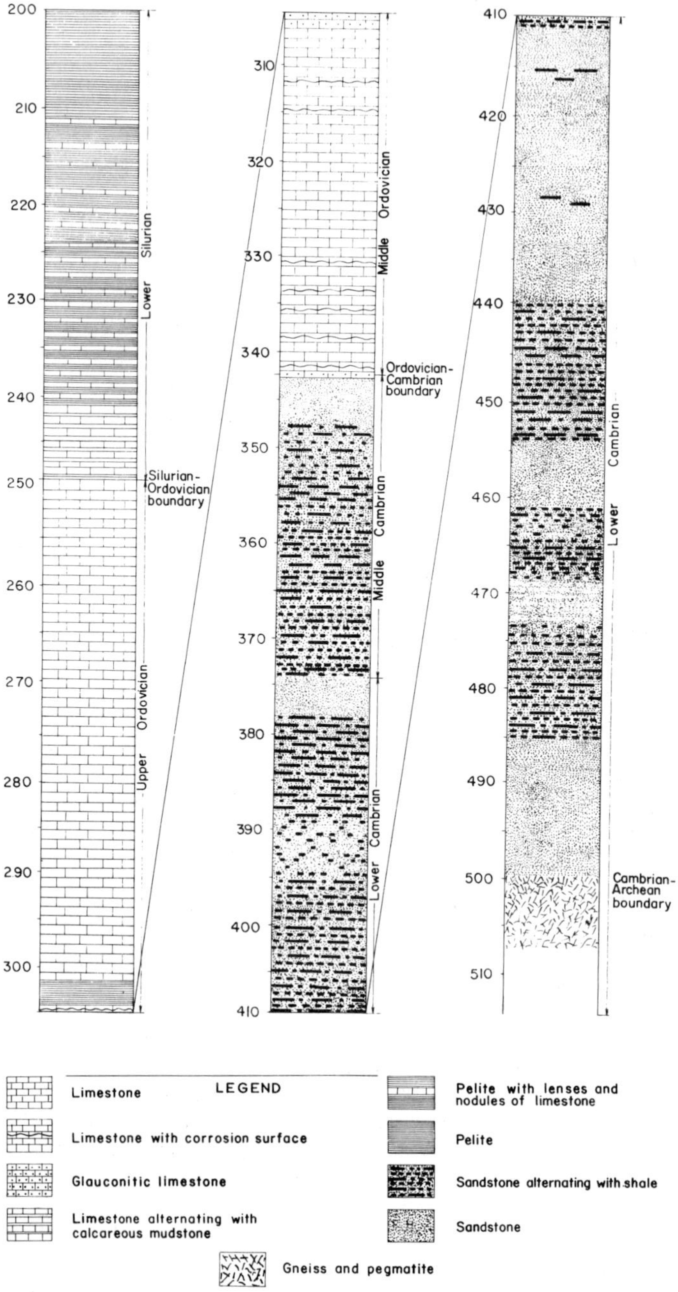
200
210
220
230
240
250
260
270
280
290
300
Lower Silurian
Silurian-Ordovician boundary
Upper Ordovician
310
320
330
340
350
360
370
380
390
400
410
Middle Ordovician
Ordovician-Cambrian boundary
Middle Cambrian
Lower Cambrian
420
430
440
450
460
470
480
490
500
510
Lower Cambrian
Cambrian-Archean boundary
LEGEND
Limestone
Limestone with corrosion surface
Glauconitic limestone
Limestone alternating with calcareous mudstone
Pelite with lenses and nodules of limestone
Pelite
Sandstone alternating with shale
Sandstone
Gneiss and pegmatite

remarkably poor in the cores; winding trails are common, but burrows occur only sparsely.

The bulk of the upper part of the cores, including Ordovician and Silurian beds, consists of limestones of different textures and compositions. Only in the uppermost, Silurian, part do argillaceous beds dominate, as this part is built up of marly pelites with intercalations of limestone, the latter diminishing in thickness and becoming more and more separated from each other towards the top of the core.

As previously mentioned, the contact of the Ordovician with the underlying Cambrian deposits is very distinct, and - to judge from the core-section of File Haidar - almost horizontal. There is no conglomerate at the base of the Ordovician, which consists of a light-grey glauconitic limestone, in which only a few rounded grains of quartz have been observed, close above the Cambrian strata. From its faunal contents it is evident that this glauconitic limestone represents a stratigraphical unit which must be correlated with the *Asaphus* Limestones in Sweden (see Table I) and with the Kunda Stage (*Vaginatum* Limestone) in the east Baltic region.

It thus appears that not only the middle and upper part of the Middle Cambrian and the Upper Cambrian are missing underneath northern Gotland, but also the lower parts of the Lower Ordovician, comprising the whole of the *Ceratopyge* Series and the Upper (*Megistaspis*) *planilimbata* Limestone and the (*Megistaspis*) *limbata* Limestone of the *Asaphus* Series.

Moreover, the available sequences in the cores of the deep borings also show signs of several stratigraphical gaps higher up in the series. The *Asaphus* Limestones in the core of File Haidar are only about 1 m thick and contain several corrosion surfaces. Thus, even the *Asaphus* Limestones of northern Gotland would very likely prove to be fairly incomplete if detailed comparisons could be made with the different zones of the *Asaphus*

Fig.10. Diagram showing the lithology of the Cambrian, Ordovician and lowermost Silurian underneath File Haidar, 8 km west of Slite and 25 km northeast of Visby. The upper 200 m were drilled by percussion drill; from this level to the bottom (507.5 m below the surface, which is 62.4 m above sea level) diamond drilling was employed. (Drawn after data from Thorslund and Westergård, 1938).

The correlation of the sediments from the drilling core with the stratigraphical units given in Table I is as follows:

—	: *Dalmanitina* Series, missing.
248.8–312.4 m	: *Tretaspis* Series.
312.4–338.5 m	: *Chasmops* Series; the fossil material obtained is not sufficient to permit an exact determination of the limits between the different zones that are distinguished within this series.
338.5–341.7 m	: *Platyurus* Limestone.
—	: *Megistaspis gigas* Limestone, missing.
341.7–342.7 m	: *Asaphus* Limestones.
—	: *Megistaspis limbata* Limestone, Upper *Megistaspis planilimbata* Limestone, *Ceratopyge* Series, *Olenus* Series, Upper and Middle *Paradoxides* Series, all missing.
342.7–ca.374 m	: *Paradoxides oelandicus* Stage.
ca.374–500 m	: *Holmia* Series.

Limestones on both sides of the Baltic Sea (Thorslund and Westergård, 1938).

Above the *Asaphus* Limestones, deposits corresponding to the (*Megistaspis*) *gigas* Limestone of the Swedish Upper *Asaphus* Series seem to be lacking, but the presence of *Platyurus* Limestone and of all four stages of the *Chasmops* Series (Table I) could again be proved.

The upper part of the Ordovician section is difficult to correlate with other areas. The limestones between 312.4 m and 289.7 m in the File Haidar boring are considered by Thorslund and Westergård (1938) as representative of at least part of the Rakvere (Wesenberg) Stage in Esthonia, which is correlated with the Lower *Tretaspis* Series in Sweden. The overlying algal limestones in this boring are correlated with the Nabala and Vormsi Stages (Lyckholm Stage) in Esthonia and the Upper *Tretaspis* Series in Sweden.

Equivalents of the Porkuni (or Borkholm) Stage in Esthonia, which is now generally considered to represent the lowermost Silurian, are missing in the borings. Thorslund and Westergård assume that a large hiatus in the profile exists at this level. This hiatus should comprise not only the *Dalmanitina* Series of the Swedish profile, but also the Lower *Rastrites* Shales.

In total, the thickness of the Ordovician sequence underneath Gotland amounts to 98 m in the Visby boring and 93.9 m in the File Haidar boring. The maximum thickness of the Silurian sequence of Gotland is about 650 m, of which about 140 m are situated below sea level. The contact Ordovician – Silurian was found in the two borings at a depth of 142.6 m and 248.8 m respectively. It is marked off by a very distinct corrosion surface. Above this surface there are no signs of stratigraphical breaks in the sequence of strata. In the lowermost part of the File Haidar core some Lower Llandoverian may be represented, though the fossil evidence is poor (cf. Martinsson, 1968). Graptolites in the File Haidar core, between a depth of 241.5 m and 217.6 m, make a correlation with the Swedish *Cephalograptus cometa* Zone likely, which means upper Middle Llandoverian. Overlying these rocks, the borings showed the presence of red clayish-marly beds, which reach a thickness of at least 54 m (Hedström, 1923d). Comparable rocks have been described by Lindström (1888a) from loose blocks found on the beach near Visby. They underly the Lower Visby Beds (*Stricklandinia* marl of Lindström) and are included in his stratigraphy as stratum a. On the basis of its fossil content he considers this stratum as Llandoverian of age. The Lower Visby Beds, the oldest rocks to be seen on the surface in Gotland, have been considered by some authors to belong to the top of the Upper Llandoverian (Munthe, 1910; Hede, 1942) whereas others claim them to be of Wenlockian age (Van Hoepen, 1910; Teichert, 1928).

THE STRATIGRAPHY OF GOTLAND: HISTORICAL REVIEW

The study of the stratigraphy of the Silurian deposits of Gotland began in 1845 when R.I. Murchison, together with M. de Verneuil, visited the island. He attacked the problem armed with an extensive knowledge of the series of strata in the Wenlock and Ludlow districts (Great Britain). The reef limestones, termed "ball-stones" by him, were correctly interpreted as local intercalations of little stratigraphical importance. As a result, the main points of his ideas about the stratigraphy of Gotland (Murchison, 1846) are in accordance with the present day interpretation, in that the strata strike about northeast and have a slight southeastward dip. Thus the strata gradually become younger when going from north to south. Later the same views about the Silurian of Gotland were expressed by Schmidt (1859, 1882, 1890, 1891). His conclusions are based both on biostratigraphical considerations and on comparisons with Esthonia and Saaremaa (Oesel).

A second approach to the unraveling of the succession of the Gotlandian strata was made by Lindström (1857, 1882a,b, 1888a). This author has been of fundamental importance for his studies on the multitude of fossils occurring in Gotland, but, unfortunately, had no correct insight into their stratigraphical use. Thus, based upon their similar lithological character, he considered all crinoid limestones as belonging to one stratigraphical horizon. The same was held for the "cephalopoda and stromatopora strata". Lindström (1888a), in his way, came to the following subdivision of the Silurian of Gotland:

(h) Cephalopoda and stromatopora stratum[1] - Upper Ludlowian
(g) *Megalomus* banks - Guelph Limestone (Canada)
(f) Crinoid and coral limestone - Aymestrian or Ludlowian
(e) *Pterygotus* stratum - basis of Ludlowian
(d) Limestones and oolite banks, with marlstones - Wenlockian Limestone
(c) Marl shale and sandstone - Wenlockian Shale
(b) *Stricklandinia* marl - Upper Llandoverian
(a) Oldest red *Arachnophyllum* shale - Llandoverian

According to Lindström these strata occur all over the island. Thus the stromatopora limestones and the crinoid limestones, attributed to his horizons f–h, are found in the north as well as in the middle and the south of Gotland. Lindström's stratigraphy is consequently a lithostratigraphy. The palaeontological difficulties which arose as a result of this lithostratigraphy are well illustrated in Lindström's view about his stratum c, marl shale and sandstone. This horizon was found to contain at least five different faunistical areas, following each other, as he stated, not in a vertical direction, but horizontally, from north to south: c_1: Visby fauna, c_2: Westergarn fauna, c_3: fauna of the central area, c_4: Petesvik – Hablingbo fauna, c_5: sandstone fauna of southernmost Gotland. It is clear, without any further comment, that Lindström's lithostratigraphy can not be maintained when the concept of guide fossils is accepted for at least part of the fossils that are found in the different faunistical areas.

With the work of Lindström and Schmidt the two main directions for further stratigraphical studies were indicated. Dames (1890), Stolley (1897) and Wiman (1897a) followed the views developed by Lindström. Their contributions, as far as these are concerned with the stratigraphy of Gotland, only give certain ameliorations of his subdivision. Other geologists, such as Holm (1901) and Munthe (1910) who both initially adhered to Lindström's views, became more and more convinced that his stratigraphy was untenable when carrying out detailed mappings in the island.

The Dutch professor of geology H.G. Jonker, being among other things engaged in the study of the origin of glacial boulders, thus came in contact with the Silurian of Gotland and arrived at similar doubts. These were

---

[1]The word stratum as used by Lindström has the same sense as defined by Rice (1954, p.393): "A layer of rock more or less similar throughout, a lithologic unit. It may consist of one or more beds, and may constitute a formation or a member or be only one of several strata in such formation or member". It will be used in this sense throughout this book. It should be noted that this definition is broader than that given in Schieferdecker (1959, p.139, cf. terms 2410 and 2411).

TABLE IV

Comparison of the Middle Palaeozoic stratigraphy of Gotland according to Hede with that of some earlier authors

Lindström (1888a)	Van Hoepen (1910)	Munthe (1910)	Hedström (1910)	Hede (1921, 1925a)
f–h (partly): Upper Limestone	South Gotland Limestone (major part)	Younger crystalline limestone (partly); *Ascoceras* layer (partly)		Sundre Limestone
d (partly): Oolite *f–h* (partly): Upper Limestone	*Phacites* Lime- stone (except for its lower parts)	Upper *Sphaerocodium* layer; *Ilionia–Spongiostroma* layer (partly)		Hamra Limestone
c_5 (partly): Sandstone; *d* (partly): Oolite	Burgsvik Sand- stone; *Phacites* Lime- stone (lower part)	Sandstone with clay; Oolite		Burgsvik Sand- stone and Oolite
d (partly): Oolite	Lauensis Marl	Lower *Sphaerocodium* layer (partly); *Rhizophyllum* reef limestone		Eke Group
c_4: Petesvik- Hablingbo; fauna; *d* (partly); *f–h* (partly)	Petesvik Marl; Nisse Limestone; Kräklingbo- and Östergarn Marl; South Gotland Limestone (partly)	Marl shale with limestone (partly); *Dayia* flags; *Ilionia* layer; *Ascoceras* layer (partly); Upper *Megalomus* layer		Hemse Group
c (partly); *d* (partly); *f–h* (partly)	Klinte Limestone (major part)	Marl shale with limestone (partly); *Ilionia–Spongio- stroma* layer (part- ly); Younger crystalline limestone (partly)	IV–VI and VII	Klinteberg Limestone
c_3; *d* (partly)	Mulde Marl (partly); Klinte Marl (partly)	Marl shale with limestone (partly)		Mulde Marlstone
d (partly); *f–h* (partly)	Fårö Limestone (partly)	Lower *Sphaerocodium* layer (partly); *Spongiostroma* layer (partly)	III ?	Halla Limestone
c_2-c_3 (partly); *d* (partly); *f–h* (partly)	Skrubbs Limestone; Follingbo Limestone; Klinte Marl (part- ly); Klinte Limestone (partly)	Marl shale with limestone (partly); Younger crystal- line limestone (partly)	IVb–VII	Slite Group
f–h (partly): Upper Lime- stone	Binger Limestone	*Spongiostroma* layer (partly)	IVa *Stromato- pora* Limestone and *Spongio- stroma* layer	Tofta Limestone
d (partly) *e*	Högklint Lime- stone and Marl; Korpklint Lime- stone	Lower *Sphaerocodium* layer (partly)	III Upper Clint Level	Högklint Limestone
c_1: Visby fauna	Hall Marl	Marl shale with limestone (partly)	II Lower Clint Level	Upper Visby Marlstone
Stricklandinia Marl	*Stricklandinia* Marl		I *Stricklandinia* Marl	Lower Visby Marlstone

N.B. The names *within* each rectangle do not indicate vertical succession.

TABLE V

Comparison of the Middle Palaeozoic stratigraphy of Gotland according to Hede with that of some later authors

Hede (1921, 1925a)	Wedekind and Tripp (1930)	Jux (1957)	Manten (this book)
Sundre Limestone	IV	Burgsvik Folgen	Hamra–Sundre Beds
Hamra Limestone	III–IV		
Burgsvik Sandstone and Oolite			Burgsvik Beds
Eke Group		Klinteberg Folgen	Eke Beds
Hemse Group	III		Hemse Beds
Klinteberg Limestone			Klinteberg Beds
Mulde Marlstone	II–III		Halla–Mulde Beds
Halla Limestone			
Slite Group		Visby Folgen	Slite Beds
Tofta Limestone	II		Högklint Beds
Högklint Limestone			
Upper Visby Marlstone			Upper Visby Beds
Lower Visby Marlstone	I		Lower Visby Beds

sustained by the well-known publication of Kiaer (1908) about the Oslo Silurian. He thereupon advised one of his students, Van Hoepen, to use the resolution of this controversy as the subject for his doctorate thesis. Based on less than four month's studies in the field, Van Hoepen (1910) presented an excellent piece of work, fully supporting the ideas of Murchison and Schmidt. When consulting the literature, however, one receives the impression that the influence of the publication by Van Hoepen on the Swedish geologists working on Gotland was slight.

In the same year Hedström (1910) published a paper on the stratigraphy of the Silurian strata of Visby and its surroundings. He clearly recognized that the reefs and their surrounding crinoid limestones are equivalent to distinctly stratified marlstones and limestones further away. As can be seen from the summarizing table (Table IV), Hedström's interpretation of the age of the rocks younger than those occurring in the coastal cliffs near Visby was disputable; it still shows some adherence to the stratigraphy of Lindström.

Munthe at this time realized more and more that a solution to the divergences of opinion would only be found when full profit was derived from the fossil contents of the sediments. "I may say emphatically, that before we can completely understand the sequence of strata in Gotland, we must have much further help from the palaeontologists" (Munthe, 1910). It was Munthe who, in 1917, when he was director of the Geological Survey, saw to it that Hede became engaged in this task. His results (Hede, 1921 and later), which will be mentioned more extensively in the next section of this chapter, definitely proved the validity of the ideas expressed by Murchison (1846), Schmidt (1890, 1891) and Van Hoepen (1910).

A third way of interpreting the Silurian sediments of Gotland, which has been proposed by Wedekind and Tripp (1930), endeavours to make a kind of compromise between the two main solutions discussed above. Palaeontologically it is based mainly on a study of coral faunas (Wedekind, 1927, 1932). The authors distinguished three marl complexes converging towards the north, and three limestone complexes. Although the limestone complexes are built up of stratified limestones with isolated reefs, they called each limestone complex a "reef". From top to bottom their stratigraphical subdivision is:

IV	Hoburgen reef
III/IV	Linde – Lau marl
III	Klinteberg reef
II/III	Slite marl
II	Visby reef
I	Visby marl

The succession of deposition, according to their opinion, should be visualized as follows:

After the circumcontinental platform had been drowned, the Visby marl was sedimentated. In this marl the Visby reef developed, and was subsequently covered by the Slite marl, while 30 km farther to the south the Klinteberg reef originated. The upper parts of the Visby reef are synchronous with the lower parts of the Klinteberg reef. The latter reef was in its turn covered by the Linde – Lau marl, in which, again about 30 km towards the south, the third reef belt developed, the lower parts of which are synchronous with the upper parts of the previous reef belt.

This view, which is not supported by the enormous amount of data collected by Hede (1921, 1925a,b, 1927a,b, 1928, 1929, 1933, 1936, 1940, 1958, 1960), did not find any favour in the geological world, until it was adapted by Jux (1957).

An attempt to correlate the earlier stratigraphical subdivisions of the Silurian of Gotland with that of Hede (1921) has been made in Table IV, whereas a comparison of the stratigraphical views of Wedekind and Tripp (1930) and Jux (1957) with the subdivision by Hede (1921) is given in Table V.

THE STRATIGRAPHY OF HEDE

From 1917 onwards, when Hede became engaged in geological work in Gotland, he has devoted almost all his energy to the study of the Silurian rocks of this fascinating island. No geologist ever had as extensive a knowledge of the sediments of Gotland and their fossil contents, as does Hede. Although he also published a number of more general papers on the

---

Fig.11. Simplified geological map of Gotland showing the distribution of the Palaeozoic sediments, according to the stratigraphy of Hede (1921). 1. Lower Visby Marlstone, 2. Upper Visby Marlstone, 3. Högklint Limestone, 4. Tofta Limestone, 5. Slite Group, 6. Halla Limestone, 7. Mulde Marlstone, 8. Klinteberg Limestone, 9. Hemse Group, 10. Eke Group, 11. Burgsvik Sandstone and Oolite, 12. Hamra Limestone, 13. Sundre Limestone.

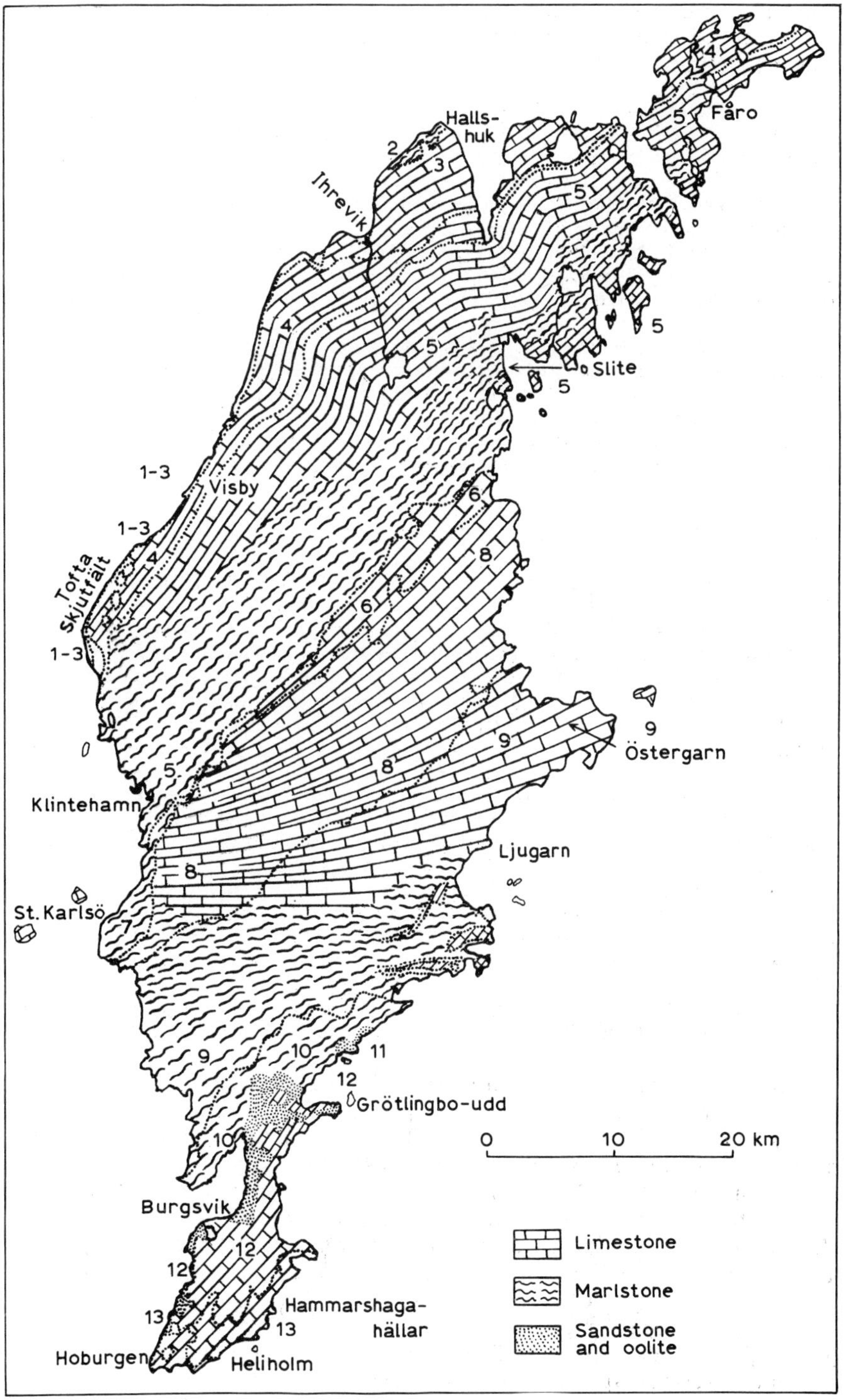
Hallshuk
Fåro
Ihrevik
Slite
Visby
Tofta Skjutfält
Östergarn
Klintehamn
Ljugarn
St. Karlsö
Grötlingbo-udd
0 10 20 km
Burgsvik
Hammarshaga-hällar
Hoburgen
Heliholm
Limestone
Marlstone
Sandstone and oolite

stratigraphy of Gotland (Hede, 1921, 1925a, 1958, 1960), the geological maps with their accompanying descriptions especially give evidence of the conscientiousness with which he carried out his work.

Hede's stratigraphy of the Silurian of Gotland (cf. Fig.11) is, from top to bottom:

(13) *Sundre Limestone*. Very fossiliferous limestones, especially rich in crinoid remains, but also in stromatoporoids, corals and bryozoans; and reef limestones.

(12) *Hamra Limestone*. At the base an algal limestone. This is overlaid by crystalline limestones and marly limestones, reef limestones and crinoid limestones.

(11) *Burgsvik Sandstone and Oolite*. The Upper Burgsvik is composed of sandstone, oolite, argillaceous shaly sandstone and claystone; the Middle Burgsvik almost exclusively of sandstone; the Lower Burgsvik of argillaceous shaly sandstone and claystone.

(10) *Eke Group*. Marlstone, passing into marly limestone and limestone in the northeastern part of its area. Reef limestones are scarce.

(9) *Hemse Group*. In the north limestones, including reef and crinoid limestones, towards the south passing into marlstone.

(8) *Klinteberg Limestone*. Dominating thin-bedded limestones, rich in stromatoporoids, corals, Algae, lamellibranchs, brachiopods and other fossils. Reef limestones and crinoid limestones.

(7) *Mulde Marlstone*. Soft, dense marlstone, alternating with harder layers of finely crystalline, marly limestone. The Mulde Marlstone wedges out towards the northeast.

(6) *Halla Limestone*. Crystalline limestones, partly finely oolitic, partly marly. A few small reefs. In the northeast in the Lower Halla the well-developed Bara oolite. Towards the southwest the Halla Limestone is gradually disappearing.

(5) *Slite Group*. Limestones, marly limestones and reef limestones in its northwestern part. Towards the southeast these limestones pass over into marl. In the most southwestern part of its range some sandy limestones and limy sandstone.

(4) *Tofta Limestone*. Limestones, partly marly, very rich in Algae, disappearing towards the north.

(3) *Högklint Limestone*. The Högklint Group is composed of crystalline limestones and marly limestones, upwards partly finely oolitic, while they often are also rich in Algae in the upper part of the group. Reef limestone and crinoid limestone are common. The lowest continuous thicker limestone beds form the base of the Högklint Limestone.

(2) *Upper Visby Marlstone*. Marlstone with intercalated limestone lenses. In the upper part of the sequence the limestone lenses grade into thin layers, which wedge out into the marlstone. The horizontal dimensions of these thin limestone layers increase in the uppermost part of the Upper Visby Marlstone. Small reefs occur.

(1) *Lower Visby Marlstone*. This unit consists of a thin-bedded, bluish-grey marl with flat lenses of harder, dense to finely crystalline marly limestone.

Since the moment of its publication, the stratigraphy of Hede has become quite generally accepted. Except for the papers by Wedekind and Tripp (1930) and by Jux (1957), later publications dealt only with minor improvements of this subdivision. For instance, Hadding (1941) carefully hinted that perhaps the Halla Limestone and Mulde Marlstone might form different lithofacies of one and the same stratigraphical unit. In 1956 the same author stated that the Tofta Limestone could be regarded as a local, extreme shallow-water facies of the uppermost part of the Högklint Group and the lowest part of the Slite Group.

In this book the Tofta Limestone will be included as a facies in the Upper Högklint Beds, the Halla Limestone and Mulde Marlstone will be united in one stratigraphical unit, the Halla-Mulde Beds, and the Sundre

Limestone will be united with the Hamra Limestone to the Hamra-Sundre Beds. For further details see Chapter XI. A tabular comparison is given in Table V.

The present author will avoid the use of lithological indications in the names of the main stratigraphical units of Gotland. Nor does he want to attach any specific rank to these units, and, therefore, following Martinsson (1962a), prefers to use the term *beds* for them. This tendency to use neutral terms for a local or regional stratigraphy has advantages insofar as it avoids confusion. Even in England, the term *beds* is now used in Silurian stratigraphy if reference is made to specific English deposits and not to the international stratigraphical standard (cf. Holland et al., 1959; Lawson, 1960).

Hadding (1941), who also did not use lithostratigraphical names, attached the chronostratigraphical rank of stages to the units in Gotland. However, this was incorrect. Since the units form the first order subdivision of the Silurian of Gotland, the name series should have been used. Moreover, the use of the word *series* would then also have been in accordance with the practice in the subdivision of the Swedish Cambrian and Ordovician (Table I). There, Hadding (1958c) also attached the rank of series to the units given in column two of Table I.

A detailed correlation with the English Silurian is still a matter of interpretation. The individual English series cannot be recognized without doubt in Gotland, because of various differences in lithology and fossil content. As will be shown in Chapter XII, only a very small fraction of the fossils found in Gotland can be regarded as index fossils. The unit boundaries in Gotland are probably not always equivalent to stratigraphical boundaries in England (for further information, see the last section of this chapter).

THE STRATIGRAPHY OF JUX

Whereas some of the older authors paid too little attention to the role played by the reef limestones and related sediments in the Silurian of Gotland, a more recent publication (Jux, 1957) offers an example of the other extreme.

Jux tried to explain all facies differences in the geology of Gotland by the occurrence of three successive reef belts. During the time of development of each reef belt, nearest to the shore a shallow-water facies was thought to be present, passing into reef debris and reef limestone, with debris again on the seaward side of the reef, now gradually replaced by marl. Three rather rapid regressions are posited by Jux, separating three "Folgen" (stages), each with a synchronous lateral development of facies belts (Visby Folgen, Klinteberg Folgen, Burgsvik Folgen).

The present author has experienced that these short periods, characterized by a rapid regression, propounded by Jux, cannot be deduced from field evidence. Instead, there were times during which the water became gradually shallower, which can be deduced both from the sedimentation and from the growth of biostromes and bioherms. The first of these periods in which the water gradually became shallower was from Lower Visby to Upper Högklint time (including the time of deposition of the Tofta limestone). After a deepening of the water during Lower and Middle Slite (Slite I–III), the second shallowing took place from Upper Slite (Slite IV) time up to and including Halla-Mulde time[1]. In Klinteberg time the water remained shallow. Hemse time began anew with deeper water that showed signs of becoming shallower until the time in which the top of the Middle Burgsvik Beds was deposited. (See also Chapter XI of this book). The waters becoming shallow, therefore, at all three times took much longer than Jux supposed for the short regressions which he posited. From Table V it follows, moreover, that the shallowing periods actually present are distributed indiscriminately over the "Folgen" of Jux. There is no field evidence either for prolonged periods with reef formation, during which the water depth should have been constant, as is assumed by that author. Instead it has been found by detailed field work that water depth generally underwent changes during the periods of reef formation (see the various discussions in Chapter XI).

It should also be remarked that the names which Jux has chosen for his three "Folgen" are unfortunate, because they can easily be confused with the Visby Marlstones, Klinteberg Limestone and Burgsvik Sandstone and Oolite of Hede (1921, 1925a) (cf. Table V).

Because of the importance, in one way or another, of the ideas of Jux for the interpretation of reefs and reef genesis, a number of points will be discussed. They show that the factual data underlying his presentation are either invalid or wrongly interpreted.

As is clear from the geological map of Gotland and the data given in Chapter XI of this book, the Slite reefs are separated very distinctly, stratigraphically, from the Högklint reefs. The division of Jux in three stages cannot account for this, since he has to assume a more or less comparable age for all reefs within his Visby Stage. Even a little field work shows the untenability of this assumption, such as a northwest–southeast traverse through northern Gotland from the northwest coast to the area of Slite.

No more are the Klinteberg and Hemse reef limestone synchronous ("Klinteberg Folgen" of Jux). A traverse from the Klinteberg to the Lindeklint could here be suggested as could some detailed field work in the east of central Gotland.

The fact that the influence of the reefs did not reach more than about half a kilometre seawards from the reefs (cf. Chapter IX) makes it impossible to consider the Slite and Hemse marlstone complexes as the distal deposits of a reef belt, as interpreted by Jux. Moreover, they are often not situated immediately behind the reefs, but only begin a few or more kilometres away from the reefs, separated from the reefs by stratified limestones containing reef debris and normally developed limestones.

The incorrectnes of Jux' views is still more evident in the case of the Eke marlstone (Jux, 1957, p.77). This sediment is also considered by Jux to be the distal deposit of reefs, which, however, are almost lacking in the Eke Beds. The Eke marlstone is geographically too far separated from the Hemse reefs to be explained as the distal deposit of these. Should, moreover, the hiatus which may occur between the Hemse and Eke Beds be of the geographical and geochronological extension as some authors suppose it (cf. p.391), there would be still further objection against Jux' conception of the

---

[1]The shallowing during Halla-Mulde time is apparent from the Halla limestones in the east of Gotland. It is in this area that the reefs from this epoch are found. Further to the west, where the limestones are replaced by the Mulde marlstone, a deepening of the water seems to have taken place. A discussion of this dichotomy within the Halla-Mulde Beds will be given in Chapter XI.

Eke Beds. However the solution of the problem of the Hemse – Eke hiatus may turn out, the fact remains that the boundary between the Hemse and Eke Beds is more distinct than that between the Eke and Burgsvik Beds, although it is the latter which forms the boundary between Jux' Klinteberg and Burgsvik Stages.

Jux (1957, pp.76–77) states that the algal limestones too, are related to reef formation, and characteristic of a shallow-water environment, at some distance from the reefs, at their landward side, where the sea-bottom relief was slight and the transport power of the water decreasing. He considers the Tofta limestone as a facies zone of his northern reef belt and the Lower Hamra algal limestone as a facies zone of the bioherms in southern Gotland. In the opinion of the present author the algal facies is essentially a shallow-water deposit quite independent of reef formation. Of course, reefs and Algae may occur together, as they do more than once, e.g., in the Klinteberg, because even a slight deepening of the sea might have caused formation of small reefs within an area in which Algae grew. This can be well observed in southernmost Gotland especially, where the basal part of the algal limestone of the Lower Hamra Beds does not contain reefs, whereas in the higher parts of this deposit the beginnings of reef development can be seen. The reverse can be observed in the Högklint Beds, where the formation of an algal limestone (Tofta limestone) occurred at the end of a period of reef growth when the water had become very shallow. Both occurrences show that the Algae do not belong to the reef facies or any of the facies that accompany the reefs. They are quite independent.

Another point to be raised is the development of the Hamra Beds in southern Gotland. The main part of the Hamra group, as defined by Hede (1921 and later) is indicated by Jux (1957, p.82, Abb.7) as a clay-marl facies ("Ton-Mergel-Facies"). Actually it is mainly formed of limestones, as is also indicated in Hede's stratigraphical terminology, as he called the group the Hamra *limestone*. Only in the near vicinity of the reefs is the occurrence of marlstone indeed locally observed. Here a calcium-carbonate percentage as low as 50 may occur. Elsewhere in the normal Hamra limestone however, this percentage amounts usually to 87–98. The insoluble fraction ranges from a small fraction to 12.5% at the maximum; a bit low for a clay-marl facies!

The Hamra limestone is not lagoonal either. It was not deposited in "einem lagunären Bereich" (Jux, 1957, p.79), in between the sandstone and the reef debris. The contrary is true. In reality the Hamra limestone was deposited at the seaward side of the reefs. Although the big black spot of Hamra limestone on the geological map of Jux (1957, Abb.7) has been described as reef limestone by Munthe (1921b) - an interpretation adopted by other authors (Hadding, 1941, p.11; Jux, 1957, p.82) - this area actually consists of thick-bedded limestone in which stromatoporoids and a few corals commonly occur. Only on the western limits of the Hamra limestone do reefs occur, e.g., at Hoburgen and Kettelsård, while indications of reefs, demolished by Recent erosion, are found on the western beach at Killingholmen in the north of Vamlingbo Parish, and in the northeastern part of Grötlingbo-udd. The Hamra reefs were thus found in a south-southeast – north-northwest belt. This means that in Jux' view the main part of the Hamra limestone ought to be re-interpreted as "Riffdetritus-Facies". Except for locally in the west, however, where narrow blankets around the individual reefs may be detrital, the Hamra limestone shows no sign of a reef-detrital character either. Moreover, this implies that the Sundre limestones further east, including the raukar fields at Holmhällar, Heliholm, Austre and Hammarshagahällar cannot be considered as a detritus facies, as is done by Jux, for they are too far away from the actual reef belt of Jux' Burgsvik Folgen. These raukar fields can, therefore, only represent the remains of real reefs (see Rutten, 1958; also this book, Chapter VIII), but of an other type than those of Högklint and Hoburgen, and comparable with those of Ljugarn and Fågelhammar in the Hemse Beds.

In as far as Martinsson's work on the succession and correlation of ostracode faunas in the Silurian of Gotland has a bearing on the geochronology of reef formation, it also does not support the theory of Jux (Martinsson, 1967, pp.381–382).

It cannot be denied that the interpretation of Jux, on first appearance, is very attractive. In fact, it represents more or less the ideal scheme we too had in mind when field work in Gotland began in 1956. However, when in the field,it becomes clear that the actual situation is much more complicated. The presence of only three main stratigraphical units, each with a standard range of synchronous facies could not be proved. After a detailed discussion of the reef types of Gotland, the picture arrived at by the present author will be presented in Chapter XV. The main objection against Jux is that his publication on a territory with a complicated palaeoecology is not supported by sufficient field observations; not much more than a number of the most beautiful exposures seem to have been visited by him and these led him to a grossly oversimplified interpretation. A further discussion of the general conclusions arrived at by Jux is consequently unwarranted.

CORRELATION WITH OTHER AREAS

In the first section of this chapter is has already been mentioned that both the *Cephalograptus cometa* Zone of the File Haidar boring (24 m thick) and the red *Arachnophyllum* shale (Lindström, 1888a, stratum a) are correlated with the Llandoverian of the English type section, while Hede (1942) also includes the Lower and Upper Visby Marlstones in this series, more especially in the Upper Valentian (Upper Llandoverian; cf. Elles and Wood, 1914). It thus appears that at least a great deal of the time covered by the Middle Llandoverian and the Upper Llandoverian is represented by Silurian sediments in Gotland.

The Högklint Beds are considered to be Lower Wenlockian of age, the Slite marlstone is correlated with the Middle Wenlockian, the Halla-Mulde Beds with the Upper Wenlockian.

The Lower Ludlowian may start with the Klinteberg Beds. The age of the Hemse Beds is especially well determined, due to its relatively rich fauna of graptolites (Hede, 1942). It belongs to the Lower Ludlowian and corresponds with the Zone of *Monograptus nilssoni*, the second of the four zones into which the English Lower Ludlowian has been subdivided. Since one specimen of *M. scanicus* has also been found, 12 m below the top of the Hemse Beds (Hede, 1919a), at least part of the third zone, too, is represented by the Hemse Beds.

Serious difficulty arises when one attempts to correlate the younger stratigraphical units of Gotland with the English type section. There is only one point on which all authors agree, viz. that the Burgsvik Beds are of Upper

TABLE VI

Correlation of the Middle Palaeozoic series of strata in Gotland with that of some other areas

Gotland	Great Britain		Scania	Oslo area	East Baltic
Hamra-Sundre Beds	? Downtonian		Lower Öved-Ramsåsa Formation	Stage 9	K
Burgsvik Beds	Upper	Ludlowian			
Eke Beds	Middle		*Colonus* Series		
Hemse Beds	Lower				
Klinteberg Beds					
Halla-Mulde Beds	Upper	Wenlockian	*Cyrtograptus* Series	Stage 8	J
Slite Beds	Middle				
Högklint Beds	Lower				
Visby Beds	Upper Llandoverian				

Ludlowian age (Hede, 1921; Säve-Söderbergh, 1941; Spjeldnaes, 1950; Jux, 1957) and presumably contemporaneous with the Upper Whitcliffe Flags. This would mean that the relatively thin Eke Beds represent not only the whole Middle Ludlowian, but also the lower parts of the Upper Ludlowian and perhaps even the upper part of the Lower Ludlowian. Munthe (1902) and Spjeldnaes (1950) found a surface of discontinuity between the Hemse and Eke Beds in the Lau district (eastern Gotland). No such indications are furnished by outcrops in the west of the island. Consequently, the problem of the age of the Eke Beds cannot yet be considered as definitely solved.

The same applies to the age of the Hamra-Sundre Beds. In 1960, Hede was still inclined to attribute a Ludlowian age to his Hamra and Sundre Limestones. However, from other sides dissentient views have been brought forth.

As early as 1846, Murchison wrote that the highest stratum of the island is "a sandy and calcareous equivalent of the Upper Ludlow rocks, with indications of a passage into the Devonian group." (Murchison, 1846, p.27).

In this connection, it should be recalled that at that time the boundary between the Silurian and the Devonian was drawn between the Upper Ludlow Group (which included the Ludlow Bone-Bed) and the Lower Old Red Sandstone.

On the basis of vertebrate remains from the Upper Burgsvik Beds, Säve-Söderbergh (1941) and Spjeldnaes (1950) correlated this horizon with the Ludlow Bone-Bed.

In 1950, White redefined the Silurian – Devonian boundary in England, and placed this at the base of the Ludlow Bone-Bed. This implied that the Upper Whitcliffe Flags would constitute the uppermost layers of the Silurian (cf. also Lawson, 1960) and the Downtonian would be the lowermost Devonian.

Spjeldnaes (1950) drew the consequences of this for Gotland and felt tempted to place the youngest Palaeozoic sediments of Gotland in the Lower Devonian (Downtonian). In the terminology of the present author these included the Upper Burgsvik Beds (about 7 m thick) and the Hamra-Sundre Beds. Thus, Murchison's view that the boundary between the Silurian and

Devonian could be traced in Gotland was rejuvenated, though it now was an otherwise defined boundary.

The International Symposium on the Devonian System, held in Calgary in Canada, in 1967, moved the Silurian – Devonian boundary up again, to the basis of the *Monograptus uniformis* Zone. This brought the Downtonian into the Silurian. The major reason for this decision was that the base of the Ludlow Bone-Bed was criticized by a number of researchers as a boundary marker, since it was said to be a facies boundary and to reflect a hiatus. It may well be, however, that the Ludlow Bone-Bed boundary is nevertheless a suitable reference level from stratigraphical and palaeontological points of view (cf. Martinsson, 1969).

Anyhow, as the situation is now, the Upper Burgsvik Beds and the Hamra-Sundre Beds of Gotland have to be considered again as belonging to the Silurian. Their correlation with the Downtonian, however, remains possible.

Kaljo and Sarv (1966) have correlated the Sundre and the Hamra Beds with the Leintwardine Beds in England. Fåhraeus (1967), however, thereupon argued that the Ludlowian of Gotland probably reaches higher in the system than anticipated by Kaljo and Sarv. The Sundre Beds and at least the upper part of the Hamra Beds (in Hede's subdivision of the Silurian of Gotland) are correlated by Fåhraeus with the Upper Whitcliffe Beds, Ludlow area, England, through the identification of the lower part of the *eosteinhornensis* conodont zone (named after *Spathognathodus steinhornensis eosteinhornensis* Walliser).

Martinsson (1965c) studied the boulders found on Hoburg Bank (southeast of Hoburg in the Baltic) and found evidence that calcareous equivalents of the Öved-Ramsåsa Beds of Scania are to be found along its northwestern parts, and the Beyrichienkalk, found in drift south of the Baltic, occurs in its higher parts. The Hoburg Bank is a longish feature along the strike of the Silurian beds and scarps of Gotland.

In connection with the movement back and forth of the Silurian – Devonian boundary, it should be noted that the line indicating this boundary, as it is drawn in Fig.3, stems from the years when the boundary was defined between the Ludlowian and the Downtonian, below the Ludlow Bone-Bed. The line indicates the situation on the floor of the Baltic basin. The course of that line, therefore, does not necessarily exclude the possibility that strata belonging to the Lower Downtonian are present on top of the stratigraphical succession in the island of Gotland.

For a summary of the correlation of the Gotlandian series of strata with that of England as well as some other areas, see Table VI.

Chapter IV

TECTONICS

PSEUDO-TECTONIC PHENOMENA

Tectonic movements of any importance have never affected the Middle Palaeozoic strata of Gotland. The whole island gives the impresssion of extreme geological quietness, as can also be said of the west Baltic as a whole. The soft marlstones, the character of the limestones, the preservation of the fossils and the very slight dip of the strata make it clear that the Palaeozoic sediments of Gotland have never been submitted to strong tectonic influences.

Notwithstanding this, there are a number of more or less localized phenomena, apart from the general dip of the strata (see the next section of this chapter), which have been explained by means of tectonics, especially by early authors. Most of these phenomena, however, can also be understood without the assumption of the occurrence of crustal movements.

First, a couple of steep cliffs sometimes are explained as fault scarps. Thus, Munthe (1902, p.269) explains the steep northwest coast of Gotland by assuming a fault, running parallel to the coast. In a later publication (Munthe, 1921b) the same assumption is made for the straight and steep coast, striking north-northeast near and north of Hoburgen, on the west coast of the southern peninsula of the island. Now that the stratigraphical subdivision by Hede has been generally accepted, implying that the Palaeozoic strata of Gotland generally have a slight dip towards the southeast, it will be understood that marine cuestas will be formed at the northwest side of the island. In the south of the island the Burgsvik Beds show a slight dip towards the east-southeast. Consequently, the direction of the coast line here is also determined by the outcrop of resistant layers. The fault or fold zone which Munthe assumes to be present between the lines Klinteberg – Gothem and Stenkumla – Bäl is explained geomorphologically by the lesser resistance of the marlstones against erosion (see also Chapter I). Finally the steep walls found between Lojsta and Östergarn (Nathorst, 1886, p.328; Munthe, 1902, p.269) can also be explained geomorphologically by differential erosion, without the assumption of the presence of dislocations.

Another topographical feature has attracted the attention of Van Hoepen (1910). This is a depression crossing the southern peninsula from Mjölhatte Träsk[1] in the west, via Halshage Träsk and Rörviks Träsk to Stockviken[1] in the east. This depression lies about 5–10 m lower than the landscape north and south of it. Van Hoepen is inclined to assume the presence of faults to

[1]Swedish: träsk = marsh; vik = bay.

explain this phenomenon. This interpretation was strengthened by the observation that the boundary between the Burgsvik sandstone and the overlying Burgsvik oolite in the Mjölhatte Träsk canal lies only 1.7 m above sea level, whereas a little more than half a kilometre northwards this boundary is situated at least 5 m higher. The present author has carefully looked for other indications which might confirm this surmise, but has not found any. On the other hand it has been observed that the boundary between sandstone and oolite is undulating (see Chapter XI). At Snäckviken, about 3.5 km south of the Mjölhatte Träsk canal, this boundary does not lie much above sea level either. Consequently, the occurrence of dislocations across the southern peninsula of Gotland is also not very likely.

Finally, the following features can also be explained without the assumption of the occurrence of crustal movements.

(1) The dislocations that Munthe (1902) felt obliged to posit in the Lau district, on the basis of the stratigraphical ideas of Lindström. These are refuted without further ado with the acceptance of the stratigraphical views of Hede (1921) (see also Van Hoepen, 1910, p.25).

(2) Three fold-like structures mentioned by Munthe (1910) from the south of Gotland. These may be re-interpreted as fossil offshore bars (see Chapter XI, pp.399–400).

(3) Saucer-like depressions which are, for instance, found very clearly developed in the Burgsvik sandstone at the foot of Hoburgen, but are also encountered in other exposures in Gotland. These are distinctly related to the high pressure exerted by the heavy load of overlying reef-limestone masses (differential compression, cf. Chapter VII, p.155).

(4) The bends found locally in the limestone layers east of Visby; also the "flexure" near the road crossing Lärbro – Storugns – Vestrume, and the strong southeastward dip near Hau (east of Bäste Träsk, Fleringe Parish; other directions of dip may be observed there as well!), etc., which were all thought by Van Hoepen (1910, pp.24–25) to be probably of a tectonic nature. In fact, these are also all genetically related to the presence of reefs.

THE DIP OF THE STRATA

After the remarks in the preceding section of this chapter, only the origin of the slight dip of the strata now remains to be discussed. This dip is generally directed towards the southeast (compare the map in Fig.11, showing the stratigraphy of Hede). The amount of the dip is about 0°30'.

Munthe (1921b) and Hede (1925a) believe that this dip has been caused by the tilting of a firm block, due to a postsedimentary crustal movement. For the axis of tilt Hede assumes a southwest to northeast direction.

From the discussion of the stratigraphy of the Palaeozoic of Gotland in some more detail (Chapter XI) it will be seen, however, that the general strike in the island is not as regular as would appear from the course of the stratigraphical boundaries, given in the maps of Hede. Variations in the strike and dip are found within some of the sequences of strata.

From these variations it follows that a tilt of the strata as a firm block through a postsedimentary tectonic influence is unlikely. Therefore, instead of a postgenetic tilt of all the Palaeozoic strata of Gotland, another picture is presented here, one of synsedimentary tilting, along an axis that has not always been in the same position, nor pointed in the same direction at the time of deposition of the Palaeozoic strata.

The regular shift of the reef belt with time towards the southeast shows that, seen as a whole, epeirogenetic uplift of the basin floor occurred in the area of Gotland. However - as will be shown - these movements, more specifically, were active over definite periods only and even alternated with periods of slight subsidence of that part of the Palaeozoic basin floor.

When analysing the average dip of the strata, it was found that it decreases from the older series of strata towards the younger ones. Thus, the average dip for the Upper Visby and Högklint Beds together amounts to about 0°31'; the average dip for the Klinteberg and Hemse Beds together is about 0°27'; the average dip for the whole of Eke, Burgsvik and Hamra-Sundre Beds is about 0°23'. These differences can only be explained when a synsedimentary origin of at least part of the dip of the strata is assumed. Owing to the prevailing downward movement of the centre of the basin, the older sedimentary strata (e.g., the Högklint Beds) already had a slight dip when the younger ones (e.g., Hamra-Sundre Beds) were deposited.

Whereas Hede (1925a, p.31) posits a general southeastward dip and a northeastward striking tilt axis for Gotland as a whole, Munthe (1921b, p.73), who has been mapping the Palaeozoic deposits in the south of Gotland, posits a north to northeast directed tilt axis in southern Gotland, in order to explain the east to east-southeast dips that prevail in that part of the island. However, the boundary between the Hemse and Eke Beds still dips approximately towards the southeast. This implies that the movements not only followed one after the other, but that they also took place in different directions. More deviations will be mentioned in Chapter XI.

For the moment, attention will now be focused on the Slite Beds. Near Bunge, the marlstone marking the top of the Slite III Beds, is very thin, and exposed at about 25 m above sea level. At about the same height the marlstone is found near Othem and Liksarve (Tofta Parish). All three places are situated close to the line of maximum extension of the Slite marlstone (Chapter XI, p.316). In Bogeklint (also called Klinteklint, about 2 km south-southeast of Boge Church) the top of the Slite marlstone lies about 10 m above sea level. Near Tjeldersholmklint the marlstone is found at about present sea level. As will be discussed in Chapter XI (p.329), the seaward retreat of the area of marl deposition at the beginning of Slite IV time, which is marked by the onset of the Slite IV limestones, took place rather rapidly. As a result, the top of the marlstone on all five places may be considered as more or less synchronous. Bogeklint is about 8.5 km southeast of the line Liksarve (Tofta) – Bunge, Tjeldersholm about 12 km. Thus over a distance of 8.5 km the boundary Slite III (marlstone) – Slite IV (limestone) descends 15 m, whereas over a distance of 12 km the descent amounts to about 22 m. In both cases this points to an angle of dip of about 0°06'. On the island of Furillen and the peninsulas Smöje-udd and Sankt Olofsholm, the boundary between marlstone and Slite IV limestone also lies only slightly above sea level. Calculation of the dip between this line and Bogeklint shows an angle of about 0°08'.

Quite another picture is manifest when observing the boundary between the Slite and Halla-Mulde Beds. Southwest of Bara Ödekyrka (ca. 3 km south-southwest of Vallstena) the base of the Halla Limestone lies about 23.5 m high. Near Bryggans Fiskläge (ca. 10 km south of Slite) the top of the Halla Limestone lies at a maximum of about 3.5 m above sea level. Since the Halla Limestone very likely attains a thickness of about 15 m here, its base will be about 11.5 m below sea level. The distance between both observation points in the direction 140° is ca. 3.5 km. Calculation of the dip of the boundary of the Slite and Halla-Mulde Beds gives an angle of about 0°35'. In central Gotland there are a number of small exposures of the uppermost Halla Limestone in the parishes of Väte, Viklau and Halla. Calculations give a dip of the Slite – Halla boundary of about 0°27'.

Calculated in a comparable way, the dip of the boundary between the Högklint (incl. Tofta) and Slite Beds is about 0°20' – 0°25'.

It thus appears that there are remarkable differences in the angle of dip when calculated from a few more or less suitable horizons. Although the present author is aware of the fact that the results of such considerations as the above should be handled with extreme care, the differences are such that they may still be accepted as an indication that moderate epeirogenetic movements did take place during the time that the Middle Palaeozoic sediments of Gotland were deposited.

The decrease in dip which presumably took place at the beginning of Slite IV time, and which resulted in a shallowing of the water, may have been caused by a downward movement of the basin centre around an about northeast-striking hinge line, situated northwest of Gotland. A new movement took place at about the transition from Slite to Halla-Mulde time. This movement was upward. The sediments of the Halla-Mulde Beds show signs of a shallowing of the water in the east of Gotland and of a deepening in the west (see Chapter XI). The axis of tilt consequently then seems to have crossed central Gotland, with an about east-northeast strike.

Combining these data it appears that in the area of Gotland neither the direction nor the location of the axis of tilt has been constant during the Middle Palaeozoic.

There is reason to assume that the same reasoning can be applied to the Cambrian and Ordovician in the west Baltic area.

Deposits of the *Paradoxides paradoxissimus* Stage occur in Öland in an about south – north belt, thinning out towards the north; the *Paradoxides forchhammeri* Stage is barely represented. Underneath northern and central Gotland both these stages are not represented at all. This suggests that during the Middle Cambrian or *Paradoxides* Epoch the direction of the coast line in the northwestern part of the basin was curved, with a direction from north-northeast bending to northeast and that this coast line crossed the area of present-day Gotland south of Visby and File Haidar.

After a temporary retreat of the basin, during the Late Cambrian or *Olenus* Epoch the sea re-invaded the area of Öland from south to north. The isopachs of the *Olenus* Series traverse the island from west to east. This may have been the direction of the coast line and also of the axis of tilt. Underneath Gotland the Upper Cambrian is missing.

The isopachs of the *Ceratopyge* Series run in southwest – northeast direction with some deviation in western direction towards the northern part of the island (Thorslund, 1960, p.96). Underneath Gotland this stratigraphical unit is not represented. Apparently, again a change in the direction of the coast line took place on the verge of the Cambrian to the Ordovician and a situation resulted with a coast direction of northeast bending to east-northeast, passing south of the present-day Visby and File Haidar.

The isopachs of the *Asaphus* Series run in Öland in about the direction of the long axis of the island. Underneath Gotland this series is also encountered, but with many hiatuses. The direction of the coast line during the *Asaphus* Epoch seems to have been about north-northeast, bending to northeast.

Too little information is available about the Middle and Late Ordovician and the Early Silurian in the west Baltic area to speculate about coast-line changes during these times, and about epeirogenetic movements of the basin floor which may have caused these changes.

JOINTS

Almost all Middle Palaeozoic sediments in Gotland show more or less well developed joints (Fig.12). They were first studied by Kaufmann (1931). Von Bubnoff (1931, 1952, p.95) has made an attempt to put the joints systems of Gotland into a more general picture of the Baltic area as a whole.

In the stratified limestones of Gotland the groups of joints are generally very regular in direction. They are systematic joints (Hodgson, 1961). The dip is vertical to sub-vertical. The dominating direction of strike is northeast, with annexed, subordinate joints generally about perpendicular to them. The criteria for distinguishing the main and annexed joints are given in Table VII. In several limestone exposures in northern Gotland this system is crossed by a second double system of joints, with its main joints striking about east-northeast – west-southwest.

In the unstratified reef limestones the strikes of the groups of joints are more irregular, their walls more uneven, fillings of calcite remarkably less common. The difference can well be established in the coastal cliff in the northwest. The crinoid limestone around the Högklint reefs is regularly jointed, the joints striking northeast. However, at the boundary with the reef limestone they terminate. On the reef limestone side of the boundary, joints are irregular and indistinct.

In the marlstones the strike of the groups of joints is the same as in the adjacent limestones, i.e., northeast, but the individual joints are generally further from one another.

In the Burgsvik sandstone joints are still more separated (6–10 m). Whereas in some exposures (Hoburgen, Kettelsård, Valar, Burgsvik) the main joints strike about northeast, in others (Öja, Fidenäs, Smiss in Grötlingbo) they show an about east – west strike. In all cases there is only one double joint system in the sandstone.

The main joints in general thus show a strike that is parallel to the supposed northwestern border of the basin and to the main strike of the Palaeozoic strata. In relationship to the bedding planes the main joints are strike joints and the annexed joints are dip joints (Badgley, 1965, p.99). The same applies to the joints in the Cambrian and Ordovician of Öland and in the Lower Cambrian in the Kalmar area, along the east coast of the Swedish mainland. The usual northeast strike is gradually replaced in the latter area, towards the south, by a dominating north-northeast strike. Kaufmann (1931) has shown that the latter joint system had already occurred during Jotnian time; it remained the main direction until the Early Cambrian. In the Early Cambrian apparently the epeirogenetic movements affecting the Palaeozoic basin of the Baltic area began, causing new directions of the groups of joints.

Of the joint systems in Gotland only the main joints often contain a filling of calcite. This indicates that only these were open when calcite

(Text continues on p.52)

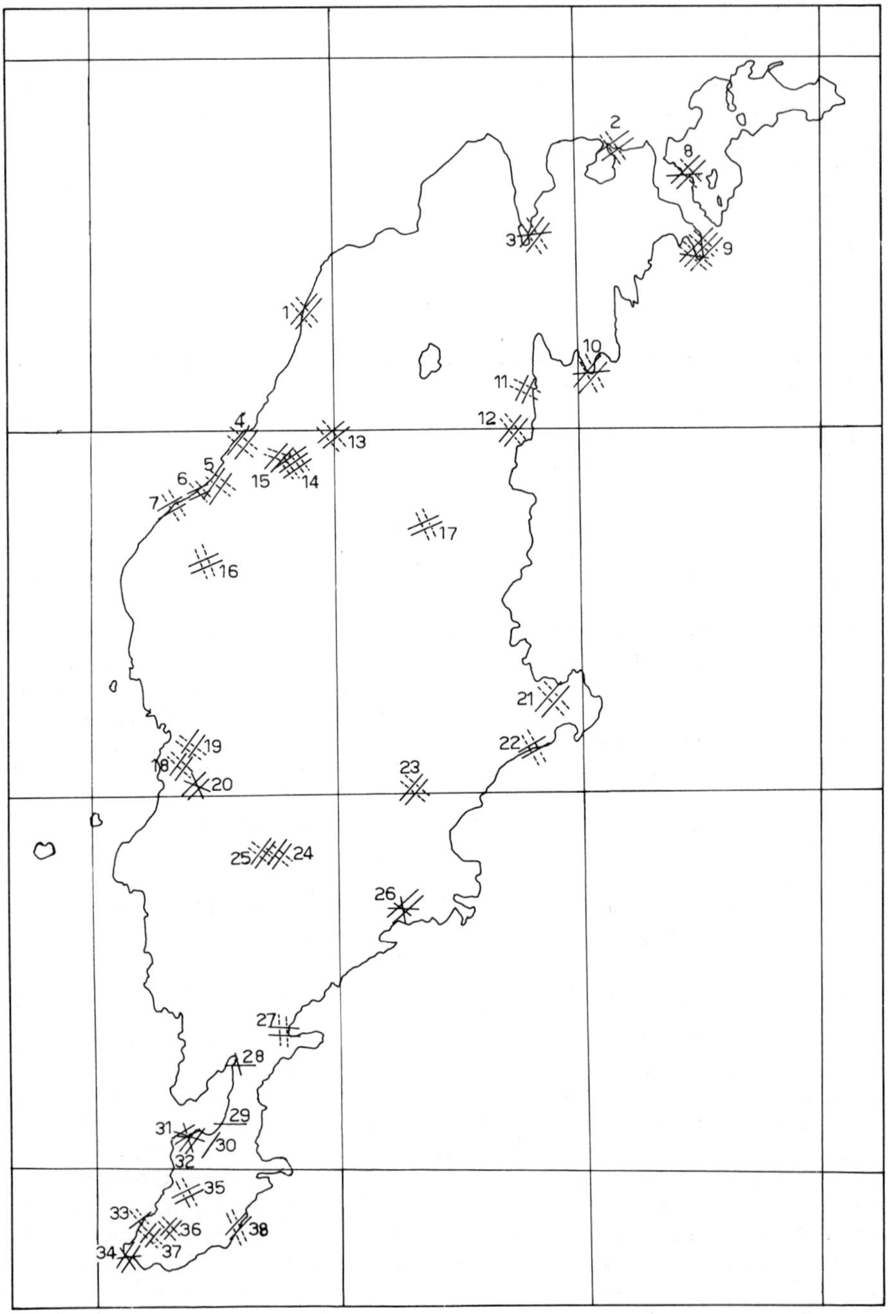
1
2
3
4
5
6
7
8
9
10
11
12
13
14
15
16
17
18
19
20
21
22
23
24
25
26
27
28
29
30
31
32
33
34
35
36
37
38

TABLE VII

Differences between the main joints and the annexed joints in limestones of Gotland (After Kaufmann, 1931)

Main joints (strike or longitudinal joints)	Annexed joints (dip or cross joints)
(1) common	(1) less common
(2) only limited variation in strike within a certain locality	(2) strike more scattered
(3) may be followed over distances up to 100 m and more	(3) often terminating against the main joints
(4) usually planar, smooth surfaces	(4) more curved, undulating and rough surfaces
(5) usually continuing downwards in the same plane	(5) at bedding planes often shifting step-wise
(6) as a rule vertical	(6) sometimes dip deviating from vertical
(7) thin (0.5–5 mm) fillings of calcite very common; occasionally thick fillings (up to 3 cm)	(7) no fillings; often exfoliation
(8) long horizontal striations rather common	(8) striations less common, short and wedge-shaped

Fig.12. Joint directions in Gotland. The following localities are represented: 1. Luseklint, coastal cliff, marlstone with interstratified limestone layers, Upper Visby Beds; 2. Vialms-udd, stratified limestone, Högklint Beds; 3. Storugns, large quarry, stratified limestone, Högklint Beds; 4. Galgberg, large quarry, irregularly stratified limestone, Högklint Beds; 5. Visby, large quarry of the former cement factory, stratified limestone, Högklint Beds; 6. Visby shooting range, small quarry, stratified limestone, Högklint Beds; 7. Högklint, cliff, stratified crinoid limestone, Högklint Beds; 8. Fårö, large quarry, stratified limestone, Slite Beds; 9. Bungenäs, large quarry, stratified limestone, Slite Beds; 10. Sankt Olofsholm, large quarry, stratified limestone, Slite Beds; 11. Solklint, Slite, stratified limestone, Slite Beds; 12. Bogeklint, small quarry, stratified limestone, Slite Beds; 13. Hejdeby, small quarry, stratified marly limestone, Slite Beds; 14. Hejdeby, small quarry, stratified marly limestone, Slite Beds; 15. Hejdeby, large quarry, stratified marly limestone, Slite Beds; 16. Stenkumla, small quarry, stratified limestone, Slite Beds; 17. Barabacke, stratified limestone, Halla-Mulde Beds; 18. Mulde-Stenstu, marlstone, Halla-Mulde Beds; 19. Klinteberg, large quarry, stratified limestone, Klinteberg Beds; 20. Alstäde, small quarry, stratified limestone, Klinteberg Beds; 21. Östergarn, large quarry, stratified limestone, Hemse Beds; 22. Sjausterhammar, stratified limestone, Hemse Beds; 23. Sigsarve, small quarry, stratified limestone, Hemse Beds; 24. Lindeklint (east), cliff, stratified limestone, Hemse Beds; 25. Lindeklint (west), small quarry, stratified limestone, Hemse Beds; 26. Burgen, large quarry, stratified limestone, Burgsvik Beds; 27. Smiss, large quarry, sandstone and claystone, Burgsvik Beds; 28. Fidenäs, large quarry, sandstone, Burgsvik Beds; 29. Öja, small quarry, sandstone, Burgsvik Beds; 30. Burgsvik, small quarry near school, sandstone, Burgsvik Beds; 31. Valar, small quarry, sandstone and oolite, Burgsvik Beds; 32. Sigreifs (Vamlingbo), small quarry, sandstone and oolite, Burgsvik Beds; 33. Kettelvik, large quarry, stratified limestone, Hamra-Sundre Beds; 34. Hoburgen, large quarry, stratified limestone, Hamra-Sundre Beds; 35. Vamlingbo, small quarry, stratified limestone, Hamra-Sundre Beds; 36. Sundre, small quarry, stratified limestone, Hamra-Sundre Beds; 37. Sundre, small quarry, stratified limestone, Hamra-Sundre Beds; 38. Vändburgsviken, stratified limestone, Hamra-Sundre Beds. (11, 17, 22 measured by the present author, all other data after Kaufmann, 1931).

deposition took place. The stress field thus has been in a direction perpendicular to the main joints, i.e., tension occurred generally in a northwest – southeast direction. This is fully in accord with the view presented in the previous section of this chapter, which posits epeirogenetic movements around a varying but often about northeast striking hinge line. In relationship to the regional structure the main joints may thus be called longitudinal joints and the annexed joints are then cross joints (Badgley, 1965, p.99).

It has been stated in the previous section of this chapter that the tilt axis has not always pointed in the same direction. During Halla-Mulde time the strike of the hinge line seems to have been about east-northeast. This same direction is indicated by the main joints of the second double system in northern Gotland. It is not impossible that this double system has been caused by the epeirogenetic movements that took place at about the Halla-Mulde time. The double system with its northeast striking main joints may have originated at that same time, but is determined by the strike of the strata. Presumably it has been intensified by post-Halla-Mulde movements.

In the limestones of southern Gotland the presence of a second system with east-northeast striking main joints can also be established, but it is much less pronounced there. The main joints in the Burgsvik Beds everywhere are in the direction of the strike of the strata (see also Chapter XI, p.407).

Chapter V

THE FOSSIL REEFS OF GOTLAND, GENERAL

Now that a general view has been obtained of the picture in which the Silurian reefs of Gotland fit, we may in this chapter begin a more detailed description of these reefs by discussing their typological, geographical and stratigraphical distribution and organic and lithological composition. Before doing so, however, a few words on the definition of the word "reef" and a few lines on reef building during the Palaeozoic in general are justified.

REEF DEFINITION

There are probably few terms in common geological use about the definition and use of which so much has been written as on the term "reef". If a conclusion can be drawn from this, it is that these organic structures vary so much in size, form and composition with local conditions in space and time that each definition of organic reefs can only be artificial and arbitrary. This is certainly true if the definition has to be applied to both Recent and fossil reefs. Consequently, there is little purpose at this point in discussing the various definitions proposed in literature or attempting to give a general definition. Instead, the present author will follow Colter (1957) by listing some of the factors that might be expected to have controlled reef formation, and some of the characteristics which might be expected to be found in reefs.

(1) Reefs have probably formed in quite shallow water because of the dependence or probable dependence of part of the constituent organisms upon light. This shallow water may lie off a gradually sloping shore, around or over an oceanic island or submerged island, or over a more or less level shelf.

(2) Subsequent growth of reefs has and does depend upon a multitude of factors, including stability, disposition and form of the original substratum, relative movements of sea level, suitability of local climatic conditions, composition, ventilation and degree of turbidity of the surrounding waters, degree of exposure to wave and current action, size and force of local waves.

(3) There must be, in each case, an association of benthonic organisms, chiefly compound organisms, building up a rigid frame, which makes up part or all of the structure. Generally, material was and is swept from this frame by wave action, forming more or less extensive fragmental deposits associated with the frame. These deposits vary in disposition with the form of the reef and the movement of the water. Some or all of the space between the frame builders is filled by this fragmental material.

(4) As a result of the intense biological activity associated with the

reef, the rate of deposition of material there has exceeded or will exceed, in some degree, that in adjacent areas of the sea floor. Given the right circumstances, the reef could or may thus grow up above its surroundings, possibly at steep angles, made possible by the frame-building organisms.

The character of wave resistance, used in several reef definitions, has been omitted from this list. This has not been done because Lowenstam (1950, p.438) rejected the structures in Gotland as reefs, since he felt that there should be no sufficient evidence that they were wave resistant (in the same journal issue, Hadding, 1950, p.405, presented evidence of "reef induced turbulence in Gotland", Lowenstam's criterion for wave resistance!) The true reason is threefold: (1) wave action is not a universal constant but depends on a variety of mainly local factors; (2) with organic structures, there is a grading spectrum from non-wave-resisting to wave-resisting depending on both environment and the variations in composition and potentials of the organic structures; (3) it is often difficult to prove that fossil reefs were wave resistant (as an example serve Lowenstam and Hadding's controversial statements about the reefs of Gotland).

The term "bioherm", introduced by Cumings (Cumings and Shrock, 1928a; see also Cumings, 1932), is considered as a (superfluous) synonym of "organic reef", since by derivation and definition[1], it means little more than "organic reef". More useful is his term "biostrome" for purely-bedded structures, such as shell beds, crinoid beds, coral beds, etc., consisting of and built mainly by these organisms and not swelling into mound-like or lens-like forms. No other word existed for this type of structure. Since 1928, the term biostrome has been widely used.

When discussing the lateral extension of the reefs in the Högklint Beds of Gotland, Hadding (1941, pp.13--14) points out that whereas some of these show the shape of an inverted cone with a curved and widely-extended base, others seem to have developed over wide parts of the sea bottom, showing a lateral extension of up to 1.5–2 km. He apparently decided that reefs of the latter type should be called biostromes. The present author is of the opinion, however, that Cuming's term is incorrectly used that way, even though it must be admitted that the distinction between a laterally extended reef or bedded outgrowths of normal reefs and a biostrome must become rather obscure under certain circumstances. The extended Högklint reefs, however, are more of a biohermal than a biostromal type of development.

The rock mass that has remained of fossil reefs, the reef limestone, is generally classified in systematic petrology as a calcareous sediment. According to the *Geological Nomenclature* (Schieferdecker, 1959), a sediment is "any part of the lithosphere that has been formed by sedimentation". In the same dictionary "sedimentation" is defined as "the process of deposition of solid lithologic materials on the lithosphere". In this sense, reef limestone is not a true sediment, since an important part of it, the reef frame, has not been deposited, but owes its existence to the building activities of organisms that created the rock on the very place where it forms part of the lithosphere. To be absolutely consistent, the reef frame could rather be called a *sedentate* and its formation *sedentation*. It is, however, so common to also use the word

---

[1]A bioherm is a reef-like, mound-like, lens-like or otherwise circumscribed structure of strictly organic origin, embedded in rocks of a different lithology.

sediment for reef limestone, that the present author will do so as well. "Sediment" should then be redefined as "any part of the lithosphere that has been formed either by sedimentation or by organisms attached to the lithosphere".

PALAEOZOIC REEF FORMATION

At this point, it may be useful to briefly sketch the picture which we have at the present time regarding Palaeozoic reef formation.

The oldest true reefs (bioherms) known, date as far back as the Middle Precambrian, and the first organisms to build such reefs were calcareous Algae. During the Cambrian, sponges also developed reef-building abilities. They were followed in the Ordovician by the stromatoporoids, corals of various kinds, and bryozoans. Four groups of organisms: stromatoporoids, corals, bryozoans and Algae, were the dominant reef builders during the subsequent periods of the Palaeozoic. They were generally accompanied by a great variety of reef-dwelling organisms. The climax of Palaeozoic reef building was reached during the Silurian and Devonian Periods. It should be noted that this Silurian – Devonian peak in reef formation is synchronous with the maximum diversification, abundance and geographical distribution of the stromatoporoids, the most dominant reef-forming organisms during these two periods.

During the Silurian Period, stromatoporoids were especially common in three large areas, viz. the North American area, the British – Baltic area and the Siberian – east Asian area. Extensive reef formation has occurred in all three areas, the reefs of North America being those most thoroughly studied. The huge extent of the eastern reef belt is shown by Lowenstam (1957, p.215, fig.1). It stretches as far north as Cornwallis Island in the Arctic, to the south of Indiana and Illinois, and to the east to the Gulf of St. Lawrence and Anticosti Island. A short review will be given in Chapter XIII. In the west of the U.S.A., reefs are found in Nevada (Winterer and Murphy, 1960). The reefs in England and the east Baltic area will also be briefly discussed in Chapter XIII. In southern Europe, graptolitic facies predominate in the Silurian deposits, possibly reflecting depths of water too great for reef growth. Unfortunately, the present author is not sufficiently informed to include a description of the reefs in the Siberian – east Asian area in Chapter XIII. No reefs seem to be present in Japan. In South America, Africa and Antarctica, no Silurian reefs are known. The Ordovician – Gotlandian reefs in southeastern Australia, indicated by Schwarzbach (1949), are now believed to be of Lower Devonian age and to be coralliferous limestones rather than real reefs (Teichert, 1952, p.35), but Talent and Philip (1956) still consider real reefs of uppermost Silurian or Early Devonian age to be present in Victoria.

Although large reefs are known to exist from the Ordovician onwards, small reefs dominate in the stratigraphical column of the Palaeozoic. Generally, several of these occur together.

HISTORICAL REVIEW

When looking at the number and character of scientific geological publications dealing with Gotland, it can be seen that initially, during the nineteenth century and the very beginning of this century, attention was focused mainly on the well-preserved organic remains of the Middle Palaeozoic of Gotland. Papers dealing with the stratigraphy were less in number, but increased rapidly due to the works of Holm (1901), Munthe (1902, 1907, 1910, 1912, 1915a, 1916), Hedström (1904, 1910), Van Hoepen (1910) and Hede (1919a,b, 1921), the discussion - following the publication of Hede's stratigraphy (Hede, 1921) - between Hedström (1923b,c), Munthe (1923) and Hede (1923a,b), and the working out of Hede's stratigraphy in the geological maps of the island (Munthe, 1921b; Munthe et al., 1925; Hede, 1925b, 1927a,b, 1928, 1929, 1933, 1936, 1940). Then work on the Palaeozoic of Gotland diminished. The total number of publications between 1931 and 1950 was only about one third of that between 1911 and 1930. Since 1950, however, interest in the world of fossils has strongly revived. In the ten years 1951 – 1960, 40 % more papers were published about the Middle Palaeozoic of Gotland than in the twenty years before. Among these, publications dealing with the reefs and their fauna again play an important part.

The first to recognize the reef character of the unstratified limestones of Gotland, was Murchison (1846). He stated that the raukar near Slite " are in fact dismantled portions of former hard coralline reefs". Since then, Dames (1890), Wiman (1897a, 1933), Hedström (1910), Bather (1914), Twenhofel (1916) and Wedekind and Tripp (1930), among others, have published material about the Silurian reefs of Gotland. A detailed historical review of work carried out before the Second World War, can be found in Hadding (1941, pp.79–94). The latter author, however, does not mention that E. Maillieux held a lecture on the reefs of Gotland for the Geological Society of Belgium in 1933. Unfortunately, only a summary has been published (Maillieux, 1933). This summary does not contain much more than a review of the stratigraphy of Hede (from which, without argumentation, the Halla Limestone has been omitted) and the distribution of the reefs. It states that there are morphological differences between the reefs of Gotland and the Devonian reefs of the Ardennes, but these differences are not described.

Hadding himself has also contributed considerably to the study of the reefs of Gotland (Hadding, 1941, 1950, 1956, 1959). With the main exception of the Tofta limestone, which he regarded as a shallow-water facies of the uppermost Högklint and the lowermost Slite Beds, he built on the stratigraphical work of his colleague at Lund University, J.E. Hede. The picture presented by Jux (1957) is quite different, being based both when treating the stratigraphy and the reefs, too much on an ideal model derived from theoretical considerations, rather than on the true situation found in Gotland in the field (cf. Chapter III).

In 1958, Rutten reported on some work done by students from Utrecht University, of which the present author was one. Following some preliminary publications (Manten, 1958, 1961a, 1962, 1966a,b, 1968, 1970) the present book is a continuation of the studies which were begun by his group.

GENERAL TYPOLOGY

All previous authors have discussed the Silurian reefs of Gotland without making any subdivision for them other than a stratigraphical one. Nevertheless, distinct differences do occur, which are very characteristic.

TABLE VIII

Main characteristics of the three reef types of Gotland

Characteristic	Upper Visby type	Hoburgen type	Holmhällar type
enclosed in a profile of	marlstone and marly limestone in alternating layers	marly limestone and limestone	rather pure limestone
most common shape	knoll; lens; inverted cone	inverted right elliptical cone; very elongated flat lens	crescent
average size of reef limestone section	less than 10 m^2	about 100 m^2	more than 1000 m^2
ratio height/length of the reefs	$1 - \frac{1}{5}$	$1 - \frac{1}{50}$	$\frac{1}{15} - \frac{1}{75}$
organic composition of reef frame	rather variable	generally variable	rather uniform
Algae	absent or very rare	common	very common
number of fossil species per reef as a whole (incl. reef dwellers)	moderate in number (10 – 60) for small reefs	many (20 – 200)	not many (5 – 40) for large reefs
dominant reef builders	corals	stromatoporoids; corals	stromatoporoids
average size of reef builders	relatively small	larger	large
average shape of stromatoporoid colonies	rather flat	lenticular	round or irregular and high
matrix	very strongly marly	strongly marly	little marly
weathered surface	conglomeratic; sometimes bed-like	generally conglomeratic; partly brecciated or bed-like	massive
surrounding sediments	narrow mantle of stratified marly limestone	reef detritus; large amounts of crinoid limestone	reef detritus; ? crinoid limestone

Therefore, the present author proposes a subdivision of the Gotlandian reefs into three main types. Table VIII gives a summary of the most important differences, as these can be observed in the field. In order to exclude, as much as possible, an element of perhaps subjective interpretation, no arguments based on palaeoecological deductions regarding these reef types have been advanced. A comparison of the last sections of the Chapters VI, VII and VIII will show, however, that the observed typological differences are, in all likelihood, actually connected with differences in the genesis of these reef types.

The Upper Visby reefs are the oldest reefs found in the island. Since they are restricted to one stratigraphical unit, there is little objection to using the stratigraphical name to also indicate the type of reef occurring in it.

Both other types of reefs, however, are found in more than one stratigraphical unit. Therefore, new names have to be introduced to characterize them. Most common are reefs of the Hoburgen type, for which the well-known cliff-complex of Hoburgen, in the south of Gotland, has been adopted as the standard example.

Less common are the large reefs of the third type. A very characteristic exposure of such a reef, which has been intensively studied, is the raukar field of Holmhällar, on the east coast of southernmost Gotland. Consequently, this type has been named the Holmhällar reef type.

Some other reef types, deviating from these three and occurring only on the small Karlsöarna (Carl Islands) off the west coast of Gotland, will be discussed in Chapter X.

STRATIGRAPHICAL AND GEOGRAPHICAL DISTRIBUTION

Reefs of the Upper Visby type occur in Gotland in the Upper Visby Beds only. Exposures are found almost exclusively in the coastal cliff in the northwest, between Nyrevsudde and Hallshuk.

The reefs of the Hoburgen type have, by far, the widest stratigraphical and geographical distribution. They are found in the Högklint, Slite, Halla, Klinteberg, Hemse, Eke, Burgsvik and Hamra-Sundre Beds. They are most abundant in the Högklint, Slite and Hemse Beds. Geographically, they are rather widely distributed over each of the three limestone areas (Fig.11; see also the enclosed geological map).

Reefs of the Holmhällar type are found in the Hemse and the Hamra-Sundre Beds. Those of Hemse age are exposed all along the east coast, between Snabben and Ljugarn. The youngest Holmhällar-type reefs are exposed in the Sundre-limestone area, mainly along the east coast of the southern peninsula of Gotland and slightly inland.

FOSSILS IN THE REEFS AND RELATED SEDIMENTS

In discussing the reef fauna and flora in the following pages, and the various reef types and stratified deposits in further chapters, one must realize that a general insight into the data presented can only be gained if they are placed against the background of their palaeoenvironment. It is necessary to recognize that all that can be said about palaeoenvironments is based upon interpretations.

These interpretations will have to be deduced from two main sources:

(1) The nature of the rocks, both reef limestones and stratified sediments. Their composition and fabric has generally been co-determined by the geographical, physical, chemical and biological environment of formation.

(2) The fauna and flora preserved in and contributing to the deposits.

The latter source again is comprised of two approaches:

(a) A comparison with taxonomically-related living representatives. Much caution is, however, required in this approach, especially since the relationships are so remote in time.

(b) A comparison with morphologically-related living organisms, with respect to any constancy of association between certain growth forms and certain environments.

Reef builders and reef dwellers

Modern reefs possess a rigid frame, built up by the activity of organisms that were firmly attached to the underlying material and who through building their calcareous structures, create and maintain an environment favourable for their growth. This framework of the reefs is generally more or less discontinuous, but the interstices are filled with sediment consisting of the remains of other organisms, of debris derived from the fragmentation of the actual reef builders and of terrigenous material trapped by the reef.

The same holds true for ancient reefs, including those found in Gotland. Therefore, the organic remains found in them can be distinguished in reef builders and associated organisms or reef dwellers. The term "reef builders" is used here to mean those organisms which actually built the solid reef frame, such as stromatoporoids, corals, bryozoans and Algae. Associated organisms may be defined as those whose remains chiefly fill in the interstices in this framework, such as crinoids, brachiopods, gastropods, cephalopods, and others. Of course, the wording of the distinction is not very sharp. As has been said before, fragments of the reef builders may help to form the matrix, and crinoid roots may add to the framework.

In the sediments directly surrounding the reefs, both fragments and intact specimens of the hard parts of the reef builders are generally present. They are found there together with remains of a great variety of other organisms which found favourable living conditions in the reef environment. Among the latter, crinoids are usually strongly dominant.

Table IX gives a collective survey of the organisms of which remains have been observed in the reef limestones and their surrounding crinoid limestones. In the following pages, the various fossil groups will be briefly discussed and their importance assessed. A more specific treatment of the organic content of each of the different reef types will be given in the respective chapters on these reef types. Although the same main taxa have contributed, chiefly, to each of the three kinds of reefs, the emphasis seems to be on different forms.

Reefs of the Hoburgen type are, by far, the richest in species. Altogether, about 265 species have been identified (of which about 50 can be classified as reef-building species) and the actual number present is even greater.

The smallest number of species known is from the Holmhällar-type reefs. Some reasons can be given to explain this: (1) the number of exposures is much smaller; (2) the rock is generally very hard and consequently difficult to sample; moreover, marl pockets have only scarcely been preserved and it is particularly in these that a very large number of reef dwellers in the Hoburgen-type reefs can be collected. These two factors, which concern the way in which the Holmhällar-type reef limestone presents itself, should certainly be taken into account. It is the author's firm belief, however, that there is still another reason of a primary nature: (3) the organic composition of the Holmhällar-type reefs has presumably always been more uniform, certainly as far as the actual reef builders are concerned.

At least 61 species are known from the Upper Visby reefs, but there, too, the actual number may be much higher; compared to the Hoburgen-type reefs, especially the reef-dwelling fauna is considerably poorer in species.

(Text continues on p. 68)

TABLE IX

Fossils found in the reef limestones and surrounding crinoid limestones of Gotland

Lithology	Reef limestone										Crinoid limestone							
		Hoburgen type							Holmhällar type									
Beds / Fossils	Upper Visby	Högklint	Slite	Halla	Klinteberg	Hemse	Eke	Hamra	Hemse	Sundre	Högklint	Slite	Halla	Klinteberg	Hemse	Eke	Hamra	Sundre
ALGAE																		
Rothpletzella sp.		+	+	+	+	+	+	+			+		+		+	+		
Solenopora gotlandica Rothpletz		+									+							
Solenopora sp.			+	+	+	+	+	+						+	+			
Unidentified calcareous Algae		+	+	+	+	+	+	+	+	+	+			+	+	+	+	
HYDROZOA																		
Actinostroma astroites (Rosen)		+																
Actinostroma sp.		+	+															
Clathrodictyon striatellum (d'Orbigny)	+		+									+						
Clathrodictyon cf. *variolare* Rosen		+																
Clathrodictyon cf. *vesiculosum* Nicholson et Murie		+																
Labechia conferta (Lonsdale)		+		+		+	+	+		+	+		+		+		+	+
Stromatopora discoidea (Lonsdale)	+	+							+	+								
Stromatoporella sp.								+										
Syringostroma sp.								+										
Unidentified stromatoporoids	+	+	+	+	+	+	+	+	+	+	+	+	+	+	+		+	+
ANTHOZOA TETRACORALLA																		
Acervularia ananas (L.)		+	+	+	+	+	+	+	+		+	+			+			
Acervularia breviseptata Weissermel		+									+							
Acervularia sp.		+	+	+	+			+				+		+				
Calostylis denticulata (Kjerulf)	+										+							
Clisiophyllum (*Dinophyllum*) *involutum* Edwards et Haime	+																	
Cyathophyllum bisectum						+		+										
Cyathophyllum sp.	+					+												
Cystiphyllum cylindricum Lonsdale		+										+						
Diploëpora grayi (Edwards et Haime)		+	+			+								+				
Dokophyllum högbomi Wedekind		+																
Entelophyllum fasciculatum Wedekind						+						+						
Goniophyllum pyramidale (Hisinger)	+																	
Hedstroemophyllum articulatum Wedekind	+	+									+	+						
Hedstroemophyllum sp.		+				+												
Holophragma calceoloides (Lindström)	+																	
Kodonophyllum truncatum (L.)		+									+							
Kyphophyllum sp.	+																	
Lykophyllum hisingeri Wedekind	+	+										+						
Lykophyllum tabulatum Wedekind	+																	
Lykophyllum torquatum Wedekind	+																	
Omphyma sp.	+	+	+									+					+	+
Phaulactis angusta (Lonsdale)	+																	
Pholidophyllum hedströmi Wedekind	+																	
Pholidophyllum tabulatum Schlotheim	+																	
Pholidophyllum tenue Wedekind	+																	
Polyorophe glabra Lindström	+	+									+							
Polyorophe lindströmi Wedekind											+							
Pycnactis sp.	+																	
Ptychophyllum truncatum (L.)							+											

TABLE IX (continued)

Lithology / Beds / Fossils	Reef limestone, Hoburgen type: Upper Visby	Reef limestone, Hoburgen type: Högklint	Reef limestone, Hoburgen type: Slite	Reef limestone, Hoburgen type: Halla	Reef limestone, Hoburgen type: Klinteberg	Reef limestone, Hoburgen type: Hemse	Reef limestone, Hoburgen type: Eke	Reef limestone, Hoburgen type: Hamra	Reef limestone, Holmhällar type: Hemse	Reef limestone, Holmhällar type: Sundre	Crinoid limestone: Högklint	Crinoid limestone: Slite	Crinoid limestone: Halla	Crinoid limestone: Klinteberg	Crinoid limestone: Hemse	Crinoid limestone: Eke	Crinoid limestone: Hamra	Crinoid limestone: Sundre
ANTHOZOA TETRACORALLA (continued)																		
Rhabdophyllum striatum Wedekind		+									+							
Rhegmaphyllum conulus (Lindström)		+	+												?			
Rhizophyllum elongatum Lindström							+											
Rhizophyllum gotlandicum (Roemer)							+									+		
Schlotheimophyllum patellatum (Schlotheim)	+	+																
Stauria favosa (L.)			+									+						
Syringaxon dalmani (Edwards et Haime)		+									+							
Zelophyllum högklinti Wedekind		+																
Zelophyllum intermedium Wedekind		+																
Zelophyllum spinosum Wedekind		+																
ANTHOZOA TABULATA																		
Aulopora roemeri Foerste						+	+								+	+		
Aulopora sp.		+	+	+		+			+			+			+	+	+	
Catenipora escharoides Lamarck	+																	
Favosites asper d'Orbigny		+	+															
Favosites gothlandicus Lamarck	+	+	+	+	+	+	+	+	+		+	+			+			
Favosites sp.	+	+	+	+	+	+	+	+	+		+	+		+	+			
Halysites catenularius (L.)	+			+	+	+	+					+			+			
Halysites catenulatus (Martini)	+	+	+	+							+							
Halysites sp.	+		+	+								+						
Milleporites madreporiformis Wahlenberg															+			
Planalveolites fougti (Edwards et Haime)	+	+								+		+						
Roemeria kunthiana Lindström	+	+				+						+						
Roemeria sp.			+									+						
Striatopora halli Lindström		+																
Striatopora stellulata Lindström		+																
Syringopora sp.	+		+	+	+	+	+		+			+			+			
Thamnopora lamellicornis (Lindström)	+	+																
Thamnopora sp.				+														
ANTHOZOA HELIOLITIDA																		
Cosmiolithus halysitoides Lindström	+																	
Heliolites barrandei Penecke			+		+	+		+				+						
Heliolites interstinctus (L.)	+	+	+	+	+	+	+	+	?		+	+			+	+		
Heliolites parvistella Ferd. Roemer	+	+	+		+	+						+		+				
Heliolites spongodes Lindström	+																	
Heliolites sp.	+	+	+	+	+	+		+	+		+	+			+			
Plasmopora calyculata Lindström	+																	
Plasmopora foroensis Lindström			+		+							+						
Plasmopora heliolitoides Lindström						+												
Plasmopora petalliformis (Lonsdale)												+						
Plasmopora rosa Lindström						+												
Plasmopora rudis Lindström			+			+												
Plasmopora scita Edwards et Haime	+			+				+										
Plasmopora suprema Lindström						+												

TABLE IX (continued)

Lithology / Beds / Fossils	Reef limestone: Hoburgen type: Upper Visby	Reef limestone: Hoburgen type: Högklint	Reef limestone: Hoburgen type: Slite	Reef limestone: Hoburgen type: Halla	Reef limestone: Hoburgen type: Klinteberg	Reef limestone: Hoburgen type: Hemse	Reef limestone: Hoburgen type: Eke	Reef limestone: Hoburgen type: Hamra	Reef limestone: Holmhällar type: Hemse	Reef limestone: Holmhällar type: Sundre	Crinoid limestone: Högklint	Crinoid limestone: Slite	Crinoid limestone: Halla	Crinoid limestone: Klinteberg	Crinoid limestone: Hemse	Crinoid limestone: Eke	Crinoid limestone: Hamra	Crinoid limestone: Sundre
ANTHOZOA HELIOLITIDA (continued)																		
Plasmopora sp.			+		+													
Propora conferta Edwards et Haime								+										
Propora speciosa Billings			+															
Propora tubulata (Lonsdale)	+	+									+							
Thecia hisingeri (Jones)	+																	
Thecia swinderniana (Goldfuss)	+	+	+	+							+	+						
Unidentified corals	+	+	+	+	+	+	+	+	+	+	+	+	+	+	+	+	+	+
ANNELIDA																		
Autodetus calyptratus (Schrenk)						+	+	+								+		
Conchicolites nicholsoni Vine	+																	
Conchicolites tuberculiferus Chapm.				+														
Conchicolites sp.		+	+	+			+					+			+	+		
Cornulites scalariformis Vine			+															
Cornulites serpularius Schlotheim				+		+	+					+			+	+		+
Cornulites sp.		+	+	+								+				+		
Spirorbis lewisi Sowerby							+								+	+		
Spirorbis sp.		+	+	+		+						+			+	+		
Unidentified annelid remains												+						
CRINOIDEA																		
Barrandeocrinus sceptrum Angelin												+						
Botryocrinus sp.			+									+						
Calceocrinus sp.												+		+	+			+
Crotalocrinus sp.								+									+	
Cyathocrinus sp.											+	+		+				
Eucalyptocrinus granulatus Lewis											+							
Euspirocrinus spiralis Angelin											+	+		+				
Gissocrinus sp.			+								+	+		+	+			
Herpetocrinus sp.												+		+				
Hypanthocrinus sp.											+							
Pisocrinus sp.												+		+	+	+		
Polypeltes sp.		+																
Promelocrinus sp.												+						
Streptocrinus crotalurus (Angelin)												+						
Unidentified crinoid remains	+	+	+	+	+	+	+	+	+	+	+	+	+	+	+	+	+	+
BRYOZOA																		
Berenicea consimilis (Lonsdale)															+		+	
Coenites repens (Wahlenberg)			+		+	+				+		+			+	+		
Coenites variabilis Hisinger				+														
Coenites sp.			+			+						+			+			
Fenestella mobergi Hennig		+	+															
Fenestella reticulata (Hisinger)		+	+			+	+				+	+			+	+		
Fenestella sp.		+	+	+		+					+	+	+		+	+	+	

TABLE IX (continued)

Lithology / Beds / Fossils	Reef limestone										Crinoid limestone							
		Hoburgen type							Holmhällar type									
	Upper Visby	Högklint	Slite	Halla	Klinteberg	Hemse	Eke	Hamra	Hemse	Sundre	Högklint	Slite	Halla	Klinteberg	Hemse	Eke	Hamra	Sundre
BRYOZOA (continued)																		
Fistulipora sp.		+	+	+		+		+										
Helopora lindströmi Ulrich	+										+							
Phaenopora lindströmi Ulrich	+	+																
Ptilodictya lanceolata (Goldfuss)	+	+	+	+			+				+	+	+	+	+	+		
Ptilodictya triangularis Hisinger											+							
Stomatopora sp.				+			+											
Unidentified bryozoan remains	+	+	+	+		+			+	+	+	+	+	+	+	+		+
BRACHIOPODA																		
Amphistrophia funiculata (McCoy)												+						
Anastrophia deflexa (J. de C. Sowerby)					+			+										
Antirhynchonella linguifera (J. de C. Sowerby)																		+
"Atrypa" phoca Salter										+								+
Atrypa reticularis (L.)	+	+	+	+	+	+	+		+		+	+	+	+	+	+		+
Atrypina angelini (Lindström)											+							
Camarotoechia borealis (Buch)	+	+	+	+							+	+	+					
Camarotoechia diodonta (Dalman)			+	+			+	+				+		+	+			
Camarotoechia nucula (J. de C. Sowerby)		+	+				+	+			+			+	+	+		
Camarotoechia sp.			+									+						
Chilidiopsis pecten (L.)			+								+	+						
Chonetes gotlandicus Hede													+					
Chonetes striatellus (Dalman)					+	+	+								+	+		
Chonetes sp.		+										+						
Chonetoidea grayi (Davidson)			+															
Cliftonia lindströmi Ulrich et Cooper		+									+							
Conchidium conchidium (L.)					+	+									+			
Conchidium knighti (J. Sowerby)						+									+			
Conchidium sculptum (Walmstedt)												+						
Conchidium tenuistriatum (Walmstedt)												+						
Conchidium sp.															+			
Crania sp.				+														
Craniops implicata (J. de C. Sowerby)						+			?				+		+			
Cyrtia exporrecta (Wahlenberg)			+									+						
Dayia navicula (J. de C. Sowerby)						+	+									+		
Delthyris elevata Dalman		+	+	+	+	+	+		+		+	+	+	+	+	+		
Dicaelosia biloba (L.)	+					+												
Dicaelosia verneuilana (Lindström)	+	+									+							
Dictyonella capewelli (Davidson)											+							
Dictyonella sp.			+	+														
Dinorthis rigida (Davidson)						+	+		+						+	+		
Dolerorthis rustica (J. de C. Sowerby)			+		+			+		+					+			+
Eospirifer globosus (Salter)			+									+						
Eospirifer grandis (Hedström)			+															
Eospirifer interlineatus (Hedström, non J. de C. Sowerby)			+		+	+											+	
Eospirifer radiatus (J. de C. Sowerby)	+	+		+							+		+					
Eospirifer schmidti (Lindström)					+	+	+		+						+	+		
Eospirifer sinuosus (Hedström)			+									+						
Eospirifer sulcatus (Hisinger)					+													
Eospirifer sp.			+	+	+		+	+				+				+		

TABLE IX (continued)

Lithology	Reef limestone										Crinoid limestone							
Beds		Hoburgen type							Holm-hällar type									
Fossils	Upper Visby	Högklint	Slite	Halla	Klinteberg	Hemse	Eke	Hamra	Hemse	Sundre	Högklint	Slite	Halla	Klinteberg	Hemse	Eke	Hamra	Sundre
BRACHIOPODA (continued)																		
Gypidula galeata (Dalman)			+	+	+	+						+	+	+				
Howellella elegans (Muir-Wood)		+		+								+						
Kozlowskiellina deltidialis (Hedström)				+														
Leptaena lovéni De Verneuil		+																
Leptaena rhomboidalis (Wilckens)	+	+	+	+	+	+	+		+		+	+	+	+	+	+	+	
Leptaenoidea silurica Hedström				+		?												
Leptostrophia filosa (J. de C. Sowerby)																	+	+
Leptostrophia impressa (Lindström)						+			?						+	+		
Levenea canaliculata (Lindström)					+	+	+		+						+	+		
Lingula lewisi J. de C. Sowerby						+												
Linoporella punctata (De Verneuil)		+						+			+	+						
Lissatrypa sp.						+			+									
Meristina obtusa (J. Sowerby)					+							+						
Nucleospira pisum (J. de C. Sowerby)							+								+	+		
Orbiculoidea pilidium (Lindström)							+											
Orbiculoidea sp.			+	+		+									+	+		
"*Orthis*" *tubulata* Lindström						+												
Orthothetes adnata Hedström				+														
Pentamerus gotlandicus Lebedev												+						
Platystrophia biforata (Schlotheim)			+	+								+	+					
Platystrophia sp.	+	+	+								+							
Plectatrypa imbricata (J. de C. Sowerby)	+	+	+								+	+						
Plectatrypa lamellosa (Lovén)			+									+						
Plectatrypa marginalis (Dalman)					+	+		+	+	+					+			+
Plectodonta duvali (Davidson)											+							
Plectodonta transversalis lata (Jones)											+							
Protoathyris didyma (Dalman)						+	+		+						+			
Protoathyris sp.											+							
Ptychopleurella bouchardi (Davidson)			+		+	+	+	+				+			+			
Resserella basalis (Dalman)		+									+							
Resserella elegantula (Dalman)			+		+						+	+						
Resserella visbyensis (Lindström)	+																	
Resserella sp.		+	+	+								+	+			+		
Rhipidomella hybrida (J. de C. Sowerby)	+	+	+	+			+				+	+			+	+		
Rhynchospirina baylei (Davidson)						+	+	+								+		
Rhynchospirina bouchardi (Davidson)				+														
Rhynchotreta cuneata (Dalman)	+	+	+	+							+	+	+					
Sphaerirhynchia wilsoni (J. Sowerby)		+	+	+		+					+			+	+			
"*Spirifer*" *insignis* Hedström						+	+											
Streptorhynchus nasutum (Lindström)				+														
Stropheodonta semiglobosa (Davidson)		+	+								+	+						
"*Strophomena*" *concinna* Lindström in museo															+			
"*Strophomena*" *orbignyi* Davidson										+							+	+
"*Strophomena*" *rugata* Lindström				+									+					
"*Strophomena*" *testudo* Lindström in museo		+									+							
"*Strophomena*" sp.			+					+				+		+				
Trimerella lindströmi (Dall)			+															
Trimerella sp.		+	+									+			+			
Unidentified brachiopods	+	+	+	+		+			+	+	+	+			+			

TABLE IX (continued)

Fossils \ Lithology / Beds	Reef limestone: Upper Visby	Reef limestone, Hoburgen type: Högklint	Hoburgen type: Slite	Hoburgen type: Halla	Hoburgen type: Klinteberg	Hoburgen type: Hemse	Hoburgen type: Eke	Hoburgen type: Hamra	Holmhällar type: Hemse	Holmhällar type: Sundre	Crinoid limestone: Högklint	Crinoid limestone: Slite	Crinoid limestone: Halla	Crinoid limestone: Klinteberg	Crinoid limestone: Hemse	Crinoid limestone: Eke	Crinoid limestone: Hamra	Crinoid limestone: Sundre
LAMELLIBRANCHIATA																		
Actinopterella sp.						+												
Conocardium sp.		+	+	+		+	+				+	+		+	+			+
Cypricardinia crispula (Lindström)							+									+		
Cypricardinia exornata Lindström in museo						+	+											
Cypricardinia sp.		+	+	+							+							
Goniophora cymbaeformis (J. de C. Sowerby)						+												
Ilionia prisca (Hisinger)						+									+			+
"*Megalomus*" *gotlandicus* Lindström														+	+			
"*Megalomus*" sp.			+															
Mytilarca acuta Lindström in museo			+															
Pterinea nodulosa Lindström in museo						+												
Pterinea sp.			+															
Rhombopteria sp.		+				+			+									
Unidentified lamellibranchs		+	+			+												
PTEROPODA																		
Conularia laevis Lindström			+															
GASTROPODA																		
Bellerophon gemma Lindström						+												
Bellerophon taenia Lindström					+	+								+				
Bellerophon sp.			+															
Craspedostoma elegantulum Lindström			+			+												
Craspedostoma glabrum Lindström						+												
Craspedostoma sp.			+	+														
Cyclonema adstrictum Lindström						+												
Cyclonema apicatum Lindström						+												
Cyclonema cancellatum Lindström						+												
Cyclonema carinatum (Sowerby)																	+	
Cyclonema distans Lindström						+												
Cyclonema perversum Lindström						+												
Cyclonema turritum Lindström		+																
Cyclonema zonatum Lindström						+												
Cyclonema sp.			+			+												
Euchrysalis lineolata Lindström						+												
Euomphalopterus alatus (Wahlenberg)	+	+	+		+						+	+						
Euomphalus walmstedti Lindström					+	+												
Holopea applanata Lindström		+																
Holopea nux Lindström						+												
Holopella minuta Lindström						+												
Hormotoma sp.			+									+						
Lophospira bicincta (Hall)			+															
Loxonema fasciatum Lindström						+												
Loxonema strangulatum Lindström					+	+												
Machrochilina bulimina Lindström						+												
Machrochilina cancellata Lindström						+												
Murchisonia attenuata (Hisinger)						+												
Murchisonia cancellata Lindström						+												

TABLE IX (continued)

Lithology	Reef limestone										Crinoid limestone							
Beds		Hoburgen type							Holm-hällar type									
Fossils	Upper Visby	Högklint	Slite	Halla	Klinteberg	Hemse	Eke	Hamra	Hemse	Sundre	Högklint	Slite	Halla	Klinteberg	Hemse	Eke	Hamra	Sundre
GASTROPODA (continued)																		
Murchisonia cochleata Lindström						+												
Murchisonia compressa Lindström						+									+			
Murchisonia crispa Lindström						+												
Murchisonia deflexa Lindström						+												
Murchisonia imbricata Lindström			+			+												
Murchisonia paradoxa Lindström						+												
Onychochilus cochleatum Lindström						+												
Onychochilus reticulatum Lindström						+												
Oriostoma acutum Lindström			+															
Oriostoma alatum Lindström		+																
Oriostoma angulatum (Wahlenberg)		+	+		+													
Oriostoma contrarium Lindström		+						+										
Oriostoma coronatum Lindström						+									+			
"*Oriostoma*" *nitidissimum* Lindström						+												
Palaeacmaea solarium Lindström						+												
Pilina unguis (Lindström)		+									+							
Platyceras cornutum Hisinger		+	+	+	+	+					+	+						
Platyceras cyathinum Lindström												+						
Platyceras disciforme Lindström								+										
Platyceras enorme Lindström			+															
Platyceras spiratum (Sowerby)					+	+												
Pleurotomaria aequilatera Wahlenberg		?									?							
Pleurotomaria bicincta (Hall)						+												
Pleurotomaria cirrhosa Lindström						+												
Pleurotomaria claustrata Lindström		+									+							
Pleurotomaria glandiformis Lindström						+												
Pleurotomaria gradata Lindström						+												
Pleurotomaria laqueata Lindström					+													
Pleurotomaria limata Lindström		+	+		+													
"*Pleurotomaria*" *linnarssoni* Lindström						+												
Pleurotomaria lloydi Sowerby						+												
Pleurotomaria marklini Lindström					+													
Pleurotomaria planorbis (Hisinger)						+									+	+		
Pleurotomaria voluta Lindström						+												
Pleurotomaria sp.	+					+												
Poleumita discors (J. Sowerby)		+	+		+	+					+				+			
Poleumita globosum (Schlotheim)		+	+	+	+	+		+			+							
Poleumita roemeri (Lindström)	+																	
Poleumita sculptum (J. de C. Sowerby)					+													
***Poleumita* sp.**			+	+								+						
Pycnomphalus acutus Lindström						+												
Subulites ventricosus Hall (according to Hedström, 1923)		+																
Tremanotus compressus Lindström						+												
Tremanotus longitudinalis Lindström														+				
Trochus astraliformis Lindström					+													
Trochus cavus Lindström						+												
Trochus gotlandicus Lindström		+																
Trochus incisus Lindström		+	+															
Trochus visbyensis Lindström											+							
Trochus sp.		+	+															
Tryblidium reticulatum Lindström		+									+							
Unidentified gastropods	+	+	+			+					+	+			+			

TABLE IX (continued)

Lithology	Reef limestone										Crinoid limestone							
		Hoburgen type							Holm-hällar type									
Fossils / Beds	Upper Visby	Högklint	Slite	Halla	Klinteberg	Hemse	Eke	Hamra	Hemse	Sundre	Högklint	Slite	Halla	Klinteberg	Hemse	Eke	Hamra	Sundre
TENTACULITIDA																		
Tentaculites multiannulatus Vine												+						
Tentaculites sp.			+				+				+	+				+	+	
CEPHALOPODA																		
Ascoceras bohemicum Barrande								+										
Ascoceras cochleatum Lindström										?								
Ascoceras cucumis Lindström						+												
Ascoceras decipiens Lindström						+				+								
Ascoceras fistula Lindström			+															
Ascoceras gradatum Lindström								+		+								
Ascoceras lagena Lindström			+															
Ascoceras manubrium Lindström						+												
Ascoceras pupa Lindström						+												
Ascoceras reticulatum Lindström						+												
Ascoceras sipho Lindström						+			+									
Ascoceras sp.			+			+			+	+								
Choanoceras mutabile Lindström			+															
Dawsonoceras annulatum (J. Sowerby)		+																
Glossoceras gracile Barrande						+												
Gomphoceras sp. (according to Hedström, 1923)						+												
Ophidioceras reticulatum Angelin			+		+	+			+	+				+				
Ophidioceras rota Lindström						+												
Ophidioceras sp.			+															
Orthoceras sp.		+	+			+		+	+	+	+	+			+			
Phragmoceras inflexum Hedström			+															
Phragmoceras praecurvum Hedström						+												
Phragmoceras sp.			+			+			+									
Unidentified cephalopods		+	+		+	+		+	+	+								
TRILOBITA																		
Arctinurus ornatus (Angelin)		+	+	+							+							
Arctinurus sp.		+																
Bumastus sulcatus Lindström			+															
Bumastus sp.		+	+	+		+					+				+			
Calymene excavata Lindström							+											
Calymene neointermedia R. et E. Richter							+									+		
Calymene spectabilis Angelin															+			
Calymene tuberculata (Brünn)		+	+	+	+					+	+							
Calymene sp.	+	+	+	+		+			+			+		+	+			
Dalmanites imbricatulus (Angelin)												+						
Dalmanites obtusus (Lindström)						+									+			
Encrinurus laevis (Angelin)	+																	
Encrinurus obtusus (Angelin)							+											
Encrinurus punctatus (Wahlenberg)	+	+	+		+	+	+		?		+	+	+	+	+			
Eophacops musheni (Salter)					+													
Eophacops sp.							+											
Proetus conspersus (Angelin)						+	+		?	+					+			
Proetus delicatus Hedström				+														

TABLE IX (continued)

Lithology / Beds / Fossils	Reef limestone										Crinoid limestone							
		Hoburgen type							Holmhällar type									
	Upper Visby	Högklint	Slite	Halla	Klinteberg	Hemse	Eke	Hamra	Hemse	Sundre	Högklint	Slite	Halla	Klinteberg	Hemse	Eke	Hamra	Sundre
TRILOBITA (continued)																		
Proetus verrucosus Lindström			+									+						
Proetus sp.		+	+	+		+			+		+	+	+		+			
Sphaerexochus laciniatus Lindström						+			+						+			
Sphaerexochus sp.			+			+									+			
Warburgella rugulosa (Lindström)											+							
Unidentified trilobites		+	+		+	+	+		+	+	+	+		+	+		+	+
OSTRACODA																		
Beyrichia sp.			+								+	+						
Craspedobolbina clavata (Kolmodin)			+									+	+					
Hemsiella maccoyana (Jones)						+									+			
Leperditia baltica (Hisinger)			+									+						
Leperditia gigantea Roemer															+			
Leperditia gregaria Kiesow						+									+			
Leperditia phaseolus (Hisinger)							+								+			
Leperditia sp.		+																
Neobeyrichia buchiana (Jones)										+								+
Neobeyrichia nodulosa (Boll)						+									+			
Neobeyrichia sp.						+									+	+		
Unidentified ostracodes		+	+	+		+					+	+	+		+			

Reef builders

Stromatoporoids

The most important reef builders of the reefs in Gotland proper are the stromatoporoids. They strongly dominate in the fauna of both the Hoburgen and Holmhällar-type reefs (see Chapters VII and VIII). Only in the smaller Upper Visby reefs are the stromatoporoids outnumbered by the corals (Chapter VI). In the reefs of the two Karlsöarna (Carl Islands), stromatoporoids are distinctly less common than in Gotland proper. This applies particularly to the lower parts of the reefs of Stäurnasar type and to the reefs of Fanterna type (see further Chapter X).

Despite the abundance and wide distribution of stromatoporoids in the Middle Palaeozoic of Gotland (and in Silurian and Devonian formations in many other parts of the world) relatively little is known about their structure and taxonomy. Some of the publications on the geology of Gotland deal incidentally with stromatoporoids. However, the group as such is still badly in need of extensive monographical treatment. The present author regrets that such a time-consuming study could not be included in his programme and consequently, only some rather general remarks can be made about this important group.

In the field, stromatoporoids are impossible to identify, with the exception of *Labechia conferta* (Lonsdale) and *Stromatopora discoidea* (Lonsdale). The former is coarsely structured and generally occurs in a thin-laminar growth form. It is quite common in Gotland, especially in reef parts with a relatively-high marl content, in marginal parts of reefs, and in some biostromal outgrowths of reefs on the surrounding beds. *Stromatopora discoidea* is usually found as thin covers over some other fossils.

In thin section, it is not very easy to study stromatoporoids in detail, owing to the recrystallization which has often affected them. Only a limited number of thin sections has been studied. Consequently the part of the hydrozoans in Table IX is not complete in either taxonomic or stratigraphical respects. The thin sections, nevertheless, indicate that the stromatoporoids of Gotland fall into two main groups.

The stromatoporoids of the first group have a skeleton composed of regular laminae and definite pillars. In this group, there are representatives of at least four genera: *Actinostroma* Nicholson (family Actinostromatidae), *Stromatoporella* Nicholson (family Clathrodictyidae) and *Stromatopora* Goldfuss and *Syringostroma* Nicholson (family Stromatoporidae). The tissue of the laminae and pillars of *Actinostroma* is compact; the pillars are strong and continuous. The tissue in *Stromatoporella* is coarse to fine porous; the pillars are not regularly superposed. In both genera of the family Stromatoporidae, the tissue is maculate and also fills to a large extent the interlaminar spaces; both genera can best be distinguished in tangential section, in which *Syringostroma* shows large, round pillars. Sections through a *Stromatopora* sp. are given by Hadding (1941, p.27, fig.20,21).

The stromatoporoids in the second group have a skeleton not composed of regular laminae and definite pillars, but of imperfect cyst plates. This group comprises the genera *Labechia* Edwards et Haime, in which the cysts occur in an imbricating manner, and *Clathrodictyon* Nicholson and Murie, in which the cysts are placed side by side or end to end, or approximately so.

Hadding (1941) discussed the different growth forms displayed by the stromatoporoids in Gotland. Roughly, the following main forms can be distinguished: (1) tabular, (2) dome-shaped or nodular, (3) spheroidal, (4) columnar. It should be added immediately, however, that there are no distinct boundaries between these four groups of shape types. The stromatoporoids exhibit an enormous variety in shape and transitions between various types of growth forms are common. The groups which Hadding distinguished are, in fact, not more than a rough expedient which aids in describing the variety that is found.

The tabular stromatoporoids spread as covers or thin lenses over larger or smaller parts of the sea floor. As covers, they are common in stratified stromatoporoid limestone. Even rather-clayey mud could be an acceptable ground for their development. In several instances such stromatoporoid covers formed a solid basis for other sessile organisms, often leading to reef building. Several such covers also commonly follow above each other, with mostly thin layers of marlstone or marly limestone in between them. A similar development but with less sediment in between, can be found in parts of some reefs, most particularly, the Gannberg variety of the Hoburgen reef type (see Chapter XI, p.357). Both the upper and lower surfaces of the covers are wavy, generally more so in the reefs than in stratified deposits, in some reefs being in part even very extreme. The thickness of an individual cover may show local variations, and generally the more so when it is strongly curved, but on the whole the thickness is rather constant per cover.

Fig.13. "Waistcoat-pocket bioherm". Boundary between Upper Visby marlstone and Högklint limestone. North side of Ihrevik. (After Rutten, 1958, fig.25.)

More common in the reefs, however, is a less regular development, with an uneven lower and upper surface, and a varying thickness. Together with the lenticular stromatoporoids, they may give the reef limestone a vaguely stratified appearance.

In the group of dome-shaped or nodular stromatoporoids, Hadding included "all the rather varying types which differ from the tabular ones by their very arched shape and from the spheroidal ones by their greater breadth". In this group he evidently also included forms shaped like thick lenses. More common, however, are thinner lenses. These are rather similar to the irregular-tabular forms, especially when occurring in large numbers as they often do. However, the thin-lenticular forms differ from the irregular-tabular forms in the fact that the successive thickenings in the plane parallel to the sea floor are not inter-connected but distinctly form separate colonies. The truly nodular stromatoporoids have a more irregular, tuberous shape than the lenticular forms. Lenticular and large nodular stromatoporoids may occur together in large numbers, constituting the bulk of several reef parts, especially of the Hoburgen type. Due to the influence of weathering and disintegration of the reef limestone, such parts generally have a coarse-conglomeratic appearance. Dome-shaped stromatoporoids and variations of that growth form are also common. They show usually well-developed latilaminae, which indicate that the colonies grew radially from an eccentric point in the fully developed colony, the successive latilaminae following more or less concentrically. The size of a colony can be large to very large, a length of 1 m being not unusual, especially in the Holmhällar-type reefs. There, as well as in the Hoburgen-type reefs and even in some stratified stromatoporoid limestones, several such more or less dome-shaped colonies may occur side by side or even joined together to a large cover with dome-shaped elevations. As Hadding also pointed out, these occupy an intermediate position between the dome-shaped and strongly wavy tabular stromatoporoids, but in contrast to the latter, they usually show pronounced latilaminae. As in other latilaminated stromatoporoids, weathering often causes the colonies to disintegrate in curved pieces, representing fragments of these latilaminae.

The spheroidal stromatoporoids often vary in size between that of a tennis ball and that of a walnut, but they may even be much larger. Usually they are strongly recrystallized and show no distinct latilaminae. They often occur in pure, massive reef limestone, which brecciates through weathering; in other cases they may contribute to a coarse-conglomeratic appearance of the rock. A single colony of spheroidal shape may form, which Rutten (1958, p.382) called a "waistcoat-pocket bioherm" (Fig.13).

Columnar or branched stromatoporoids have only sporadically been observed in the fossil reefs in Gotland.

Presumably both environmental factors and specific differences have contributed to the variety in growth forms. The former factor probably was the most important, but the author has not been able to clearly prove this. The subject of different stromatoporoid growth forms will be further touched on in Chapter XII.

Corals

Compared to the stromatoporoids, fortunately much more systematic literature is available on the corals that are represented in the Middle Palaeozoic of Gotland. No doubt this is due mainly to the fact that the external form of the corals is generally well preserved and characteristic and also the microstructure in the majority of specimens is still well recognizable. At least 36 species of tetracorals, 13 species of tabulate corals and 17 species of heliolitids have been found in the reefs. Most likely, however, the actual number of species present is distinctly higher still.

It seems that the solitary corals suffered comparatively less competition from the stromatoporoids and compound corals in a muddy environment. Solitary corals are abundant in the Upper Visby Marlstone and also, e.g., in the Lerberg Marlstone of Stora Karlsö (Chapter X). In the reefs they are commonest in the Upper Visby-type reefs and the lower parts of the Hoburgen-type reefs in the Lower Högklint Beds. Generally, however, even within the group of the anthozoans, the solitary corals are unimportant as reef builders, being overshadowed by the compound forms.

These compound corals play a prominent part in the reefs of Karlsöarna. In the lower parts of the reefs of Stäurnasar type *Halysites* particularly is a dominant reef builder. Compound corals are generally also common in the upper parts of these reefs and in the reefs of Fanterna type. Comparatively, the Karlsöarna reefs are much richer in corals than those of Gotland proper.

Compound tetracorals, although fairly common, have only a scattered distribution in the reef limestones of Gotland; comparatively speaking, *Acervularia ananas* (Linnaeus) is the most frequently found and occasionally in very large colonies. In the Hjännklint, a reef part of 4 m wide and 6 m high consists almost exclusively of *Acervularia* colonies, which are also common in the environment of that reef. A similar, but somewhat smaller occurrence of this coral was found in Storburg, Hoburgen.

Unbranched, massive compound corals are common in most of the

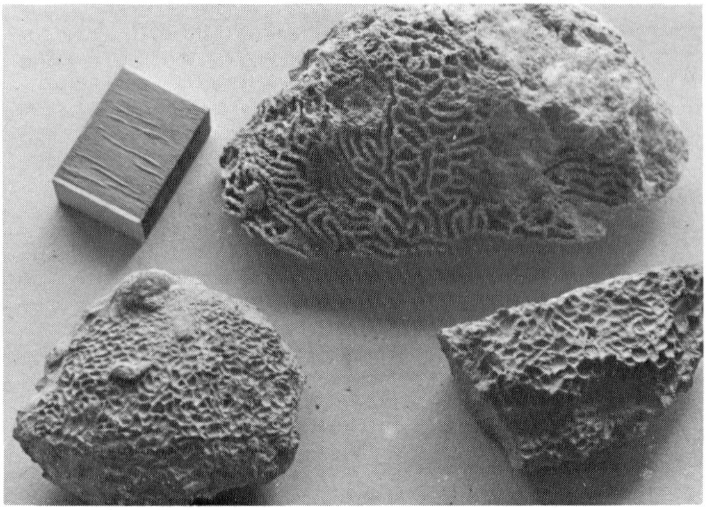

Fig.14. Two species of *Halysites*. Above *H. catenulatus* (Martini), below parts of two colonies of *H. catenularius* (Linnaeus). Both species are commonly found in the Upper Visby Beds, in both the reefs and the stratified sediments. They are also found in several reefs of Hoburgen type, relatively more frequently in the lower parts thereof than in the upper parts. (After Manten, 1962, fig.8.)

---

reefs of Gotland proper. They have been found to be most abundant in the Upper Visby reefs and least common in the reefs of Holmhällar type. Although they can commonly be found in all parts of the Hoburgen-type reefs, they are often relatively most frequent in the lower parts, where it is not unusual that they form the solid substratum for other reefbuilders. *Halysites* (Fig.14) is, of course, also well represented there, both by intact colonies and fragments, but the massive forms among social corals often outnumber it even in these reef parts. Among the latter, there is in the first place, *Favosites gothlandicus* Lamarck in its various growth forms, followed by a range of heliolitids, lead by *Heliolites interstinctus* (Linnaeus), but also comprised of some other *Heliolites* species, as well as species of the genera *Plasmopora* and *Propora*. The size of the colonies is generally small or moderate; in the lower reef parts, they are often bun-shaped or tabular, whereas larger and more roundish colonies are most common in the more marginal parts.

Thin covers of *Thecia swinderniana* (Goldfuss) are also often found, characteristically in the marginal parts of the reefs, although they occur elsewhere in the reefs too, as well as in the stratified sediments directly around the reefs. *Thecia* also appears in a branched form, the shape apparently being influenced by environmental conditions. The same holds for the laminar and branched growth forms both found with *Heliolites parvistella* Ferd. Roemer. More about this and other aspects of coral palaeoecology will be said in Chapter XII.

Branched corals are present mainly in the more central reef parts.

Where they are found, they often developed in a pool in the reef surface surrounded by stromatoporoids, sometimes together with some massive coral colonies.

Bryozoans

The number of bryozoan colonies observed in the reefs is very much smaller that that of the stromatoporoids and also less than the number of corals. Since they are, moreover, generally of small size, bryozoans usually do not contribute much to the total volume of the reef limestone. The only common exception are the reefs of Fanterna type in Karlsöarna.

The branched and fenestrate bryozoan colonies probably relied generally on the more protected parts of the reefs. Even there they may from time to time have suffered serious destruction, as their broken remains are commonly found in several reefs and the surrounding sediments. Occasional colonies which are found overgrown by stromatoporoids are well preserved in the reefs of Gotland proper. In Karlsöarna intact colonies are much more common.

Locally encrusting bryozoans are found, again most commonly in the more central parts of reefs, where they cover both reef builders and reef sediment; probably most of these encrustments belong to the genus *Fistulipora*.

Algae

Because one is apt to overlook the Algae in the Silurian reefs of Gotland, their contribution to reef formation has often been underestimated. Hadding, for instance, was guilty of this in his major contribution on the reefs of Gotland (Hadding, 1941), but later corrected himself (Hadding, 1950, p.407). Once having recognized the Algae, they are found to be quite common, even abundant in parts.

The role of the genus *Solenopora* was rather similar to that of the modern coralline Algae. The genus is fairly well represented. In places, its porcellaneous remains characterize substantial parts of the reef limestone. This is particularly true for the Högklint reefs, and also in Hoburgen-type reefs in the Slite Beds and elsewhere where it is found rock-forming, thereby encrusting other reef builders in places. The shape of *Solenopora*, as that of modern coralline Algae, varies and is generally very irregular. In cross-sections, growth laminae can be seen, generally less than 1 mm thick. These laminae are roughly concentric around the point of attachment, becoming increasingly irregular with greater distance from it and usually finally splitting up into a number of different lobes.

There has been much discussion on the true nature of the algal remains which have long been known as *Sphaerocodium*. The starting point has been the algal limestone in the Lower Hamra Beds. In the older literature, the algal material constituting the balls occurring frequently in the deposit is called *Girvanella problematica* Stolley, 1893, and the deposit consequently *Girvanella* limestone. In 1891, Rothpletz introduced the genus *Sphaerocodium*, on the strength of material from the Alpine Triassic. In 1895, Seward had already shown that the genus diagnosis of Rothpletz was dubious and insufficient (cf. Seward, 1898). Nevertheless, Rothpletz stuck to his genus and in 1908 described material from Gotland as *Sphaerocodium gotlandium*. Since then, the name *Sphaerocodium* has frequently been used in literature on the Silurian of

Gotland. This was even more so as algal remains with a similar structure to that of the Hamra Beds were also found in several other beds, including reef limestones, on the island. Pia (1926) stated that *Sphaerocodium* is only an intergrowth of several species of *Girvanella*. Hadding (1933) also noted that the "*Sphaerocodium*" balls are only seldom homogeneous, but that often two or more Algae contributed to the formation of these limestone concretions, whereas other organisms could also have aided in this. In 1941, he used the name *Pilothrix* for layers or thin coatings of *Sphaerocodium* material as these are found in the reefs (Hadding, 1941, p.38). Although the present author did not go deeply into the problem of the nature of these questioned algal remains, the impression gained is that in the reef limestone the filamentous Algae seem to belong mainly to the genus now known as *Rothpletzella*, with probably some contributions by *Girvanella*. In the algal balls, these two again are present, the first one probably being the more important, but there are also some layers of *Spongiostroma*-like material; even encrusting bryozoans have taken some part in the formation of a number of the balls. A detailed taxonomic study of the Silurian Algae of Gotland would be very useful.

The algal crusts found in part of the reefs cover both reef builders and reef sediment and their contribution may thus have been both in strengthening the reef frame and in binding the loose sediment filling the interstices in that frame. In the reef limestone of the Högklint, characteristic alternations of *Thecia* and algal layers were found in some marginal reef parts, similar to those described by Hadding (1941, p.40). Particularly important are algal crusts in the reefs of Holmhällar type.

It is the presence of Algae in many of the reefs in Gotland that gives the most evidence of the lowest limit of the water depth in which these reefs grew, as these must have been confined to parts of the sea floor within the reach of sunlight.

Associated organisms

As was said before, in addition to the true reef builders, a great diversity of other organisms was present on, in and around the reefs. Those which lived on the reefs comprise both forms which were attached to the reef and free-living forms. Attached were, e.g., crinoids and several solitary corals, and not attached such animals as crustaceans, molluscs and arthropods. Mixed with debris of reef builders, calcareous mud and terrigenous detritus, their remains constitute part of the interstitial material of the reefs. Together with reef builders, debris derived from the fragmentation of them and the remains of organisms living around the reefs, associated reef organisms are also found embedded in, or even may form the major part of, the sediments surrounding the reefs.

Crinoids

There is hardly any reef in Gotland in which remains of crinoids cannot be commonly observed, macroscopically as well as microscopically. In the latter case, their optical behaviour and usually characteristic shape make them recognizable in thin sections. They must have been present in large numbers on the outer parts of the reefs and probably also directly around them. The ease with which they disarticulated made them by far the major source of bioclastic debris. Consequently they contributed considerably to the formation of the stratified deposits around the reefs, particularly in the Hoburgen and Holmhällar-type reefs. Although several crinoid roots were seen still attached to colonies of reef builders, the number of these is insignificant compared to the extremely abundant stem fragments. The number of identifiable calyces found was even smaller.

Some aspects of the palaeoecology of the crinoids found in Gotland will be discussed in Chapter XII.

Brachiopods

Brachiopod shells are commonly, often even abundantly, found in the reefs of Gotland. Where the matrix is marly, and especially in marl pockets, they can be extracted and identified. In many other cases, this is virtually impossible. The latter holds true particularly for the Holmhällar-type reefs, but also for several Hoburgen-type reefs or parts of them. The contribution of the brachiopods to the total reef limestone is merely a passive space-filling one.

Molluscs

Of the classes belonging to the phylum of molluscs, that of the gastropods is most commonly represented. Gastropods occur particularly in the reefs of the Hoburgen type, but are also well represented in reefs of the Upper Visby type. More of their remains were found in the reef limestone than in the crinoid limestone, the commonest form in the latter being *Platyceras cornutum*. Fragments of gastropods, whose structure suggests that they have recrystallized from rhombic aragonite, are not uncommon in thin sections of reef limestone. A comparable type of recrystallization is shown by remains which may be those of lamellibranchs. The latter, as molluscs, might also have contained aragonite. It is significant that brachiopod remains never show a similar type of recrystallization.

An occasional pteropod has been observed.

Lamellibranchs occur in reefs of Hoburgen type, but are distinctly fewer in both number of species and number of specimens than the gastropods. The relative scarcity of lamellibranchs in the reefs of Gotland is notable since in several modern reefs, including the Great Barrier reef (Maxwell, 1968), the lamellibranch species are possibly as varied as and certainly more numerous than the gastropods. The commonest lamellibranch in the Hoburgen-type reefs of Gotland is the genus *Conocardium*.

In several reefs of the Hoburgen and Holmhällar type, one or several specimens of cephalopods were seen, but only very few were observed in the surrounding stratified deposits. Cephalopods are not particularly rare in the reefs of Gotland in contrast to other areas, such as the English Wenlock, where they are only very rarely found in association with reef-bearing beds (Colter, 1957, p.27).

Arthropods

The easiest recognizable arthropod remains are those of the trilobites. They are represented in each of the three reef types and are rather common both in the reef and the surrounding stratified sediments. At least 17 species are known from these rocks in Gotland.

Remains of ostracodes are also easily recognized. Their parts are usually found throughout the reef matrix, but are best preserved in marl pockets, where both articulated and disarticulated valves can be seen.

Sponges

Sponge spicules occur fairly commonly in several of the reefs, but they are relatively inconspicuous and hardly contribute to the bulk of the reef rock. The silica of the spicules is usually replaced by calcite. It has not been possible to identify the isolated spicules in genera. Both hexactinellid (triaxonid) and lithistid sponges are represented, the latter being comparatively most common in the more central parts of the reefs, which might suggest a preference for a more sheltered environment.

Protozoans

In thin sections of both reef limestone and surrounding sediments, peculiar bodies can be seen, which in sagittal sections are flask-shaped (Fig.15 Similar bodies from the English Wenlock have been described by Colter (1957, pp.28–30), who was struck by their resemblance to certain Mesozoic and Recent Tintinnia (sulphylum Ciliophora). Whereas Tintinnia have long been thought not to occur in deposits older than Jurassic (Pokorný, 1958), they have recently been described from the Devonian of the Sahara (Cuvillier and Sacal, 1963) and the Silurian of the Betic Cordilleras in Spain (Hermes, 1966). This increases the likelihood that these flask-shaped fossils from Gotland and from the English Wenlock are also Tintinnia.

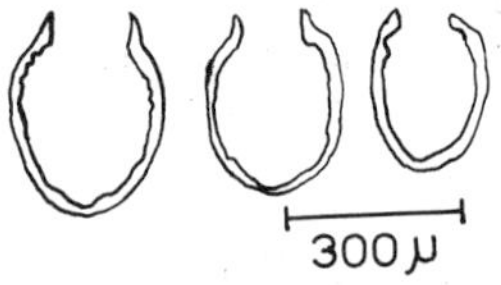

Fig.15. Flask-shaped bodies, resembling Tintinnia. Reef limestone, Högklint, Högklint Beds.

Soft-bodied animals

In addition to the true reef builders and the associated organisms of which hard remains have been preserved, there is no doubt that several more animals have been present on and in the reefs at the time of their formation. However, since these did not have hard parts which could fossilize well, no trace of them is generally left. Exceptions are, in some cases, infilled borings and burrows and faecal pellets.

Pellet-like bodies are found in some places in the matrix and since they consist of massive marly limestone, there is reason to assume that these are faecal pellets. Occasionally, traces of borings are also found in the reef limestone, transversing reef builders and filled by either calcite or sediment; borings are, however, not nearly as common as in many Recent reefs.

THE MATRIX OF THE REEFS

As the reefs grew upwards, they became subject to progressively more severe erosion. Debris thus derived from the fragmentation of the reef builders was mixed with remains of other organisms, with calcareous mud and with terrigenous detritus. This mixture fills in all the interstices in the reef frame and is partly bound by calcareous Algae. Together they constitute the reef matrix.

Detailed study of this reef matrix shows that there is great variation from one place to another, even within one reef. Four more ore less pure types may be distinguished, but all possible transitions between these occur, while the transitional or complex matrix forms are more frequently found than the "end" forms of matrix which are: (1) inorganic mud, with some small reef debris and remains of associated organisms scattered through it; (2) fine-grained or dense limestone with scattered small fossils; (3) a matrix composed almost entirely of remains, broken or otherwise, of reef builders and associated organisms with crinoids often dominating; (4) vaguely stratified, procellaneous limestone of a stromatolitic nature.

Matrix forms (1) – (3) speak for themselves. Type (4), however, needs some further discussion.

The stromatolitic type of matrix has several characteristic properties, both macroscopic and microscopic. In polished surfaces, it is commonly visible as strings of pinkish or creamy porcellaneous limestone. The vague stratification may be horizontal, but is often wavy, surrounding other reef builders (Hadding, 1950, plate 1). The layers may even be vertical, which proves that they were not formed by normal sedimentation. Using a sample of reef limestone from the Galgberg, Högklint Beds, Colter (1957, fig.10,11) demonstrated that the material was coherent at all stages in its formation and that the successive layers represent temporary outer surfaces during its outward growth. He concluded that many tabular bryozoan colonies evidently found the horizontal, dipping and vertical layers all suitable for attachment.

In thin sections, the wavy layering in the stromatolitic limestone is even more apparent. The boundary with a normal limestone matrix of usually greenish, sometimes brownish colour, is often distinct, but there are also gradual transitions. Transitions also occasionally occur from stromatolitic towards distinctly filamental algal material. The number of enclosed fragmental remains of other organisms is generally small; there is some embedded inorganic material of clay or sometimes silt size, either in isolated grains or in small pockets. Rarely does a sample show a large number of imperfect tubules, 0.05 – 0.10 mm in diameter, as are also reported by Colter (1957, p.40) from similar matrix material in the English Wenlock reefs.

It follows that although some of the characteristics mentioned point towards an algal origin, there is no absolute proof that the material is an algal or otherwise organic limestone. The same conclusion, however, would result from an independent study of several Recent calcareous deposits of undoubtedly algal origin.

If the assumption that stromatolitic reef matrix is formed by Algae is warranted, one must conclude that it has been the Algae themselves which have precipitated calcium carbonate and which were not merely sediment binders. Evidence supporting this conclusion includes the observed transitions to filamentous *Rothpletzella* material; the porcellaneous appearance of the sediment, so similar to that of known algal limestone with *Rothpletzella*; the different colour with respect to the normal limestone; and the small amount of inorganic detrital sediment in it, as compared to the other kinds of matrix.

Locally a small interstice in the reef frame seemed to have remained unfilled at the time of reef development, because it generally had only a small opening. These are now found filled with calcite. The bottom of such filled cavities is usually rather smooth, evidently as a result of fine sediment that has been sieved into the cavity; the walls and roof are commonly irregular and determined by the shape of the surrounding reef builders. The calcite filling the cavity is granular and similar to that filling cracks in the reef limestone, suggesting that it was deposited well after the time of formation of the reefs. No fibrous calcite has been observed, such as is described from the Permian reef in the central southern U.S.A. (Newell, 1955, p.308), and which seems to have been deposited synchronously with reef formation and prior to the infiltration of fine sediment. Nor has fibrous calcite been observed by Colter (1957, pp.45–46) in the Wenlock reefs in England. He suggests that its absence may be connected with the size and form of the reefs, or their depth below the surface of the sea in which they developed. The reef which Newell studied is much larger than those found in Gotland and England, and Colter quoted Newell, who mentioned that in modern reefs, aragonite is formed by the passage of warmed sea water, supersaturated with calcium carbonate, through the reef at ebb-tide, after it has lain for some time on the extensive reef flat.

Chapter VI

THE UPPER VISBY REEF TYPE

INTRODUCTION

One of the most conspicuous characteristics of the west coast of Gotland, over a distance of about 60 km, between Nyrevsudde and Hallshuk, is the difference between the lower and the upper part of the coastal cliff. The lower part consists of bluish grey marlstone, slaty to thin bedded, which alternates with thin, elongated lenses of marly limestone. Upwards, these limestone lenses increase in number and thickness, although most of them are not thicker than 5 cm. Small reefs occur. This part of the cliff is built by the Visby Beds, mainly the Upper Visby Beds. The upper part of the cliff is built by fossiliferous limestones, thick bedded, varying from a grey to whitish or reddish colour, and with intercalated large reefs. These sediments form part of the Högklint Beds. The boundary with the underlying Upper Visby Beds is not well defined. Usually the first thick and continuous layer of limestone is taken to be the basis of the Högklint Beds.

From the coastal cliff in an inland direction, the Högklint limestones are generally found at the surface. Only occasionally does some Visby marlstone outcrop in low-lying spots.

In the coastal cliff, exposures of Upper Visby sediments are common. Nevertheless, the range of exposures is sometimes interrupted. This is especially the case at those places where at present the sea hardly ever reaches the foot of the cliff, and where parts are covered by scree. Beautiful sections, however, are not rare either, especially south of Visby.

GEOGRAPHICAL AND STRATIGRAPHICAL DISTRIBUTION OF THE REEFS

A point of orientation in the south is the Korpklint, which bounds the Tofta Skjutfält in the north. This cliff, whose name is confusingly used twice in the area, is smaller than the true Korpklint (Västerhejde Parish), which is found a short distance further northwards. The upper part of the section, with sediments of Högklint age, occurs more inland in the case of the southern Korpklint. The Upper Visby cliff along these two Korpklintar can best be studied starting from the Ygne Fiskläge. The next access to the beach is at Fridhem. If one wants to study the steep Högklint cliff from the side of the beach, Fridhem is the best starting point. Going northeastwards from there, one may, via Axelsro, get close to Kneippbyn. Northeast of Kneippbyn the beach is interrupted for a short distance. At Kneippbyn Fiskläge the beach can be reached again.

Not only the reefs of the Högklint Beds but also those of the Upper

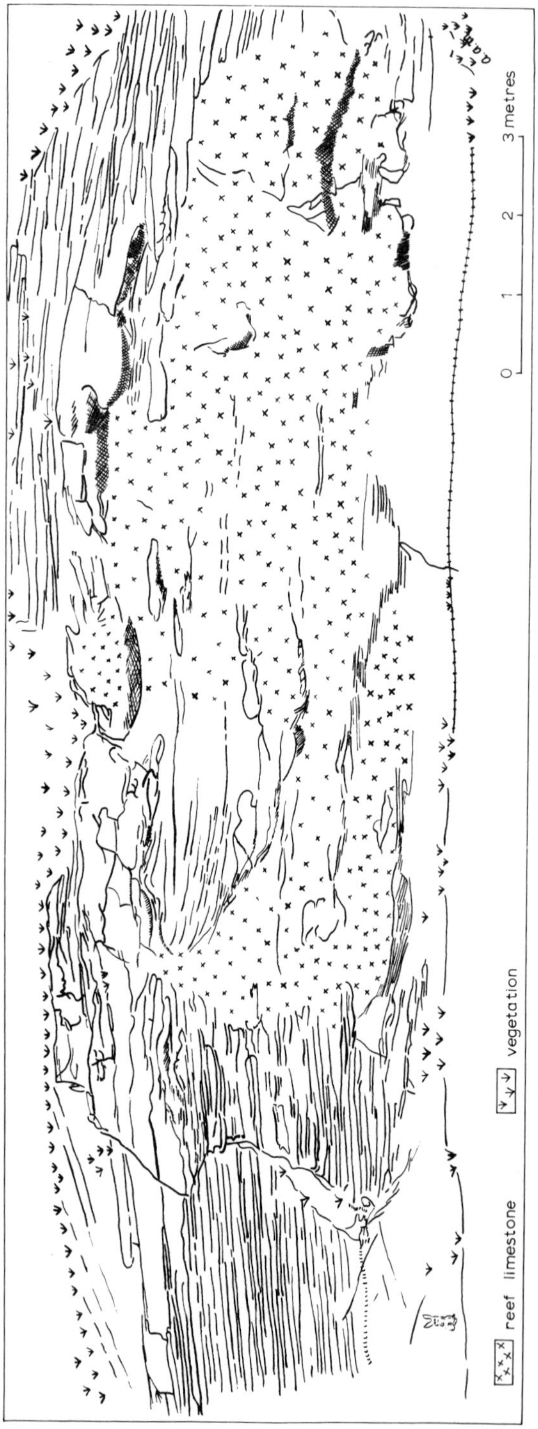

Fig.16. The southernmost of the three reefs exposed along the road southwest of Snäckgärdsbaden Hotel. The reef contains favositids, heliolitids, tetracorals and stromatoporoids, together with several kinds of reef dwellers, in a voluminous marly matrix. Around the reef a mantle of stratified limestone which passes gradually, but within a rather short distance, into the normal Upper Visby interstratification of marly limestone and marlstone.

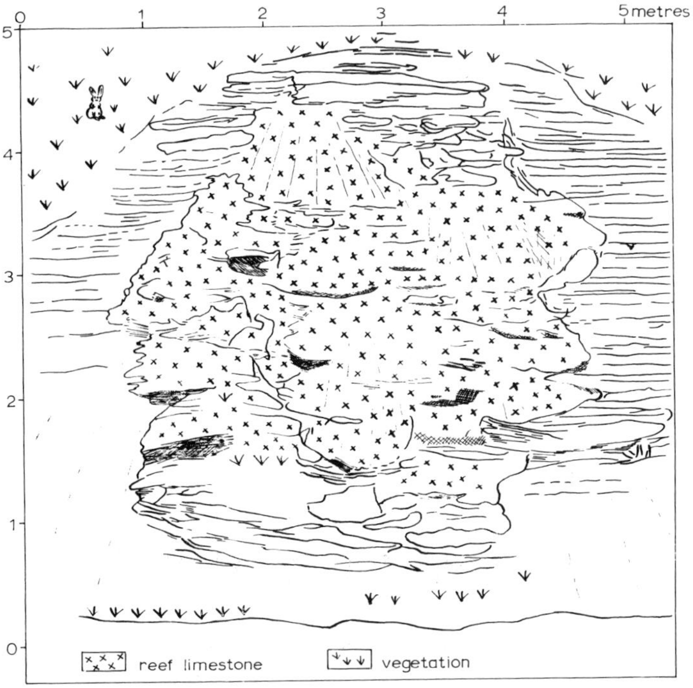

Fig.17. The central of the three Upper Visby reefs exposed along the road southwest of Snäckgärdsbaden Hotel. It is about equally as thick as the reef of Fig.16, but much less laterally extended. The character of reef limestone and surroundings are similar to those of its northern neighbour.

Visby Beds are very common and beautifully exposed south of Visby, in the cliff between Nyrevsudde and the Visby shooting range. The coast along this shooting range has not been studied by the present author. Further south there is a second such range (Tofta Skjutfält), of which the coastal stretch Nyrevsudde – Blåhäll – Stavsklint – southern Korpklint forms part. Thanks to the kind cooperation of the military authorities, the present author was allowed to include this area in his studies. It showed several interesting exposures.

The number of Upper Visby reefs observed per kilometre of cliff varies south of Visby from about 18 (Kneippbyn) to about 12 (Tofta Skjutfält). That the latter number is lower depends, however, mainly on the degree of exposure of solid rock in the coastal cliff. Altogether almost one hundred Upper Visby reefs have been recorded south of Visby, in varying sizes and degrees of clearness. Consequently these reefs are certainly not as rare as was assumed by some earlier authors (e.g., Hadding, 1933, p.60).

The reefs south of Visby occur in the upper two-thirds of the Upper Visby section, albeit with some variation. (Thus, in the area of the Tofta Skjutfält, the oldest reefs occur a few metres lower than at Kneippbyn.) However, comparatively the most reefs are found in the uppermost part of the Upper Visby Beds. Unfortunately, as a consequence, they are often situated rather high in the cliff wall, with the result that a part of them is nearly or completely inaccessible. It is also unfortunate that almost all sections through the reefs follow the general direction of the coastal cliff, which is about southwest – northeast.

North of Visby, three rather well-known Upper Visby reefs are found south of Snäckgärdsbaden Hotel (cf. Hede, 1940, pp.16–17). They lie along a road which is regularly treated with calcium chlorate. The dust from this road covers the exposures like plaster so that they are, therefore, no longer as beautiful as at the time when the photographs in the publication by Hede were taken. For a detailed study, many of the reefs exposed in the coastal cliff, mentioned earlier, are now more suited than these three at Snäckgärdsbaden. Drawings of two of the Snäckgärdsbaden reefs are given in Fig.16 and 17.

Northeastwards, between Snäckgärdsbaden and Lundsklint, the coastal cliff is generally hidden behind scree material. North of Lundsklint the Upper Visby sediments are again better exposed. The average number of reefs, present per kilometre, is distinctly lower there than south of Visby. Further northeast, moreover, it can be observed that reef growth started continually later during Late Visby time. In the environment of Lickershamn, the oldest Upper Visby reefs occur only comparatively close to the boundary with the overlying Högklint Beds, into which they often continue and in which they generally reach their maximum extension. At Hallshuk, in the extreme north of the area which shows the Visby Beds, no more reefs occur in the Upper Visby Beds. The reefs, recorded north of Visby, which began growth during Late Visby time, total about forty in number. Some of these are completely enclosed in Upper Visby sediments, but some only have their lower parts enveloped by the Upper Visby Beds.

PALAEOGEOGRAPHICAL DISTRIBUTION OF THE REEFS

In the previous section it has been shown that the Upper Visby reefs are most common south of Visby, where they also start at a distinctly lower level than in the north. Circumstances favourable for their growth, however, seem to have extended gradually. Other data from the Upper Visby Beds (Chapter XI) indicate a gradual decrease in the depth of the water during Late Visby time. Reef development makes certain demands as to water depth. Therefore, it could be assumed that a relationship between the first occurrence of reefs at a certain locality and water depth at that time existed.

This suggests that, when reef growth started in the south, the sea further northeast was still too deep. It is further assumed that the depth contours were generally parallel to the contemporaneous shore line. Thus, a slight angle has to be presumed between the Upper Visby shore line and the present coastal cliff. Since the latter trends about northeast – southwest, the direction of the Upper Visby shore line may consequently be assumed to have run about north-northeast – south-southwest. This conclusion is important for the explanation of a number of observations which will be dealt with in later sections of this chapter. The Upper Visby reefs developed in a belt about parallel to the shore line. With the shallowing of the sea, this reef belt gradually moved in the direction of the basin centre.

GENERAL CHARACTER OF THE REEF LIMESTONE

The Upper Visby reef limestone generally shows a nodulose and unstratified or an indistinctly stratified structure. The nodulose or conglomeratic character is caused by the reef builders. These consist of grey, recrystallized limestone, which is harder than the marly reef matrix. Consequently, they protrude somewhat in not-very-fresh exposures. The marly matrix generally constitutes quite a large percentage of the total reef volume; often in the order of 40–50%. Probably as a result of so much marl sedimentation, the reef builders, on the average, did not reach the sizes of those which are found in reefs of the Hoburgen or Holmhällar type. The fact that most of the compound reef builders (predominantly colonial corals) have distinctly larger horizontal than vertical dimensions, is often one of the causes of the vague stratification that can be seen in some of the reefs. For another part, this may be due to the large matrix volume, leading to intercalated short and thin marlstone layers. An example of a vaguely stratified reef is that occurring about 0.15 km northeast of the staircase close to the southern boundary of the Visby shooting range. Many of the reef builders in that reef show a more or less similar thickness (4–6 cm), and they also seem to have had some preference for certain horizons within the reef. Their lower and upper sides, bounded by the marly matrix, thus bring out approximately horizontal lines, which do not correspond, however, to bedding planes outside the reef limestone. As will be discussed in the section on coral palaeoecology (Chapter XII), variations in the supply of continental clastics may have strongly influenced the development of the main reef builders. Thus, temporarily greater supplies of mud not only affected the volume and character of the reef matrix, but presumably also the vitality of the organic frame.

The dominating flat shape and horizontal position of the reef builders may also help to explain why the reef limestone, though generally unorganized in character, nevertheless offers a less disorderly impression than that of several reefs of Hoburgen type. The percentage of reef builders which are found in their orientation of growth, is comparatively high (see further Chapter XII). Several (presumably mutually dependent) factors may have contributed to this, such as the shape of the colonies, the small dimensions of the reefs and the consequently smaller internal pressures, and relatively less water turbulence.

In some of the larger reefs a slight decrease in marliness is found higher in the reef. This is the case, for instance, with three large reefs about 0.1 km north-northeast of Nyrevsudde (Tofta Skjutfält), of which the southern one (more than 6 m thick) and the middle one (about 5 m thick) are well exposed. A slight decrease in marliness may in some cases also be observed when the amount and character of the matrix are studied in horizontal direction from the periphery towards the centre of the reef.

Despite the apparent loose construction of the reefs and the high percentage of marly matrix, the reef limestone is more resistent to erosion than the nearby normal stratified sediments of the Upper Visby Beds (marlstone interbedded with harder marly limestone). In many cases, the reefs protrude from the cliff wall. The three reefs at about 0.1 km north-northeast of Nyrevsudde are also a good example of this. All three rest upon a kind of ledge, consisting of stratified marly limestone, which is, in this form, limited to the direct environment of the reefs. This limestone is thickest underneath the northern reef, which is apparently sectioned through its periphery. The ledge dips there towards the cliff. The lowermost portion of that reef is talus-like in character and contains many fossils lying awry. The limestone underneath the reefs will be further discussed in a later part of this chapter.

REEF-FORMING ORGANISMS

In general the same organisms are found in the Upper Visby reefs as in the stratified sediments of this stratigraphical unit, though in different frequencies. Compound corals play the most important part in the reefs, mainly with three important groups: Favositidae, Heliolitida and Halysitidae.

Of these three groups, the favositids are the most striking. The shape of their colonies varies greatly, from thin, undulous plates to almost-globular specimens, presumably depending mainly on the environment in which they developed. A soft substratum may have promoted horizontal expansion; increasing sedimentation could cause the death of some corals, but may also have stimulated the upward growth of others.

A *Favosites* sp. can easily be recognized, but the subdivision of the genus is difficult. Part of the species ascribed to this genus in the literature may be dubious. Many of the favositids in the reefs under discussion may presumably be best assembled in one species. This then becomes *Favosites gothlandicus* sensu lato. In this sense, this species shows a great range of variations and includes such other "species" as *F. forbesi* Edwards et Haime and *F. clausus* Lindström.

It can easily be decided whether a *Favosites* sp. is lying in its orientation of growth or not: its upper side is characterized by wide, open tubular calyces, the lower side by closed, round cells.

The heliolitids or "sun" corals, also played an important part in the Upper Visby reefs. As in the case of the favositids, there is a great variation in shape. The growth of the heliolitid corals was apparently less rapid. Under the influence of strong marl sedimentation, colonies growing from a solid substratum, among which many reef-building specimens, generally developed rather-globular forms (cf. Chapter XII). Some colonies, showing the shape of a mushroom with a short stem, may have grown in even more turbid water. The latter are found occasionally in the reef limestone, but are somewhat more common in marlstone layers of the stratified Upper Visby succession. As in the favositids, some flat heliolitid colonies may have also adapted themselves to increased marl sedimentation by the formation of more or less knobby upper surfaces. Colonies exhibiting this phenomenon have also been observed in both the marlstone layers and in the marly limestone, but particularly at transitions from a limestone to a marlstone layer and in reefs (e.g., in two reefs about 0.15 km south of Axelsro). There, the knobs on the corals were on the average 1–3 cm high, but a *Heliolites* sp. has been observed with upward outgrowths as high as 4.5 cm. The heliolitids of the Upper Visby Beds belong to a number of species from the genera *Heliolites*, *Plasmopora* and *Propora*, which are very difficult to distinguish on the basis of their external appearance.

The chain-pore coral genus *Halysites* occurs in two species, one with a fine network, *H. catenularius* (Linnaeus), and one with a coarse network, *H. catenulatus* (Martini) (Fig.14). They consist of upright lamellae, which may branch, and which enclose substantial interspaces, in which all sorts of foreign material, e.g., marl, could be stored. Thus, they could live even in very turbid water. Consequently, they are found even more commonly in marlstone than in limestone and reef limestone. In the latter, but occasionally also in the stratified sediments, they are partly broken up into a heap of sharp sherds. The reason for their fragmentation may have been that the interspaces were not yet completely filled when the colony was put under pressure, or that the filling material was not yet consolidated enough at that time.

Several species of solitary corals are also well represented in the reef limestone. In general, however, they were quantitatively rather unimportant as reef builders.

In general, stromatoporoids are strongly subordinate to corals in the Upper Visby reefs. Only one small reef has been found, which was mainly formed by stromatoporoids (see p.108). As the reef matrix contains more marl, so generally the number of stromatoporoids in the reef is proportionately smaller. The shape of the stromatoporoids seems to have also been strongly influenced by the environment. Most common are specimens with the shape of a flat disc. In two reef remains about 0.1 km south-southwest of Stavsklint, the largest colony measured 19 cm in horizontal diameter, with a maximum height in the centre of 5 cm. Larger colonies are rare in the Upper Visby Beds. Occasionally stromatoporoid colonies are also found with a mushroom-like form, moreover, partly showing a knobby upper side. Sometimes stromatoporoid colonies are covered with a network of *Aulopora* sp.

In the field it is very difficult to identify the various genera of stromatoporoids. An exception is *Stromatopora discoidea* (Lonsdale), which shows a surface pattern with small "suns"; in young specimens just one, in older ones more than one. Generally this species occurs as thin covers over other fossils. Also found encrusting, are some small specimens of *Clathrodictyon striatellum* (d'Orbigny).

Some corals may, in the field, easily be confused with stromatoporoids. As Colter (1957, p.255) also experienced, some specimens which externally resemble stromatoporoids, may on sectioning be found to be corals such as *Planalveolites* spp., or off-centre sections through the large solitary coral *Schlotheimophyllum patellatum* (Schlotheim).

In most cases, the stromatoporoids are found unattached to any kind of substratum. It remains an open question in how far this always represents the original condition. Occasionally it can be established that colonies of stromatoporoids or corals - both in the reefs and in the stratified sediments - began their growth by attachment to an older colony, a solitary coral or any other solid fragment.

In several reefs, fragments of bryozoans have also been observed, among which remains of *Fenestella* sp. and *Helopora* sp. Their contribution to reef building as a whole has been slight.

In several places within the reefs, the marl forms small pockets. In these it often shows a very thin stratification. Both in the matrix and in these pockets, solitary corals, brachiopods and molluscs (mainly cephalopods and gastropods) are commonly found.

Although in and around the Upper Visby reefs, crinoid remains have been found, nowhere are these rock-building. This is in contrast to the occurrence of crinoids in the Högklint Beds and younger stratigraphical units.

A more detailed survey of the fossils known from the Upper Visby reefs is given in Chapter V, Table IX (pp.60–67).

SHAPE OF THE REEFS

Somewhat schematized, the most common shapes found among Upper Visby reefs are the knoll, the lens and the inverted cone. In both the latter cases, a section through the reef in the present coastal cliff is generally not symmetrical, but shows a steeper southwest than northeast side.

The knoll shape

An example of a knoll reef is found about 0.4 km south-southwest of Stavsklint, in the area of the Tofta Skjutfält (Fig.18). It measures about 1 m in both length and height. The reef is extremely marly. At the top of the reef

Fig.18. Small reef, about 0.4 km south-southwest of Stavsklint, Upper Visby Beds. The length and height are both about 1 m. The matrix of the reef is extremely marly. The water flowing at the boundary between the cover of weathered rock and the underlying solid rock shows the impermeability of the Visby marlstone to water. (After Manten, 1962, fig.3.)

the stratified sediments arch over it; otherwise the reef seems to have exerted little influence on the position of the surrounding sediments. Immediately against the reef, a little more stratified limestone is present than elsewhere in the succession[1]. This limestone is very fossiliferous, especially in relatively-small solitary corals (1–3 cm long) and in brachiopods.

A knoll-shaped reef, which is somewhat vaguely developed, is present about 0.45 km southwest of Blåhäll (Tofta Skjutfält). This one too, is about 1 m long and high. It rests upon a saucer-shaped swelling of stratified limestone, which in its centre is about 10 cm thick, but increases in thickness towards the margin. Similar limestone also builds a 0.5–1 m-broad mantle at the right side of the reef, faintly interfingering with the almost vertical southwest side of the reef. At the northeast side of the reef, which rises outwards at an angle of about 35°, there is a very poorly developed limestone mantle. Around the top, the mantle is even lacking completely. The reef itself shows a vague stratification, which is caused by the voluminous marly matrix, together with reef builders of a flat-lenticular or tabular form.

Most knoll-shaped reefs occur comparatively low in the reef-carrying part of the Upper Visby Beds. They are small and very marly (cf. also p.97).

The lens shape

There is a gradual transition from the knoll-shaped to the lens-shaped or lenticular reefs. The difference between the two is mainly in the ratio between the dimensions: knoll-shaped reefs are about equal in length and thickness; lens-shaped reefs are distinctly longer than thick. Moreover, the latter are usually larger than the knoll-shaped reefs. Thus in the area of the Tofta Skjutfält, where the mentioned examples of the knoll reefs occur, a lenticular reef is found about 0.2 km north-northeast of Nyrevsudde which measures about 3 m long and 1.5 m thick. This reef also, possesses a rather strongly marly matrix, but it does not show stratification. Especially at both sides of the reef, but also underneath it, the succession of stratified sediments is distinctly richer in marly and relatively hard limestone (Fig.19). Many lens-shaped reefs found elsewhere are still significantly larger (see also p.95).

Both the larger size and the better-developed limestone mantle of the reef suggest that the lens-shaped reefs grew under somewhat more favourable conditions than the knoll reefs. There is no essential difference in fossil content between the smaller and larger reefs within the Upper Visby Beds.

Two other, approximately lenticular reefs are exposed closely together in the southwest – northeast-orientated coastal cliff about 0.15 km south of Axelsro. The northern one of these is shown in Fig.20. Both reefs are well-developed and occur just below the boundary between the Upper Visby and Högklint Beds. Both distinctly show the asymmetry, mentioned at the beginning of this section. At the southwest side of both reefs, there is a very well developed mantle of hard, stratified limestone. The first reef, which is about 9 m long, shows a rather steep boundary with this limestone mantle.

[1]Where the dimensions of Upper Visby reefs are mentioned in this book, the mantle of stratified limestone, as found with several of them, is not included, unless otherwise indicated.

Fig.19. Upper Visby reef about 0.2 km north-northeast of Nyrevsudde.

Fig.20. The northern of the two Upper Visby reefs south of Axelsro (below fallen tree). Much stratified limestone is found at the southwest side of the reef (right), less at the northeast side (left). There is no difference in the reef fauna between the reef core and the reef margins.

At the northern reef, about 6 m long, this contact is even vertical and slightly interfingering. At the northeastern side, the limestone mantle is only very poorly developed. The lower lateral contact between reef limestone and stratified sediments there rises gradually over the latter. At the second, northern reef, a talus-like zone, of about 1 m thick, is intercalated between the reef and the mantle of stratified limestone.

It is rather usual in the Upper Visby reef-limestone exposures to find that the southwest side of the reef limestone is steeper compared to the lower boundary at the northeast side, as was observed in the two reefs described above. Between Högklint and Visby, this is true for 75% of the well over forty larger Upper Visby reefs which were observed there. Only two exceptions to this rule have been observed. In the remaining little over 20% of the exposures, there is no distinct difference in the dip of the two lateral boundaries between reef limestone and stratified sediment.

An explanation for the observed asymmetry may be found if the angle between the directions of the Upper Visby reef belt and the present coastal cliff is taken into account (see p.82). This implies that the northeastern side of the reef-limestone exposures is more related to the original seaward side of the reef, and the southwestern side more to the original landward-directed side. The east side of the reef was presumably in direct contact with the open sea and received a constant supply of water rich in food and nutrient salts. Reef expansion, therefore, will have been mainly in this direction, thus causing the lateral reef boundary at that side to rise over continuously younger surrounding deposits. It is true that the seaward side of a reef is also subject to the greatest demolition, but with the type of reefs such as those found in the Upper Visby Beds, which only protruded slightly above their surroundings, this seems not to have significantly influenced horizontal expansion. On the other hand, organisms detached from the reefs were deposited mainly at the landward side. Those still alive then, came into relatively more turbid and less well-refreshed water. With reefs as these, developing under marginal living conditions, this may, in several localities, have been fatal for at least some of these detached reef builders, even though other, non-reef-building organisms were able to maintain themselves at those places.

A further good example of an asymmetrical reef, and one which rather convincingly suggests that expansion of the Upper Visby reefs indeed took place mainly in a seaward direction is the reef exposed about 0.6 km southwest of Blåhäll. This reef is about 4 m long and has a southwest side, about 3 m high, which is almost vertical and which interfingers slightly with a narrow limestone mantle. Some stratified limestone is present also underneath the northeastern side of the reef, which rises at about 25° over the surrounding layers. The lowest point of the reef-limestone mass as a whole is at the southwest. Reef expansion as revealed by this section, seems to have taken place only towards the northeast, in reality perhaps towards the east, that is in the direction of the open sea.

The inverted-cone shape

In cases where both seaward and landward expansion of a reef took place rather slowly, a reef-limestone mass developed in the shape of an inverted cone (Fig.21). In general the available data suggest that cone-shaped reefs developed under somewhat less favourable circumstances than lens-shaped reefs, but under conditions more favourable than those for the knoll reefs. Thus, for instance, the conical reefs are generally larger than the knolls. On the other hand, the rocks of the inverted-cone reefs show a less-distinct reef character than is general in the lens-shaped reefs.

From the various reefs with the shape of an inverted cone, two will be selected as examples.

The first example is the reef about 0.15 km northeast of the staircase close to the southern boundary of the Visby shooting range, already mentioned in the discussion of the general character of the reef limestone (p.83). In cross-section, it shows a V-shape with a distinctly convex upper side. Its maximum length and height are both about 2.5 m. Both the northeast and the southwest side of the reef limestone rise outwards at an angle of about 55–60° against the stratified sediments. The mantle of stratified limestone is somewhat more strongly developed at the southwest than at the northeast side of the reef, but the difference is not great. With this mantle included, the reef is about 7 m long. The general character of the reef limestone gives grounds to the surmise that reef growth was not very prosperous (see p.83).

Another reef with the shape of an inverted cone is exposed about 0.1 km northeast of Kneippbyn Fiskläge. Its maximum height and length are about 4 m;

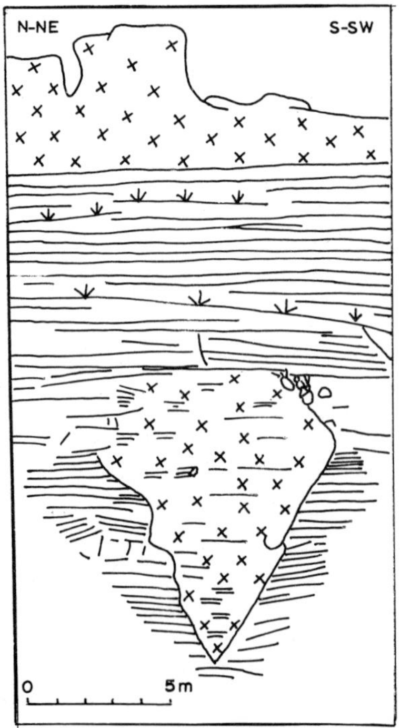

Fig.21. Upper Visby reef with the shape of an inverted cone. Southern part of the Högklint. The reef is enclosed in the uppermost Upper Visby Beds and the lowermost Högklint Beds. Internal stratification of the reef makes its reef character somewhat vague, but a mantle of stratified limestone identifies it, nevertheless, as a reef. At the top of the succession, Högklint reef limestor

Fig.22. Upper Visby reef between Axelsro and Kneippbyn, as seen from the southeast. At its base a 2.25 m thick "attempt at reef growth".

---

the southwest side is a little steeper than the northeast side; the upper side is rather flat. The reef limestone is vaguely stratified. At both the northeast and southwest side of the reef a well-developed mantle of stratified limestone is present. The normal Upper Visby stratification sags underneath the reef.

If a reef with the shape of an inverted cone is cut through its periphery by the present cliff, it may give the impression of being a large knoll reef. Although the difference between such an exposure and a true knoll reef is but slight, if only one cross-section is available, it remains useful to restrict the indication "knoll reef" within this stratigraphical unit to the small reefs discussed in the beginning of this section.

An interesting exposure, showing a reef of real intermediary position between an inverted cone and an exceptionally large knoll shape is found about half-way between Axelsro and Kneippbyn (Fig.22). This reef is rather large and well developed. Since it is exposed rather low in the cliff wall, it is reasonably accessible. It illustrates formation under less favourable conditions than seems to have been the case with large reefs of a lenticular shape.

In the succession of stratified sediments underneath this reef, a portion about 1–1.5 m broad and 2.25 m thick is present. This portion is very fossiliferous and seems to represent unsuccessful incipient reef growth. Most common are flat-lenticular coral colonies (*Favosites*, Heliolitida, *Halysites*), whilst a few thin stromatoporoid covers are also present. Of the characteristic interstratification of the Upper Visby Beds, the marlstone layers generally continue through this part, though often undulously. The limestone layers, however, are often interrupted by the occurrence of the fossil colonies, some of which appear not to be in their orientation of growth.

At the southwestern side of this "attempt at reef growth", the limestone layers strongly increase in thickness. The marlstone layers, still present in the fossiliferous part discussed above, gradually disappear in a transition, about 1 m broad, which lies between this fossiliferous part and the complex of thickened limestone layers. This transition is eroded to a depth of about 0.5 m into the cliff wall, causing a shallow niche (Fig.22). About 1.5 m towards the southwest, the limestone complex (right below in Fig.22), passes again into the normal Upper Visby stratification. The thickened limestone layers are very fossiliferous, especially with solitary corals and brachiopods, but also with small coral colonies, a few thin stromatoporoid covers, crinoid remains and bryozoan fragments. The presence there and the character of the limestone complex may indicate that north of the niche still more coralliferous rock has been present.

Above the "attempt at reef growth" there is the true reef, about 6 m thick and in the middle also about 6 m long. If the unsuccessful "attempt at reef growth" is excluded, the shape of the reef is thus that of a large knoll. With the underlying coralliferous rock included, it approaches rather the inverted-cone shape. A surrounding mantle of hard, stratified limestone is present around the true reef, but is less well developed than is often the case. Especially at the northeastern side, the reef limestone may in some places lie directly against the normal Upper Visby sediments. At the steeper southwestern side, the reef limestone interfingers with the mantle of stratified limestone. At both sides of the reef, in its direct environment, several coral colonies are found. These are not in their growth positions. The sediments there are relatively richer in solitary corals than is the reef limestone. In the reef limestone, by far the most coral colonies are in their orientations of growth, although some colonies in oblique positions and upside down are also found. The matrix is very marly; especially close to the margins, it shows some vague stratification; at several places in the reef, it forms marl pockets.

Compared to the sediments lower in the cliff section, the stratified sediments in the normal Upper Visby succession at the level of the reef, but which are at some distance from it, do not show distinct evidence of relative increase in the cumulative thickness of the limestone layers. This is in contrast to what has been found in the environment of several other reefs, e.g., the two reefs near Fridhem (see the section on specific levels of reef development, p.107) and the reefs south of the Visby shooting range (see the section on the coastal cliff north of Kneippbyn, p.108). Such an increase may point to a more general improvement of living conditions for limestone-forming organisms. That such an increase in limestone deposition is lacking here is even more remarkable when we realize that the reef is situated rather high in the Upper Visby Beds. Presumably its top is only about 2–3 m below the Upper Visby – Högklint boundary.

Influence of the open sea

In the preceding paragraphs, emphasis has been laid on the influence of the open sea in explaining the shape of the reefs. To support this presumption, two more examples will be discussed in this section. Both deal with two

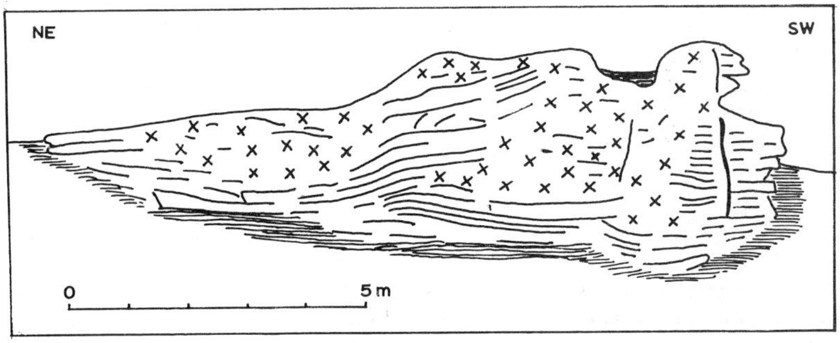

Fig.23. Two Upper Visby reefs, about 0.5 km southwest of the southern Korpklint. The northeastern reef has probably been initiated by one or more reef builders washed off the southwestern reef.

inter-related reefs, of which the youngest in its relationship to the older reef may show influence of the open sea.

The first example is found in the area of the Tofta Skjutfält, about 0.5 km southwest of the southern Korpklint (Fig.23). Two reefs rest here upon an irregular limestone swelling. At the southwest side of the southwestern reef there is a broad limestone mantle against the rather steep reef wall. In between the two reefs there is an area, about 2 m long, which is also occupied by stratified marly limestone. In this limestone, the layers dip away from the southwestern reef, but towards the northeastern reef. This suggests that the latter reef, which is smaller, started growth later than the southwestern reef. Perhaps its growth was initiated by one or more reef builders which were loosened from the southwestern reef, but found a suitable place for further development on the limestone mantle of this mother reef at its seaward side. It is the only case of a kind of flank reef observed in the Visby Beds. At its northeastern side, the younger reef also, is surrounded by a mantle of stratified limestone.

The other example lies about 0.5 km northeast of Axelsro. There two reefs are found, whereof part of the one closely overlies the other. The lower one is small and possesses a very poorly developed limestone mantle (see also the section on the limestone lateral to the reefs, p.104). At its base, the stratified limestone passes very gradually into the reef limestone. Immediately above this reef, and also further northeast, there is an elongated, solid complex of stratified fossiliferous limestone. It forms part of the limestone mantle of the second reef. A section of the latter can be seen above these limestone layers, which sag slightly under the reef. This reef is much larger than the lower reef; in the section, it shows a length of about 15 m. Unfortunately, its top is no longer exposed, but it seems likely that the upper side will have been rather flat.

There are no indications of a direct relationship between these two reefs. Presumably they began their growth independently. Development of the smaller, western reef may, however, soon have been hindered by that of the other. This western reef is very marly. Together with the many flat-lenticular reef builders, this causes some vague stratification. The second reef, apparently, thrived much better. It probably developed in more direct contact with open water. This may imply that the western reef became continuously more screened, until it finally died and was covered by the limestone mantle of the expanding eastern reef. This latter reef shows no sign of an internal stratification, and its matrix is also comparatively less marly.

Both examples may thus indicate that reefs which were in direct contact with the open sea had greater vitality.

Exceptions to the rule

As was mentioned earlier, two reefs have been observed south of Visby, in which the steepest side was not the southwest side of the reef but the northeast side. The first exception can be observed about 0.55 km northeast of Axelsro, high in the coastal cliff. There a lens-shaped reef is found, which is about 15 m long and 1–3.5 m thick. It shows large but dominantly flat-lenticular reef builders and a very marly matrix, which, as usual, occupies quite a large volume. In between the reef builders, there are a number of thin and undulous marly layers, some of which may be followed for more than 1 m; they are relatively the most common in the lower part of the reef. The reef also contains a few intercalations of stratified marly limestone. Even compared with other Upper Visby reefs, this one is not a model of vigorous reef growth. The greatest thickness of the reef limestone is found relatively close to the northeastern side of the exposure. Towards the southwest, the lower lateral boundary between stratified sediments and overlying reef limestone gradually rises. This suggests that, although reef growth at the seaside continued, horizontal expansion took place mainly in a southwest- or westerly direction, that is more coastwards than seawards.

The second exceptional case that should be mentioned here, is found in the cliff directly southwest of KneippbynFiskläge. At that place, two reefs are exposed next to each other in about an east-northeast – west-southwest section. The maximum dimensions of the southern reef are about 3 m long and 1 m thick. The east-northeastern boundary of this reef, with the underlying stratified sediment, rises at about 60°, the west-southwestern boundary

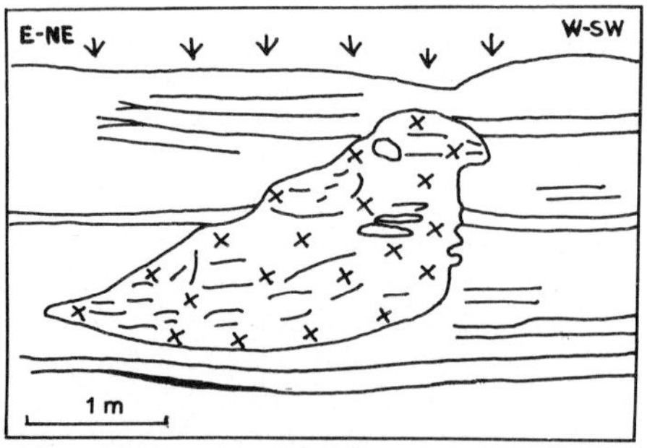

Fig.24. The northern of the two reefs found in the coastal cliff southwest of Kneippbyn Fiskläge. The cross-section suggests that the centre of reef growth gradually moved from the middle towards the west-southwest. There is hardly any mantle of stratified marly limestone around this reef. Conspicuous limestone layers are drawn. The remaining parts of the reef surroundings are taken up by the normal Upper Visby alternation of layers of marly limestone and marlstone.

is approximately vertical and interfingers slightly with the stratified rocks. This is all in agreement with the general picture. However, the northern reef shows quite another picture (Fig.24). It is about 3 m long and 2 m high. The deepest place in the reef base is about in the middle. The west-southwest boundary is approximately vertical and interfingers slightly with the surrounding sediments. At the other side, the stratified sediments soon regained ground on the reef. The exposures thus suggest that following an initial expansion, the living surface of the reef gradually became more and more restricted to the west-southwestern part. These two Kneippbyn reefs are, moreover, remarkable in the almost complete absence of a mantle of stratified limestone. At the base and at the west-southwest side, a few comparatively thicker limestone layers still occur, but at the east-northeast, these are lacking, with the exception of a few at the very top. The reef limestone is of the normal Upper Visby type.

Instead of a more or less stationary reef growth at the landward side and reef expansion at the seaward side, the latter reef and its enveloping sediments thus reflect the less vigorous situation of death and retreat at the landward side and only stationary growth at the seaward side. The example fits into the general picture as far as the relatively greatest vigour at the seaward side is concerned, but differs from other reefs insofar that it was the seaward boundary which, as a consequence of the inability of the reef to extend itself during most of its life, developed about vertically.

DIMENSIONS OF THE REEFS

As has already appeared from the examples described in the previous pages, the dimensions of the Upper Visby reefs are generally modest. Many knoll reefs are not much longer and thicker than 1 m. The largest lens-shaped reefs are of up to about 15 m in length and 4–5 m in thickness. Specimens with the shape of an inverted cone may have reached a maximum thickness of little over 6 m. The average of all Upper Visby reefs may be taken as 3–4 m in length and 2–3 m in thickness. These figures do not include the dimensions of a number of generally large reefs, which began their growth in Late Visby time, but reached their greatest extension during the Högklint Period.

Since it has already been mentioned that lens-shaped reefs generally occur higher in the Upper Visby Beds than the knoll reefs, the above figures show an increase in the dimensions of the reefs upwards in the stratigraphical succession. Relatively large reefs only occur high in the Upper Visby Beds. Small reefs occur throughout the reef-carrying part of the Upper Visby Beds, but are most common in the lowest part.

The following example, observed in the coastal cliff close to Kneippbyn, gives an impression of how this increase in size may appear in the field.

Southwest of the Kneippbyn "Fiskläge" (only a few small fishermens' barns), two small reefs occur, which have already been mentioned at the end of the previous section of this chapter. The reef limestone is exposed in a cross-section of about 2.4 and 2.0 m^2, respectively.

Immediately underneath the fiskläge a third reef is exposed (Fig.25A). It shows a faintly developed mantle of stratified limestone, is about 1 m thick, 1.5 m long and has an exposed reef-limestone surface of about 1.2 m^2. The southwest boundary of the reef is about vertical. The northeast boundary is $<$-shaped; the reef limestone rising

A B C D

Fig.25. Some of the Upper Visby reefs found near Kneippbyn Fiskläge. A. Small reef found underneath the fiskläge. B. Larger reef, about 15 m northeast of the previous reef, photographed from the west-southwest. The reef occurs in the protruding rock part and shows a mantle of stratified limestone at both sides. A swelling of marly limestone underneath the reef sags somewhat under the pressure of the latter; on the photograph, this is partly hidden behind vegetation. C. Vague reef, found in between those in B and D, resting upon a slight swelling of marly limestone above the hammer. D. Larger reef about 0.1 km northeast of the fiskläge, seen from the west-southwest; for comparison of size see the hammer at the foot of the cliff. The reefs of B and D are found at a higher level in the Upper Visby Beds than those of A and C, illustrating an increase in reef size upwards in the succession.

initially at an angle of ca. 25° outwards, and afterwards retreating at an angle of about 35°.

A fourth reef is exposed about 15 m further northeast (Fig.25B). This one is about 2 m long, 1.25 m thick and measures in cross-section about 2 m^2. The reef possesses a solid mantle of stratified limestone, at its base sagging under the reef limestone, at its top arching over it.

After a small, indistinct reef, with a surface of less than 1 m^2 (Fig.25C), a sixth reef is encountered about 0.1 km northeast of Kneippbyn Fiskläge (Fig.25D). It is about 4 m long and thick and 9 m^2 in the cross-section. At the northeast side a very-well-developed mantle of stratified limestone is present, dipping reefwards; a comparable mantle is present at the southwest side.

Of these six reefs, the first, second, third and fifth have their bases all at about the same height in the Upper Visby Beds. The average surface of reef limestone exposed per reef is less than 1.75 m^2; none of them possesses a strongly developed limestone mantle. The fourth and sixth both begin at a level, which is about 1 m higher than the previous level of reef initiation; of these two reefs, the one only slightly exceeds the average size of the reefs of the lower level, the other one is considerably larger; both show a well-developed limestone mantle.

A good example of greater geographical extension, which shows increase in reef size upwards in the Upper Visby Beds, is found in the coastal cliff between Stavsklint and Korpklint (Tofta Skjutfält). About 0.2–0.3 km northeast of Stavsklint, six round knoll reefs are found, on the average about 1 m in diameter. They occur close to each other, with their bases at about the same level. As a rule, they rest upon locally thickened layers of harder, marly limestone. Moreover, a few other local limestone swellings occur, which did not later lead to reef formation. The reef knolls are very marly and show a vague stratification, which, however, cannot be correlated with the alternation of layers in the surrounding stratified sediments.

A short distance southwest of these knolls, on a somewhat higher level, an almost 3 m long, flat reef is exposed. At its base, stratified limestone is found, about 0.2 m thick, which is also present, in greater thickness, at the right and left side under the reef. A few more reefs occur on this same level, northeast of the above-described knoll reefs.

About 0.5–0.35 km southwest of the southern Korpklint, five relatively large Upper Visby reefs occur within a distance of about 150 m; their maximum length is 6 m, and maximum thickness is 5 m. Four of them have their base at more or less the same level. This third level is in its turn about 1 m higher in the Upper Visby Beds than the second. The basis of the four reefs is now about 9 m above sea level. The growth of the fifth and smallest of these reefs started at a level about 2 m lower. Because of selective Quaternary erosion, the reefs protrude as humps from the cliff.

DISTINCTION BETWEEN REEF LIMESTONE AND STRATIFIED SEDIMENTS

Although the majority of Upper Visby reefs are easily recognizable, there are also cases where it is more difficult to distinguish between the stratified sediments and the reef limestone of the Upper Visby Beds. South of Visby, this especially applies to reef-like formations in the lowermost 2 m of the reef-carrying part of the Upper Visby succession. Thus it is doubtful whether a few reef-like formations appearing between the Korpklint (southwest of Högklint, Västerhejde Parish) and the Korpklint (at the northern boundary of the Tofta Skjutfält) represent fossil reefs or not. A good example is a small reef-like formation about 1 km south-southwest of the northern of these two Korpklintar, the Västerhejde Korpklint. In general appearance, there is only a slight difference from the stratified sediments. The voluminous marly matrix in places gives rise to a vague stratification. However, the succession of marlstone and limestone found in

Fig.26. Upper Visby reef at the south-southwest side of Högklint.

the stratified sediments does not continue through the reef-like formation. Moreover, a few limestone layers sag slightly under it or arch over it. Type and colour of weathering in the reef-like formation are about the same as those in the stratified sediments.

Such resemblance in weathering and colour with the surroundings is also found in several other reef-like formations. In fact, the interruption in the normal stratification - or, if the formation shows a vague stratification, the lack of correlation between this and the normal stratification - and the relatively higher number of coral colonies, are the only indications that the reef-like formations in question presumably represent elementary reefs.

As will be discussed in the next section, reef growth was often preceded by a gradual improvement of the environmental conditions, resulting in the deposition of a local complex of stratified marly limestone or varying thickness. This limestone may be overlaid by a reef. Sometimes the remains of an unsuccessful attempt at reef development are found on top of a limestone swelling. Some examples are found in the environment of Axelsro. In places above a local swelling of limestone, which may or may not bend slightly downwards, more coral colonies are present than elsewhere in the surrounding sediments. *Halysites* spp. often play a major part, but often other genera also are present, whereas even stromatoporoids may contribute. Failure of reef growth can be established from the continuation of the Upper Visby stratification through these rock parts, albeit sometimes with a thinning of the marlstone layers. Exposures like these show that even the criteria for reef growth, given in the preceding paragraph, have to be used with care and only in combination with each other.

Although indistinct reef-like formations, as discussed in the first paragraph of this section, are generally found in the lowermost part of the reef-carrying part of the Upper Visby Beds, some also occur higher up. This is illustrated south-southwest of Högklint. A reef exposed there is vaguely stratified, strongly marly and shows the shape of an inverted cone. It has developed so late during the Late Visby Period that its uppermost part is found surrounded by Högklint sediments. The Upper Visby deposits pass, close to the reef, into thicker layers of hard, marly limestone. Both this limestone and the Högklint limestone on top of it, terminate mainly against the reef, or partly bend under it or over it. Directly south of this reef, a second one is present, with its base still a little higher in the Upper Visby Beds (Fig.26). This reef is somewhat smaller and more lens-shaped. Also this reef has a voluminous marly matrix and is vaguely stratified. It rests upon a small swelling of stratified limestone. Laterally of it, hardly any limestone mantle is present; the reef limestone is bound almost directly by the normal Upper Visby sediments. The top of the reef is surrounded by Högklint limestone.

The poor development of these reefs may presumably have been caused by the presence of one or more other reefs, which were more to seaward and which influenced their environment.

In all cases mentioned above, a high content of marl contributed to a vague stratification in certain reefs. One reef, however, is a notable exception to the rule. It is exposed at the foot of Stavsklint (Tofta Skjutfält), some tens of metres northeast of the southwestern end of this cliff. In a large part of the section, this reef shows some stratification, caused not by marl but by a high content of stratified limestone of a character similar to that of the sediment which forms the mantle around the reef. The reef builders occur in between these short limestone layers, sometimes as isolated colonies, usually in heaps together, corals, stromatoporoids and bryozoans intermingled. The uppermost limestone layers show some arching. Because of the close connection between stratified limestone and reef limestone, the exact dimensions of the reef are difficult to determine; its length is about 4 m, its thickness 2–3 m. The upper side of the reef is still about 9 m below the boundary between the Upper Visby and Högklint Beds.

LIMESTONE UNDERNEATH THE REEFS

The development of the Upper Visby reefs only became possible after a local change in the environmental conditions had taken place, which was favourable for the potential reef-building organisms. This involved a relative increase in the deposition of marly limestone in such localities, at the cost of the marlstone.

As was already mentioned in passing, in the previous section of this chapter, many Upper Visby reefs came to rest in this manner over one or more thick (generally 10 or more cm) and local layers of hard, marly limestone. Whereas, however, such limestone swellings needed to be mentioned earlier in order to be better able to interpret that which was found on top of them, a few pages will now be devoted to these limestone deposits themselves.

The local improvement in environmental conditions causing the limestone swellings did not always last long enough to lead to effective reef growth. In several places within the Upper Visby Beds, local lenses of hard, marly limestone which are not covered by reefs, can be seen. Sometimes such a limestone swelling, comprising one or more layers, occurs rather isolated in a section. In other cases, some swellings are found above each other, each passing over a short distance again into the normal Upper Visby alternation of layers of marly limestone and marlstone, which may thereupon

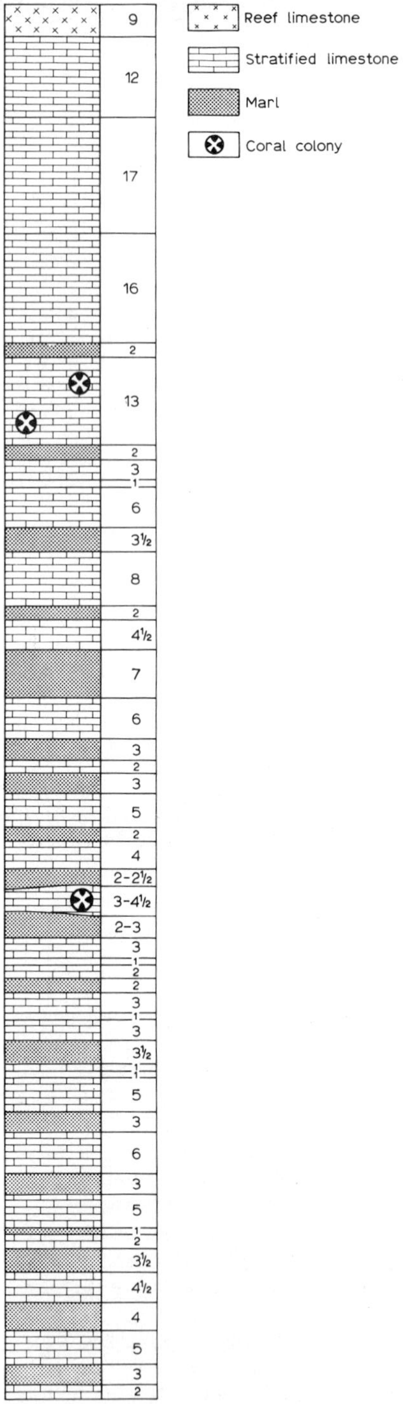

Fig.27. Detailed section, showing the Upper Visby interstratification underneath an Upper Visby reef of about 2 m long and 1 m thick. Note the increase in (marly) limestone towards the base of the reef. The numbers indicate thicknesses of the successive layers in centimetres. About 0.2 km south-southwest of Stavsklint (Tofta Skjutfält). (After Manten, 1962, fig.6.)

be followed by another limestone lens. There are also examples where two or more swellings occur side by side, as local thickenings of what are elsewhere not very typical, thin layers of the usual marly limestone.

Examples of such local limestone lenses are well-exposed in, among other places, the higher parts of the Upper Visby Beds at Axelsro. The high fossil content points to a mainly organogenic formation. Generally brachiopods and solitary corals dominate, but compound corals are also found. However, the difference in the conditions under which formation of the normal Upper Visby sediments took place was only gradual, since coral colonies occur also in the normal Upper Visby succession. The larger ones (up to 20 cm in horizontal section and generally between 1 and 10 cm thick) though are usually either fully enclosed, or have only their bases, in a fossiliferous layer of marly limestone within the succession of strata. Apparently a richer development of solitary organisms was often a good preparation for the development of compound potential reef builders.

Some of the local limestone accumulations represent a kind of transitional form towards what has earlier been called "attempts at reef building which failed". Thus, slightly more than 0.5 km southwest of Blåhäll, some swellings of vaguely-bedded, very fossiliferous limestone are found, which are on the average 0.5 m thick and not quite 1 m long. They are roughly semi-globular, with the convex side downwards. The underlying sediments sag slightly under the limestone swellings.

The improvement in environmental conditions which can be deduced from the cliff closely southwest of Blåhäll, continued somewhat further. High in this cliff, in three places, a very local swelling of layers of hard limestone can be observed. In each of the three occurrences, the layers together have a thickness of about 2 m. The fossil content consists predominantly of brachiopods and solitary corals, with in between these a number of coral colonies. In the lower parts of the swellings, the limestone layers are separated by thin layers of marlstone, some of which distinctly arch over the limestone. On top of the limestone swellings, in each of the three cases, a small accumulation of true reef builders is found; they are predominantly rather flat forms, and are separated by marlstone. Within the accumulations there is no stratification. Immediately above the accumulations, the normal Upper Visby alternation continues. Reef development, thus, only lasted a very short time.

There are many examples where the limestone thickenings are overlaid by real reefs. At this place, a reef exposed about 0.2 km south-southwest of Stavsklint has been selected. It is about 2 m long and 1 m thick. Hard limestone is found both at its base and on its sides. The latter rock is restricted to the direct environment of the reef. At a very short distance from the reef, the limestone layers already start decreasing in thickness. Marlstone begins to occur in between the layers, and over a distance of some metres the deposit passes into the normal Upper Visby alternation. Some of the limestone layers continue as thin layers of marly limestone within the stratified succession; the majority of the layers thin out entirely.

In view of the fact that reef development was preceded by an improvement in the local conditions, it would be interesting to know how far down in the sedimentary succession this local change in environmental conditions first began to make itself perceptible. In this connection, the thicknesses of the alternating layers of hard, marly limestone and of marlstone directly underneath the above-mentioned reef south-southwest of Stavsklint were measured (Fig.27). It is clear that upwards, towards the reef, the limestone layers began increasingly to dominate over the layers of softer, generally thinly foliating marlstone. Some of the higher limestone layers are already rich in coral colonies, up to 20 cm long. The majority of these are in their orientation of growth, a few are upside down. The shape of the colonies varies from rather flat to almost globular. If from the reef limestone downwards, the total thickness of the limestone layers for each successive 25 cm is counted, the following figures are obtained from the section shown in Fig.27: 25, 23, 19, 15.5, 15, 14.5, 17.5, 13.5. For this part of the Upper Visby Beds, a ratio of 15 cm marly limestone/10 cm marlstone may be considered as being normal in localities where there are no reefs. It therefore appears that the improvement in local environmental conditions, which finally resulted in reef growth, started when the sediments some 75–80 cm underneath the reef were laid down.

In a number of cases, several reefs within a certain length of coastal cliff (up to a few hundreds of metres) are seen to have started growth at about the same level. This level, then, is characterized by the presence of limestone layers which are thicker than the average at that height in the Upper Visby succession. Underneath the reefs these layers are often even more thickened. With the aid of coloured-chalk markings, such levels can easily be traced in the field. The greater dominance of the limestone layers can, from time to time, be checked by determining the total limestone thickness per 25 cm, as was done in the above example. The level on which the increase in limestone started, appears from such counts to fluctuate somewhat. Underneath reefs the increase in limestone deposition generally began from one to a few decimetres lower. Some examples of reefs which started growth more or less simultaneously will be given in a later section of this chapter.

It should be noted here that the development of reefs which started at the same level, did not end at the same time, not even within a restricted area. Some reefs reached a thickness of less than 1 m, others are several metres thick and some even continue until well into the Högklint Beds.

Reef development in Late Visby time was not always preceded by preliminaries of such long duration as is reflected by the local limestone swellings or the development of a substratum zone of larger extension. There are also Upper Visby reefs which are hardly underlaid by extra-thick limestone. They apparently represent cases where soon after an improvement of local conditions began, an early attempt at reef building was already successful.

As most reefs are surrounded by a mantle of stratified limestone (for details see the next section of this chapter), the possibility that some exposures show a very peripheral section through a reef should also be taken into account. In such a section, a lens of stratified limestone seen underneath the reef limestone is in fact a deposit formed laterally to an older part of that reef. The number of cases in which reef growth was preceded by increased limestone deposition may, therefore, be somewhat less than one would think when working in the field.

LIMESTONE LATERAL TO THE REEFS

As already mentioned above, increased limestone deposition, as compared to areas without reefs, also took place laterally to many Upper Visby reefs. An example was already given in the previous section - the reef about 0.2 km south-southwest of Stavsklint. Especially at the southwest side of the reefs (which because of the general orientation of the coastal cliff are usually sectioned in a northeast – southwest direction) this limestone deposit is often found to be well developed. At the northeast side, as exposed, the limestone mantle is generally thinner, especially with the smaller reefs. Overlying the reefs a local limestone zone is found in some cases only. Otherwise, there is the normal Upper Visby alternation of marly limestone and marlstone. A cover of limestone is present mainly in those reefs, which end relatively close to the Upper Visby – Högklint boundary, e.g., the three reefs of Snäckgärdsbaden. In an upward direction the limestone of such a reef cover becomes more-thinly stratified.

The limestone lateral to the reefs is, for instance, clearly developed around the two reefs found about 0.15 km south of Axelsro, and mentioned already as examples of the shape of the Upper Visby reefs (p.87). Here also, more stratified limestone is present at the exposed southwest side than at the northeast side or even underneath the reefs. Only a very few marlstone layers alternate with the limestone. Southwest of the more southern of these two reefs, the limestone can be followed, in a total thickness of almost 2 m, over a distance of more than 10 m. In one place it is even developed as a kind of transition form between the reef-surrounding stratified limestone and the reef limestone. Southwest of the more northerly of the two reefs (Fig.20), the stratified limestone, about 2 m thick, extends over a distance of about 7 m to near the southern reef. The normal succession of Upper Visby sediments sags under the reef-surrounding limestone, though less than under the reef limestone.

In the majority of cases, the stratified-limestone mantle of Upper Visby reefs is no more than a few metres wide (also at the southwest side of the reefs), and it is often even narrower.

Like the stratified limestone found underneath several reefs, the limestone lateral to the reefs is almost always very fossiliferous. Solitary corals and brachiopods are abundant and there are also several compound corals, bryozoans and crinoid remains, and some stromatoporoids. In a general way, this limestone can be compared to the stratified limestone with reef debris and crinoids which is found around reefs of Hoburgen type. The latter, however, take up much more space around those reefs, particularly the crinoid limestones. This may be connected with the generally larger size of the Hoburgen-type reefs and also with the much more abundant crinoid development around them. Although in the higher parts of the Upper Visby Beds, and particularly in the reef-surrounding limestones, an increase in the number of crinoid remains can be noted, nowhere the crinoids notably outnumber the other fossils. None of the Upper Visby reefs possesses a crinoid limestone around its higher parts which, in any way, could be compared to that around many of the younger reefs in the area of Gotland (e.g., "Hoburg marble", "Karlsö marble").

As was also mentioned earlier, in some cases of rather vague reefs, the presence of a limestone mantle may help to characterize such a marly and somewhat-stratified formation as a fossil reef. An example is found in the southern part of the Högklint cliff (Fig.21). The very-marly matrix of this reef shows a faint stratification, which, however, does not correlate with the stratification in the normal Upper Visby succession of limestone and marlstone layers, as found in the close vicinity. The reef is bound from these sediments at both sides by a narrow mantle of marly limestone, which is distinctly stratified and very fossiliferous.

The differences in the amounts of stratified limestone as found southwest and northeast of the exposed reefs may probably be explained in a similar way as the asymmetrical shape of most of the reefs (p.89). As was shown there the northeast side of the exposures may give an impression of the situation that existed at the side directed more towards the open sea, whereas the southwest side of the exposures may represent the situation more as it developed at the landward-directed side of the growing reefs.

There is a second indication in support of this explanation. The cliff coast about 0.1 km north-northeast of Stavsklint has an orientation about north-northeast – south-southwest. Three large Upper Visby reefs are exposed there. Both underneath and on both sides of the northern of these three reefs, much more surrounding stratified limestone is found than around the other two reefs. To judge from the present local situation, most of the reef limestone of the northern reef is still hidden in the cliff. The exposure thus gives an impression of the situation at the west side of the reef, and thereby suggests that at this west side, during reef development the strongest limestone deposition took place; this was at that time the landward side of the reef.

An indication that the least limestone deposition took place at the original seaward-directed side of the reefs, is probably presented by a small and comparatively vague reef, exposed about 0.5 km northeast of Axelsro. It rests over a thick layer of limestone. A small cave is present at the southwest side. In the cave it can be seen that at the lateral and upper sides of the southern part of the reef limestone, only a thin mantle of marly limestone is present, rapidly passing into the normal alternation of marlstone and marly limestone. These rocks also bound the reef at the northeast side.

In the few cases in which there are no signs of extra amounts of limestone around a reef, one should be cautious in deciding that these represent possible sections through the east-southeastern periphery of the reef. The angle between the direction of the present section and that of the coast line at the time of formation of the reef should also be taken into account. If this angle is not rather small, the exposure cannot solely represent the original seaward side of the reef. Other factors must also have played a part.

As a general rule (but one that has exceptions), it can be said that the limestone mantle has developed most strongly in those cases where the environmental conditions for the growth of the reef itself also seem to have been most favourable. This holds for the succession from the oldest to the youngest reefs found within the Upper Visby Beds, but also where, within one level, very distinct and vaguely developed reefs alternate. Under favourable conditions a reef probably rose more rapidly to well over its surroundings, with the result that it was also more subjected to the demolishing forces of the water. The presence of the reefs may have caused increased water turbulence, preventing part of the terrigenous debris from settling. At the same time, under favourable conditions a reef occupied a larger area and thus contained a greater quantity of reef builders and consequently also a greater quantity of potential reef debris. And, fourthly, the more favourable environmental conditions may also have attracted many organisms to live in the direct environment of the reef.

As is usual with stratified sediments around a reef, those around the Upper Visby reefs also often arch somewhat over, and sag somewhat under the reefs. The main causes of this are the effects of greater rates of development on the reef, compaction of the stratified rocks at the side of the reefs and depression of those beneath due to the weight of the reefs.

SPECIFIC LEVELS OF REEF DEVELOPMENT

In some localities in the coastal cliff examples can be found of reefs which apparently started growth at more or less the same time.

The part of the coastal cliff which starts about 0.2 km northeast of Stavsklint will be considered as a first illustration of this phenomenon.

The first occurrence of reef limestone found there, belongs to a reef, about 3 m long, and flat. Its lowest exposed part rests over a succession, about 20 cm thick, of some layers of hard limestone. A comparable rock, in greater thicknesses, is also present right and left of the reef centre, underneath the lateral parts of the reef.

Directly northeast of this reef, the first occurrence is found of six reef-like, round knolls, with an average diameter of just over 1 m. They follow shortly after each other, all at about the same level in the cliff wall and

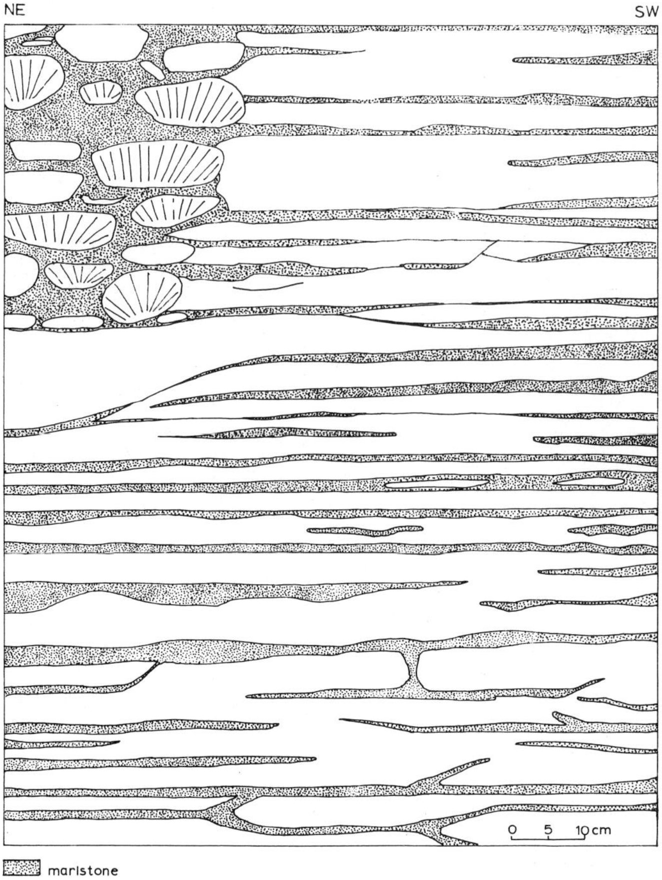

Fig.28. Distribution of marlstone and limestone layers underneath and lateral of the northern one of the two Upper Visby reefs, which are found near Fridhem, northeast of Högklint.

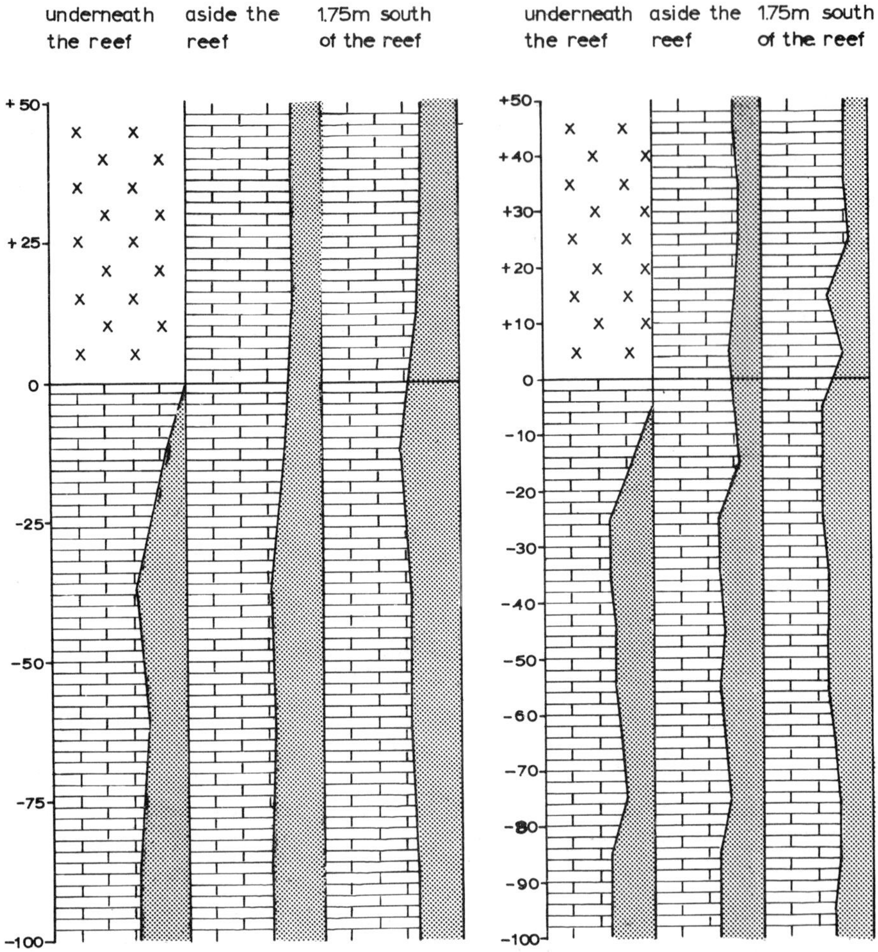

Fig.29. Graphical representation of the total thickness of the limestone layers per successive units of 25 cm (left) and 10 cm (right) in vertical successions underneath and lateral to the northern Upper Visby reef, found near Fridhem, northeast of Högklint. See also Fig.28. Both sets of graphs present the same picture. Working with 25 cm units emphasizes somewhat more the general trend, by eliminating some fluctuations.

---

somewhat lower than the level at which the first-named reef occurs. The knolls possess a strongly marly matrix and show a vague stratification, which, however, cannot be correlated with the stratification found in the surrounding stratified sediments. The knolls mainly rest upon one or more thick layers of hard, marly limestone. There are also some platforms of this limestone, generally a few metres long, that occur at the same level, but on top of which no reefs developed.

Going further north, the level of preference, on which the six reef knolls started growth, disappears again as such. Some reefs found there are vague and, with one exception, have their bases at a higher level.

In order to gain some understanding of how specific levels of reef growth originated, a second example is still to be mentioned.

Near Fridhem, north of Högklint, two small Upper Visby reefs occur, with a distance of about 7.5 m in between them. They both rest over the same layer of marly limestone, which thickens to 10–20 cm underneath the reefs, but in between the reefs is only 3 cm thick and does not differ there from the other layers of marly limestone which occur in the Upper Visby Beds (Fig.28). The reefs are small in size. The southern one is at a maximum 1 m long and thick, the northern one is at a maximum 1 m long and 0.75 m thick. Both show the normal characteristics of the Upper Visby reefs.

Some sections were measured in the vicinity of these reefs. On the basis of the thicknesses measured, the contribution of the limestone layers to each successive 25 cm was calculated and the same was done for each successive 10 cm. The results are shown in Fig.29; both graphs present the same picture. Underneath the reef, the limestone layers together take up about 65% of the total section. The thickened limestone layer below the reef again represents an improvement in the environmental conditions, which at a certain moment made reef growth possible. This improvement in conditions can also be traced in the environment of the reef, but began there at a later stage. It first became clear there at the level at which reef building started. At the height of the reef, the limestone layers in the reef environment take up about 75% of the total succession of strata. A temporary increase in marl deposition is represented by a marlstone layer, about 4 cm thick, not far below the top of the northernmost of the two reefs. Such a thick marlstone layer is rather unusual at this height in the Upper Visby Beds. It is not impossible that this increased supply of terrigenous debris caused the death of the northern reef, or of both reefs. Above the reefs, a ratio of the cumulative thicknesses of the limestone layers against those of the marlstone layers of 3/1 is restored again. The steepness of the cliff wall prevented detailed measuring. About 3 m above the base of the reefs, the boundary between Upper Visby and Högklint Beds is exposed.

The comparatively thick limestone layer underneath the reef, together with the distinct increase in the amount of limestone at some distance from the relatively small reef, are clear indications that the greater deposition of limestone cannot be attributed to the reef and the debris which it produced, but represents a real alteration of the environmental conditions.

The differences in the height within the Upper Visby section, at which the increase in limestone formation is notable for the first time, suggest that the improvement in conditions started locally and spread from there over the direct environment. Reef growth occurred particularly at those places where the improvement in environmental conditions was strongest. Generally, these were also the places where the improvement of the environment first started. If these places were very local, the growth of adjacent reefs started at different levels; if the environmental alteration involved a somewhat larger area, more reefs could start growth at about the same level.

THE COASTAL CLIFF NORTH OF KNEIPPBYN

Good examples of Upper Visby reefs are found north of Kneippbyn (about 3.7 km southwest of Visby). They illustrate several of the phenomena described in the previous pages.

From Kneippbyn Fiskläge northwards, a few reefs can first be seen, which have already been described in the section on the dimensions of the Upper Visby reefs (pp.95–9 There is no particular reason to describe the five lens-shaped reefs which follow next. The territory of a shooting range soon follows. The reefs to be dealt with here are found in the environment of the boundary between the free beach and the shooting range.

About 25 m southwest of this boundary a small reef is exposed, which is the only Upper Visby reef in which stromatoporoids dominate. Including the stratified limestone at its base, this reef is in its centre about 0.9 m long and about 0.7 m thick (Fig.30). The base occurs about 4 m above present sea level. The succession starts with a horizontal, 12 cm-thick layer of hard limestone which consists, for a very substantial part, of brachiopods and solitary tetracorals. This layer is overlaid by a marlstone one, 3 cm thick. On top of this, a lens of limestone occurs, which is 40 cm long and in its centre 5 cm thick, and again very fossiliferous. Over the lens a layer of marlstone follows, on the average about 2.5 cm thick. On top of this layer, six large stromatoporoids are present, of a semi-globular shape, on the average 75 cm long and in their centre 10 cm thick. These stromatoporoids cover each other in a kind of right-left alternation, and are separated by layers of marlstone, generally less than 1 cm thick. The lowermost stromatoporoids have settled upon solitary corals; except for the lower side, they envelop these corals almost entirely. Also an occasional *Favosites* colony, only a few centimetres large, is completely surrounded by one of the lower stromatoporoids. In between the stromatoporoids some colonies of *Heliolites* are

Fig.30. The only Upper Visby reef in which stromatoporoids play a dominant part. North of Kneippbyn.

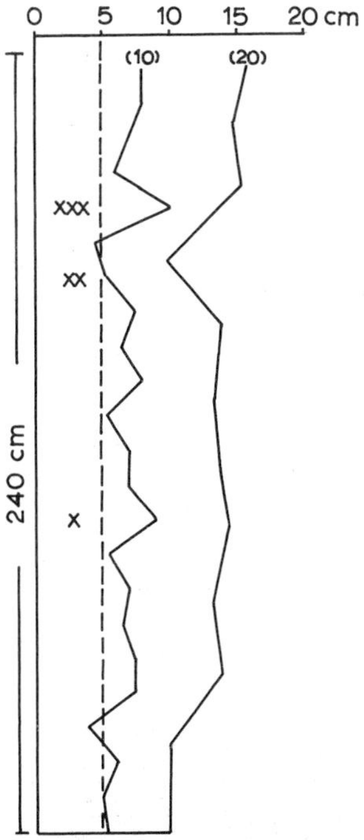

Fig.31. Graphical representation of the cumulative thickness of the layers of marly limestone per successive units of 10 cm (left) and 20 cm (right); about 7.5 m southwest of the stromatoporoid reef of Fig.30. Although the succession is seemingly a normal Upper Visby section, the levels on which reef growth started can be well distinguished. The development of the stromatoporoid reef was preceded by a long improvement in local conditions. The * horizon corresponds to the 12 cm-thick limestone layer underneath that reef.
A temporary increase in marl deposition (**) probably caused the death of this small reef. Soon conditions became more favourable again and the *** horizon corresponds to the beginning of the limestone thickenings underneath the reef directly northeast of the stromatoporoid reef. The three special horizons can be followed over several tens of metres. The other limestone layers all thin out after a shorter distance and other ones appear. The total general picture of the ratio between the thickness of the marlstone and limestone layers, however, remains the same.

---

present in the reef. Upon the margins and the top of the reef, some compound corals and some solitary corals have settled.

The stratified Upper Visby sediments underneath the above-described succession show no signs of sagging. Laterally, the layers surrounding the lower 20 cm of the reef come to a dead end against it. The higher layers dip upwards against the reef, with an increasing angle, which is at the maximum 45°. The uppermost surrounding layers arch over the reef. Only 20 cm above the reef the normal, horizontal stratification is restored. In the stratified rocks closely above the reef, relatively many coral colonies are present.

A layer of limestone, 5 cm thick, which is very rich in brachiopods and solitary corals, is found at the southwest side of the reef, about half-way up. Apart from this layer, there are no extra amounts of limestone to surround the reef laterally. The boundary between reef and stratified sediments shows a similar pattern at both sides.

At about the height of the top of the stromatoporoid reef, just over 1 m further northeast, the first of a succession of locally-thickening limestone layers is found. The succession reaches a thickness of 0.5 m. The thickened limestone layers are separated by marl layers of 1–3 mm thick. On top of the local limestone deposit, a reef is exposed about 1.5 m long and less than 1 m thick. It is, unfortunately, mainly hidden behind recent scree. The southwestern boundary of this reef lies above the stromatoporoid reef, described before. Both at the southwest and the northeast side of the reef a well-developed limestone mantle is present, but the one at the northwest side thins out much more gradually. Similar to the limestone layers underneath the reef, the layers of the mantle in the northeast are knobby and rugged and very fossiliferous (many compound and solitary corals and brachiopods). The marl layers in between the limestone layers are thin, 1–3 mm, particularly those close to the reef.

About 5 m southwest of the stromatoporoid reef, and with its base at the same level, there is an unsuccessful attempt at reef building, which is about 0.75 m wide and thick. At the base there is a limestone layer, locally thickened to about 8 cm. Also a number of limestone layers show an increase in thickness there. Some coral colonies are present; slightly higher in the section there are even quite a number of them, up to 15 cm in horizontal diameter and up to 8 cm thick. The layers of marlstone continue,

Fig.32. Upper Visby reef in the coastal cliff north of Kneippbyn, about 25 m north of the southern boundary of the Visby shooting range. At the base of the reef the normal Upper Visby interstratification of marly limestone and marlstone. Stratified limestone occurs lateral to the reef. This limestone is best developed at the southwest side (at the right on the photograph). (After Manten, 1962, fig.4.)

Fig.33. Upper Visby reef, north of Kneippbyn, in the south of the Visby shooting range. The reef shows a well-developed limestone deposit at its southwest side, gradually passing into the normal Upper Visby interstratification of marly limestone and marlstone. (After Manten, 1962, fig.5.)

albeit thinner, around the fossils or arching over them. Although their number and thickness decrease upwards, the locality could not fully differentiate itself from the normal sedimentation pattern of the Upper Visby Beds.

In the section exposed at a distance of about 7.5 m southwest of the stromatoporoid reef, the ratio of the amounts of marlstone and marly limestone was measured (Fig.31). The picture obtained gives an average of about 7 cm of limestone per unit of 10 cm thickness. This section is reasonably representative of the Upper Visby Beds as exposed in the coastal cliff in this area.

Directly north of the southern boundary of the shooting range, a reef, about 1 m long and thick is exposed. The limestone mantle of this reef is especially well developed at its northeast side, which is an exception to the general rule. The reef started growth during the first improvement in environmental conditions shown in Fig.31.

About 25 m over the boundary of the shooting range, near a staircase which is built against the cliff wall, a reef is present which has maximum dimensions of about 5 m long and 4 m thick (Fig.32). It started development at about the same level as the stromatoporoid reef. Underneath the reef there is hardly any extra limestone. The reef mantle of local limestone layers is more strongly developed at the southwest than at the northeast side. At both sides the boundary between reef limestone and surrounding stratified sediments rises outwards at an angle of about 60°. The reef is built mainly by compound corals. In comparison with other Upper Visby reefs, it is somewhat more unorganized, with a greater number of colonies which lie obliquely or upside down.

In between the above and the next major reef, a small reef is exposed, at about the level of the stromatoporoid reef. It possesses a rather well-developed mantle of limestone layers, separated by very thin layers of marlstone.

About 50 m past the staircase, the next Upper Visby reef is exposed (Fig.33). The reef is comparatively large for this reef type. It clearly belongs to the second level of reef development. In between the predominantly flat reef builders, mainly compound corals, several small pockets of very thinly stratified marl are found. Against the southwest boundary of the reef, which is very steep and in the upper part almost vertical, a zone of stratified limestone, at the minimum 2.5 m wide, is present. The northeastern reef boundary, which rises at an angle of about 25° outwards over the stratified sediments, also has a mantlè of stratified limestone, which is on the average 1.5 m wide, and particularly well developed against the lower half of the reef. At both sides the layers of the limestone mantle dip slightly towards the reef, caused by subsidence of the entire complex of reef and reef mantle in the normal Upper Visby deposits. The limestone layers therein are less thickened underneath the reef than under many other reefs of this type.

Following three small, vague reefs, another larger reef is found about 75 m past the above mentioned. Also this one started growth in the second level. It is found slightly lower in the present cliff wall than the previous reef, but this is caused by a sagging, here even stronger, of the entire Upper Visby Beds under the pressure of the reef and its surrounding sediments. The reef is about 4 m thick, and is about 5 m long without the surrounding limestone mantle, and about 10 m long with it.

Especially at the southwest side of the reef quite a large amount of limestone is present, measuring about 4 m in thickness, with a width of about 2 m at the base and about 4 m at the top. In between the limestone layers of the mantle practically no marlstone layers are present. At a height of 2–3 m above the base of the reef a notable zone, about 1 m thick, is present in the reef surroundings. In this zone several thick limestone layers extend much further southwestwards than the actual limestone mantle of the reef. At a distance of 7 m away from the reef, several can still be clearly recognized as extra-thick layers, but they gradually become thinner and the Upper Visby interstratification of limestone and marlstone regains more and more its normal appearance. At the height of the uppermost metre of reef limestone such long, thickened limestone layers are less common.

Below the deepest part of the reef, limestone layers are present to a total thickness of about 0.5 m, with in between them a few thin marlstone layers. Northeastwards, this basal limestone complex increases to an average thickness of about 1 m and a local maximum of 1.5 m. Southwestwards underneath the reef, the increase is even stronger (up to 2 m) and the marlstone layers disappear there almost completely.

The boundary of the reef limestone at the northeast side rises for the lowermost 1.5 m of reef thickness over the stratified limestone with an angle of about 60°; the angle decreases further upwards to as little as 15°, after which it increases rapidly again. The southwest boundary is much steeper, being about 65° in the lowermost metre and upwards from there about vertical. Also the boundary between the limestone mantle and the more normal Upper Visby succession is steeper at the southwest side (about 60° outwards, without considering here the zone with the extended limestone layers at 2–3 m above the reef base) than at the northeast side (about 30°). The upper surface of the reef is faintly convex.

The reef about 0.15 km northeast of the staircase has already been discussed earlier in this chapter, as an example of the inverted-cone shape (p.90) and as one illustrating the general character of the reef limestone (p.83). Past this reef there are only two vague reef-like developments. The beach then is interrupted by a coastal area in which the sea directly bounds the coastal cliff.

THE REEFS NORTH OF VISBY

The reefs which are exposed north of Visby, particularly along the coast, present relatively little new information compared to the reefs which have been described from south of the capital of Gotland.

One of the sites showing an Upper Visby reef is near the Luseklint, the first major cliff northeast of Nyhamns Fiskläge. The cliff is built predominantly by Högklint reef limestone. An Upper Visby reef is present there, low in the section, in the southwest. It has a well-developed mantle of stratified limestone, which is somewhat wider at the southwest side than at the northeast side. The boundary between reef limestone and mantle at the southwest side is rather steep. At the northeast side, both the boundary between reef and mantle and the boundary between mantle and the normal Upper Visby succession advance outwards with an angle of about 20–25°. Above the reef, the lowermost stratified Högklint Beds are to be found.

Further north, but still at the base of the high cliff, two more reefs are found which are mainly enclosed by the uppermost Upper Visby Beds.

Directly north of the cliff, two reefs, belonging to the Upper Visby Beds, are found rather closely together. The sediment in between these reefs is stratified limestone. Of both reefs only the uppermost parts outcrop, all the rest is hidden behind scree. The southern reef is better exposed. It is about 2.5 m long and thick and strongly weathered. It shows the normal nature and composition of the Upper Visby reefs. The northern reef probably was somewhat larger. The mantle of stratified limestone around it is well exposed. The boundary between reef and mantle is very steep at the southwest side and rises more gradually over the stratified sediment at the northeast side.

Rather characteristic for the exposures north of Visby is the gradual transition from the Upper Visby to the Högklint Beds. This is also reflected in the reefs. Several reefs have their bases in the uppermost Upper Visby Beds and their upper parts enveloped by the Högklint Beds. Some of these even gave rise to reef masses which are comparatively large for Gotland.

A good example is given by a large cliff, about 0.7 km northeast of Lundsklint, which can easily be recognized in the field, because of the enormous accumulation of loose blocks at its foot. Fig.147 shows part of this complex, about 180 m long, of Högklint reef limestone. At the very left, an isolated Upper Visby reef can be seen. Its base is about 6 m below the Upper Visby – Högklint boundary. Including the mantle of stratified limestone and also the thickened limestone layers underneath, the reef is about 6 m long and about 5 m thick. The reef protrudes from the cliff wall; its upper part is covered by travertine and scree.

In between the reef mentioned above, and the large complex of Högklint rocks, the Upper Visby Beds are relatively rich in marly limestone. The Högklint reef complex is distinctly rooted in the Upper Visby Beds. The lowest point is about 7 m below the stratigraphical boundary. The reef complex consists of a large number of lenses, separated by some limestone layers.

A similar development can be found with several other Högklint reefs which occur low in the stratigraphical succession. Others, however, show greater unity in the total amount of reef limestone which is exposed. An example is the reef about 0.28 km northeast of Lundsklint (Fig.44), which also clearly has its root in the Upper Visby Beds. Underneath these reef complexes the Upper Visby Beds show a distinct local increase in the thickness of the limestone layers.

In the lower coastal cliff in between the two large cliffs mentioned in the preceding paragraphs, there is a distance of about 0.4 km in which only two small reefs are found, completely enclosed by the Upper Visby Beds. Some other reefs started growth at about the transition from Visby to Högklint time, or slightly earlier. This illustrates that the development of Upper Visby reefs in the area north of Visby was less common and also started later than that further south.

SYNTHESIS

The phenomena described in the previous pages were often based on a generalization of specific observations. The more such generalizations cover a wide subject, the more the risk inherent in such generalizations usually increases. The author is aware of this, but nevertheless, if a general understanding of the Middle Palaeozoic geology of Gotland is to be reached, this has to be done. Therefore, the discussions of the reef types found in Gotland will be concluded by an attempt to synthesize the collected data. The same will be done with the information from the various stratigraphical units (Chapter XI). Finally, all these views will be brought together to give the total picture of Chapter XV.

In the course of Late Visby time, conditions on the sea floor apparently changed gradually. More marly limestone and less marlstone was deposited. The situation did not alter uniformly over the entire sea floor. In some restricted or very restricted parts, the modification in environmental conditions was ahead of the general trend. More organisms populated these areas and contributed to extra limestone formation. Also, they provided hard objects upon which larval forms of the potential reef builders could settle. The early colonies themselves again provided a suitable place for the settling of further larvae. If the environmental conditions remained reasonably favourable, this could lead to the formation of small reefs, which are often surrounded by mantles of stratified marly limestone. However, small alterations in the environmental conditions could still strongly influence the development. Consequently, several attempts at reef formation failed. Where the pioneering organisms did manage to found a reef, environmental conditions in combination with differing growth rates in different individuals led to slightly different forms of reefs, such as knolls, reefs with the shape of an inverted cone or lenticular reefs.

Corals, in a rather great variety of forms, strongly dominate the organic element in the reefs. Apparently conditions were still too unfavourable for a really substantial contribution by the stromatoporoids, and this applied even more for the calcareous Algae. Crinoids also found the environment still too unfavourable for abundant development.

With a continued gradual improvement of the environment towards the end of Visby time, reef development was increasingly favoured. The foundations were thus laid for much richer reef growth in the next, Högklint Period. A number of Högklint reefs root in the Upper Visby Beds. With an increase in reef size, the organic composition of the reefs also became more varied.

The difference in abundance of reefs north and south of Visby suggests that reef growth was restricted to a certain depth zone. The direction of that reef zone was probably parallel to the direction of the coast line at that time.

Chapter VII

THE HOBURGEN REEF TYPE

INTRODUCTION

The Upper Visby Beds, in the northwest of Gotland, are overlaid by the Högklint Beds. These also carry reefs, some of which already started growth in the uppermost Upper Visby Beds or around the lower boundary of the Högklint Beds. Thus there is a continuation of reef formation from the one stratigraphical unit into the other. Nevertheless, the reefs in the two units are different. The reefs in the Högklint Beds are not of the Upper Visby type, but of the Hoburgen type. The main differences between the two are shown in Table VIII. More or less similar reefs are present in several other stratigraphical units in Gotland: the Slite, Halla, Klinteberg, Hemse, Eke, Burgsvik and Hamra-Sundre Beds. The latter beds include the well-known (in Gotland) locality of Hoburgen, in the very south of the island. There reefs of this nature are well exposed and this site has, therefore, been made the type locality for this kind of reef.

Although Högklint also is a well-known locality, and one which consists largely of reef limestone of the type now to be described, it has purposely not been selected as the locality after which the reef type is to be named. The major reason is that Högklint has already given its name to a stratigraphical unit, and, as mentioned above, reefs similar to those found in this unit are present also in other stratigraphical units. Calling the reef type the Högklint reef type could easily lead to confusion as to the stratigraphical distribution of this kind of reef. Secondly, the exposures in Högklint reveal less detail about various aspects of this kind of reef as do some other sites, such as Hoburgen.

In the following pages the Hoburgen reef type will be described in some detail, using information obtained from Hoburgen but also from various other localities and other stratigraphical units, in order to reach the best possible picture of the reef type in all its variations.

The reef limestone is more resistant to erosion than any of the stratified sediments and often reef limestone appears as hillocks or cliffs, frequently with steep walls at the side which was or is affected by post-Glacial or Recent wave erosion.

THE HOBURGEN COMPLEX

Since Hoburgen has been made the type locality of the most-common reef type of Gotland, it is appropriate to include in this chapter a short description of that locality (Fig.34).

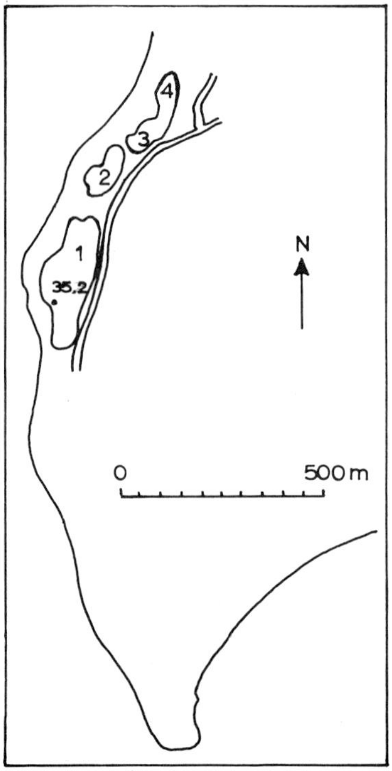

Fig.34. Sketch map of Hoburgen, showing the position of the four hillocks ("burgar").

Hoburgen, which means "the high citadel", is located in the southwest of the southern peninsula of Gotland. It consists of four hillocks, called "burgar" (singular "burg"). Together they have a length of about 750 m and a breadth averaging 150 m. Of these hillocks, the most southern one, the Hoburg proper, also called Storburg (Swedish: stor = large) or the first burg, is the most important. To the north follow successively the second, third and fourth burgar. The last two are very-narrowly connected on the east side, but separated in the west.

As said above, the Storburg is the largest hillock and also the highest. Its summit is 35.2 m above sea level. In the southwest it shows the following sedimentary succession:

Around and a little above sea level Burgsvik sandstone is found, with in its uppermost part a characteristic and very fossiliferous layer. On top of it is 60–80 cm oolite, in which a little sandstone is interbedded. This is overlaid by limestone in a somewhat reef-like development (max. thickness 1 m), which passes westwards into a synchronous crinoid limestone (max. 60 cm thick). Above this reef-like deposit there is limestone, very rich in algal balls (see also the section on the Hamra limestone in Chapter XI, pp. 410–411). Next is crinoid limestone again, passing into reef limestone in the east. This reef limestone is mainly built by stromatoporoids and corals in a marly matrix. Also present are bryozoans, crinoids, brachiopods and some lamellibranchs. From there upwards, almost the whole west cliff is composed of reef limestone. In the uppermost part, finally, there is the so-called "Hoburg marble", a red crinoid limestone with thin marly intercalations.

The south cliff of Storburgen (about 20 m high) consists at the bottom of marly Hamra limestone, and has at the top 8–10 m of "Hoburg marble". The latter sediment also occurs in the east cliff.

The upper surface of the Storburg is somewhat undulating; the low tops are in all likelihood connected with reef centres. Between the reef limestone, exposed locally on top of the Storburg, "Hoburg marble" is still found at several places in between the reef builders. Where we imagine this non-reef material to be absent, the reef should show a surface comparable to a field covered with freshly-lifted large potatoes.

The north side of the first burg, as well as the south side of the second one, is composed of marly limestone, which presumably in former times also took up the space between the two burgar.

The second burg, too, shows sandstone and oolite in the lowermost part of the west cliff. These are overlaid again be reef limestone, which in the middle of this wall takes up almost the full height, but to the north and south passes into crinoid limestone. In the rather poorly exposed east side of this burg, mainly "Hoburg marble" outcrops.

The top of the second burg shows in the west a landscape with several higher parts, presumably caused by the presence of fossil reefs. In the east, on the other hand, the surface is rather smooth, which leads to the presumption that no reefs occur there.

Also for the indentation between the second and third burg, there are indications that stratified limestone has been present there, which has only rather recently been removed by erosion.

The third burg shows in the west crinoid limestone at its bottom; while the rest is composed of reef limestone. In contrast to these well exposed west cliffs, the east side is nearly completely overgrown with vegetation.

The west cliff of the fourth burg is almost exclusively composed of reef limestone. Here also the east side is covered with vegetation. The upper surfaces of the third and fourth burgar are uneven, but show an east-southeastward dip everywhere.

FAUNA, FLORA AND MATRIX OF THE REEFS

In fact the essential facts about the organic remains and matrix in the reefs of Hoburgen type have already been given in Chapter V. Some additional information, more specific for the reefs of Hoburgen type, will be added in the following paragraphs.

Stromatoporoids, by virtue of their large numbers and frequently large size, form the most-conspicuous part of the fauna of the Hoburgen-type reefs. The contribution made by the corals varies from reef to reef and also between the various stratigraphical units in which the reefs occur. Massive compound corals are generally well represented, although they are in almost every reef distinctly subordinate to the stromatoporoids. Branched corals are present, sometimes in large colonies or groups of colonies, in the more-central parts of some of the reefs, but may be almost absent in other reefs, even in close vicinity. Solitary corals, occurring in so many specimens and species in the Upper Visby reefs, are not very abundant in most of the exposures of Hoburgen-type reef limestone. It is quite usual if only a few are observed over an exposed surface of 1 m^2. Comparatively the highest numbers of solitary corals are found in the reefs which follow in stratigraphical age directly after the Upper Visby reefs, those in the Lower Högklint Beds. The most-striking figures are presented by the reef limestone drawn as part B of Fig.151. In the major part of the exposed reef area, about 200 solitary corals per 1 m^2 can be seen, whereas in the lowermost 0.5–0.75 m of the reef, this number can be as high as about 750 per 1 m^2.

Bryozoans are present more as isolated branches than as intact colonies, but are common in that fragmented form. Their contribution to the bulk of the reef limestone, however, is generally negligible.

Coralline Algae of the genus *Solenopora* are fairly common in some parts of the reefs.

The distribution of the various reef components over the reefs varies. One has to imagine that the reefs, at the time of their development, showed a great variety of living and dead material. From place to place there were differences in vigour and vitality of the reefs. At one spot there was hardly a space as large as a human hand which was not covered with living reef-building organisms, while in other places, perhaps only a few metres away, there were only a few reef builders, separated by much matrix material, fragmented organic remains and organisms out of their growing positions. Some rough general lines can be subtracted from the variation seen.

Generally, the reef builders in the lowermost parts of the reefs are smaller than higher upwards (Fig.35). The same holds for the reef builders in the uppermost parts in various reefs.

Massive tabular stromatoporoids and corals are frequently found in the basal parts, but they are generally largest and relatively-most abundant in the marginal regions. The stromatoporoid *Labechia* and the coral *Thecia* may locally form expansions of a biostromal nature from the reefs out on to the surrounding sediments.

Generally the reef builders are embedded in a matrix, which may vary from marl to rather pure limestone. In places, pockets or stratified intercalations of matrix material may also occur, both marginal and central in the reefs. In the larger reefs, it can, in several instances, be established that the volume of the reef limestone taken up by the matrix is higher at the original-landward side than at the seaward side. In some reefs, however, also the latter side may be relatively rich in matrix material.

This latter situation is illustrated in, e.g., the east-eastsoutheast wall of the Spillingsklint (Othem Parish, Slite Beds). In the southern part of this wall, reef limestone is exposed which has been formed at the southeast, seaward side of the reef. At a level of 4.5–2 m below the top of the reef, at least ten pockets of marly sediment occur, on the average 25 cm

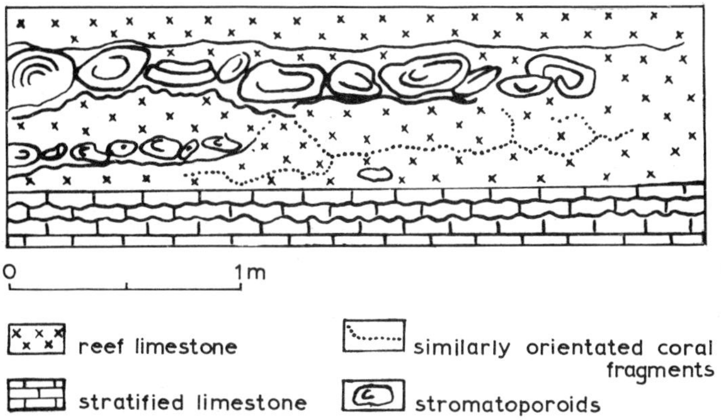

Fig.35. Detailed section, showing the base of the reef limestone in the Hjännklint, about 1.5 km southwest of Halls Fiskläge. Högklint Beds. The contact with the underlying stratified fragment limestone is remarkably smooth. The lowermost reef limestone consists of stromatoporoids and some corals in a matrix of very finely crystalline to dense limestone. In this matrix, remains of several other fossils can be seen, but they are strongly recrystallized and not or almost not identifiable. Two well-recognizable stromatoporoid horizons are notable. In the lower of these, the colonies are on the average 15 cm long; in the upper horizon they measure up to 40 cm. The horizons are bounded by irregular lines. Some of these lines are caused by fragments of branched corals, lying almost side by side in an orientation which is about north-northwest — south-southeast. The other lines are caused by weathering of somewhat softer matrix material from between the reef builders. The reef limestone above the upper stromatoporoid horizon shows a conglomeratic structure.

---

long, 15 cm deep and 10 cm high. In this zone also a cave was developed, and a horizontal niche, about 6.4 m long, 0.9 m high and a few decimetres deep; both originated most likely through erosion of stratified sediment. Both the cave and the niche are covered by reef limestone. South of the niche, a few similar notches were found, 1–3 m long, which, however, are overlaid by stratified crinoid limestone with reef debris, intercalated in reef limestone.

In the Solklint (Slite, Slite Beds), it was found that at the original-seaward side of the reef most local intercalations of layers of stratified marly limestone have a direction of dip towards the reef centre. At the original-landward side the directions of dip are somewhat more irregular. The dip directions, particularly those at the seaward side, suggest that most of the local stratified intercalations were formed at the lee side of reef parts which had risen over the surrounding reef surface.

As said, not only marginally, but also in the more central parts of many Hoburgen-type reefs, pockets and, though less frequently, larger intercalations of stratified sediment do occur. This is generally marl-stone, limestone with reef debris or crinoid limestone. Deposition may have taken place at quieter places in between the reef builders. In a very few cases crinoid calyces were observed in marl pockets.

In part of the occurrences of stratified sediment, deformations can be observed, which are due to differential compression and to displacements that must have taken place within the reef-limestone bodies. In older exposures, such stratified portions are often partly or completely weathered away, leaving holes and niches of various sizes and forms. For more details about intercalations of stratified sediments in the reef limestone the reader is referred to the descriptions of the various localities in Chapter XI (e.g., pp. 288, 295, 296, 320, 324, 359, 362, 366, 369).

In addition to debris of the reef frame, animal remains found in the matrix include crinoids, brachiopods, molluscs, arthropods, sponge spicules and small, tintinnid-like bodies.

The crinoids probably lived more on the sides of the reefs than on the reef surface (see also the section on crinoid palaeoecology in Chapter XII, particularly pp. 448, 452; cf. also pp. 461, 464).

Frequently brachiopod shells are found in small pockets. The author is unable to answer the question whether these shells were generally transported there together or whether the animals had a certain preference for sheltered patches. In some of the pockets brachiopod shells occur together with remains of other reef dwellers, in other cases they are the only fossils in such pockets. The local brachiopod assemblages often show high size and shape variability. Several of the reef brachiopods were attached during their life to the reef by means of a muscular stalk or pedicle protruding from an opening under the beak of the pedicle valve.

Gastropods are regularly found in the reef limestone, with many species, but they are only occasionally abundant; most common among them is *Platyceras cornutum* Hisinger. Lamellibranch species are less common and certainly less varies than the gastropods, in contrast to modern reefs in the Great Barrier Reef province (Maxwell, 1968, p.168). Cephalopods are distinctly less common than the two previous molluscan groups.

In places the reef matrix is of the stromatolitic type described in Chapter V (p. 77).

In general, field investigations of the composition of the reefs are easier in the comparatively-younger exposures than in several of the older, inland exposures, where observation is hindered by the patina which covers the weathered rock and in part also by the brecciation which has taken place.

STRUCTURE OF THE REEF LIMESTONE

The general appearance of the reef limestone is to a large degree dependent on the shape of the reef builders and the total volume, composition and distribution of the reef matrix. In his well-known study of the Gotlandian reef limestones, Hadding (1941) distinguishes two types of reef-limestone structures: *massive* and *bed-like*. Massive are those reef limestones which are built up of elements with varying or indeterminate orientation. Two subtypes are distinguished: conglomeratic and brecciated. The bed-like structure is subdivided according to whether the structure is due to embedded loose material or to tabular reef builders.

The present author accepts the classification of Hadding, in general, but with an addition and a modification.

The addition is necessary because not all occurrences of massive reef limestone can be classified as either conglomeratic or brecciaceous. Also sections are found showing massive reef limestone with relatively smooth surfaces. Some parts with a stromatolitic matrix may show such exposed planes. Smooth surfaces may also be found in strongly-recrystallized reef limestone. Some raukar, which are the remains of the most-resistant parts of a reef, for instance, show such smooth surfaces, but also exposures along joint planes. Reef-limestone exposures of this type may be described as having a *solid structure* or being massive sensu stricto. A solid reef-limestone surface, however, is not always a stable situation. If it has been exposed to weathering for a long time, it may pass into a brecciaceous appearance (see below in this section).

The modification in Hadding's classification of reef-limestone structures is proposed, because the present author has objections to the term *bed-like structure*. A bed is generally defined as being a division of a stratified complex, marked by a more-or-less-distinct divisional plane from its neighbours above and below. Bed-like means: having the appearance of a bed. In fact, however, it is meant that in the reef limestone more-or-less-distinct beds can be distinguished, that is to say, that the rock shows a (vague) bedding or stratification. Therefore, the terms *stratified* and *vaguely stratified* are preferred to bed-like to describe this type of reef-limestone structure.

Massive structure

In reefs which show a brecciated structure, the matrix of the reef limestone has generally undergone stronger recrystallization or cementation and hardening than in those with a conglomeratic structure. The matrix may thus form an equally-resistant part of the reef as the reef builders. An example of reef limestone with a brecciated structure is that of the Bogeklint (Slite Beds, p. 320).

Between the reef builders and the hard matrix, marly films may be found. In other cases there is no sharp boundary between them. In such cases even colonies of reef builders can be found which through recrystallization gradually pass into the matrix limestone. Elsewhere only a few fossils are recognizable in the brecciaceous reef limestone. In general, more fossils are visible in brecciaceous reef limestone than in the strictly massive (solid) type, but less than in conglomeratic or vaguely stratified reef limestones.

The pieces into which the brecciaceous reef limestone breaks apart, generally measure a few centimetres in diameter, but may vary from less than 1 cm to large fragments of 20–100 cm in diameter.

The brecciated reef-limestone structure may, as Hadding (1941, p.51) suggested, have originated already in an early stage, because the rock could not carry its own weight but broke to pieces. Thereupon the pieces were dislocated in relation to each other. As a rule, movements were small and the cracks insignificant, but the rock nevertheless obtained a characteristic brecciated structure. The cracks can best be seen in thin sections. They are irregular, pass through reef builders and matrix alike and are partly filled with calcite. Broken shells commonly present evidence of small displacements.

As has been said, reef limestone with a solid structure may pass into brecciated limestone as a result of weathering. One of the reasons for this may be the presence of many intact or nearly intact stromatoporoid colonies, with a characteristic latilaminar structure. In weathering, the rock easily splits along these latilaminar planes. In the Hallshukklint, brecciation was also found to occur where branched coral colonies, in a little-more-marly matrix than elsewhere, are the main reef builders. In other cases, the explanation may be that parts of the apparently solid reef limestones are, in fact, brecciated in structure, but are stronger and rather-uniformly recrystallized, because their matrix was relatively pure. In the freshly exposed rock, the pieces may still adhere to each other, but they become detached when the rock has long been subjected to weathering. This latter explanation is supported by the observation that, in the reef limestone of Hoburgen type, parts with a rather-solid structure occur in several places enclosed in reef limestone with a brecciated structure, into which these parts also gradually pass.

In general, reef limestone which is predominantly built by corals was stronger of construction than rock very rich in stromatoporoids. This appears from the observation that massive coral limestone is generally not brecciated, but solid in structure. Brecciation was neither observed in a reef part which consists of an intimate interoccurrence of colonies of corals and stromatoporoids and was found in the south of the Hjännklint (Hall Parish, Högklint Beds). Apparently, coral colonies have greater carrying power than stromatoporoids. Neither do sections through large coral colonies generally brecciate through weathering.

Also in places where the reef matrix was relatively pure over larger parts of a reef, a solid reef-limestone structure may have formed which does not brecciate through weathering. Presumably the total reef mass has been solid there since the time of deposition of the reef matrix, through authigenic sedimentary formation of calcite (see also below in this section) or a stromatolitic type of matrix. Only slight dislocations took place within it afterwards. These may have occurred along erratic lines which can be seen in several of the larger exposures of solid reef limestone. They generally divide an exposed wall into spool-shaped parts of usually about 1 m length. Their origin may be connected with zones of weakness in the reef, e.g., surfaces containing the upper sides of several reef builders. Many of the lines are weathered out as thin grooves. Some are filled with calcite, some contain marl. In the south of the Hjännklint, one groove contains a great many coral branches which lie horizontally and are orientated about northwest – southeast (Fig. 35). These branches up- and downwards pass into structureless calcite. Along part of the lines solution also seems to have taken place and some took the character of stylolites.

If in weathering, the matrix of the reef seems to be less resistant to erosion than the embedded reef builders, and still more, if in addition the latter are more or less bun-shaped, the rock presents a conglomeratic appearance. If we only look at the exposed section, the term nodulose (=knotty or having nodes; cf. Rice, 1954, p.273) can also be used to describe the general appearance of the rock. Examples of reef limestone with a conglomeratic structure are found, among many others, in the Galgberg (p. 291), near Herrvik (p. 363), in Torsburgen (p. 363) and in the Lindeklint (p. 371).

Stratified structure

In several reefs a number of horizons are found which are richer in matrix than the remainder of the reef. This may cause a more or less vague stratification. Generally these horizons do not occur through the reef as a whole but only in limited parts of it, more often at the periphery and in the lowermost or uppermost part of a reef than in the centre.

Also the reef builders themselves may aid in giving the reef limestone a stratified character, by occurring dominantly in flat-lenticular or tabular colonies.

These two causes of a more or less vague stratification in the reef limestone are generally not independent. Increased sedimentation, which in itself already produces stratification, will in its turn also have influenced the shape of the reef builders.

As has been discussed in the previous chapter, stratification is found in quite a number of the Upper Visby reefs. In the basal Högklint Beds stratified parts within the reef limestone are also common. Higher in the reef-carrying Högklint, too, reef parts with some stratification are regularly encountered. Examples are drawn in Fig.36 and Fig.153. The raukar at Lickershamn, too, show several intercalations of a stratified character. For the Upper Visby and Högklint Beds together a vertical succession of stratified – conglomeratic – brecciated, solid and conglomeratic – conglomeratic with some stratified, is a perceptible rule, though one with several exceptions.

As was said four paragraphs earlier, in several reefs a tendency to a stratified structure can be observed in the top parts. Often this is caused by comparatively more matrix and/or debris. It suggests that at least in many cases reef growth ended not abruptly but was preceeded by a period of decline. Examples are found in the Högklint Beds in Högklint (p. 285), in the Slite Beds in the Bogeklint (p. 320), in the Hemse Beds in the Lindeklint (p. 371) and in the Hamra-Sundre Beds in Hoburgen (p. 415). Vaguely stratified reef limestone is also common in the lowermost parts of some reefs, for instance in the Klinteklint (Hemse Beds, p. 366).

In the light of the above observations it is easy to understand that in small reefs (e.g., those of the Klinteberg or Eke Beds) a (vaguely) stratified structure is comparatively more common than in large reefs.

Not included in this classification are reef-limestone parts which generally consist of fossil fragments, small fossils and marl or marly limestone, often in random intermingling. These parts occur in a number of reefs as intercalations in the reef limestone. They will be discussed in the section dealing with debris-filled depressions.

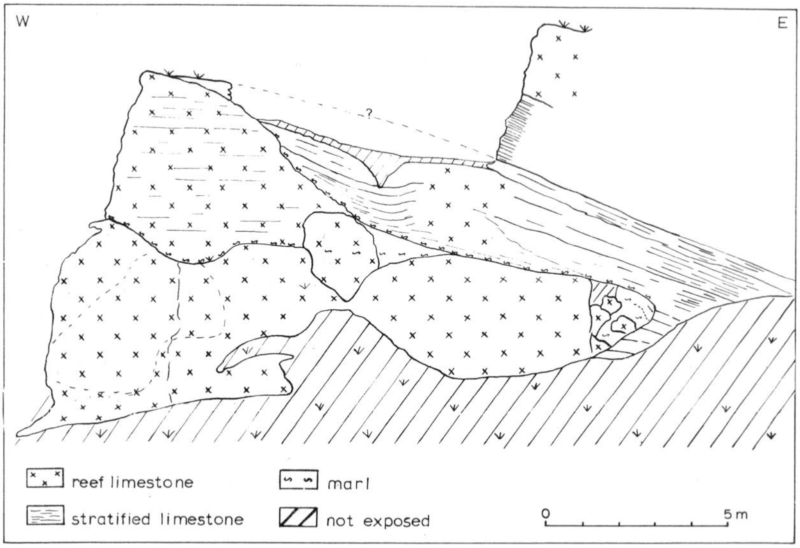

Fig.36. Reef and stratified limestones, belonging to the Högklint Beds, as found approx. 0.9 km south of Sigsarvebodar. The lower part of the reef limestone is massive. Within the dashed line at the left, it is of solid structure and light grey in colour, fossils are macroscopically hardly recognizable, except in marl pockets; there are several nests of calcite crystals. Elsewhere the lower part of the reef limestone is of brecciaceous appearance and grey in colour; the wall there is less steep; some stromatoporoids can be distinguished, of most other fossils at the best "ghosts" are present (spots of slightly differing colour). The higher part of the reef limestone at the left is vaguely stratified. It brecciates through weathering. The weathered surfaces show several fossils. The colour of the weathered rock is yellowish grey.

The reef is traversed with a few cracks. Of these the one at the boundary between massive and vaguely-stratified rock presumably follows a plane where reef growth has been interrupted; it is covered by a thin layer of marl, which continues eastward even after the crack has ended in another, roughly circular crack. At the upper boundary of the reef limestone, there is also such a thin marl layer. In the east, the reef limestone is somewhat crushed in appearance. The overlying stratified limestone is coarsely crystalline and fossiliferous. Intercalated in it is a small occurrence of stromatoporoid reef limestone. The younger stratified limestone has thinner layers, is finely crystalline and less fossiliferous. It is overlaid again by reef limestone.

Recognizability of fossils

The Hoburgen-type reef limestones can also be classified in another way, viz. as reef limestones which macroscopically still show the fossils which contributed to their formation and reef limestones in which

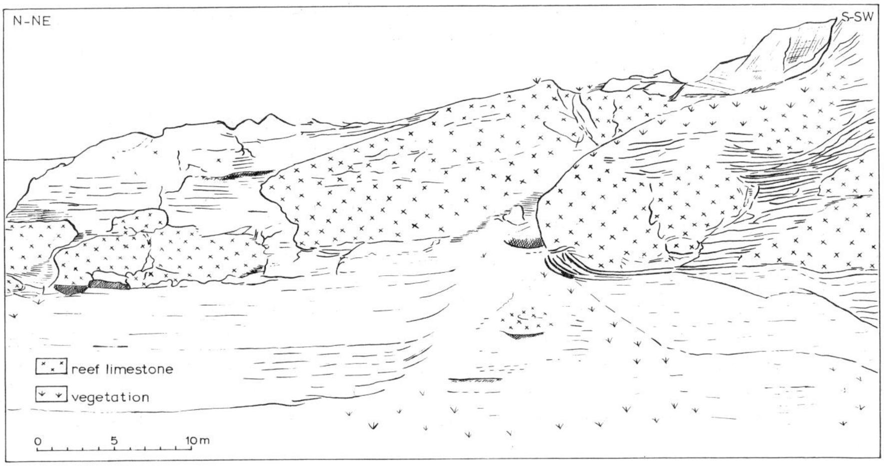

Fig.37. Detail of the northern part of the reef limestone complex about 0.7—0.9 km north of Lundsklint. At the right, the characteristic subdivision into smaller reef-limestone bodies, as found also in a number of other reef-limestone massifs in the Högklint Beds. At the left, so much stratified limestone is found between the reef-limestone bodies that they appear as separate reefs. At the base, a badly exposed small reef, belonging to the Upper Visby Beds. On top of this reef, the Upper Visby succession contains more limestone than north and south of it.

---

diagenetic recrystallization of the total mass has gone so far that fossils are only occasionally recognizable in the field. The variation which the Hoburgen-type reefs show in this respect is greater than with that of the Upper Visby and Holmhällar-type reefs.

In general reef limestones of the first group show a conglomeratic structure, or are more or less vaguely stratified, but some have a solid or brecciated structure. Reef limestones in which fossils are hardly recognizable generally show a solid or brecciated structure.

In all cases in which the fossils are still readily recognizable, they are found to have nevertheless been recrystallized. There are, however, differences in the degree of recrystallization. If the matrix is strongly marly the fossils are less changed. Apparently the clay content of the sediment reduced the permeability to water and thus hindered the recrystallization process. This is further indicated by the fact that the fossils found in the stratified marl deposits in Gotland are least recrystallized. On the other hand, there is also a general rule that the purer the limestone of the matrix, the stronger the recrystallization of the reef builders and also the harder the matrix itself. In such cases the external shape of the fossils may still be distinct but the finer internal structures are often lost. Hadding (1941, p.46) rightly warned that as a result of this, tabular corals may be mistaken for stromatoporoids or vice versa and thus care is required in determining the fossil content of a reef. In the Högklint Beds, but also elsewhere, examples have been seen where, in addition, the external shape has partly disappeared. The one part of a large fossil, especially a coral colony, may then still show its outline, whereas in the other part even this has been lost and the fossil has passed into solid reef limestone.

It is not impossible that hardening of the calcareous mud which constitutes the matrix of the reef limestones in which fossils are only vaguely distinguishable ("ghosts"), took place contemporaneously with deposition, assuming that calcite also crystallized simultaneously from the no-doubt-saturated solution in which the mud was deposited (cf. Hadding, 1941, p.44). Together, thereafter, they underwent secondary, diagenetic formation of calcite. If the matrix was almost pure, it may even have become the most-resistant part of the reef limestone, which then often shows a solid structure.

Nowhere in the reef limestones of Gotland does dolomitization seem to have taken place. At any rate, there is no large-scale dolomitization, such as has affected the Silurian reefs in, e.g., the Great Lakes area of North America.

No distinct reef frame preserved

Contemporaneously with reef growth, the reef was attacked by destructive forces, which detached reef builders and caused their falling over and rolling or fragmentation. Furthermore, in most cases the reef limestone has to a greater or lesser degree been compressed by the weight of the higher reef parts and the overlying rocks, which has led to dislocations within many reef-limestone occurrences. During the Glacial Period the sediments were, moreover, subjected to the pressure of the inland ice. Where the rock in Recent times has been cut by cliffs, sliding may also have taken place (cf. Fig.139; another good example can be observed west of Stuguklint, see p. 296).As a result, even in those exposures where fossils are still generally recognizable, parts are found only occasionally which still seem to have something left of the original calcareous skeleton of the reefs. In the Högklint Beds such restricted parts are found in the Kinnertorpklint, in the south cliff at Brissund, and in the Hjännklint. Also a few parts with large and dome-shaped stromatoporoids in the Högklint and Galgberg may be fairly intact. At Hoburgen (Hamra-Sundre Beds), a relatively-large part of apparently-original character is found in the south of the west cliff of the second hillock. It consists almost entirely of stromatoporoids.

SHAPE AND DIMENSIONS OF THE REEFS

The shape of the reefs of Hoburgen type varies rather, from a single spheroidal stromatoporoid colony (Fig.13) to an inverted right-elliptical cone of up to about 2 km in its longest horizontal direction and a thickness of up to 30m, or to generally thin but extensive stretches of patch-reef nature. In defining the form of the reef in more detail one meets with more difficulties, because there seems to have been quite a lot of variation and only two-dimensional exposures are available of many of the reefs, whose sections, moreover, do not generally go exactly through the centre of the reef.

Inverted-cone-shaped reefs

Although the point of origin of the reefs is generally not satisfactorily exposed, one may assume that the majority of them began growth at a more-or-less-central point and from there in the course of time spread out. In the case of the inverted-cone-shaped reefs, such as the one shown in Fig.38, an inverted conical base developed. The plan form, however, is not generally roughly circular, as an early description by Hadding (1941, p.13–14) suggests, but more elliptical (cf. Hadding, 1956, fig.1, 4; this book Fig.75, 74). Consequently it is more accurate to describe the generalized form of these reefs as inverted elliptical cones. The reef surface usually slopes in a roughly-radial way outwards from the centre. This means that the central axis of these conical reefs is generally approximately vertical; hence this reef shape may be described as an inverted *right*-elliptical cone. Stratified sediments often arch over the reef; the boundary is generally more-or-less sharp; the covering sediments may be compacted into irregularities in the reef surface.

The ratio between the thickness of a reef and its horizontal dimensions may have been mainly determined by the depth of the water and alterations therein during the time of formation of the reef. In the Högklint Beds, which developed in water that gradually became shallower, the base of conical reefs is usually shallow or very shallow; for the Hoburgen reefs (Hamra-Sundre Beds), development of the reefs in deepening water is assumed and there, with roughly the same thickness of the reefs, the horizontal extension is much smaller.

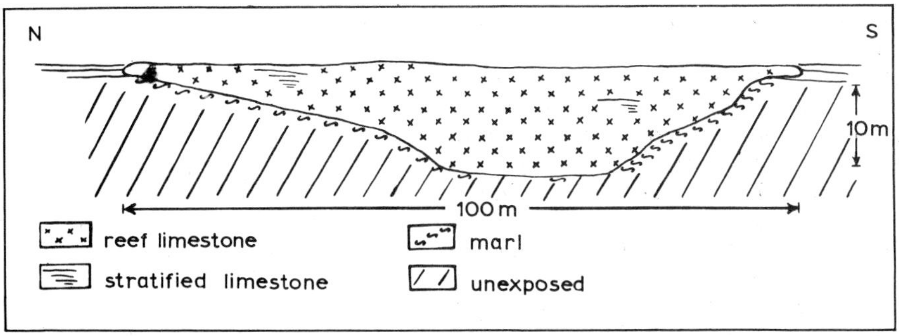

Fig.38. Schematic drawing of the Häftingsklint, approx. 5.5 km north-northeast of Ihrevik Fiskläge. The lower part of the reef is enclosed in marlstone with interstratified layers of marly limestone; the top is surrounded by stratified limestone. Within the reef, two large parts, consisting of stratified limestone. Högklint Beds.

Patch reefs

Water depth was probably also a determining factor leading to the formation of patch reefs. Reef development also in these cases may have started at a certain point, but from the very beginning the reef expanded more laterally than in an upward direction, causing the formation of a reef basin with a low angle of slope (cf. Fig.66). Such reefs frequently show a length of up to several hundred metres. Other exposures of this length or even more, present evidence that they formed through lateral fusion of two or more such patch reefs (see further on in this chapter). In the Högklint Beds, the patch reefs are generally thinner than the inverted-cone reefs. Hadding (1941, p.14) was of the opinion, that these are not patch reefs, but peripheral sections through inverted-cone reefs. However, he produced no evidence to prove this. It would be illogical, however, to assume that almost always the central part of such reefs has been removed by erosion, and also the fact that some of the large reef-limestone exposures show laterally fused reefs of this elongated type seems to support their classification as patch reefs. Several more patch reefs are found in the Slite and Hemse Beds.

Lenticular and irregular reefs

Where lateral expansion was less strong than in the patch reefs, a more lens-shaped (or lenticular) or an irregular reef form developed. The thickness is generally about 5–10 m; some are thicker, but never more than 25 m. In some of these, the horizontal extension is hardly more than the reef thickness, in others the length measures a few times that of the thickness. Generally, the base is less flat than that of the patch reefs and the same, to a lesser extent, may apply to the surface. Whereas most of the other reefs generally may have risen at low angles from the surrounding sea floor, some of the reefs in this group present evidence that they have risen at quite steep angles. This is indicated particularly be laminar reef builders, usually the stromatoporoid *Labechia*, whose latilaminae at the reef margin may be inclined at high angles, flattening off higher upwards, thus suggesting a reef margin which was convex outwards. As the adjacent stratified sediment is unaffected, there is no reason to assume slumping, at least not after their deposition took place. Should slumping have occurred before their sedimentation, a steep reef edge still has to be assumed. The lateral extent of the successive stromatoporoid covers, however, rather points towards successive growth on a strongly-inclined surface. Displaced material, on the other hand, are, undoubtedly, blocks of reef limestone with steeply-inclined reef builders which are included in reef talus in some localities, and which also indicate that the parent reefs must have risen well over their direct vicinity. These blocks will be further discussed later in this chapter, when dealing with the reef talus.

The Klinteberg reefs

The reefs found exposed in the Klinteberg, and also elsewhere in the Klinteberg Beds, are on the average much smaller than those in the

Fig.39. Photograph taken at the north side of the Klinteberg, northwestern part. At the base, crinoid limestone; higher in the wall, reef limestone. The latter deposit contains, as main reef builders, colonies of stromatoporoids and corals (*Halysites, Favosites, Acervularia*). The stromatoporoids are almost all flat lenticular in shape. The matrix is strongly marly and there are many marl pockets with crinoid fragments. The reef limestone is unorganized and vaguely and irregularly stratified. The boundary between crinoid limestone and reef rises from left to right in the photograph, passing along the top of the niche, and is not very distinct.

Högklint, Slite or Hemse Beds. Some representative examples are given in Fig.39–43. The largest reef observed in the Klinteberg reached a thickness of about 9 m, but generally the reefs in the Klinteberg Beds are more-or-less lenticular in general outline, with a thickness of not more than a few metres. The matrix of the reefs is generally strongly marly. Stromatoporoids are the main reef builders, but compound corals are also well represented. Most of the stromatoporoids in the Klinteberg reefs are lenticular.

The distinction between a reef and the surrounding stratified limestones is in several instances rather blurred. The surrounding rocks may also be very fossiliferous, the reef limestone may be vaguely stratified and contains marl pockets and intercalations of stratified limestone. Both the reef and stratified limestones may brecciate through weathering.

The crinoid limestones around these reefs are often cross-bedded. The bedding planes are often rugged, which may be due to wave action. Algae are commonly found in certain areas between the reefs, elsewhere large coral colonies are common. These are some of the indications that the reefs in the Klinteberg Beds probably developed in shallower water than the great majority of Hoburgen-type reefs in the other stratigraphical units of Gotland.

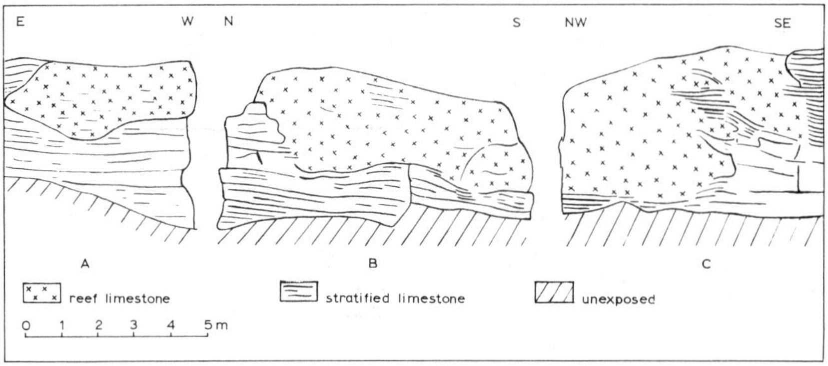

Fig.40. Three sections through one reef in the Klinteberg Beds. The reef is exposed in the south of the Klinteberg, about 325 m south-southwest of the three-fork of roads close to Klinte Church. Section C forms one of the walls of an old quarry. This quarry is mainly excavated in stratified marly limestone of brownish grey colour. In its lower exposed parts, this limestone is thickly bedded, with layers up to 1 m thick, which through weathering may appear to be composed of a number of thinner layers, occasionally cross-bedded. Higher up the layers of the limestone are on the average 3–15 cm thick. The sediment is very fossiliferous, crinoid fragments being especially abundant and there are several coral colonies, generally in their orientations of growth. Further, there are solitary corals, stromatoporoids and many brachiopods.

In the right of section C it can be seen how this stratified limestone interfingers with marly reef limestone. Most reef-building colonies are flat lenticular in this reef, with an average length of 20–30 cm. The origin of reef growth was probably close to the corner between sections C and B. In the lower and upper parts of the reef, the reef builders are flattest but in the middle rounder forms also occur. Within the reef limestone north of this reef centre, local intercalations of stratified limestone occur. Northwestwards, the reef thins out again between the stratified sediments, as can also be seen in section A. The reef limestone enclosed between the three sections probably represents about one quarter of the original reef, which then may have had a diameter in the order of 20 m and a thickness of not much more than 5 m.

INTERRUPTIONS AND FLUCTUATIONS IN REEF GROWTH

No doubt a great variety of factors, generally very complexly interconnected, has operated in the moulding of reef topography. It is most difficult and inappropriate here to analyse each of these factors; an attempt to do so will be made in Chapter XIV. What has to be done first is to describe all the effects they brought about in the reefs and to attempt to understand their origin. This will be done in this and the next two sections of this chapter, and has in fact to some lesser extent also come up already in the previous two sections.

Fig.41. Photograph, showing the southern 4 m of section B in Fig.40. Klinteberg.

Interruptions in reef growth

Although in several places in Hoburgen, the exposed reef limestone is traversed by horizontal or subhorizontal lines or cracks, for almost all of these a recent origin has to be assumed, as a result of weathering. There is, in fact, only one case where a line might follow a primary structure of the reef, probably a plane of interrupted reef growth. This line is exposed over a rather-long distance in the fourth hillock (Fig.217). Unfortunately it occurs at such a height in the steep wall that over most of its extension it cannot be directly studied if one has no special appliances available. There is no difference in fossil content above and beneath the line, but the volume of marly matrix is higher at this level than anywhere else in the reef, whereas also a few lenses of marl, with an average length of about 25 cm, were found intercalated in the reef limestone at this level. In these marl lenses, a few crinoid fragments were found but no large fossils such as stromatoporoids or coral colonies or even fragments of these. This supports the theory of a temporarily-quieter environment which led to increased deposition of terrigenous debris.

Whereas interruptions in reef growth are rare in Hoburgen, several can be found in reefs of Hoburgen type in the Högklint, Slite and Hemse Beds. An example where a thin layer of marlstone traverses the entire section through the reef limestone is shown in Fig.36. Generally, however, an interruption in reef growth affected only part of the reef (see the left

Fig.42. Photograph of the most northern part of section B in Fig.40. Klinteberg.

part of the reef drawn in Fig.44. They may be revealed as intercalations of some extent and generally-restricted thickness, which consist of stratified sediment, as discussed in the section dealing with the reef organisms and matrix. Other local interruptions in growth have led to the formation of depressions or pools in the reef surface, which will be discussed in a later section in this chapter.

Erosion of the reef surface

In the younger parts of the Högklint Beds, some reefs are found which apparently suffered from marine erosion at the time of their formation. In the environs of Gustavsvik, for example, reefs are found which are rather-flatly sheared off at their top and are overlaid by horizontal layers of limestone rich in Algae and crinoids (Hede, 1940, p.26, fig.8).

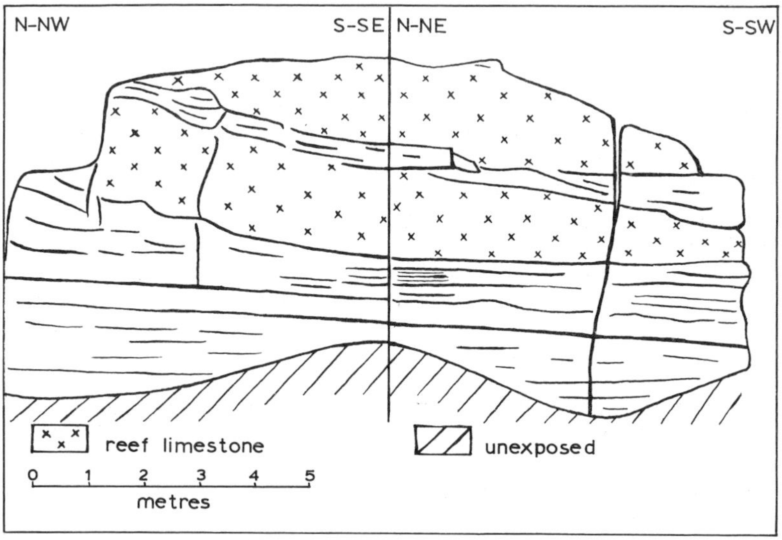

Fig.43. Section through the west-northwestern part of a reef in the Klinteberg, about 20 m north of the reef shown in Fig.40. At the base, crinoid limestone, with upwards an increasing amount of crinoid fragments and also reef debris. This deposit only faintly sags under the overlying marly reef limestone. The latter is intersected by a zone of stratified crinoid limestone.

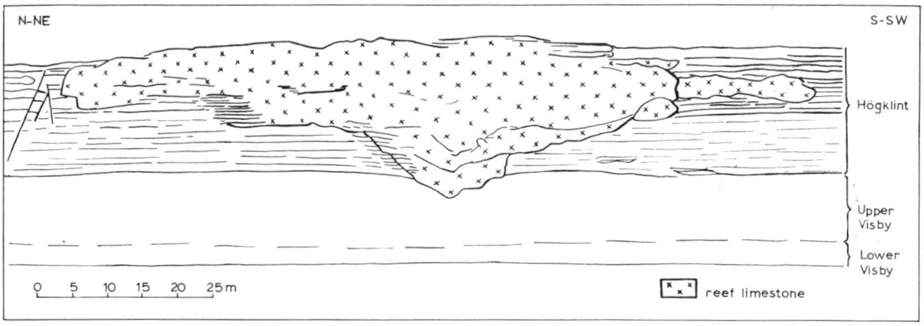

Fig.44. Högklint reef, 0.28 km north-northeast of Lundsklint. The reef began its development already during Upper Visby time. From its root, the lower surface of the reef rose gradually and probably in all directions. In this way, a reef-limestone body was formed with roughly the shape of an inverted elliptical cone with a widely extended base, of which the longest axis was probably more or less parallel to the direction of the coast line during the time of reef development. Reefs of this form are also found in several other places in the Högklint Beds and in other stratigraphical units containing reefs of the Hoburgen type. North of the above-figured reef, the stratified limestone shows some local faults along which small displacements have taken place. Local occurrences of stratified limestone within the left-hand half of the reef indicate that in places of the reef its growth has been temporarily interrupted. There is, however, no distinct subdivision of the reef into separate smaller bodies, as can be seen in several reefs in the Högklint Beds.

This phenomenon can be observed in a much-more-striking way in some exposures in the Hemse Beds, particularly in the north of the Gannberg (Fig.177), where some very elongated, but thin patch reefs may follow over each other, separated by thin zones with stratified sediment rich in reef debris. Since the surfaces of the reef-limestone bodies are notably horizontal over long distances, it is not likely that variations in the extension of one and the same reef, as will be discussed next, have caused the exposed pattern. More likely is reef development in very shallow water, with presumably some variations in depth, in which recolonization of the sea bottom by reef builders took place very quickly.

Fluctuations in the extension of the reefs

In several exposures, especially in the Hemse Beds, there are alternations of reef limestone, reef debris, crinoid limestone with reef debris, and other related deposits. These suggest that at least part of the reefs have had periods of lateral expansion which alternated with periods in which the reef retreated again. The deposits surrounding the reefs moved backwards and forwards with the reef. Such fluctuations are relatively most common at the landward side of the reefs.

An example where this phenomenon is exposed in the Hemse Beds is the Kaupungsklint, in which close to Ardre Ödekyrka, the following section is found:

0.54 m +	Greyish white to light brownish-grey crinoid limestone, irregularly bedded, marly, strongly recrystallized, contains several small nests of calcite; carries "*Megalomus*" *gotlandicus*.
1.00–1.30 m	Stromatoporoid limestone, grey, greyish-white or light brownish-grey, finely crystalline, strongly recrystallized, faintly marly, rather structureless, conglomeratic in appearance.
0.65 m	Light grey to light brownish-grey crinoid limestone, more-or-less-distinct and somewhat-irregularly bedded, faintly marly, finely crystalline, contains markedly less stromatoporoids than the rock underlying and overlying it; "*Megalomus*" *gotlandicus* is present.
0.47 m +	Stromatoporoid limestone of the character as above.

When going north-northeastwards it appears that the above sediments can replace one another also in a horizontal direction. Behind the farm, about 0.2 km north of Ardre Ödekyrka crinoid limestone is exposed to a thickness of about 2 m, partly rather rich in reef debris. Reef limestone is found there over a length of about 10 m and a height of up to 1.4 m.

About 0.1 km north of the farm, the wall, which is about 2 m high there, consists of stratified marly limestone with beds 0.2–6 cm, bedding planes are irregular. At the base of the wall the fossil content is high, especially in stromatoporoids; upwards within a short distance, but nevertheless without a distinct boundary the number of stromatoporoids decreases very strongly. Some tens of metres further north-northeast, this stratified sediment ends over most of its height against reef limestone, of very disorderly appearance, with mainly small reef builders, of which several are not in their orientation of growth, and a marly matrix.

In the Visneklint, also belonging to the Hemse Beds, the following section has been observed:

210 cm +	Grey stromatoporoid reef limestone of Hoburgen type, with also corals, bryozoans and crinoids; inserted some nests of stratified limestone, generally less than a few square decimetres.
18 cm	Reef talus
17 cm	Vaguely stratified limestone
36 cm	Reef talus

30 cm — Vaguely stratified limestone, brownish light grey to grey and finely crystalline

80 cm + — Talus-like deposit consisting of a matrix of marly limestone with a great number of fossils and fossil fragments, none of which is in a position of growth; upwards many intercalations of stratified limestone occur, the number of reef builders decreases strongly and within a short distance there is, at the top, a gradual transition into more-general limestone, not particularly fossiliferous, but with some stromatoporoids and crinoids.

These rock types also occur elsewhere in this wall, but in varying succession and thickness. Thus, reef talus can also be found with reef limestone both overlying and underlying it. The alternation of rocks gives the impression of a stratified deposit and this may have caused Hede (1929, p.47) to speak about stratified stromatoporoid limestone. Doubtless, however, the rocks originated at the periphery of a reef with slight fluctuations in its extension.

INTERRELATIONS BETWEEN REEFS

Where the sea floor and water conditions were particularly favourable for reef growth, it is not uncommon to find that several reefs began development close to each other and almost synchronously. With continued growth of these reefs, they became ever-more serious competitors. The result would be either fusion of adjacent reefs or the end of one or more of the competing reefs.

Natural selection among reefs

Particularly in Hoburgen, it can be seen how in many cases reef-limestone masses have been formed by a number of reefs, with sometimes different growth rates, which grew close to, against and over each other. A good example is provided by the seaside wall of the third burg (Fig.45).

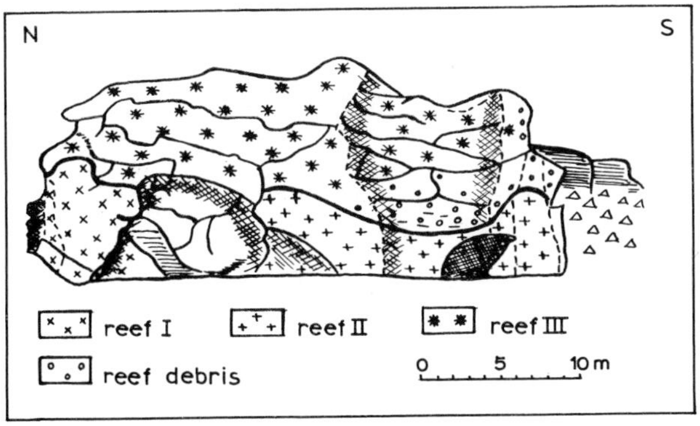

Fig.45. Section shown in the west side of the third hillock, Hoburgen. Hamra-Sundre Beds. Reefs I and II are separated by stratified limestone, in which a cave has been formed by the *Littorina* sea. The cave is known as "Hoburgsgubbens Matsal" (the dining room of grandfather Hoburgen). A smaller cave is found in the southwest. This cave has two openings, of which the western one is drawn at the right; it seems to have been excavated by erosion of a softer part of reef II. The upper part of the section is taken up by the third and largest reef.

This wall contains a large cave, known as "Hoburgsgubbens Matsal" ("the dining room of grandfather Hoburgen"). On the left as well as on the right side of this cave, a reef is exposed with stratified limestone in between. The presence of the latter perhaps explains why a cave could be excavated at that place by the waves of the *Littorina* Sea. Both reefs have expanded over the stratified limestone. On the left end in this wall in some places, traces of stratified limestone are still preserved, which contain a great abundance of crinoid fragments, several brachiopods, but only rather few reef builders. The difference between the unstratified and stratified limestone is quite distinct.

In the middle of the back wall of the cave, the stratification in the bedded limestone shows a kind of basin-like depression; on the left side this interfingers with the reef limestone, and on the right side the stratified sediment thins out against the reef limestone within a short distance. It seems that south-southeast of the cave, a third reef was present, which grew rather quickly and expanded also in a north-northwestward direction, into the passage between the two reefs mentioned earlier. The third reef soon collided with the reef on the right side. In the deeper part of the cave, it can be seen how this led to a further narrowing of the passage between the reefs, which now became confined by the left reef and the protruding third reef. Continued obtrusion of the latter reef northwards increasingly narrowed the zone in which stratified limestone could be deposited, until that reef also collided with the left reef and expanded over it. It is the third reef which forms the roof of the cave.

It is thus likely that the two small reefs died as a result of being buried with the reef outwash from the adjacent third reef. It is important to note that this latter reef developed at the east, or seaward, side of the other two. Apparently it was the most vigorous and vital of the three.

Lowenstam (1950, p.474), writing about the reefs in the Great Lakes area of North America, reported a similar phenomenon, but one where the conquering reef earlier had also led to the genesis of the smaller reefs, because the zone of reef-induced turbulence in the environment of a greater reef prevented settling of fine terrigenous clastics and favoured the development of secondary reef centres. The greater the distance from the primary centre, the greater the probability that the secondary reefs would attain wave resistance, whereas more-closely-spaced building sites had less chance of survival because of encroachment of excessive reef debris.

It is doubtful whether Lowenstam's description is also applicable to the situation found in the third hillock of Hoburgen. The stratigraphical succession at that locality makes it most unlikely that the third reef could have been significantly older than the two which it covered. It is more likely that its mere position at the seaward side of the other two caused its more-favourable development; otherwise the three may have started growth under more-or-less-equal conditions.

Completing the description of the exposures presented by Hoburgen's third hillock, one finds on both the left and right side of a second cave, with two openings, southwest of Hoburgsgubbens Matsal, reef limestone of the right-hand reef, with only a slight thickness, caused by the early death of that reef. The reef limestone is thinnest in the south-southeast, being somewhat thicker right and left of the cave, thus forming a saucer-like depression in the reef surface. In this depression, initially material midway between stratified and reef limestone was deposited. The protruding third reef provided an increasing amount of debris to fill the depression;

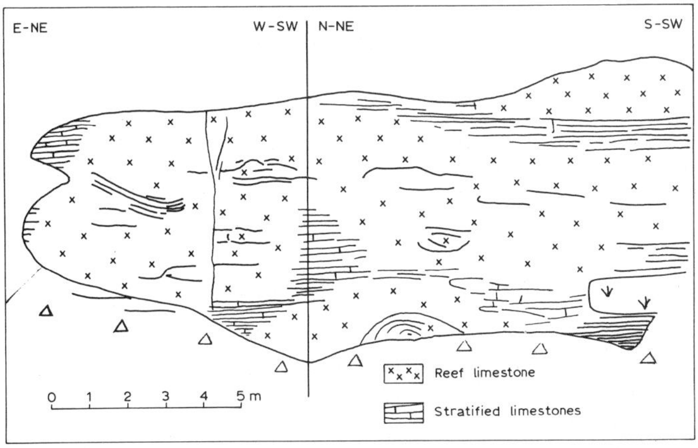

Fig.46. Section about 2 km east-northeast of Jungfrun. Högklint Beds. The cliff, which further northeastwards lies parallel to the present coast line, has bent inland here. The section shown is that of the second exposure, going inland. Three reef-limestone parts are sectioned, overlying each other. The reef limestone exposed at the base is strongly marly. A cave about 2.5 m long and 0.7 m high has been formed in it. Reef-building colonies are relatively small. The matrix is rich in reef debris and solitary corals. This reef limestone is overlaid by rather coarse crinoid limestone with reef debris. This limestone is regularly stratified. The middle reef limestone presumably covered the lower one from the south-east, but never extended much further northwest than the present exposure. This can be deduced from the presence of stratified limestone in the northwest corner (centre of the drawing). It could not be established whether this middle reef limestone belongs to the same reef as the reef limestone at the base, or whether it represents a second reef which developed a little to the seaward of the first, and therefore under more favourable conditions, causing the death of its landward neighbour, and finally overgrowing it. A second complex of stratified limestone and a third reef limestone occur on top, to which the same reasoning is applicable. (After Manten, 1962,fig.20.)

---

the dips in the deposit gradually increased, as did the coarseness of the deposit. Finally the reef extended over its debris, to build the upper part of the cliff, at the same time reaching there its maximum extension in this direction, as is shown by the preserved remains of a talus zone, which will be discussed later in this chapter.

The second hillock of Hoburgen presents a few further examples where the reef at the seaward side conquered in the competition with other reefs. These will be described when dealing further with the exposures in Hoburgen in Chapter **XI** (pp. 415, 417, Fig.215).

Some similar situations are also found in the Högklint Beds (see Chapter **XI**, pp. 296, 299, 301 and Fig.46).

The reef of which the root is found exposed in the Lithberg Grotta (Storburg, Hoburgen) shows that in the later stages of its development, conditions for its growth turned out to be less favourable. Whereas initially the reef expanded at all sides, the exposed wall makes it clear that at a certain level, the growing reef surface was forced to withdraw to its central part (Fig.59). Close to a small cave, above the Lithberg Grotta, a vaguely-stratified and very marly intercalation is present in the reef limestone; at about the same level some comparable indications of decreased reef growth are found in the reef south of it. In the centre of the surface of the Lithberg reef and its direct southern neighbour growth still continued for some time, but not as long as the growth of some other neighbouring reefs, such as the one northeast of it (partly exposed left of the Lithberg reef) and the reef some 10 m south of the two smaller reefs. On the left above the Lithberg Grotta an isolated part of the reef talus is preserved, in which nearly all fossils occur in an almost vertical position, suggesting that the reef has probably not extended much further northwest than the place of the present cliff wall, which thus presents a peripheral cross-section, but also that it stood well out over the surrounding sea floor. The exposed reef limestone of both central reefs does not reach the top of the wall; the reef limestone of the reefs northeast and south of the two does. If also the amount of talus material around the two central reefs is noted, the impression is obtained that they developed slightly more coastwards than the passage between the two other reefs; this passage was gradually narrowed because of expansion of the northeast reef, which in a later stage of its development even managed to override the two.

The examples given suggest that with several young reefs developing closely together, the reefs at the coastward side had the least chance to reach large size. If they were located behind other reefs, the expansion of these decreased water turbulence around the more coastward reefs; if the position was slightly behind a passage, water turbulence and erosion strongly increased; both could lead to the death of the affected reefs. Competition did not necessarily conclude with the surviving reef over-riding one or more others. Especially in the first of the two possibilities outlined above, it may well have been enough that the more-seaward of the competing reefs grew faster upwards and sideways, whereby the created lee-side environment with decreased water movement and supply of oxygen and food, and increased sedimentation was already in itself sufficient to retard and finally end growth of other reefs present at that lee side.

Fusion of reefs

Rapid colonization of a part of the sea floor by reef builders is revealed by a reef base which is close to the horizontal. Such rapid lateral expansion may have taken place immediately after the beginning of reef growth, but also at some later stage. When it happened later, the local root of reef formation can be seen to be surrounded by sediment layers older than the ones which directly underlie the reef base. It now happens that some of the patch reefs in Gotland of which the base is exposed for a length of some importance, do not show only one, but two or more such local wide-angle deepenings of reef limestone into the underlying stratified sediments; these deepenings may each represent a section through roots of reef formation. This suggests that long patch reefs may have started growth on more than one place and as a result of fusion attained the length which they show. It may be assumed that with the rapid and flat coloniza-tion that took place, little reef debris was produced and consequently the ends could join without a debris deposit in between, so that it is hard to discover where the actual fusion took place, except that slight variations in the slope of the reef base might give some rough idea as to approximately where two reefs fused. Such slight variations in the course of the reef base may, as they are found now, however, also have had other causes and in themselves, they cannot be taken as proof of the lateral fusion of reefs. An example of a patch reef which may have originated through fusion of some earlier smaller patch reefs, is shown in Fig.154.

It is not impossible in some cases that smaller reefs of a more-lenticular or irregular shape, which started growth closely together, may have also fused. This is suggested by exposures such as the one drawn in Fig.70.

Compound reefs

In the Högklint Beds it can often be seen that several more-or-less-lens-shaped reef-limestone bodies occur very closely together, to form a large reef-limestone massif. The size of the individual bodies varies largely, between only one metre or so to several tens of metres in horizontal extension and several metres in thickness. The combined massif may be up to 120 m long and about 20 m thick. The problem in such situations is whether the individual bodies are to be regarded as separate reefs, which by their close proximity and growth have joined or fused, or whether the larger massif should be regarded as one reef. Between the reef-limestone bodies, mantles of stratified limestone are found. The transitions from obvious reef limestone to apparently stratified crinoid limestone take place with bewildering rapidity, both laterally and vertically. In some cases, these separating mantles of stratified sediment are only thin and/or they may after some distance even wedge out, making the boundary between adjacent reef-limestone bodies indistinct (Fig.37, 48, 147). In other cases the mantles are thicker (Fig.46, 152), but in a horizontal direction, they may after a short distance abut abruptly against another reef-limestone body

(Text continues on p. 142)

Fig.47. Example of a raised cliff wall, predominantly consisting of reef limestone in lengthy patch-reef development. North of Brissund. Högklint Beds.

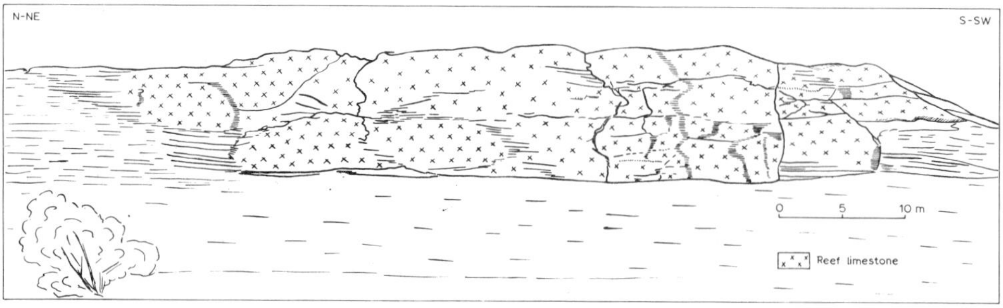

Fig.48. Lundsklint, about 17 km north-northeast of Visby. At the base, interstratified marly limestone and marlstone of the Upper Visby Beds is exposed. The upper half of the section shows reef limestone and stratified limestone, belonging to the Högklint Beds. The left part of the reef is subdivided into some smaller units by means of narrow mantles of stratified limestone. In the right part, local and more or less horizontal planes intersect the reef limestone. These planes appear in the section as weathered-out grooves. Also narrow, subvertical occurrences of stratified sediment are present. The stratified limestone north of the reef limestone is a coarse-spathic crinoid limestone, also containing solitary corals and washed-in stromatoporoids and halysites. (After Manten, 1962, fig.12.)

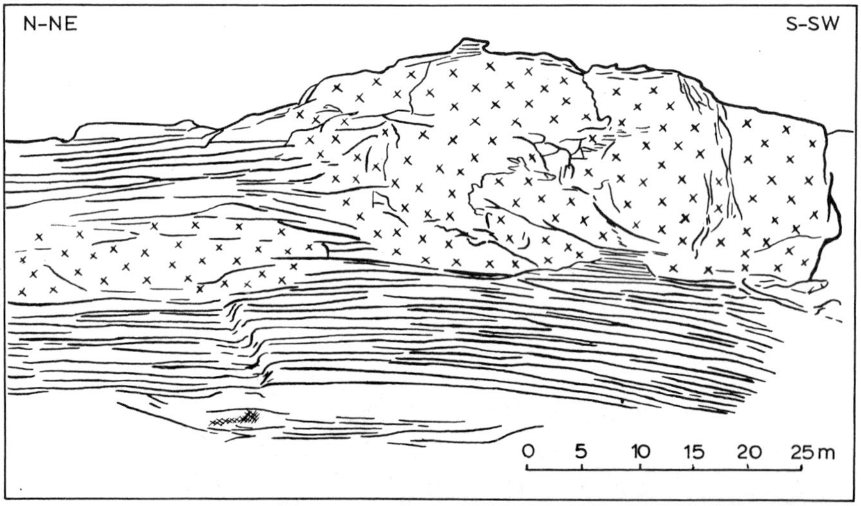

Fig.49. Section about 0.65 km north of Lundsklint. The main reef limestone occurrence belongs to the same reef as the reef limestone shown in Fig.50. North of it is a smaller reef. Both belong to the Högklint Beds. They overlie the Upper Visby interstratification of marly limestone and marlstone, which shows a flexure-like deformation, caused by differential compression under the influence of the overlying rocks.

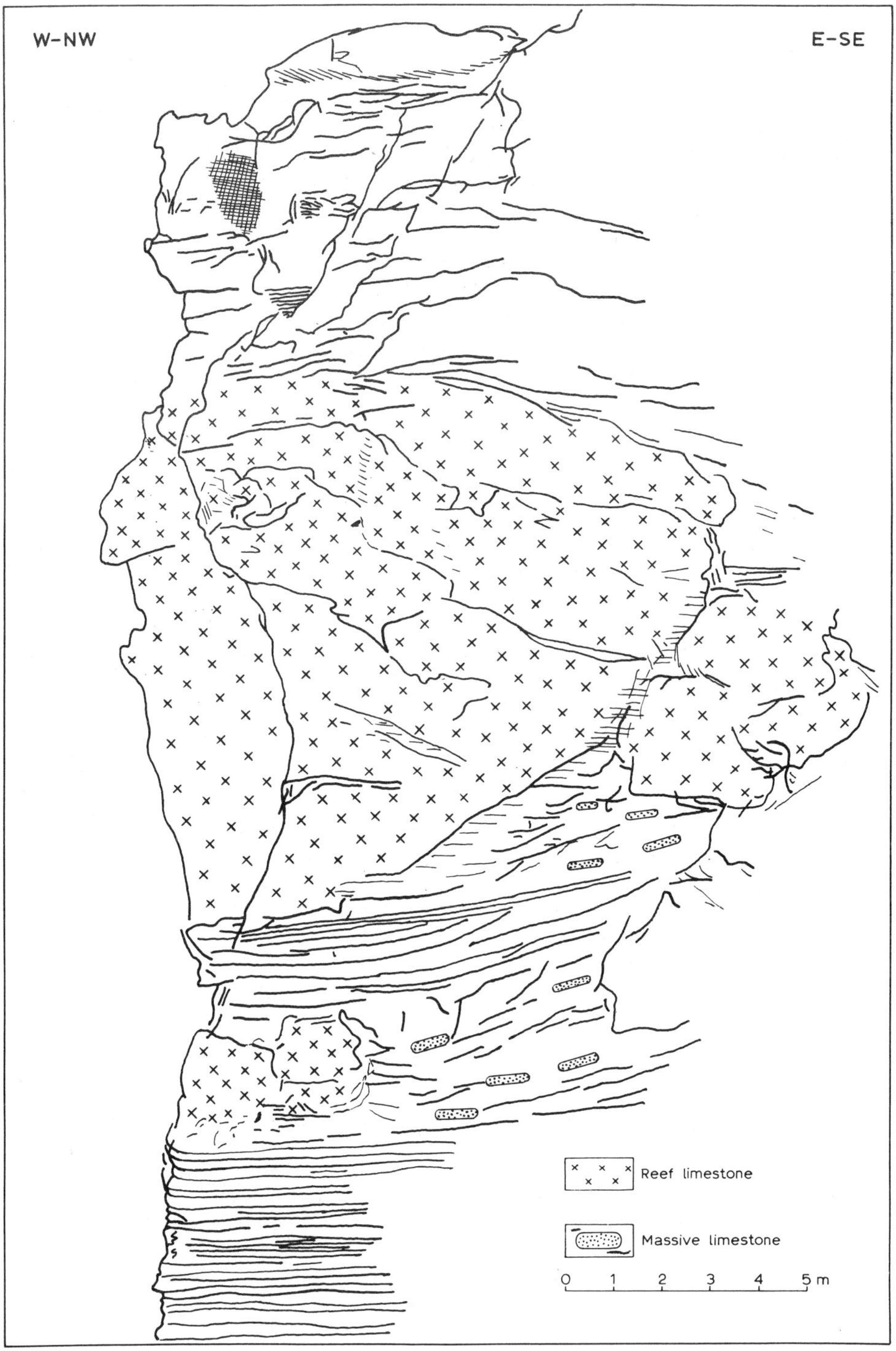

Fig.50. Legend see p. 142.

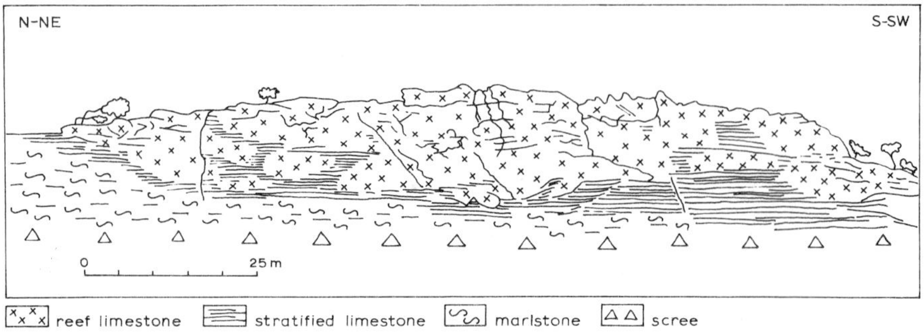

Fig.51. Reefs belonging to the Högklint Beds in the Luseklint, approx. 16 km north-northeast of Visby. Note that the subdivision of the reef-limestone mass into smaller units is distinct at the north-northeast and south-southwest sides of the exposure, but not in its central part.

(Fig.51), showing that deposition of stratified sediment at one place was synchronous with reef formation in close proximity. In some exposures the mantles are so well developed that the bodies do indeed appear as separate reefs (Fig.49,50; compare also the southern and northern parts of the cliff section presented in Fig.136).

In between reef-limestone massifs which show such a subdivision into smaller bodies, others occur in which no stratified intercalations are found at all. Compare Luseklint (Fig.51) and Lundsklint (Fig.48) with the more than 100 m long and 20 m high reef-limestone mass, which is located only 0.28 km north of Lundsklint and does not show any subdivision (Fig.44). Its development is comparable to the reef in the Häftingsklint (Fig.38). See also the reef exposed about 0.5 km north of Lundsklint (Fig.52).

The exposed wall of the Luseklint (Fig.51) shows clearly that a development with a number of individual smaller reef-limestone bodies is distinct in the peripheral parts of the reef-limestone mass, but not in its more central parts. The same pattern is suggested by the cliff walls of Snäckgärdsbaden (Fig.133, 136), where no individual smaller bodies are visible in section BC and in the central and southern parts of section AB, whereas these bodies are clearly present in the northernmost part of section AB. The examples suggest a process with a comparatively rapidly

Fig.50. Section about 0.62 km north of Lundsklint. At the base, stratified marly limestone and marlstone and a small reef, belonging to the Upper Visby Beds. The local deposits of hard and massive limestone at the right of this reef, which are enclosed in stratified marly sediment, are remarkable; they seem to be restricted to the close environment of the reefs. Higher in the cliff wall, reef limestone of Högklint age is found. This is overlaid by stratified Högklint limestone. At the right of this reef is a narrow intercalation of stratified limestone, which divides the reef into a smaller and a larger unit. (After Manten, 1962, fig.15.)

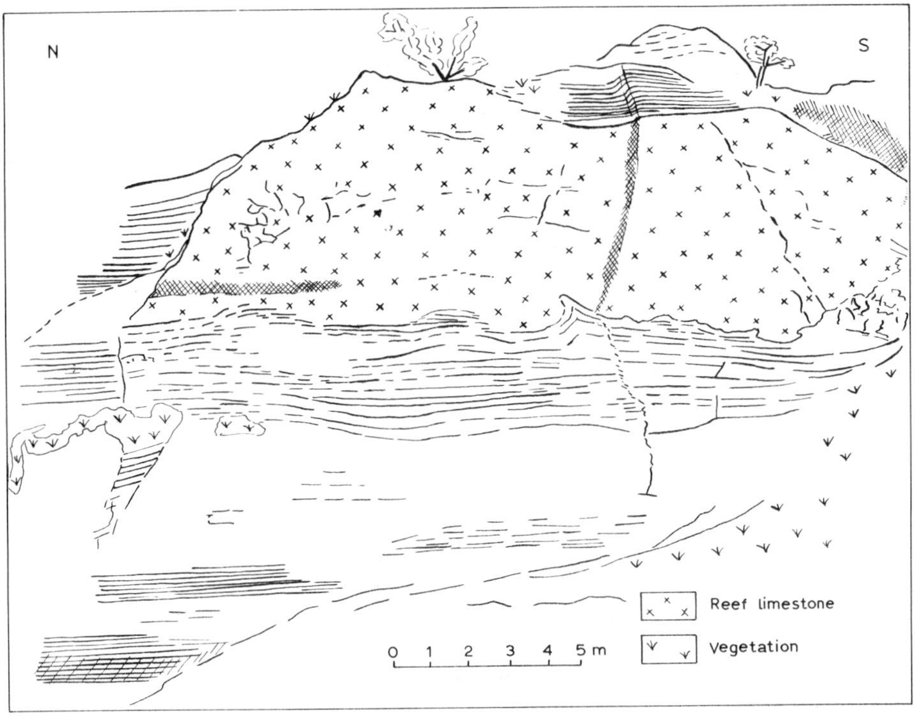

Fig.52. Reef about 0.5 km north of Lundsklint. Högklint Beds. The reef is an unstratified mass, predominantly built by stromatoporoid colonies. The rock is strongly recrystallized. It is surrounded by stratified limestone with crinoid fragments. (After Manten, 1962, fig.19.)

---

expanding reef of the inverted-cone form, which at its margins joined or fused with other, smaller reefs. See also the situation pictured in Fig.146; had the main reef expanded somewhat further south-southwestwards, it would have covered or fused with the adjacent small reef. The situation shown in Fig.46 has presumably developed through a process described before as natural selection among reefs, thus also through the contributions of more than one reef.

It has already been described how within a reef, intercalations of stratified sediment can be commonly found. It is not impossible, and for some exposures even likely, that these may also have contributed in causing the compound nature of some reef-limestone masses.

DEPRESSIONS

Two main types of depressions can be distinguished, those which developed as basins between some closely-neighbouring reefs, and those which originated within one reef when a certain part of the reef surface did not grow upwards at the same rate as the areas around it. In the latter case, reef debris could at a certain moment start to accumulate in the depression, thus building the floor of the depression again up to a level approaching that of the surrounding reef surface, after which reef builders generally colonized the area again. But it could also happen that organisms other than the reef builders common on the reef surface profited by the sheltered conditions in the depression and thus a different fauna developed there. In the latter case, the author prefers to call such a lower area of the reef surface a *pool*. An intermediate form is that in which the common reef builders themselves flattened out temporary unevennesses in the reef topography; examples of this have been seen in the Lännaberg (Slite Beds, p. 323).

Interreef basins

In several places in Hoburgen, passages or basins developed between small reefs that grew close to each other. They could be filled with reef debris or stratified limestone containing a varying amount of such debris.

A good example of an interreef basin is found in the second hillock of Hoburgen (Fig.53), a little north of the rich stromatoporoid development pictured in Fig.214. At both the left (northwest) and right (south) side of the depression, reef limestone is found, while from the shape of the layers of stratified material in the basin it seems very likely that a reef has been present also west of it; or perhaps the reef at the northwest side once extended a little farther to the south.

The lower part of the basin is filled with talus material, which most likely came mainly from the reef south of it. This is overlaid by stratified limestone with only a limited content of reef debris. Thereafter in the basin a tongue of another reef invaded from the east. The weight of this extending reef further contributed to give the layers of the stratified limestone in the basin a saucer-like shape. At the left side of this reef tongue, a narrow, almost vertical zone of talus is found, not more than 0.5 m thick. Also at the right of the reef tongue there is some talus, over which the reef expanded southwards. Thereafter the invading reef came into contact with the reef south of the basin, but soon afterwards its growth terminated.

Slightly south of the middle of the west wall of the second burg of Hoburgen, there is another basin, which developed between at least two reefs. It is filled with very marly sediment in which reef outwash is embedded, mainly stromatoporoids that got loosened from the reef, but also coral colonies; further crinoid remains, several solitary corals and a few brachiopods are found in it. Crinoids are nowhere as richly abundant in the interreef basins in Hoburgen as they are in the crinoid limestone ("Hoburg marble") which is normally found around the reefs. Higher in the wall the interreef basin and the two exposed reefs flanking it are overlaid by stratified limestone of which the layers show a slight downward buckling where they cover the basin sediments. Recent erosion there has led to the formation of a cave.

In the majority of cases the reef debris and crinoid material in interreef basins is coarser than elsewhere around the reefs. However, in some instances more sheltered places were created. There the deposited material had little to suffer from water movement. In such places even

Fig.53. Hoburgen, detail of the southwestern part of the second hillock. A basin is shown which developed in between three or perhaps even four reefs. The reef limestone part which is visible at the right, belongs to the reef very rich in stromatoporoids, part of which is shown in Fig.214. This reef also supplied the debris on which the hammer lies. On top of this debris, stratified limestone has been deposited. Finally, the reef at the west side expanded to such an extent that it penetrated into the basin, causing a deflexion of the limestone layers which it overlies. Hamra-Sundre Beds.

intact crinoid calyces can be found, as appeared, e.g., in the south cliff of Brissund (Fig.140). An interesting observation made in the latter locality was that on top of one of the crinoid calyces, over the anal opening, a specimen of *Platyceras cornutum* Hisinger was seen in its living position.

A rather peculiar interreef basin is the one shown in Fig.54, 55, 56, presenting almost a transitional form to the filled depressions within a reef, to be described next.

Filled depressions within the reefs

The reef limestone at the south side of the interreef basin in the middle of Hoburgen's second hillock, described above, appears to have been formed by one of a few centres of growth into which a reef, which initially developed as a whole, at a certain stage has fallen apart. Apparently the individual growth centres differed much in vigour. Between the part bounding the interreef basin and the one south of it, a depression in the reef

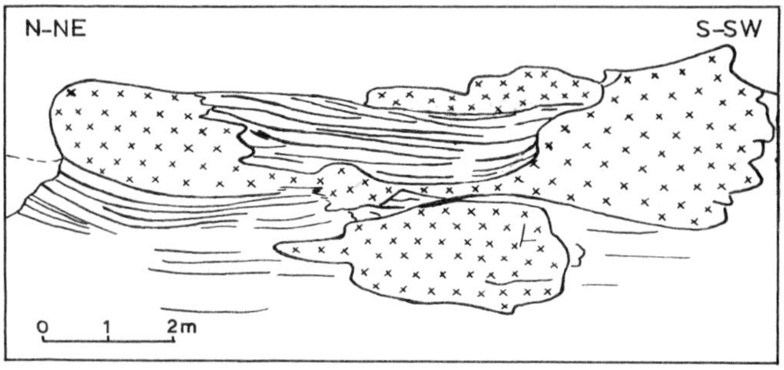

Fig.54. Exposure about 0.8 km north of Lundsklint, showing three reefs. The lowermost of these is the least well exposed and is for the major part still surrounded by Upper Visby sediments. The upper two reefs are, at their base, connected through an approximately 4 m long isthmus of reef limestone. These reefs are illustrated in more detail in Fig.55 and 56. They belong to the Högklint Beds. The isthmus is underlaid by limestone with interstratified thin marly layers and is overlaid by stratified, hard limestone with reef debris.

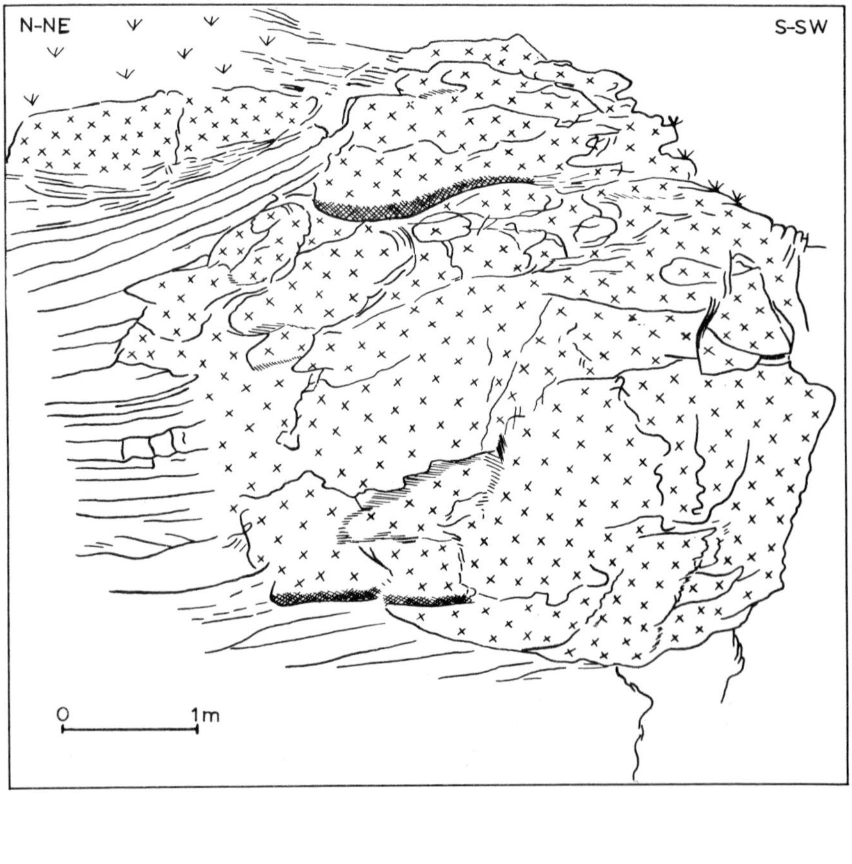

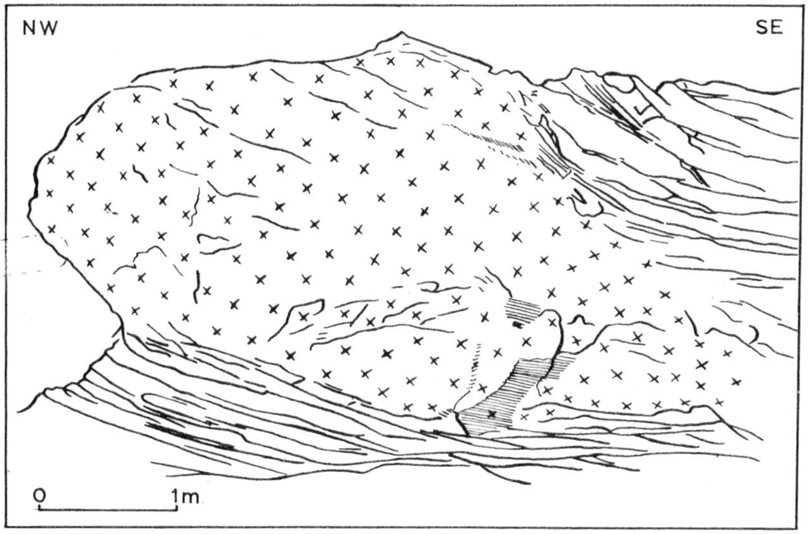

Fig.56. The northern reef of Fig.54. The reef rests over an alternation of limestone layers, about 20 cm thick, with marly layers, about 5 cm thick. The highest layer is marly and is ruggedly overlaid by the stromatoporoid reef limestone. At the right hand top is a mantle of sediments, about 0.5 m thick, that forms the transition from the reef to the normal stratified limestone. It is somewhat marly and contains reef debris.

surface initially developed, filled with stratified limestone and reef debris, until the southern growth centre, which was larger and apparently also more vigorous, expanded over the depression. Periods of vigorous growth apparently alternated also in this growth centre with times of lesser growth. In a small indentation in the wall it can be seen how within this reef part, another depression began to develop, though it was soon covered again by reef limestone. Elsewhere it shows a distinct increase in matrix volume in some places, locally even leading to a vague stratification within the reef. Nevertheless this southern growth centre managed to survive the other ones; it is the only one of this reef which extends to the plateau of the second hillock, but while standing on this plateau it can be seen how from all sides the stratified limestone expands over the reef limestone.

Fig.55. The southern reef of Fig.54. Conglomeratic reef limestone, due to stromatoporoid colonies with rather round shapes, which are separated by a marly matrix. The upper part of the reef shows several crinoid remains. At the left in the drawing is stratified limestone ending against the reef. The rock is whitish in colour and relatively hard. The layers are on the average 10–15 cm thick, with only little or hardly any marl in between. They show only a few recognizable fossils.

The example illustrates, as do the data presented earlier in this chapter about fluctuations and interruptions in reef growth, that the reefs of Hoburgen type were liable to comparatively rapid changes in form. It seems probably that they hardly ever remained perfectly stationary but generally slowly advanced or retreated. Every movement they made affected in some degree the water circulation around them, thereby influencing favourably or unfavourably the growth of the organisms further on the reef. They may be compared to huge living pulsating organisms, slowly stretching out an arm here and withdrawing one there, in some places showing youth and vigour, in others disease and death; capable of withstanding the rough buffetings of storms and surf and yet extremely sensitive to some ostensibly much slighter changes in the environmental conditions.

In part of the exposures of Hoburgen-type reef limestone, cracks are found. Of these particularly the horizontal and subhorizontal ones deserve special attention. Good examples are found in the raukar of the Lännaberg (Slite, Slite Beds).

Two types of horizontal and subhorizontal cracks can be distinguished there. Those of the first type often cut indiscriminately through both matrix and colonies of stromatoporoids and corals. The reef limestone above and below the crack is of similar character. Cracks of this type are also found in all possible other directions, up to vertical. These cracks have undoubtedly formed after induration of the rock.

Directly above some other cracks the fossil content of the reef limestone is different from that underneath them. Thus reef builders may be much smaller and thinner there; or corals may temporarily outnumber the stromatoporoids. These are cracks of the second type. They can be followed for distances of 1–7 m and in some cases in more than one rauk. Three-dimensionally, they often show the shape of a basin, with rising edges which gradually fade away in the reef limestone. These edges vary in height between one and a few decimetres. It seems likely that these cracks are connected with original phenomena in the reef-limestone structure, presumably parts which for some time were lower than the surrounding growing reef surface. In a few instances it has been observed that a crack indicated the base line of a debris-filled depression, containing crinoid fragments and reef debris which were overlaid again by reef builders in situ. On one occasion, four basin-shaped cracks were found above each other. In the Lännaberg, most of these cracks of the second type were found in the north, that is, more at the landward side of the reef.

In fact there is no sharp distinction between pockets of stratified sediment or reef debris, found between reef-building colonies, and filled depressions in the reef surface. It is mainly a matter of size. In part of the larger and thicker filled depressions, phenomena of compaction and buckling can be observed similar to those often found in the stratified sediments underneath the reefs.

If the debris, filling a depression, is very coarse and comprises many complete reef builders, it is sometimes difficult to distinguish it from the surrounding reef limestone. In the Solklint (Slite, Slite Beds) a debris-filled depression was recognized only because out of ten coral colonies which were found embedded closely together in some reef debris, not less than nine were upside down.

The distinction from the true reef limestone can also be blurred when in the part of a reef surface that developed into a depression, reef growth never completely came to an end. The marl or fine debris deposited in such a depression may then still enclose some reef builders in their growing positions and the result may vary from an intercalation of stratified material with some reef builders to only-vaguely-stratified reef limestone.

Pools in the reefs

From the foregoing subsection, it will be clear that also no sharp boundary can be drawn between debris-filled depressions and pools characterized by a different faunal composition. Lower portions in a reef surface were always susceptible to debris accumulation. Local conditions may have been determinative as to whether, in the depression, organisms could develop or not; and where the first applied, such organisms in the one case settled directly on the floor of the depression, in another on a basal layer of debris. Furthermore, in the one case, the organisms had little to suffer from an influx of reef debris; whereas in another pool the remains of these organisms are mixed unsystematically with debris washed into the pool from the surrounding sea floor.

Since the organisms in a pool grew in a relatively-sheltered part of the reef, generally not many colonies in pools are found detached and out of their normal positions. Where many colonies in a depression are not in their life orientations, these are generally massive colonies rather than the branched ones which often characterize the pools, and these will then more likely be washed in from the surrounding parts of the reef. Also fragmental debris is then generally found. But what about those massive colonies that are found in their correct orientations? These may either also have been swept into the pool or have developed there. Which of the two applies is often difficult or even impossible to determine.

THE ROOTS OF REEF FORMATION

In some places in Hoburgen it can be observed how a reef began to develop on the sea bottom and from its centre of origin gradually expanded laterally over progressively-younger layers of stratified sediment.

A beautiful example of a reef root is found in the middle of the west wall of Storburg. It is exposed below the overhanging part with Hoburgsgubbens Trappa (Fig.77), at the left side, in a cave. It is a strikingly-untidy mass, because many of the fossils occur in orientations other than those of growth; there are also many fragments of fossils, and the marly matrix occupies a rather-great volume of the total reef rock. As reef builders, several compound and solitary corals are present, together with many stromatoporoids; among the latter a few large specimens are found, but the majority are small-lenticular ones. In the matrix some crinoid remains occur. The developing reef very soon produced a thin layer of debris around it, over which it expanded. The weight of the developing reef led to the formation of a saucer-like depression in the sediments beneath (Fig.57). The sediment in the direct environment of the reef root is very poor in fossils and consists of an alternation of layers of argillaceous marly

Fig.57. Storburg, Hoburgen. Hamra-Sundre Beds. Buckling of stratified sediments underneath the root of a reef, as exposed in a cave in the middle of the western cliff.

Fig.58. Storburg, Hoburgen. Hamra-Sundre Beds. Alternation of argillaceous marlstone and hard, marly limestone, close to the root of a reef. The photograph was taken on the right at the back of the small cave visible on Fig.57.

material with layers of hard and splintery limestone, which are a few centimetres thick on the average (Fig.58).

Considering the general picture presented by the central part of the west wall of Hoburgen's Storburg (Fig.77), it may be that the reef root described has been the beginning of a secondary centre of reef growth, which after a short time fused with the parent reef. No distinct boundary could be established with the reef limestone of the large and evidently-older reef exposed north of it.

In Lithberg Grotta (Lithberg Cave), in the north of the Storburg, another reef root is found, surrounded by stratified sediment. It is not impossible that in this case reef growth began at two different places, but close to each other, because at the left side in this cave the boundary between reef and stratified limestone is seen to be a little lower than in the middle of that cave. The reef root is relatively rich in fossils; although flat reef builders dominate, there are also several rounder ones; stromatoporoids are in the majority. The stratified sediment in the back of this cave, especially close to the reef root, is rather marly and shows a vague alternation of softer marly layers with harder ones of marly limestone, but the alternation is less well developed than in the exposure north of Hoburgsgubbens Trappa. As far as can be seen, the reef that grew out of the root in Lithberg Grotta expanded all around. Southwards, it extended over its own talus until it soon abuted against the talus of a presumably slightly older reef (Fig.59).

An outcrop of lesser quality occurs in the south of the second burg, close to the reef part rich in stromatoporoids, shown in Fig.214. This reef root also shows a rather-untidy building, with a great many fossil fragments as well as fossils that are not in their life orientations; the whole is embedded in a matrix of strongly-marly limestone. There, too, in the very beginning of reef growth, the stromatoporoids already outnumbered the compound corals. Also solitary corals, crinoid remains and brachiopods were found.

A little south of Hoburgsgubbens Trappa, in the Storburg of Hoburgen, the lowermost part of the cliff shows a rock which in all likelihood was deposited very close to the base of a reef, which later on expanded over it. It is a marly sediment very rich in fossils, among which especially the numerous thin and faintly wavy tabular stromatoporoids are notable. It is overlaid by reef talus (Fig.60). More information on the general development of the sediments underneath reefs will be given in a later section of this chapter.

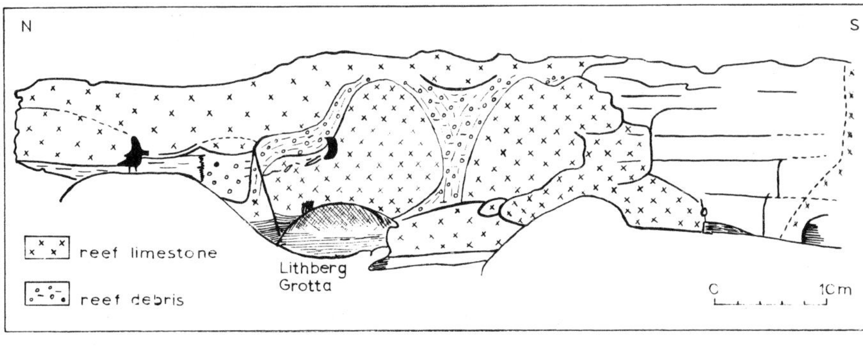

Fig.59. Northern part of the west wall of Storburg, Hoburgen. In the centre, two smaller reefs surrounded by a talus mantle. The reefs are overridden by a third and larger reef which had its point of origin about northeast of the other two and the peripheral section of which is exposed in the left of the wall. At the right of the two reefs, stratified fragment limestone is found, after a short distance followed again by reef limestone (cf. Fig.77).

Fig.60. Sediment very rich in flat stromatoporoids, deposited very close to the base of a reef, and overlaid by reef debris. Middle of the western wall of Storburg, Hoburgen. Hamra-Sundre Beds.

Exposures in other stratigraphical units confirm the observations in Hoburgen that in the majority of cases, already in and around the roots of reef formation, stromatoporoids predominate. There is no general faunistic zonation in reefs of Hoburgen type.

FISSURES

In some exposures of Hoburgen-type reef limestone, especially in the Slite Beds, but also in a few reefs in the Högklint and Hemse Beds, fissures are found traversing the rock.

The reef limestone exposed in Hejnum Hällar (see p. 325) shows several vertical cracks, up to 1 m deep; some are up to 15 cm wide, others just a few centimetres. There is a strongly-dominant 120° direction (E 30°S). The fissures are fairly straight and there are no indications that they were originally filled.

Also in the reef limestone of Gisslauseklint, about 3 km east of Othem (p. 325), several vertical and subvertical cracks are found. Some of these, with northeast or southeast directions, almost certainly are joints. A few, however, with northeast directions are characterized by weathered surfaces covered with enormous amounts of fine reef debris. Fragments larger than 2 cm hardly occur, the majority being much smaller. There are indications that once the entire cracks were filled with such debris. The cracks traverse reef-limestone parts which are relatively poor in large reef builders.

These fissures are found only in reefs of large horizontal extension. Therefore, it seems likely that their origin is connected with reef expansion and was due to strain set up when younger reef parts progressed over unconsolidated sea-floor deposits which underwent compaction under the weight of the growing reef.

Since such fissures are more characteristically found in reef of Holmhällar type, a further discussion on their nature is postponed until the next chapter (pp. 200–205).

STYLOLITES

In some reef limestone exposures, stylolite seams can be observed. They are much less common there than they are in several exposures of stratified, finely-crystalline limestone. In these stratified limestones, especially in the Slite and Hemse Beds, the stylolite seams generally, seen on a large scale, are parallel both to the orientation of the bedding planes and to the other stylolites observable in the exposure, even though they apparently run independently of each other. In the reef limestones, on the other hand, several of the stylolites strongly deviate, even up to 90°, from a horizontal direction. They may even cross each other, or divide into two separate branches. Stylolites in uneven-grained stratified limestones take an intermediate position. No stylolites have been observed in the coarse crinoid limestones.

The formation of the stylolites can be explained by a process of differential chemical solution in indurated sediments, under pressure, and with deposition of insoluble residues in situ. This is the solution-pressure theory of Stockdale (1922, 1943). Two other theories on the origin of stylolites, advanced by Shaub (1939, 1949, 1955) and Prokopovich (1952) are inconsistent with the available data on stylolites as observed in Gotland (Manten, 1966b, 1968).

The greater variance in the orientation of stylolite seams in the reef limestones may be explained by inhomogeneities in this sediment, which led to local anomalous directions of the pressures, during solution.

Of the several examples which illustrate that solution of hard rock has actually taken place, an especially explicit one has been reproduced as Fig.61.

REEF-SURROUNDING SEDIMENTS

The stratified sediments enveloping the reefs can be separated into those underlying the reefs, those occurring lateral to the reefs and those overlying the reefs. Such a division simplifies their description, but it should always be kept in mind that there are no sharp boundaries or even distinct patterns in the distribution of the various sediments. The sediments laterally surrounding the reefs can be further subdivided into the talus mantle and stratified limestone with crinoid remains and reef debris; the talus mantle is often absent, but where present, it passes outwards without sharp boundary into the crinoid limestone with reef debris.

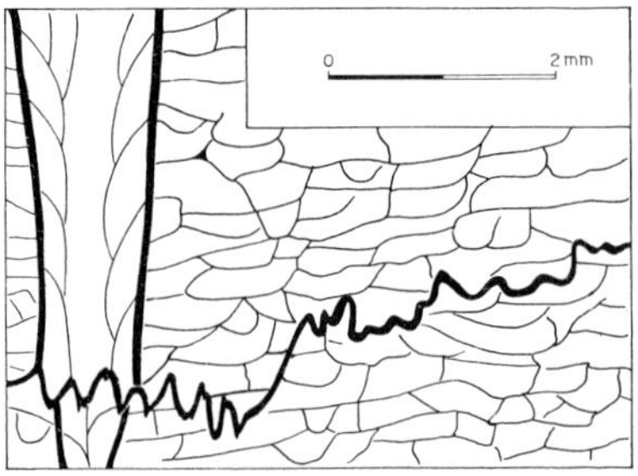

Fig.61. Vertical section through a stromatoporoid colony (*Stromatoporella* sp.), crossed by a stylolite seam. A cf. *Syringopora* corallite, enclosed by the stromatoporoid, shows that solution has caused a shortening of at least 2 mm. Reef limestone. Hamra-Sundre Beds, Hoburgen. (After Manten, 1966b, fig. 3.)

Stratified sediments underneath the reefs

Normally approximately the same stratified sediments are found underneath the lowest point of a reef of the Hoburgen type as elsewhere in the same beds where there are no overlying reefs. This is generally a dense or finely crystalline, somewhat marly limestone. The bedding planes are often covered by a film or thin layer of marl, and in some cases there is even an alternation of layers of marl and limestone. In some localities a coarser and more fossiliferous limestone is found. Whether deposition of coarser and fossiliferous limestone or of fine limestone poor in fossils did take place may possibly have been determined by the intensity of current action over the sea floor.

In the stratified limestone, small coral colonies are found locally, often in their position of growth. Their development was somewhat faster than the velocity of sedimentation of the limestone, with the result that they soon extended slightly over the surrounding sea floor. As they got older, their velocity of growth decreased and some extra deposition of mud might easily have caused their deaths. Especially marl deposition may often have been fatal, since several coral colonies are covered by a marl film such as found on several bedding planes.

Solitary corals, and also bryozoans in places, are generally more common in the sediment underneath the reefs than colonies of corals and stromatoporoids, which indicates that the environmental requirements of the latter were somewhat higher. However, they were present, and slight improvements in growth conditions caused them to increase in number and to start reef building. In a few localities, small lumps of limestone in biohermal development represent attempts to form reefs which failed at a very early stage, but in most cases, the reef growth succeeded.

Some of the Lower Högklint reefs originated in the uppermost Upper Visby Beds or at the Visby – Högklint boundary (Fig.44, 54, 147). In

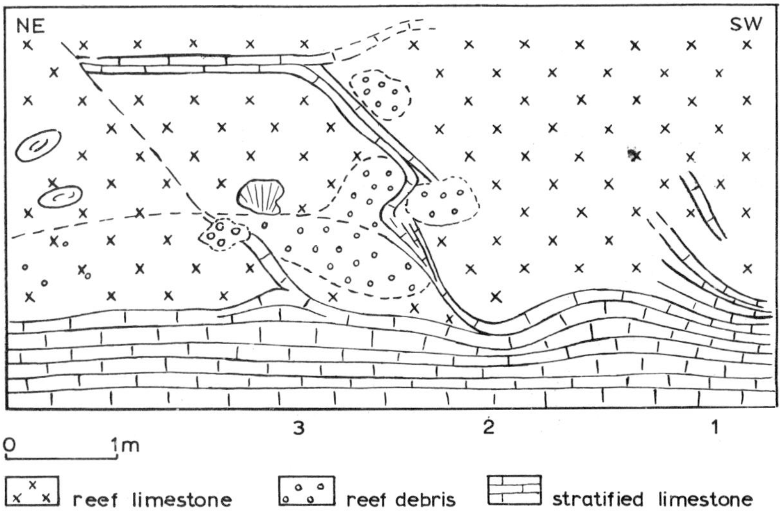

Fig.62. Detailed section, showing the contact between stratified limestone and reef limestone in the Hjännklint, about 1.45 km southwest of Halls Fiskläge. Högklint Beds. The basal stratified limestone is fossiliferous and of varying coarseness. At three places, the stratified limestone penetrates into the reef limestone. All three penetrating wedges are finely crystalline and poor in macroscopically-recognizable fossil remains. The reef debris is especially rich in fragments of solitary and social corals in a calcareous matrix. The boundary between reef debris and reef limestone in between the stratified wedges 2 and 3, can be followed northeastwards in the reef limestone. It separates the basal reef limestone with small stromatoporoids and reef debris from the overlying reef limestone in which larger stromatoporoids are the main reef builders.

---

contrast to the general situation with Hoburgen-type reefs, around the roots of these early reef formations, local and thick limestone layers are usually found, similar to those occurring underneath the Upper Visby reefs which have been described in the previous chapter.

The actual boundary between stratified sediment and overlying reef limestone generally is somewhat irregular. In some cases, however, especially underneath the extensive flanks of patch reefs, the boundary can be remarkably plain (Fig.35). In a few other cases the stratified sediment penetrates in wedges the basal reef limestone (Fig.62, 156).

In many exposures, the stratified limestones have sagged underneath the reef-limestone masses overlying them. The phenomenon which caused this could best be called "differential compression" (different weights on the same material), rather than "differential compaction" (the same weight on different materials). The dips in the sagging sediment, underlying and bordering a reef, can be up to 35°, but are generally much less. The steepest dips are found where close underneath the reef-carrying layers a marlstone complex occurs. Where compression was very strong in such

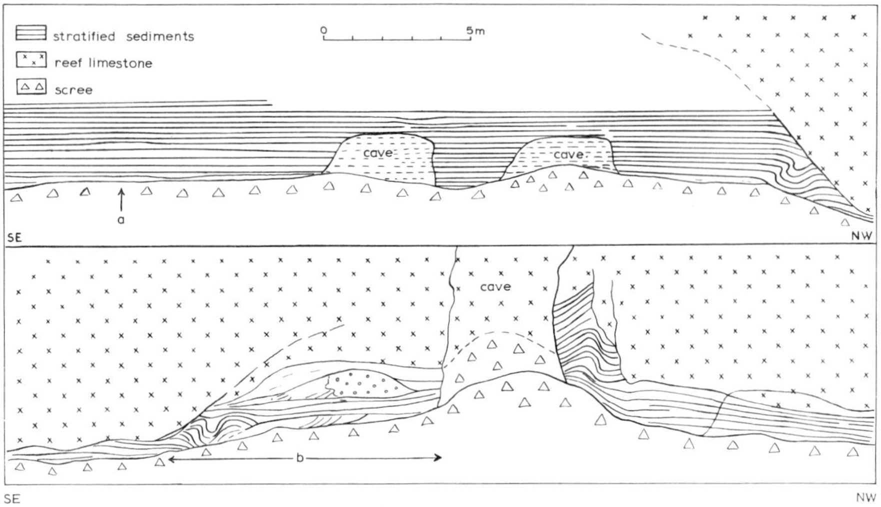

Fig.63. Hallshukklint, about 0.27–0.21 km southeast of the Hallshuk lighthouse, showing the location of Fig.73 (a) and Fig.64 (b). In the southeast, 26 m of the section is formed by stratified limestone with a number of interstratified marlstone layers. In the rest of the section, the stratified limestone is overlaid by reef limestone. At only one place does the stratified limestone at the base reach a height of about 5 m. A cave has developed there, indicating that this sediment is less resistant to erosion than the reef limestone. The stratified material was most probably deposited in between two reefs. At the basal lateral contacts between reef and stratified limestone, the stratified sediment is folded, due to the squeezing out of stratified sediment from underneath the reef, as a consequence of the heavy burden placed upon it. All sediments belong to the Högklint Beds.

---

cases some layers may even have been squeezed out from underneath the reef. Also fold phenomena are occasionally found (Fig.63, 64).

Differential compression is probably also mainly responsible for the rather undulant upper boundary of the Upper Visby Beds along the northwest coast of Gotland. These beds generally reach higher in places where the overlying Högklint Beds developed as stratified sediment. A similar phenomenon is found in the Eke Beds. The stratified limestones underneath some Eke reefs is thin and overlies the Hamra marlstone; the limestone layers show sag dips up to 35°.

A stromatoporoidal deposit as shown in Fig.60 is not a common phenomenon underneath a reef.

Underneath the more peripheral parts of the reefs, fossiliferous limestone is generally found (Fig.65). This deposit is usually coarse or medium grained, but is sometimes finely crystalline. The coarseness of the sediment and the number of fossils contained in it can vary without any apparent regularity in both the vertical and horizontal directions within a given layer, as well as between successive layers. As a rule, the sediment

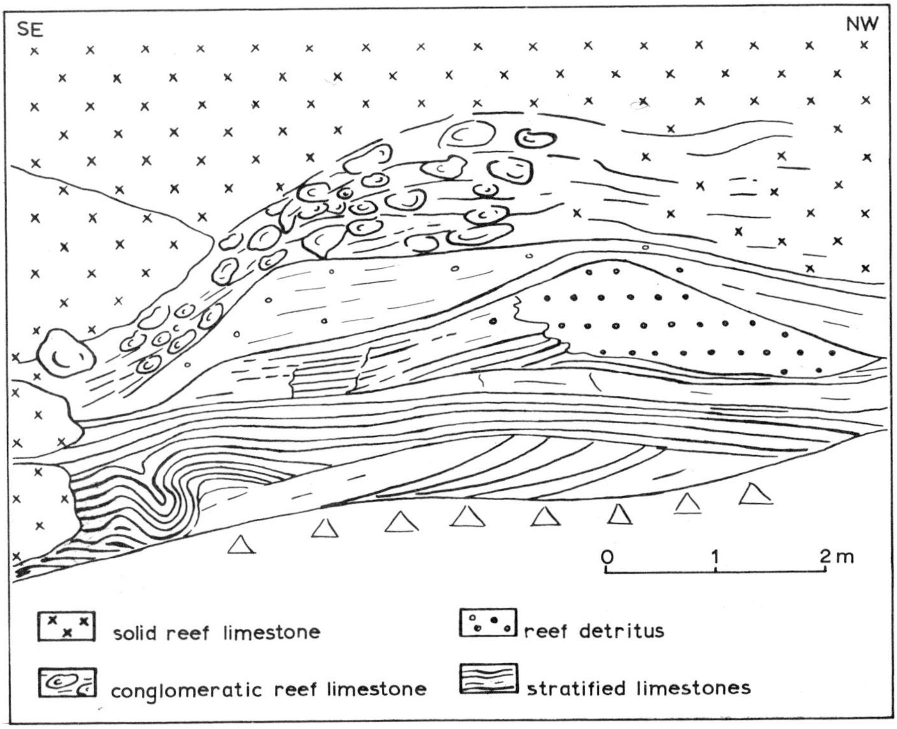

Fig.64. Detailed section, showing the contact between stratified and reef limestone at b in Fig.63. Hallshukklint. Högklint Beds. The stratified limestone encloses a body of reef debris, against which the surrounding layers end. The layer on top of it is extremely rich in fossils, particularly solitary corals; the number of fossils decreases northwestwards. The basal reef limestone, overlying this layer, has a conglomeratic structure, with colonies in rather a marly matrix. The remaining reef limestone shows a solid structure.

---

is especially rich in crinoid remains. Favourable conditions for reef growth were usually also favourable for an abundant crinoid development; crinoid limestone could be formed even in an early stage of reef growth. The colour of the rock varies; it may be yellowish or brownish-grey, locally light greenish-grey, grey to whitish-grey, and often red-mottled.

In addition to the crinoids, especially stromatoporoids and thin coral colonies are generally present; also solitary corals, brachiopods, bryozoans and a few gastropods are usually found. In comparison to the stromatoporoids in the reef, those in the underlying sediment are as a rule distinctly smaller, with a horizontal diameter averaging less than 7 cm in several localities; some are lenticular or dome-shaped, but most are tabular. As far as can be ascertained, most of the flat stromatoporoids and corals are in their normal orientation, and only a small fraction are upside down. It can be seen that some of the reef builders enclosed in the lower-

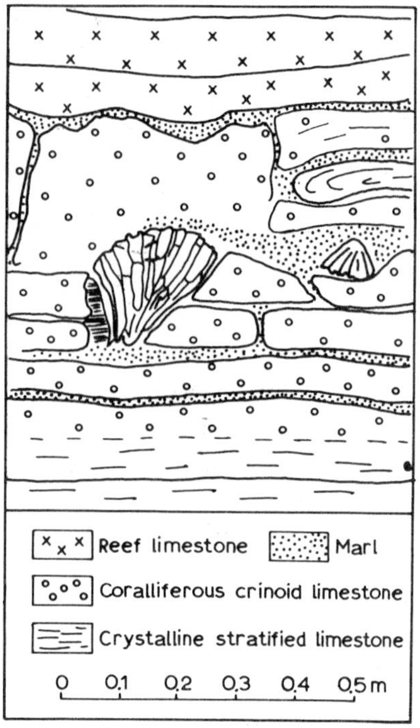

Fig.65. Detailed section of the contact between stratified limestone and reef limestone at b in Fig.70. Hallshukklint. Högklint Beds. At the base is crystalline stratified limestone, which passes upwards into an extremely fossiliferous limestone, largely built by solitary corals and crinoid remains. The layers enclose a coral colony which seems to be in the same position as during growth. Thinner layers of marlstone lie between the limestone layers. The thickest of these contains a small coral colony, lying upside down. Above it lies a stromatoporoid colony. The overlying reef limestone contains marl in small pockets. (After Manten, 1962, fig.23.)

---

most reef limestone started life on stratified sediment deposited laterally to a lower and earlier part of the reef. Branched corals or bryozoans are represented by fragments, although occasionally a large piece of up to 20 x 20 x 10 cm is found in its original coherent form. Algal balls with an average diameter of 2–3 cm and occasionally up to 10 cm are found in several localities.

The thickness of the crinoid limestone varies. In one place, there may be only one layer of a few to some tens of centimetres thickness, while in others there is a complex of layers up to about 1 m or more thick. The layers, often about 2–5 cm thick, may wedge out over short distances or show a tendency towards current bedding. In many cases, films or thin layers of marl occur on the bedding planes. As a rule, there is less marl the closer one gets to the reef base.

An example of the upward decrease of marl in a section through sediments underlying reef limestone was measured underneath the northern part of a Högklint reef about 0.5 km south of Sigsarvebodar, where the reef limestone is mainly exposed in raukar. The succession from the reef limestone down is given in Table X.

The decreasing occurrence of marl upwards to the reef limestone is even more apparent when we summarize the thickness of the sediment types per each successive 25 cm (Table XI).

In the limestone underneath the peripheral parts of a reef, there generally is a distinct upwards increase in the amount of reef debris, which may even become the dominant constituent in the uppermost part of it, as will be shown while discussing the reef talus.

A notable section through stratified limestones underneath a reef was found in the Hjännklint (Hall Parish) (Fig.66). In the south, there is a coarse and fossiliferous limestone at the base, which is especially rich in crinoid remains. It is overlaid by a limestone layer, the lower and upper side of which are both strongly hummocky, so that the layer as a whole resembles a string of very large beads. The layer can also easily be followed because of its brownish colour, which contrasts with the yellowish grey of the underlying limestone and the greyish white of the overlying limestone.

The "beads layer" is covered by a complex of limestone layers which very irregularly interdigitate. The weathered rock falls apart into angular pieces of varying size and thickness. The limestone is coarse, and rich in such fossils as stromatoporoids, crinoids, bryozoans, branches of branched coral colonies, and small massive coral colonies; all fossils are strongly recrystallized and sometimes they are in a beautiful way partly liberated on bedding planes. Part of the fossils is probably reef detritus. The rock has a distinctly stratified appearance, even though the 1–7 cm thick layers cannot as a rule be followed more than 0.3 m. Three layers are exceptions to this (see Fig.66). Layer 1 is a very thin layer of marly, scaly-weathering material occurring closely above the "beads layer". Layer 2 consists of very fine limestone which contains no fossils and is thinly stratified; its thickness is about 3 cm, and its colour is yellowish grey. In the very south of the sketched section, this layer is not particularly distinct, but after some metres, it is well developed. Layer 3 is similar to layer 1.

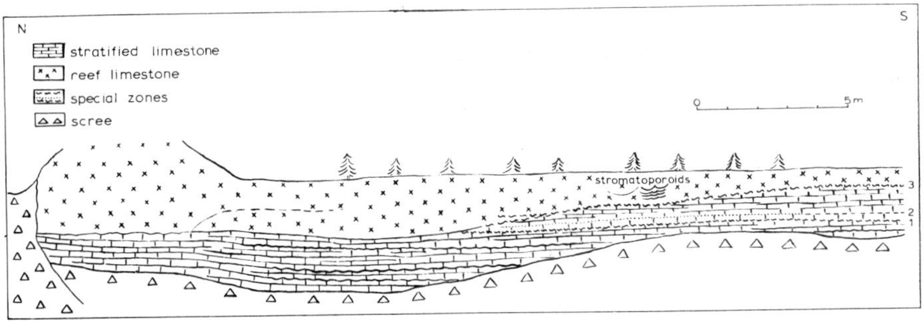

Fig.66. Section exposed in the Hjännklint in north Gotland, about where a small road connects the roads Hallshuk – Hall and Hallshuk – Norrby. Högklint Beds. For a detailed explanation, see the text.

TABLE X

Stratigraphical succession underneath a Högklint reef about 0.5 km south of Sigsarvebodar

		reef limestone
23	cm	crinoid limestone
6	cm	crinoid limestone
9	cm	reef limestone
1	cm	marlstone
1.5	cm	crinoid limestone
0.25	cm	marlstone
4.75	cm	crinoid limestone
0.2	cm	marlstone
3	cm	crinoid limestone
< 0.1	cm	marlstone
2.5	cm	crinoid limestone
5	cm	marlstone
3.5	cm	crinoid limestone
0.1	cm	marlstone
1	cm	crinoid limestone
0.5	cm	marlstone
2	cm	crinoid limestone
1.5	cm	marlstone very rich in fossils, especially crinoids
1.8	cm	crinoid limestone
2	cm	marlstone
2	cm	crinoid limestone
0.1	cm	marlstone
1.5	cm	crinoid limestone
< 0.1	cm	marlstone
2	cm	crinoid limestone
2.5	cm	fossiliferous marlstone
1	cm	crinoid limestone
0.8	cm	marlstone
1.7	cm	crinoid limestone
0.5	cm	marlstone
2.5	cm	crinoid limestone
0.7	cm	marlstone
1	cm	crinoid limestone
8	cm	very fossiliferous marlstone
21.5	cm	crinoid limestone

TABLE XI

Stratigraphical succession of Table IX summarized per 25 cm

Reef limestone	Crinoid limestone	Marlstone
25	–	–
–	25	–
9	14.45	1.55
–	15.1	9.9
–	13.1	11.9

The part of the wall in the south of Fig.66 from the scree up to layer 3, forms a rock shelter. The sediment above it is reef limestone which, about 8–10 m from the place which is represented at the right margin of the drawing, shows a beautiful example of wealthy stromatoporoid development.

The stratified limestone forming the lower part of the wall in the central and northern parts of the drawn exposure differs from that in the south because coarse fossiliferous limestone irregularly alternates with fine limestone poor in fossils in both the horizontal and vertical directions; a few "beads layers" are interbedded. At the base of the reef limestone, several stromatoporoids and other fossils are found embedded in a matrix of finely crystalline to dense limestone.

The great variety of deposits found in the Hjännklint underneath the reef limestone suggests that the local conditions under which these were laid down have varied both in space and in time.

Also an exceptional development is the occurrence of lumps of hard and massive limestone enclosed in stratified marly sediment lateral to the Upper Visby reef and underneath the Högklint reef of Fig.50.

Talus mantle

By far, not all reefs of the Hoburgen type are surrounded by a mantle of reef talus. There is a talus mantle around some of the reefs exposed in Hoburgen itself, whereas around some other reefs at Hoburgen such a mantle is absent. In reefs of similar type in other stratigraphical units, a talus mantle is an exception rather than the rule. In the absence of such a mantle, the reef limestone is generally directly surrounded by crinoid limestone with reef debris. Directly against the reef, the content of debris may sometimes be higher than that of crinoid remains, especially at the landward side of the reef, but the abundant crinoids, the approximately horizontal stratification, the direct transition in crinoid limestone with reef debris and the absence of reef blocks were criteria which caused the author not to call this true reef talus. It is admitted, however, that the distinction between talus mantle and surrounding crinoid limestone with reef debris is arbitrary.

As was mentioned earlier, most of the Hoburgen-type reefs (with the exception of those which developed in Hamra-Sundre time, and which include the reefs in the Hoburgen area) developed during periods of on average slowly falling sea level. It may be assumed that as a sea becomes very shallow, turbulence and currents become so strong that they prevent the local accumulation of a proper reef-debris mantle, and scatter the debris over the surrounding sea floor. The general distribution of reef debris in the reef-surrounding sediments will be discussed further in Chapter IX. At this place only the talus as found directly aside the reef limestone will be described.

As said, the talus supplied by reefs of the Hoburgen type can well be studied in some places in the Hoburgen complex itself.

The first outcrop to be discussed here is found in the south of the seaward wall of the third hillock of Hoburgen (Fig.45). There is a depression on top of reef II which is filled with limestone rich in reef debris. This is overlaid by the talus of reef III, which was deposited while this reef expanded westwards. When the reef was still at a certain distance from this locality, only small debris material was deposited there, but with the gradual approach of the reef, the deposit became coarser and included larger complete colonies of stromatoporoids and other reef builders. The talus directly around the reef even contains reef-limestone blocks of 1 m^3 or more, in which all the flat stromatoporoids show about the same orientation. Perhaps these loosened as a whole from the reef and rotated while moving down the reef.

This phenomenon of large dislodged blocks can also be studied while standing on top of the third hillock, as well as in it west-southwest cliff where it is beautifully exposed (Fig.67). This cliff section passes almost entirely through reef talus. It shows several of such tumbled blocks in which almost all of the fossils occur in the same orientation, which can be at all angles to the horizontal. Fig.68 is a photograph of part of such a reef-limestone block enclosed in the reef talus of this west-southwest cliff. Note that almost all of the flat stromatoporoids at the left are roughly parallel to each other but with an orientation about perpendicular to the only slightly dipping stratified limestone at the right.

Another place in Hoburgen, where Hamra reefs are exposed together with their surrounding talus is in the north of the west wall of the Storburg, close to the cave called Lithberg Grotta (Fig.59). Two reefs occur there; the southern one presumably began its growth somewhat earlier than the northern reef. The southern reef shows a well-developed steep talus, especially at its north side. It also has a depression at the top which is filled with reef debris. The reef does not reach the top of the cliff and is covered by a thin zone of younger reef limestone. The northern of the two reefs is surrounded by talus both at its north and south sides. Parts occur in the talus of this reef which in all likelihood can be interpreted as being large blocks which were torn loose as whole pieces from the reef by wave action.

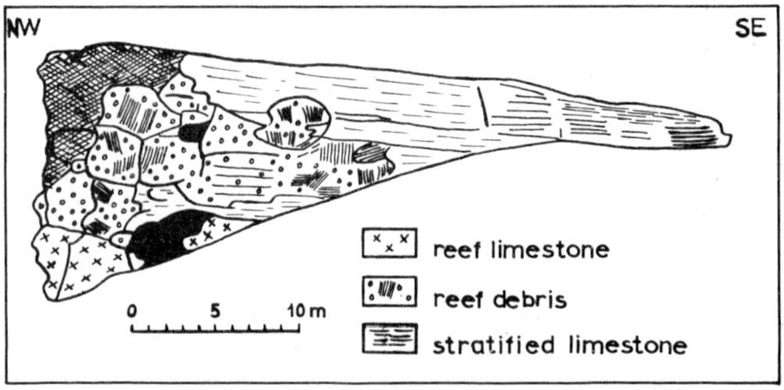

Fig.67. Hoburgen, west-southwest cliff of the third hillock. Hamra-Sundre Beds. At the lowermost left side, part of reef II of Fig.45 is exposed. This reef is overlaid by the talus of the larger reef III which developed east of reef II and expanded over it. South-southeastward in this wall, the talus deposit wedges out in stratified limestone. In the course of time, these stratified sediments are seen to have progressed over the talus and again levelled the sea floor. Above the exit of the cave at the left (drawn black), there is a thin remnant of stratified limestone overlying the talus material; it contains a rather high amount of reef debris; immediately behind it is still the reef talus deposit.

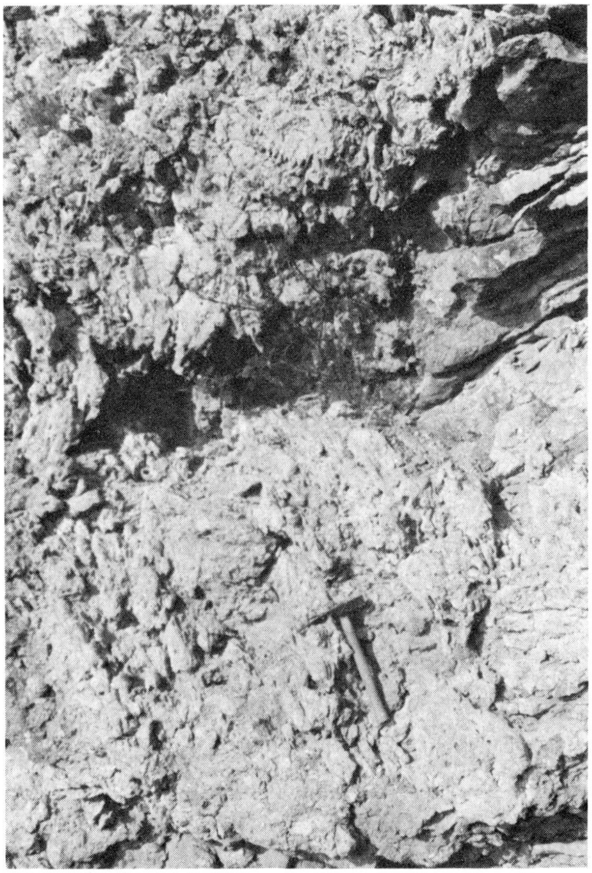

Fig.68. Reef talus. Note that almost all of the stromatoporoids are perpendicular to the stratified limestone at the right of the photograph. They probably became separated from the reef all together in a huge block. West-southwest wall of the third hillock, Hoburgen. Hamra-Sundre Beds.

In a talus deposit in the northeast of Klinteklint (Gammelgarn Parish, Hemse Beds), there is a block 1.10 m high and 1.30 m broad in which a great many flat-lenticular and tabular stromatoporoids up to 0.70 m long occur parallel to each other in a very orderly manner, with a dip of about 20° to the south-southwest. It is likely that this is also a block which was detached from a reef as one piece.

In this same locality, other parts of the talus deposit also show parallel flat-lying reef builders, but these were probably not brought there as blocks. In itself, it is not remarkable to find that fossils became generally parallel orientated in places where hardly anything else besides marl and very flat fossils were deposited. Deviations from this pattern occur when rounder fossils are intermixed because the flat fossils then were not deposited on a smooth bottom and consequently show dips in various directions, according to the unevenness of the underlying material, bottom currents and the like. Indeed, such parts are found in the northeastern Klinteklint as well. Parts in which the fossils and fossil fragments were likely deposited one by one can be distinguished from the large blocks by the greater number of fragments in the first, and also by the abrupt ends of some of the flat-lenticular and tabular colonies with no indication in the direct environs of where the detached peripheral parts have gone.

In both the examples from Hoburgen and Klinteklint, the blocks are found in talus which was likely deposited more at the coastward side of the reefs. Observations of present-day reefs have shown that storms approaching the reefs from the landward side may cause major destruction of the reef edge. Large portions of the reef edge may then become dislodged and are carried down the slopes. Two factors make the landward side of Recent reefs more vulnerable than the seaward side: the edge at that side is often steep and overhanging and the reef there is often less solid than at the seaward side. For the reefs of Gotland, the second factor may also have played a part. In the many cases where the reefs stood only a few metres above the surrounding sea floor, steep overhanging edges would have been less common than with the thick reefs of the present day. But it is notable that nowhere are tumbled blocks so common as in Hoburgen, and the Hamra Beds are the only distinct case where reefs of the Hoburgen type developed in water of increasing depth. The Hamra reefs on the average have a greater thickness in proportion to their horizontal extension and may have grown upwards at a higher rate than most of the other reefs, leading to relatively steeper edges and perhaps even less-solid reef frames.

A notable phenomenon found in a small southwestern cliff near the southern end of the second hillock of Hoburgen should also be described here, since it is likely to be connected with the occurrence of a talus slope of a reef (Fig.69).

At the very left at the bottom of this cliff, a small remnant of reef limestone overlaid by stratified limestone occurs (not visible in Fig.69). The whole remaining exposed lowermost part of the cliff is composed of reef limestone very rich in tabular stromatoporoids. These stromatoporoids all dip distinctly in the north-northwest direction, but nevertheless they are in all likelihood in their positions of growth (lower half of Fig.69). Because of their flat shapes, they give the rock a somewhat irregularly stratified appearance. Presumably, this reef limestone developed over a talus slope previously formed by the same reef. At the left, the stromatoporoid reef limestone abuts against stratified sediment; at that left side, a few lenses of marly clay are intercalated in the reef limestone. The dip of all the stromatoporoids makes it likely that they extended from the top downward. If the reef limestone had developed from the sea bottom upwards, one would expect to find them in horizontal positions with each new stromatoporoid extending a little farther to the right than its predecessor, at the edge. The centre of the reef must have been more into the hill than the corner between this southwestern cliff and the westward-facing cliff south of it. The topography in the corner between these two walls shows a slope parallel to the dip of the stromatoporoids, but the direction of slope turns southwards from northwest to south-southwest, thus giving the area the shape of a quarter of the area of a cone with its top in the northeast corner. Perhaps this topography is determined by the talus slope over which the stromatoporoid reef later expanded westward.

Stratified sediments were deposited on top of the dipping reef limestone at the right side; these do not extend more than a few metres westwards. From the west, the reef limestone with the dipping stromatoporoids was overgrown by a younger reef which expanded eastwards. At the left, this reef directly overlies the older reef limestone; at the right, it extends over the intercalated stratified sediments.

Fig.69. Detail from the southwest of the second hillock, Hoburgen. Hamra-Sundre Beds. At the base, a dipping stratification in the reef limestone, caused by tabular stromatoporoids. Overlying it is a younger reef, part of which can be seen in the top left hand of the photograph.

Stratified sediments lateral to the reefs

Crinoid limestones are present underneath the more peripheral parts of several reefs of the Hoburgen type, but are generally much better developed laterally around the reefs and their enveloping talus and partly also on top of the reefs. They most commonly appear as light-coloured limestones with a more or less well-developed stratification. They are so enormously rich in crinoid remains that in some instances the name crinoid breccia would be appropriate. The crinoid limestones contain a varying amount of reef debris, which is most commonly made up of fragments and entire colonies of stromatoporoids, corals, bryozoans and locally also calcareous Algae. Solitary corals are regularly encountered; other fossils, such as brachiopods, cephalopods, gastropods and trilobites, are seldom abundant. Compared with the bioherms, the variety in species is less. Where the reef limestone is surrounded by a talus deposit, this talus usually gradually passes into the crinoid limestone with reef debris. Where there is no talus mantle, as is the case in several reef exposures, there may be a rather sharp boundary between the reef limestone and the surrounding sediment rich in debris and crinoids, but in some instances, this boundary is a narrow gradual transition zone. There are also examples where the contact is interfingering, especially in the lower part of a reef (Fig.70, 71). In cases of a sharp boundary, occasionally this boundary is

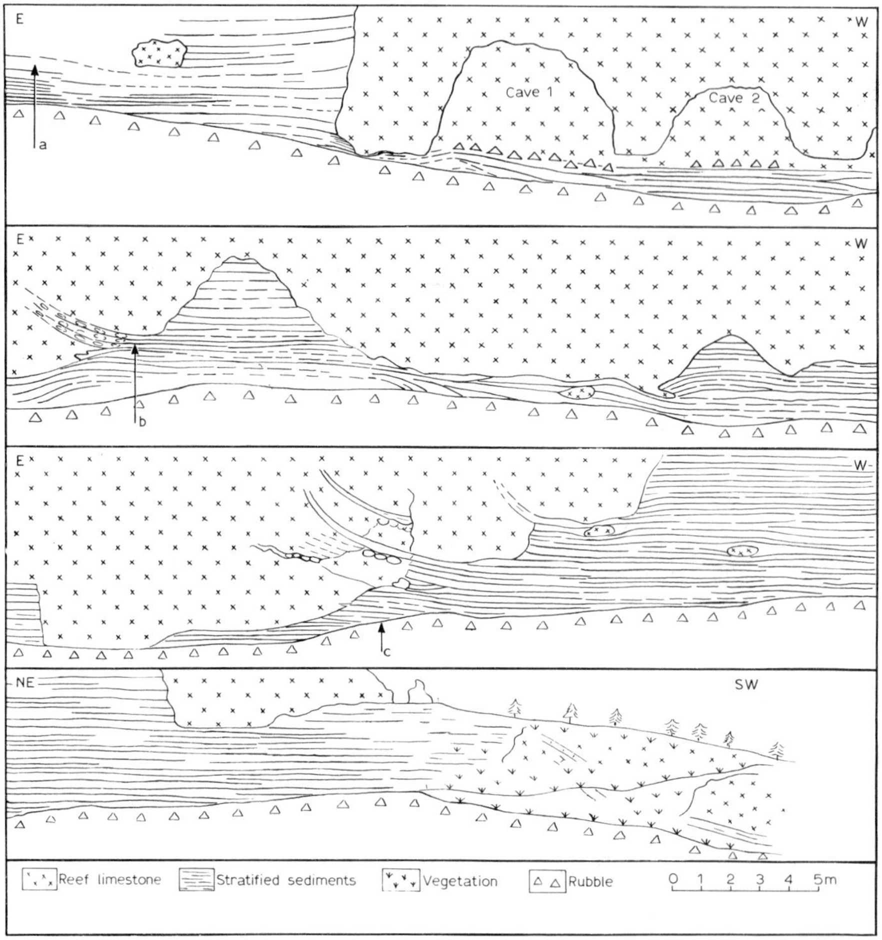

Fig.70. The lower part of the Hallshukklint, southeast of the lighthouse. a, b and c indicate the locations of Fig.72, Fig.65 and Fig.70, respectively. Note that the boundary between stratified limestone and reef limestone is generally more distinct at the east side of the reef than at the west side. At the west side, layers of stratified limestone may penetrate into the reef limestone. Högklint Beds. (After Manten, 1962, fig.21.)

very steep and rather straight; in such instances one likely is dealing with an eroded reef edge against which the crinoid limestone was deposited.

In general, the crinoid limestone with reef debris is widely variable in composition, texture and structure where reefs lie closely together.

Although the crinoid limestones present abundant evidence of current or wave action in the form of displaced corals and stromatoporoids, the degree of rounding and sorting is usually low, though both are seen. The impression gained, therefore, is that except for in some narrow passages

(Text continues on p. 168)

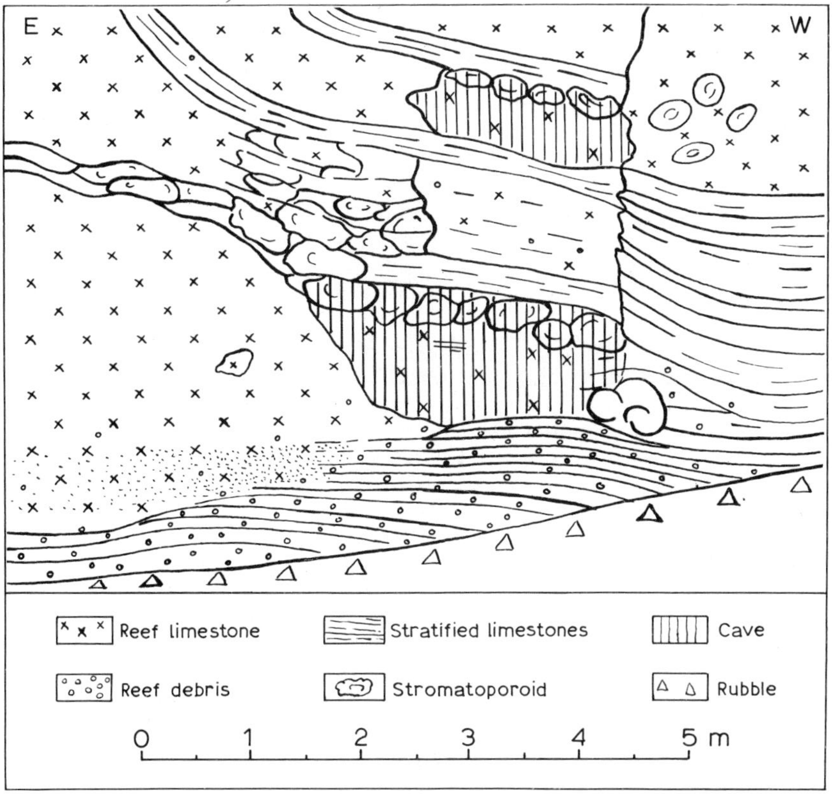

Fig.71. Detailed section of the contact between stratified limestone and reef limestone at c in Fig.70. Hallshukklint. Högklint Beds. At the base, stratified limestone is seen, containing reef debris. In the east, it is overlaid by reef limestone. At about 0.2 m above the stratified limestone, the reef limestone shows a deviating horizon, with fewer and smaller stromatoporoids (6 colonies per square metre, and on the average 15 cm long), in rather a marly matrix with abundant small (about 1 cm) fossils and fossil fragments. At the base of this horizon the fossils are somewhat larger, but fewer in number. Thin, finely-crystalline layers occur there, which can be followed in the stratified limestone further west, where they form the bottom of the lower cave. Towards the east this horizon is already indistinct at the left margin of the drawing, and it disappears about 5 m east of the cave. Upwards, the horizon passes into normal reef limestone with rather a brecciated structure.

At the top of the cave, there is a horizon with rather large stromatoporoids. It is overlaid in the east by a comparable rock, which at its westward side ends against a reef-debris like deposit. The latter, in its turn, is distinctly bounded against the normal stratified limestone to the west. Eastwards, the conglomeratic stromatoporoid limestone passes into brecciated reef limestone.

The upper cave shows a layer of finely-crystalline stratified limestone both at its base and top; they penetrate into the reef limestone. (After Manten, 1962, fig.24.)

between neighbouring reefs, the crinoid gravels associated with the reefs were not subjected to continual washing backwards and forwards.

Close to the reefs at the coastward side, the crinoid limestone tends to be more marly than elsewhere around the reefs and marl may be found on the bedding planes there.

In some sheltered basins between a number of neighbouring reef-growth centres the crinoid remains and reef debris were relatively little influenced by water movement. The deposits found there consequently strongly contrast to those found in narrow passages. In such sheltered places even intact crinoid calyces may be found. An example of such a deposit is in the southern cliff of Brissund (Fig.140).

While discussing above the alternation of reef expansion and retreat, it was pointed out that crinoid limestone can be found alternating with reef talus and reef limestone in one vertical succession.

An increase in the content of crinoid remains and especially of reef debris in a vertical section through a succession of stratified sediments, is a reflection of expansion of a nearby reef. Examples of such a situation are the exposures in the Hemse Beds close to the crossing of the three roads to Etelhem, Garde and Lye (p. 368).The expansion of a reef, as reflected in this manner, can be either laterally or upwards. Without other exposures or indications about reef behaviour, it is difficult to establish which of the two reasons caused the higher debris deposition in a certain locality, although generally a combination of the two may be supposed.

Since there is usually a gradual transition in character from the limestones at the side of a reef to the stratified limestone on top of a reef, a few words may also be said here about the latter deposit. A good example of crinoid limestone overlying reef limestone is the "Hoburg marble" exposed on the plateau of Storburg, Hoburgen; reef debris can be observed immediately over the reefs there, indicating that up to the moment that the entire reefs were buried, wave action worked loose parts of the reef surfaces, which were then embedded in younger sediments. The size of the debris material is much smaller than in the talus directly surrounding the reefs of Hoburgen. Within a vertical distance of only a few decimetres, the size of the debris decreases rapidly, as well as the quantity. The sediment about 30–50 cm above the top of the reefs is a nearly normal crinoid breccia or limestone.

In several cases, it can be seen that the uppermost lateral layers and also some higher layers arch over the reef. This may be due to a stronger settling of the stratified sediments than of the unstratified reef limestones. However, the phenomenon may also be partly synsedimentary. The dead reef formed an elevation on the sea bottom and may have been covered by the younger layers as with a blanket.

In places where the sedimentary complex was cut by cliff walls in Quaternary times, the heavier reef limestone may have caused thrust phenomena in its surrounding sediments. An example of this is shown in Fig.139.

With increasing distance from the reefs, the crinoid limestone generally passes gradually into stratified limestone which is often finely crystalline and not very rich in fossils; the thickness of the layers varies

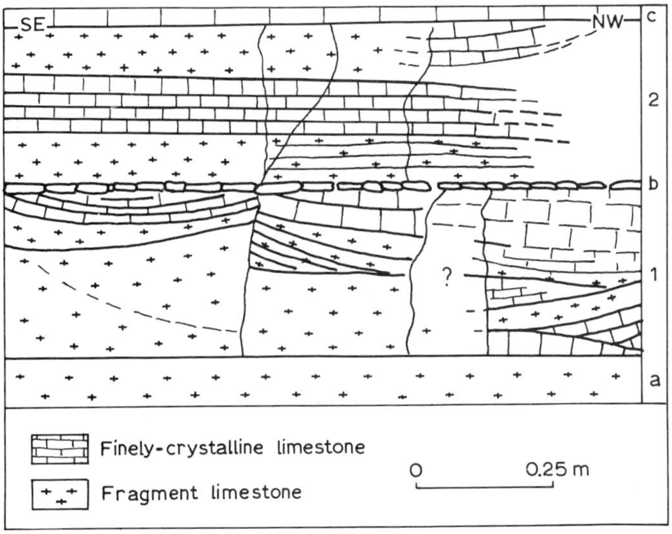

Fig.72. Relationships between finely-crystalline or micro-crystalline stratified limestone and fragment limestone at a in Fig.70. Hallshukklint. Högklint Beds. At the base, fragment limestone (a) occurs. This is overlaid by two mixed layers (1, 2) separated by a thin, crumbled layer of very finely-crystalline limestone (b). At the top, a layer of very finely-crystalline limestone occurs again. The two mixed layers show cross-bedding. Almost all lamellae of these layers differ in appearance, varying from coarse fragment limestone with fossils and fossil fragments up to 2 cm large to micro-crystalline limestone. Macroscopically, the latter shows hardly any or completely no fossil remains, but microscopically it is found to contain some small fragments, and traces of microcoquina, completely recrystallized. About 10 m northwest of this detailed section, a large reef-limestone mass occurs. Only about 4 m northwest of the section, a small reef-limestone exposure is found (cf. Fig.70). This suggests that reef limestone is also present directly behind this section, within theklint. (After Manten, 1962, fig.22.)

(Fig.72, 73). In several localities, films or thin layers of marl are found between the limestone layers.

The picture is somewhat more complicated when several reefs together form a kind of barrier (Fig.74). In the passages between the reefs, coarse, well-worn calcareous fragments were often deposited, especially where the passage was relatively narrow, on the order of 0.1 km or less. Such a limestone deposit in a passage is usually practically free from clay and the calcium carbonate content may exceed 99%. The layers are thick and the bedding planes are rugged.

The limestone at the seaward side of the reefs is over some distance of the same character as that in the passages, but is less even-grained. It passes into stratified limestone showing no reef influence.

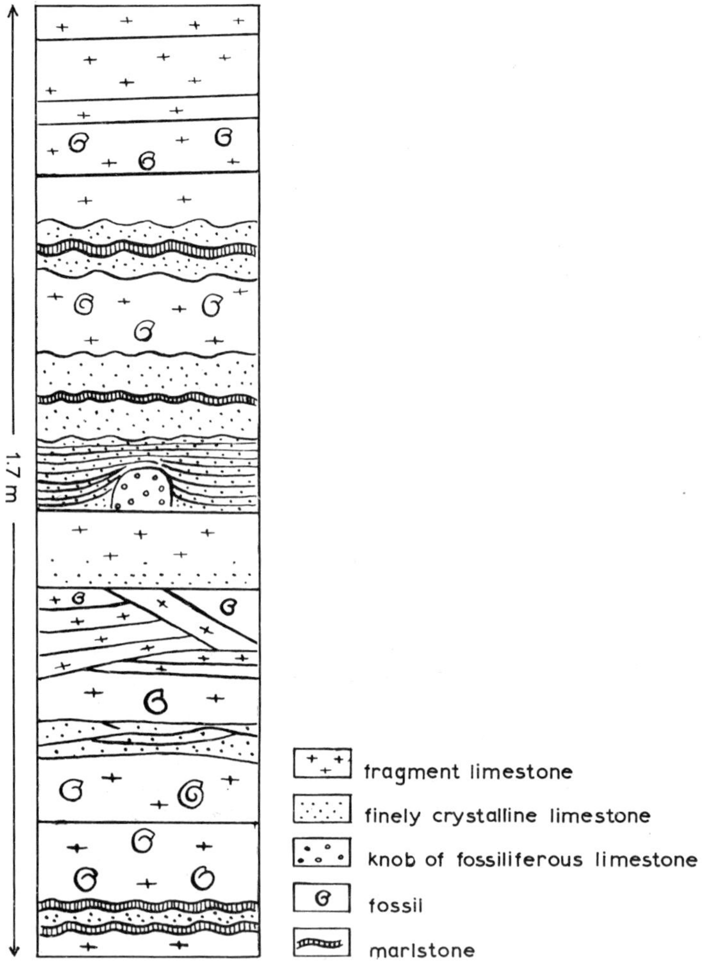

Fig.73. Detailed section, showing the character of the stratigraphical succession found at a in Fig.63. Hallshukklint. Högklint Beds. Coarse and generally fossiliferous fragment limestone alternates with finely-crystalline to micro-crystalline limestone which is less fossiliferous. The finely-crystalline limestone in the middle, encloses a knob of very fossiliferous limestone, which apparently has been deposited there as a whole. It may have been washed off a reef which is still hidden behind the façade of stratified sediment. Note that several bedding planes of the finely-crystalline limestone are rugged; also that marl is generally found in connection with the fine limestone.

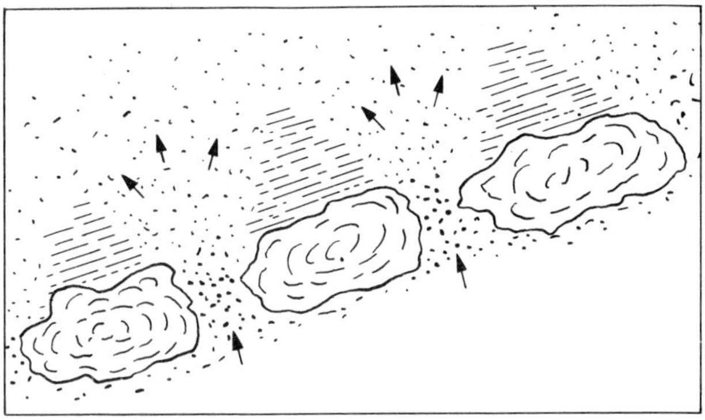

Fig.74. Schematic sketch of the distribution of fragments and marly sediments at the reefs of Smöjen, 9 km east-northeast of Slite, as established by core drillings. Slite Beds. The reef bodies, which are drawn, measure 0.2–0.3 km in diameter and the intervals between them are less than 0.1 km. The limestone complex in which the reefs are enclosed, overlies marly shales. The change from marlstone to limestone deposition was caused by a shallowing of the water. Together with other reefs, those which are drawn form a barrier, in a northeast – southwest direction. The distribution of sediments around the reefs is more regular than at Kappelshamn (Fig.75). In the passages between the reefs, coarse, well-worn calcareous fragments were deposited; crinoids, bryozoans, corals, calcareous Algae, and shells included. On the northwestern (lee) side, behind the reefs, marly layers are found with interbedded limestone layers. In the extensive field at the northwest side (inside) of the reef barrier, there is a moderately coarse or fine fragment limestone which here and there is marly. (After Hadding, 1956, fig.1).

Directly behind the reefs at the landward side, marly layers may be found with interbedded limestone layers, but most of the area shows moderately coarse to fine fragment limestone, which here and there is marly. The stratification is more marked and the layers are thinner than in the coarse limestone between the reefs.

The situation of several reefs occurring in an area together, but randomly distributed, is even still more complicated (Fig.75). These reefs caused irregular current action and consequently a more irregular deposition of coarse and fine material.

In a number of localities with reefs a hundred metres or more apart, thick layers of medium to fine material occur between thinner layers of finely crystalline limestone. In a few cases, the thicker layer shows an internal current bedding (Fig.76).

The stratified sediments between neighbouring reefs take more and more the character of those in reefless areas as the distance between the reefs increases. Stratified sediments deposited at greater distance from the reefs and showing little or no reef influence will be described in Chapter XI.

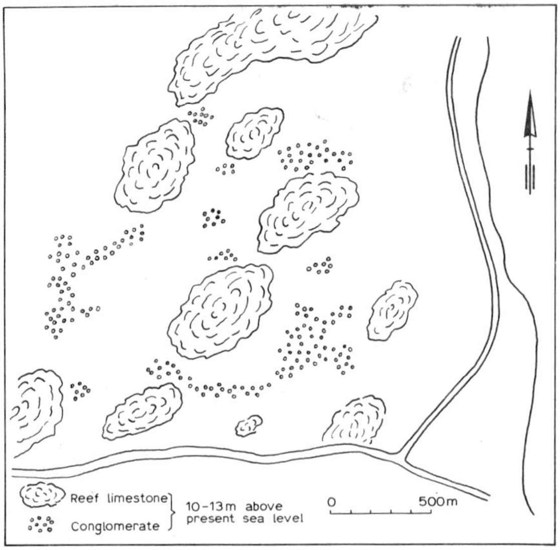

Fig.75. Occurrence of reefs and conglomerates at Kappelshamn. Högklint Beds. Due to the random distribution of the reefs, current action was complicated, and the distribution of coarse and fine material was, therefore, more irregular than with reefs occurring in a single row (Fig.74). The conglomerates, whose distribution has been mapped, appear as bands of rather insignificant thickness in layers of fragment limestone. These bands also occur elsewhere and approximately at the same stratigraphical level in the Upper Högklint Beds. Their formation may be connected with a general elevation of land. (After Hadding, 1956, fig.4.)

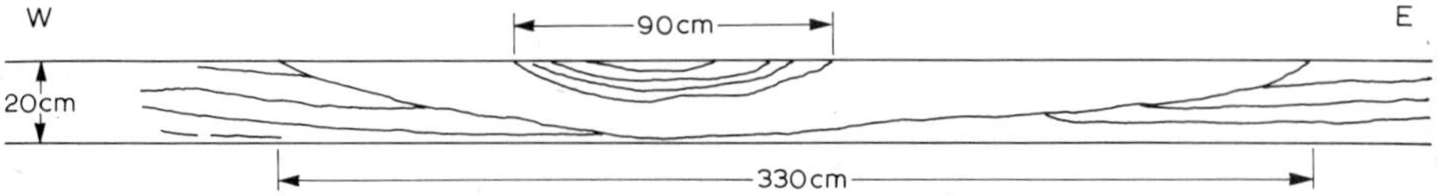

Fig.76. Current bedding within a thicker layer of fragment limestone in an inter-reef deposit. Such a thicker layer is generally interstratified between thinner layers of finely-crystalline limestone. Sections of this character are found in some areas where Högklint reefs occur some hundred or more metres apart. Towards the reefs, the sediments change to crinoid limestone with reef debris. The above example was observed east of Ihrevik. It is found at about the same stratigraphical level as the conglomerate bands mentioned in Fig.75. (After Manten, 1962, fig.27.)

Distinction between stratified sediments and reef limestone

Also something must be said here about the degree of difference between the reef limestone and the normal succession of stratified sediments. In order to make this point clear, the sediments underneath the reefs will be shortly considered again as well.

It has been noted already that sediments underlying the root of a Hoburgen-type reef generally do not differ much from those found elsewhere at about the same stratigraphical level. This is in contrast to what was found for the Visby Beds, where in the deeper-water marly Lower Visby Beds no reefs are found, and where in the Upper Visby Beds distinct local improvements of the environment often seem to have preceded reef growth. Apparently conditions for reef development became better in shallower water with limited deposition of terrigenous debris. In Högklint, Slite or Hemse times, a lesser change in environmental conditions was needed to lead to reef formation.

Once colonial corals or stromatoporoids had established and built a platform for further reef development, other organisms apparently found conditions there more favourable than on the surrounding sea floor. The result is that there is a distinct difference between reef and stratified limestones in most stratigraphical units in Gotland.

It has also been noted already that the difference between stratified and reef limestone in the Klinteberg Beds is less than elsewhere in Gotland. In parts of the Klinteberg, the stratified limestones are very rich in coral colonies, often of large size and generally in their positions of growth. The stratified limestone between Hästings and Hallbjens, northeast of Guldrupe Church, is very rich in stromatoporoids and Algae. East of Bringes (southwest of the former railway station at Bjerges, Vänge Parish), a zone of marly limestone about 1 m thick is found, which is very rich in stromatoporoids, bryozoans and corals; it may even be somewhat reef-like in appearance, but is distinctly bedded, generally with layers of 2–10 cm thick. The overlying finely oolitic limestone and the underlying *Spongiostroma* limestone are also relatively fossiliferous. Between Krasse and Västerby (Guldrupe Parish), north and south of the (former) Buttle Station, northwest of Ovre Lundsmyr, southeast of Hejde Church, and in other localities, the stratified Klinteberg Beds also approach a reef-like character, without loosing their stratified nature, due to a relative increase in corals, stromatoporoids, bryozoans and crinoids.

The general impression is that in the shallow sea of Klinteberg times, conditions were favourable in many places for the development of a rich and varied organic life. Where conditions were relatively more favourable, potential reef builders became more and more common in the stratified sediments; these sometimes gradually developed a semi-reef retaining a rude stratification, and sometimes passed finally into a true unstratified reef. Where conditions were most favourable, the reef builders in a very early stage developed an unstratified reef surrounded by crinoid limestone with reef debris. All intermediate stages between general stratified limestone and unstratified reef limestone can thus be encountered in the Klinteberg Beds.

The boundary between reef and stratified limestone

After comparing in the field a number of exposures of reefs and reef-surrounding sediments in Gotland, one usually experiences no difficulties in drawing the boundary between reef limestone and stratified sediments. In a few instances, however, there may be uncertainties.

In the first place, there may be doubts in some cases where the exposed cliff walls are very steep and boundaries have to be determined from a distance. The sediment surrounding the reef is not always very distinctly stratified close to a reef and, on the other hand, intercalations of stratified sediment, tabular reef builders and cracks caused by weathering may give the reef limestone a somewhat stratified appearance. As soon as close observations can be made, the problem is usually less great.

Secondly, the boundary may be difficult to fix in older and partly weathered outcrops, particularly in cases where there seems to be little difference in composition between reef limestone and stratified limestone. An example of such a situation is the weathered reef mass (h) in the western cliff of the Storburg of Hoburgen, north of the overhanging part which is called Hoburgsgubbens Trappa (c in Fig.77). The southern boundary, which is in the lower part of the cliff with stratified limestone at the other side and at the top of the cliff with another reef, is still reasonably traceable. But in the north, the reef limestone gradually passes into a thick-bedded limestone. The peripheral part of the reef is not as fossiliferous as reef limestone normally is, and among the fossils present are many crinoids. The adjacent stratified sediment contains several fossilized potential reef builders. The stratification does not end abruptly, but gradually fades away. Nevertheless, this is virtually the only indication of where the boundary has to be sought.

Locally in some exposures, especially in the Hemse Beds, where reef limestone gradually passes into crinoid limestone with reef debris, the exact boundary between the two is sometimes difficult to determine very precisely; nevertheless, it can generally be established to within one or a few decimetres in larger exposures. The distinction sometimes remains a problem where only small and scattered exposures are available, e.g., as in the Ausarveklint (Hemse Beds), where in some small outcrops stromatoporoids dominate but are embedded in vaguely stratified limestone rich in crinoid remains. Without knowing the nature of the sediments around the exposure, it is hardly possible to tell at which side of the boundary between reef and surrounding sediments the exposed rock was formed.

The lower boundary of a reef sometimes is indistinct, either because the underlying sediment is also very fossiliferous, or because the lowermost unstratified part with the appearance of reef limestone shows too few reef builders to account for the different nature of the rock by their frame-building activity. Thin sections can then help to fix the boundary. In such detailed sections, the fossiliferous limestone underneath a reef clearly demonstrates how the sediment was compacted under a somewhat irregular reef base. The matrix has often been distinctly squeezed between the solid fossils and fossil fragments, and individual particles are usually closely sutured into their neighbours. In such sections, the lower reef boundary can then be fixed at the place where the fossils start to be embedded in a structureless matrix which shows less indication of great compaction and where the individual rock components show no suturing.

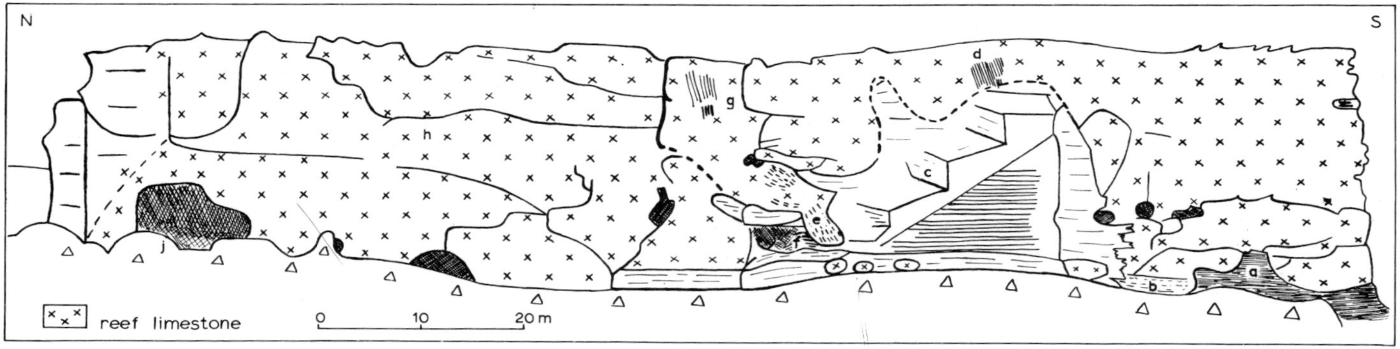

Fig.77. The central part of the west wall of Storburg, Hoburgen. In the north and south, reef limestone; in between these, stratified limestone is exposed in an overhanging part of the wall, reminiscent of the underside of a staircase and therefore, in accordance with other phenomena in Hoburgen, called here "Hoburgsgubbens Trappa". a = stratified clayish marlstone; b = stratified marly sediment, probably deposited very close to an early stage of reef growth, the rock is very fossiliferous and among the fossils present, there are many thin and slightly wavy stromatoporoid covers, overlying it is some reef debris, followed by about 22 m of reef limestone; c = stratified limestone, forming "Hoburgsgubbens Trappa"; d = small occurrence of reef debris, containing many flat stromatoporoids which all dip very steeply in the same direction, suggesting that they form part of a block of reef limestone which loosened as a whole from the reef; e = reef limestone, or more probably reef talus, very rich in strongly-dipping stromatoporoids; f = cave in which a reef root is exposed, north and in the roof of the cave, stratified limestone is found; g = small occurrence of steeply-dipping reef talus; h = reef limestone, which in the north shows no distinct boundary with thick-bedded limestone; at the base of the reef some crinoid limestone with generally small and thin stromatoporoids is exposed; in the lower part of the reef limestone, the matrix is very marly and includes several marl pockets and lenses; j = cave, above which the reef contains a substantial part which is built almost exclusively by the coral *Acervularia*.

Another aspect of the boundary between reef limestone and surrounding sediment that deserves to be mentioned here is the occurrence in some places of sharp boundaries between massive and stratified sediment where it seems there is very little else to distinguish the reef from the neighbouring stratified rock. The peripheral part of such reefs may contain only a few reef builders, certainly not enough to make it likely that together they had once formed a continuous rigid framework. Sometimes crinoid remains are even the most prominent fossils in the reef limestone. Nevertheless, the unstratified peripheral reef rock seems to have been practically incompactable, while the adjacent stratified limestone is distinctly compacted. What determined the difference in behaviour of the rocks?

Newell et al. (1953, p.82, fig.39H), in an attempt to explain similar sharp boundaries found by them in the Permian reef complex in southern North America, supposed that compaction of unconsolidated sediments around the reefs resulted in a lateral squeezing out of some sediment in the direction of the reef, where compaction was less.[1] This would explain the apparent rigidity and resistance to compaction of the massive lateral reef limestone as well.

The present author agrees with Colter (1957, p.54) that this mechanism seems unlikely to explain all sharp lateral boundaries of the nature as described above. Locally in Gotland, the limestone and especially the thinner interbedded marlstone layers which may neighbour the massive rock contain pellet-like bodies and ostracodes with valves intact, and it is unlikely that either of these would survive a squeezing process. In agreement with this no such pellet-like bodies or intact ostracodes have been observed in sediments underneath reef limestone which have been subjected to squeezing out in directions away from the reef centre.

SYNTHESIS

It will be apparent from the data presented in this chapter about the Hoburgen reef type, that the reefs must have developed in shallow water. Water depth was less than that in which the Upper Visby reefs formed. The following data support this conclusion:

(1) The stratified sediments in which the reefs are intercalated are limestones and marly limestones, but no marlstone. Directly around the reefs, crinoid limestones are generally present, sometimes showing cross-bedding, rounding, or other shallow-water characteristics; a talus mantle may separate reef and crinoid limestone and the latter almost always contains amounts of reef debris.

(2) The Högklint Beds present abundant evidence of decreasing water depth during their time of deposition; they contain many Hoburgen-type reefs and these beds conformably overlie the Upper Visby Beds.

(3) Algae are fairly common in the Hoburgen-type reefs; stromatoporoids, not corals, are the dominant reef builders.

---

[1]This process thus affected the stratified sediments aside of the reefs and should not be confused with the squeezing out of stratified sediment from underneath a reef, due to the heavy burden placed upon it (cf. p. 156 and Fig.63, 64).

(4) Erosion of some reef surfaces has taken place. The common occurrence of patch reefs suggests that shoaling in several instances prevented further upward expansion.

When compared to the Upper Visby reefs, the reefs of Hoburgen type developed under generally much more favourable conditions. This is apparent from the distinctly lesser deposition of terrigenous sediment, the larger size of the reefs and the much greater diversity in organic composition.

In their distribution the reefs show relationships to the contemporaneous coast line. Although they often present an irregular distribution within a certain zone (Fig.75), the orientation of that zone is parallel to what on the basis of other information may be assumed to have been the direction of the shore line. Moreover, the orientation of the individual reefs shows the longer axis to have that same direction (cf. also Chapter XI, e.g., pp. 289, 322, 323, 366, 407, and the enclosed geological map of Gotland). Where reefs occurred very closely together, the seaward ones developed under more favourable conditions than their close neighbours on the landward side.

That the width of the reef zones often amounted to several kilometres supports the belief that the Silurian basin was a large, flat and shallow-bottomed sea, where comparable conditions prevailed over extensive areas.

The distribution and size of the reefs, combined with influences exerted by wind and current directions, determined to some extent the depositional pattern of the stratified sediments found around the reefs, but not to such a strong degree as is the case around the large reefs of the present day. This subject will be somewhat further touched on when speaking about the reef debris in Chapter IX.

Chapter VIII

THE HOLMHÄLLAR REEF TYPE

DISTRIBUTION OF HOLMHÄLLAR-TYPE REEFS

Along the eastern shore of the southern peninsula of Gotland reef limestones are found that differ in character from those described in the two preceding chapters (cf. Table VIII). They are predominantly exposed in picturesque erosion-forms, the so-called "raukar" (singular "rauk") or stone giants (Fig.78). Just like the reef limestones of the Hoburgen type, these reef limestones have a greater resistance to erosion than the surrounding stratified sediments. The latter, for the most part, have disappeared, leaving the reef limestones as promontories or small off-shore islands. The best outcrops are found at Holmhällar, near Austre, on the island of Heliholm, at Hammarshagahällar and along the southeastern shore of Faludden (see also Chapter XI). All of these belong to the Sundre limestone, as defined by Hede (1921). The Sundre limestone is included by the present author in the Hamra-Sundre Beds.

Furthermore, reef limestones of similar character also occur in the Hemse Beds, in the raukar fields at Ljugarn, Fågelhammar South and North,

Fig.78. Rauk (stone pillar, stone giant), consisting of reef limestone of Holmhällar type. Heliholm. Hamra-Sundre Beds.

and Sjausterhammar (see also Chapter XI). Reef limestones which show a tendency towards a transition from the Holmhällar type to the Hoburgen type are found between Snabben and Sandviken (Östergarn Parish) (cf. pp.183, 372, 374). These also belong to the Hemse Beds. All localities mentioned are situated along the east coast of Gotland.

In older publications the sediments from the above-mentioned localities have been taken together with some other deposits elsewhere in Gotland, which show reef limestones of Hoburgen type, under the names of *Ascoceras* limestone or Etelhems limestone (Munthe, 1910). The resemblance between these Hoburgen-type reef limestones and the reef limestones in the raukar fields named above is mainly in the common occurrence of red colours. After Hede (1921) had shown that the *Ascoceras* limestones belong to stratigraphically separate units, Munthe (1921b) introduced the more neutral name of red-brown marble reef limestone, or, in shorter form, marble reef limestone. The present author is of the opinion that there is no need for these names, which may even cause confusion, because their usage may still wrongly suggest a strong community of, e.g., character, palaeoecology and/or age. There is no need whatsoever for a special name, even if only adopted for descriptive purposes, for reef limestones which correspond in a characteristic of so subordinate importance as a common red colour (cf. also p.188). Therefore, these names will not be used in this book.

Because the present author started his investigations in southern Gotland, Holmhällar was the first reef of this type that he intensively studied (see also Rutten, 1958). Later field work by him confirmed that the raukar field at Holmhällar was one of the best exposures of this kind and, therefore, there is no objection to making Holmhällar the type locality of this reef type.

The accompanying map (inserted loose at the end of this book) gives the distribution of the raukar over this locality. In the southeast the raukar rise up to 3.5 m above present sea level; in the west the highest raukar reach up to about 6.5 m above sea level. A few raukar found scattered in both the west and northeast have been omitted, as they are strongly weathered and overgrown by lichens and did not allow such detailed studies as were carried out in the mapped part.

Several of the characteristics of the Holmhällar reef will be discussed in the following pages.

FAUNA, FLORA AND MATRIX OF THE REEFS

Reef-forming components

The Holmhällar-type reef limestone can be characterized, in short, as a pure, compact and hard limestone, which is unstratified, has a greenish grey to reddish brown colour, is very rich in large stromatoporoids, which are generally strongly recrystallized, and is of a relatively homogeneous nature. On the average, the reef limestone consists to the extent of approximately 70% of the remains of animal reef builders and reef dwellers, with 30% composed of a calcareous matrix and calcareous Algae. Among the reef-building animals stromatoporoids are most evident; they produced approximately 60% of the total rock volume. The reefs, therefore, are stromatoporoid reefs. Other animal reef builders played a very minor part. Corals take the second place, with an average of only about 3%. Almost nowhere do bryozoans show high percentages. Moreover, when found they are often only fragmental and are dispersed among the other fossils.

In the matrix remains of several other animal groups may be observed. Among these, cephalopods are quite common. Gastropods are less well represented. The same is generally true for brachiopods and lamellibranchs. Whereas the cephalopods are obviously more common (cf. p.120), particularly the latter two groups are apparently less common in these reefs than they are in reefs of the Hoburgen type. Trilobites, too, are only scarcely represented. The reefs show only slight evidence of destructive work by boring organisms. Except for the pronounced recrystallization, little seems to have been altered after the death of the reef builders and before Quaternary erosion began.

In comparison with the fauna of the Hoburgen-type reefs, the Holmhällar reefs appear to be considerably poorer in species (see Table IX). This may partly be due to the fact that fewer reefs of this type are exposed and that their massive construction and marked recrystallization greatly hinder the collecting of fossils. But it is the impression of the present author that primary circumstances have also played a part in this.

Calcareous Algae are well represented in the Holmhällar and similar reefs. Macroscopically, however, it is generally very difficult or even impossible to distinguish them from the calcareous matrix, as Hadding (1950) also experienced.

A striking difference as compared to the Hoburgen reefs is the great regularity in organic composition of the Holmhällar reefs. This aspect will be elucidated in the subsequent paragraphs of this chapter.

Method of inventarization

It is difficult to give exact data about the percentage volumetric composition of the reef limestone for a great number of observation points, especially if these data have to be collected in the field. Nevertheless, it was important to have more information about the distribution of the stromatoporoids, corals, bryozoans, crinoids, and calcareous matrix and Algae throughout the reef. Therefore, an estimation method was designed by which the quantities of the various components were indicated by means of the figures 1–7. The rough definitions pertaining to these figures are given in Table XII. Estimations were made in Holmhällar in 300 observation points, which are marked on the accompanying map; and where this appeared useful, one or even more additional observations have been added afterwards. Although the figures obtained are rather subjective in character, in the field they supplied useful information. As a follow-up, it was decided to look for a more exact basis for the estimations that had been made. This has been done by careful mapping of 25 reef-limestone surfaces, 1/4–1/12 m^2 large, on millimetre paper, with a scale varying between 1:5 and 1:1. By calculating the total surface taken up by each of the reef-limestone components which have been distinguished, percentages could be fixed for their distribution over these two-dimensional parts. These have been compared with the estimation figures (cf. column one in Table XII) listed earlier for the same parts. The results are given in the columns three and four of Table XII. With the aid of the average values in column four, the graphs in Fig.79–82 have been drawn. As a result of the way in which the graphs were constructed,

TABLE XII

Estimation method for the composition of Holmhällar-type reefs

Quantification figure	Definition	Percentages	Average percentage
	For stromatoporoids, corals or bryozoans:		
1	reef limestone almost exclusively built up by the fossil group in question.	76–96	84
1–2	. . . .	68–81	73
2	reef limestone built up for a large part by the fossils in question	51–63	57
2–3	. . . .	48	48
3	number of colonies rather large, but almost all of these are separated by the matrix	26–42	31
4	several colonies present, but not very common	21	21
5	only a limited number of colonies	3–14	5
6	just one or a few colonies.	0.2–8	3
7	not observed	–	–
	For crinoids:		
1	extremely numerous	26–72	42
2	numerous	17–29	23
3	very common.	12–14	13
4	mainly concentrated in a number of "pockets" but also dispersed in the matrix	6	6
5	only a few "pockets", also scattered fragments.	3	3
6	only scattered in the matrix	0.4–2	0.7
7	not observed	0–0.2	–
	For calcareous matrix and calcareous Algae together:		
1	constituting a very large portion of the reef limestone	79–87	83
1–2	. . . .	66–79	71
2	abundant throughout the limestone	58	58
2–3	. . . .	–	45(?)
3	mainly concentrated in thin films between the animal reef builders and in a few other places.	21–39	30
4	not much, only locally more common	–	25(?)
5	rather small amounts	6–11	8
6	very small amounts.	3–9	5
7	almost absent.	0–6	2

they are not very exact but nevertheless give a reasonable picture of the variations which occur in the distribution of the main reef builders and the matrix over the reef limestone.

Stromatoporoids

As has already been stated, stromatoporoids are by far the most important constituent of the Holmhällar reefs. They generally occur close to each other, separated only by thin layers of calcareous matrix or Algae. The shapes of the individual colonies vary: some are more or less spheroidal or dome-shaped. Other colonies are firmly united so that the boundaries between them are indistinct. Both types may be large. In other reef portions most colonies are tabular; this may have led locally to a kind of pseudo-stratification. A great diversity of other stromatoporoid forms also occurs, some of which are quite extreme. A few random examples are drawn in Fig.226.

Several observations on the palaeoecology of stromatoporoids are presented in Chapter XII. Included there are also data collected in Holmhällar and related reefs.

From the graphs in Fig.79–82, it can be seen that the percentage of reef limestone consisting of stromatoporoids distinctly decreases from the central part of the reef towards the ends of the crescent. Near the margins of the reef, moreover, more colonies show a tabular shape and are smaller in size than in the centre, where huge specimens are found. Aberrant growth forms, too, are more common in the marginal reef portions. In those areas where many tabular colonies cause a vague stratification, as in some places near the margins of the Ljugarn reef, the colonies at the original seaward side can sometimes be seen dipping moderately reef-downwards.

Most of the stromatoporoids in the Holmhällar-type reefs are pronouncedly recrystallized. This has made reef portions with a high stromatoporoid content more resistant to erosion. As a result of this, most raukar have a pedestal of rock which is extremely rich in stromatoporoids. In higher portions, which suffered comparatively less from erosion - since they are not attacked by the present wave action - similar reef limestone may occur, but there less-massive reef limestone is also preserved locally. Such less-massive portions were undoubtedly also present lower down, but are now eroded. The impression that the stromatoporoid content is higher in the lower portions of the reef than in the upper ones consequently is thought to be due to the Recent influence of selective demolition.

The colour of the stromatoporoids on a fresh fracture is usually whitish to light grey. In recrystallization the latilaminar structure has often been preserved, and on weathering the colony may break along the latilamination planes. Locally, this leads to a somewhat crumbly demolition, especially where the reef limestone is no longer affected by wave action.

The preponderance of large stromatoporoids in the reefs of Holmhällar type is fairly general, with only one regional exception. Between Snabben and Sandviken, along the east coast of central Gotland, Östergarn Parish, the remains of several reefs are found. Although many of the reefs in this region show large and even very large stromatoporoid colonies, part of the reef limestone does not give the same impression of compact stromatoporoid rock as does the Holmhällar reef. There may be more smaller colonies, the

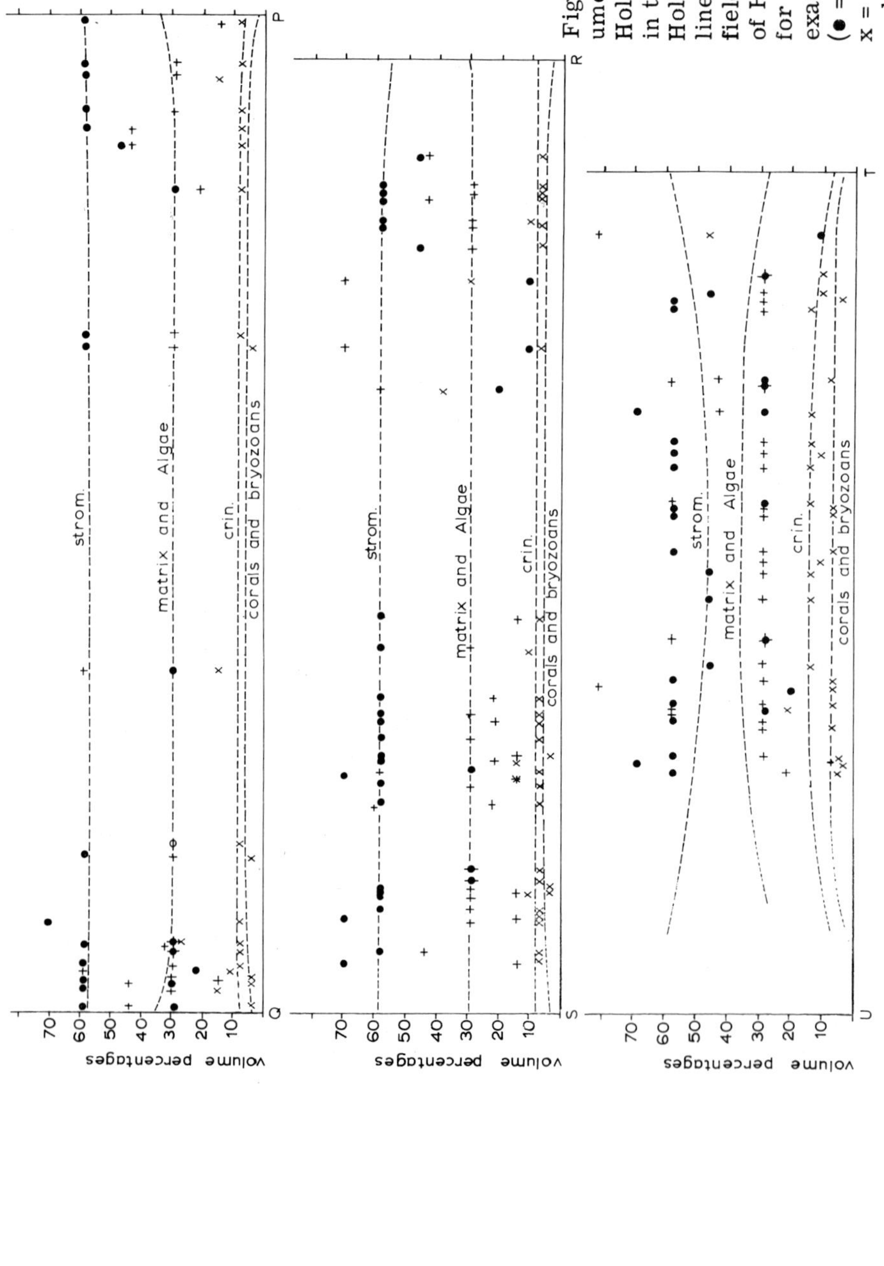

Fig.79. Approximate volumetric composition of the Holmhällar reef limestone in the central part of Holmhällar, along three lines across the raukar field (see the enclosed map of Holmhällar). The curve for the crinoid remains is exaggerated by a factor two. (● = stromatoporoids, x = crinoids, + = matrix and Algae).

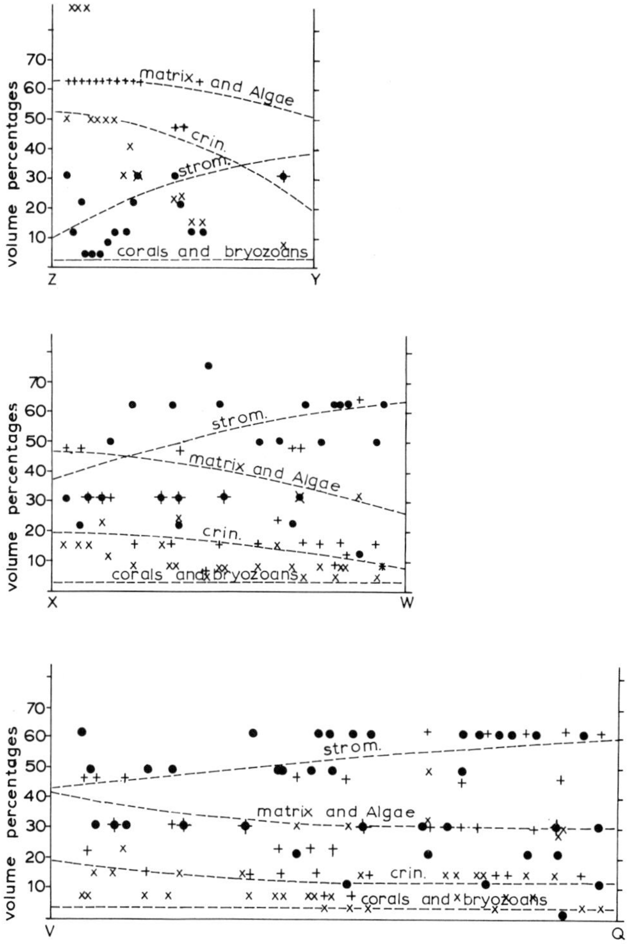

Fig.80. Approximate volumetric composition of the Holmhällar reef limestone in the western part of Holmhällar, along three lines across the raukar field (see the enclosed map of Holmhällar). The curve for the crinoid remains is exaggerated by a factor two.

---

number of tabular stromatoporoids may be larger and some may even be of laminar character; also the matrix percentage may be higher. As mentioned in the beginning of this chapter, in this region there seems to be some tendency towards a transition to reef limestone of Hoburgen type.

Corals

Corals do not, in general, constitute an important part of the reef limestone. Solitary corals were only found at little more than half of the observation points. Although compound corals are somewhat more common,

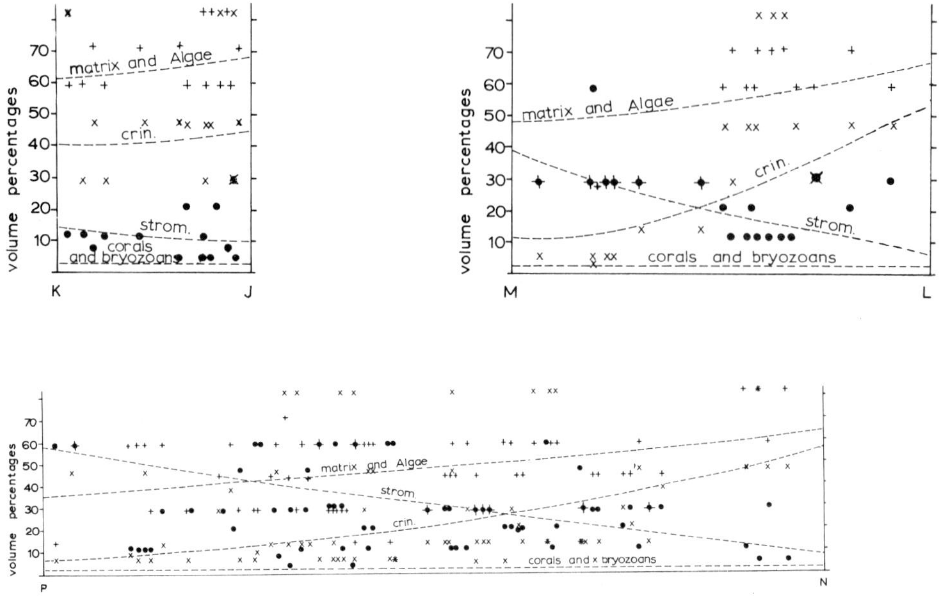

Fig.81. Approximate volumetric composition of the Holmhällar reef limestone in the northern part of Holmhällar, along three lines across the raukar field (see the enclosed map of Holmhällar). The curve for the crinoid remains is exaggerated by a factor two.

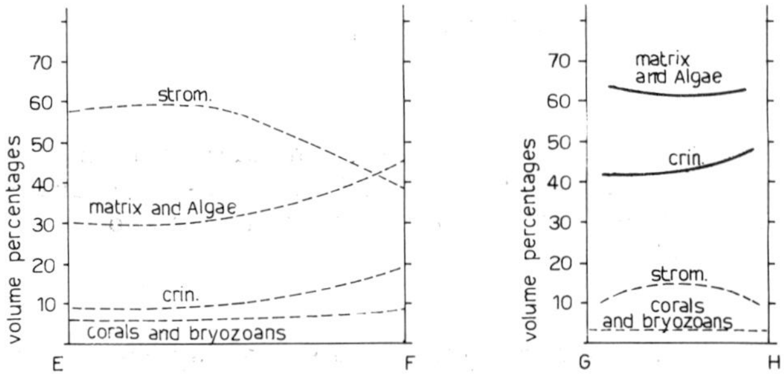

Fig.82. Approximate volumetric composition of the Holmhällar reef limestone along two lines about perpendicular to the longitudinal axis of the reef. Left: in about the central part of the reef, right: at the northern end. The curve for the crinoid remains is exaggerated by a factor two.

their total volume is only about 3% of the reef and only locally does their contribution occasionally surpass 5% of the total reef-limestone mass. Both massive and branched colonies are present. Massive colonies are relatively most common in the north of Holmhällar. Branched colonies are seldom found intact; most have disintegrated. In Holmhällar no increase in the total number of corals towards the reef margins has been established; such a tendency seems to be evident in Ljugarn, but there too, the differences between the coral percentage in the reef centre and at the margins are only slight.

A distinctly higher percentage of corals has been found in pools in the reef surface (see later in this chapter), where apparently the vitality of the stromatoporoids decreased as a result of a stronger deposition of calcareous mud.

Crinoids

Crinoid remains are anything but rare in the Holmhällar reef and in most other reefs of similar type; the smaller stem and crown fragments are particularly common. However, some large stem fragments and root remains have also been observed, embedded in calcareous mud. Generally, the higher percentages in crinoid material are found near the assumed edges of the reef. As is shown in the accompanying graphs (Fig.79–82), the percentage of crinoids increases both in the north and west and in the southeast of the Holmhällar raukar area. These increases go together with an increase in matrix volume and a decrease in the total stromatoporoid volume. This is one of the arguments in favour of the theory that crinoids grew more on the reef sides and probably also in the immediate surroundings of the reef than on the upper surface of the reef (see further Chapter XII).

Large accumulations of crinoid material within the reef are found in several of the debris-filled depressions (see later in this chapter). Such portions are of a much less massive character than the reef limestone proper.

Matrix

In between the reef builders, there is a certain amount of calcareous mud. Macroscopically, this matrix is often hardly if at all distinguishable from the reef-building Algae, so that an exact percentage is difficult to give (see also the paragraph on the Algae, p.188), but an average of 10–20% of the total volume seems to be a fairly safe estimate. Close to the reef margins and in debris-filled depressions and in pools, the matrix percentage is generally much higher, locally up to as much as 70% of the reef-limestone volume.

As a rule, the matrix of the Holmhällar reef is hard and dense. In a few cases, small pockets of softer material, greenish-coloured and marly, are intercalated. In a few other cases, the cementing limestone was found to be somewhat porous.

The matrix of the Holmhällar reef occurs in two varieties: a red to reddish brown one and a greyish green one. Generally, the latter colour predominates. In the north of Holmhällar, however, the red colour is found more frequently. In most cases, the two colour varieties are clearly separated from each other. Then, in the raukar, the red variety generally overlies the greyish green one. In a few cases, especially in the north of Holmhällar, both varieties are intermingled. Nevertheless, it can also be established in many of these cases that the grey colour is more common in the lower parts of the reef limestone than in the higher parts (e.g., observation points 3, 10, 33, 37, 45). Also, if the reef limestone is predominantly grey, some red-coloured spots may be concentrated in the top portion of a rauk (e.g., observation point 31).

Presumably the red colour is mainly secondary in character, caused by an infiltration of Fe^{3+}-containing solutions. In the red-coloured reef-limestone parts, iron oxide has been deposited on the outer surfaces of the fossils and along very fine cracks, which intersect the sediment irregularly. The interior of the fossils in the reddish-brown reef-limestone variety is generally white in colour.

Locally, there is a breccia-like mixture of the red and the light-coloured variety of the Holmhällar-type reef limestone. This is interpreted as an indication that slight disturbances have taken place within the reef limestone, presumably caused by compaction. This compaction may have taken place when the reef became thicker and perhaps even more so when it became buried under younger sediments. Small slickensides, encountered in several places in the reef limestone, provide further evidence of compaction. Relatively large disturbances of reef portions are marked by the fissures which will be described later in this chapter.

It is not unlikely that a certain fraction of the calcareous matrix found in reefs of the Holmhällar type originated in situ by disintegration of the skeletons of reef builders and reef-dwelling organisms. Much of it, however, must have been stirred up by water turbulence and was redeposited afterwards, as is suggested by the distinct relation between calcareous-mud deposition and stromatoporoid growth (see also pp.197–198). This mud may have had its origin either on the reef itself or in its immediate environment. Stromatoporoids, with their massive colonies, are bad potential sources of small-grained bioclastic debris. These colonies constitute the main part of the reef. Around the reef, however, enormous amounts of crinoids grew. Their skeletons easily disintegrate post mortem into bioclastic sands and sediments which are even finer than these sands. The representatives of all other phyla were probably greatly subordinate to the crinoids as matrix-mud suppliers. It should be noted that in the graphs (Fig.79–82), there is some similarity between the curves for the matrix and the crinoid remains.

Algae

A large portion of what in the field seems to be matrix of the reef, on closer examination - especially in thin sections - appears to consist of calcareous Algae. As already stated, a distinction between true calcareous matrix and Algae is often difficult or even impossible to make. Therefore both have been taken together when estimating the composition of the reef limestone at the more than 300 observation points. More detailed examination of samples from a number of observation points revealed that in the reef centre a very large proportion of the "matrix" consists of Algae; this is also true, to a great degree, for the area close to the southeastern margin. Towards the north and west end of the reef, the percentage of Algae decreases as does that of the stromatoporoids. In pools and debris-filled depressions, most or all of the matrix consists of calcareous mud. Therefore, it may be concluded that stromatoporoid development was more strongly influenced by mud deposition than by algal growth.

Conclusions

Apart from such reef portions as pools and debris-filled depressions, which are still to be described (pp.191–196), the Holmhällar reef, and also the other reefs of this type, shows a striking regularity in organic composition, with stromatoporoids as the dominant reef builders and with corals and bryozoans playing only a very subordinate part. In Holmhällar (as well as in other similar reefs) stromatoporoid development was most vigorous in the reef centre and decreased towards the north and west flanks, in which direction mud and crinoid remains gain in importance as reef-limestone components.

SHAPE AND DIMENSIONS OF THE REEFS

On the basis of what has been said in the previous section about the distribution of fossils within the Holmhällar reef, it can be assumed that the most vigorously developing part of the reef was in the centre, that is in the southeast. Both towards the north and the west, the vigour of reef growth decreased. This is especially evidenced by the percentage of stromatoporoids in the reef limestone.

A similar distribution of different degrees of growth vigour over a reef is found in modern reefs which grow perpendicularly to the dominating wave or current direction. At both edges of such a reef deposition of debris takes place, over which the reef may expand. In this way, a reef develops which possesses the shape of a crescent or a horse-shoe. At the inner curve, growth is less, possibly due to slighter water movement and the consequently smaller supply of food and nutrient salts. The opening of the crescent is generally directed towards the coast. The dimensions of such a reef seem to be related to the depth of the water.

It should be noted that the distribution of the raukar in Holmhällar shows a crescent-shaped pattern. It is unlikely that this is due to the abrasive action of the present sea, since the southeastern part of the raukar field forms the head of a promontory and is most severely attacked by wave action.

The conclusion seems warranted, therefore, that the Holmhällar raukar field represents a fossil crescent-shaped reef, with the opening directed towards the northwest. A similar reef shape can be deduced from the raukar fields of Hammarshagahällar, north of Holmhällar, and Heliholm, south of Holmhällar, as well as from a number of raukar fields in the Hemse Beds.

One of these reefs in the Hemse Beds still shows part of the surrounding sediments. This reef lies about 1.2 km north-northeast of Sjausterhammar. There a northwest – southeast-orientated raukar zone reaches the beach, where the outer boundary of the raukar field is turned towards the south. After a few tens of metres, this outer reef-limestone boundary again retreats westwards, via a section with a northeast – southwest direction. Around the place where the raukar group reaches the present shore in the northeast, a few of the raukar show reef debris overlying the true reef limestone. The preserved debris blanket is up to a few decimetres thick and dips slightly towards the northeast. Where the reef-limestone margin in the southeast begins to retreat inland, the reef limestone is underlaid by stratified, strongly-recrystallized crinoid limestone, dipping towards the centre of the reef. Similar stratified limestone is also exposed further south of the reef. From these data, it appears that this reef possesses a semi-circular outward boundary. At the inward side, however, no stratified sediments are present, but some reef limestone very poor in crinoids is exposed. The reef thus shows a semi-circular rather than a crescent shape.

Most of the other reefs of Holmhällar type in the Hemse Beds were presumably also semi-circular or crescent-shaped, rather than elongated like the Hoburgen-type reefs.

Along the northwestern shore of Sandviken, however, some remains of reef limestone are found, which probably formed part of relatively small reefs, which were perhaps elongated reefs, roughly narrow-elliptical in plane. The exposures, unfortunately, do not permit any reliable conclusion.

It has been argued in Chapter IV, that the slight dip in the Gotlandian strata is due to syn-sedimentary tilting. Though deviations have occurred, the main direction of dip is towards the southeast. It is likely that the direction of the depth contours, and also that of the coast line, has been more or less parallel to the tilt-axis, viz. in general approximately northeast – southwest. In the southern peninsula of Gotland, reef limestone of the Holmhällar type is found from Faludden in the north to Heliholm in the south, that is, in a roughly northeast – southwest-directed belt. Combining these data, it appears very likely that the crescent-shaped reefs of Holmhällar and environment, just as more recent reefs, had their opening directed towards the coast which existed during the time of reef formation.

The original shape and orientation of the Holmhällar reef, as deduced from the detailed field analysis, are indicated on the accompanying map by means of a dashed line.

In the Hemse Beds, reefs of the Holmhällar type occur in the Snabben – Sysne-udd area in a north-northeast – south-southwest arrangement, with their openings directed roughly west-northwest. A similar orientation of the opening is found in the Ljugarn, Fågelhammar and Sjausterhammar reefs. Therefore, in the Hemse Beds too, a relationship may be assumed between the form and orientation of the Holmhällar-type reefs and the direction of the coast line at that time.

Measured in a straight line, the Holmhällar reef had a base approximately 650 m long. The greatest breadth and height were attained in the centre. The highest raukar are still found there.

A horizontal extension comparable to that of the Holmhällar reef was reached by the Ljugarn reef of the Hemse Beds. That reef, as measured between the two ends, shows a chord length of about 550 m. The Fågelhammar reefs are somewhat smaller. Still smaller are the reefs in the Sjausterhamma area.

Several reefs, probably on the order of fifteen, must have been present in the Hemse Beds between Snabben and Sysne-udd. Most of these were quite small, with a chord length of 50–150 m. About half-way along the southwestern coast of the Sysne peninsula, the remains are found of a reef which probably measured not more than 10 m in chord length. Of this small reef only a nucleus of about 3 m in diameter is exposed. It is surrounded at a distance of one to a few metres by reef-detrital limestone. This limestone dips about 20° away from the reef nucleus and overlies the margins of the reef.

In the Hemse Beds, the reefs of Holmhällar type thus obviously decrease in size in a northeastward direction. A tentative explanation of this observation will be given in the discussion of the Hemse Beds in Chapter XI (p.386).

The thickness of the reefs of Holmhällar type exposed in the Hemse Beds is only slight. The raukar only rarely reach over 4 m above sea level. Only in Fågelhammar South are higher raukar, up to 6 m, found. It is not known how much reef limestone is still below sea level in Fågelhammar and Ljugarn. In the surroundings of Sjausterhammar, the underlying sediments of part of the reefs are exposed at around sea level. But since Fågelhammar is only about 6 km southwest of Sjausterhammar and the general dip of the strata of Gotland does not deviate much from a southeasterly direction, it seems unlikely that a great thickness of reef limestone is still hidden below sea level. Presumably the reef limestone in the centre of the large reefs had a maximum thickness on the order of 6–8 m (see also the section on the Holmhällar-type reef limestones of the Hemse Beds in Chapter XI), whereas the smaller reefs, e.g., in the Snabben – Sysne-udd belt, may have been even thinner.

DEPRESSIONS IN THE REEF

In the Holmhällar raukar field, several indications are found that during development of the reef its surface in places was rather uneven. Parts with the characteristic stromatoporoid fauna alternate locally with parts in which reef and/or crinoid debris dominates or in which a different fauna is present. Just as with the Hoburgen-type reefs described in Chapter VII, those parts in which debris could accumulate will be termed *filled depressions*, and the ones in which a different fauna developed will be called *pools*. It is admitted that this distinction is not a sharp one. Pools, too, may contain a certain amount of debris and, on the other hand, depressions may have developed upwards into pools. Nevertheless, the distinction facilitates a description of the reefs.

Normal reef portions, depressions and pools not only alternated synchronously, but they also succeeded each other in time. Thus, debris-filled depressions usually overlie reef limestone rich in stromatoporoids. Debris accumulated in them until, in a later stage, reef builders may again have expanded over the filled depressions.

Arkell (1935, pp.98–99), in a study on Jurassic reefs from the vicinity of Oxford, postulated that in these reefs growth might have been mainly confined, after a while, to the margins. In this way, outgrowths from the margins might have enclosed sheltered embayments, in which reef limestone of different faunal and sedimentary facies could accumulate.

A comparable mode of origin may have led to the formation of depressions in the surface of reefs of the Holmhällar type in Gotland. In places with the best supply of water, rich in food and nutrient salts, reef builders may have grown faster and larger than elsewhere, thus leading to an uneven topography, with sheltered depressions behind the faster-growing reef parts. In these depressions debris could accumulate, or a different fauna could develop.

No apparent regularity has been found in distribution of the debris-filled depressions and pools over the reefs. They seem to have developed in all parts of the reefs, certainly not only behind outgrowths from the reef margins. This may have been connected with the probably seaward-sloping upper surface of the reefs (cf. Chapter IX, p.223), which may have caused not only the reef margins but also many other parts of the reef surface to be in regular contact with circulating water. But in the lower parts of the raukar, remains of filled depressions and pools are rarely found; this, however, is assumed to be due to selective Recent erosion (see the section on the formation of raukar, later in this chapter, p.208).

Debris-filled depressions

Among the debris deposited in depressions there are, of course, remains of stromatoporoids, corals and bryozoans, but also crinoid fragments, orthoceratids, trilobites and occasional brachiopods, gastropods or lamellibranchs. Whereas crinoid stem fragments are extremely abundant in some of the depressions (e.g., observation point 73), they are much less numerous in others (e.g., points 131, 173A). In general, there seems to be a relationship between the amount of crinoid remains in a filled depression and the distance of this depression from the reef margin. This is one of the indications which suggest that crinoids grew more abundantly on the flanks of the reefs than on the upper surface (see also Chapter XII).

Other factors too, however, must have influenced the filling of these depressions, such as fluctuations in the wind direction and in the strength of wave action. The latter can be illustrated with the facts observed in point 78 (Fig.83). This point forms part of a large depression, a remnant of which is still left of dimensions of over 2 m high, about 5 m long and about 2 m broad. The debris filling consists in very large part of red-coloured crinoid remains. They give the sediment as a whole a red colour, even though the calcareous matrix in between them is mainly greyish green. However, this matrix is a less important element of the sediment than are the crinoids. The sediment shows a distinct stratification, with layers dipping towards the southeastern edge of this rauk. Presumably, the depression was deepest in this vicinity. In between the crinoids, numerous remains of stromatoporoids and corals are present. These are mainly concentrated in a number of special layers, which alternate with layers of a more standard crinoid content. Within these layers of coarser material, the stromatoporoid and coral remains occur in all possible positions, intermingled at random with crinoid fragments. Towards the top of the depression, larger fossils decrease in number and in the highest part of the rauk, only a few are present. The presence of larger fossils in some layers and their absence in others cannot be accidental but must have been caused by an agent such as occasional severe storms that attacked the developing reef.

Distinct stratification is also found in the depression seen at observation point 85. Crinoid remains are numerous there too; stromatoporoids and corals are principally represented by fragments. The individual layers vary in thickness between one and a few centimetres. The stratification evidently

Fig.83. Holmhällar, observation point 78. Depression in the reef surface, at the bottom filled with crinoids and fragments of stromatoporoids and corals, at the top mainly with crinoid remains. Hamra-Sundre Beds.

is caused by a sorting according to size of the crinoid and other fossil remains, whereas slight variations in the matrix volume may also have contributed somewhat. Fluctuations in the intensity and direction of wave action may have been the main factor leading to this stratification.

The boundary between the actual reef limestone and an overlying debris-filled depression is usually distinct, due to the sudden decrease in the number of stromatoporoids and the simultaneous increase in fragments of fossils. From this, one can surmise that several of the depressions have developed from lower parts within the reef surface in which reef growth was ended within a short time, perhaps due to the washing-in of great amounts of debris. A gradual smothering of reef builders by washed-in material would have led to less distinct boundaries and to transition zones in which the autochthonous fauna was more gradually replaced by the debris.

The debris material in the depressions is generally softer than the stromatoporoid reef limestone. It is, therefore, interesting to note that no filled depressions have been preserved around present sea level, where erosion was strongest. The only examples found are higher up in the raukar, resting on a pedestal of stromatoporoid limestone (cf. p.183). Where larger depressions occur at that level, the lesser resistance to erosion is often demonstrated by an inward curving of the rauk surface (e.g., observation points 191–192, 137).

An example of a debris-filled depression is found in Holmhällar, at observation point 137. The depression overlies reef limestone very rich in branched corals, bryozoans and stromatoporoids. The presence of the first two groups suggests that, prior to the debris accumulation, this locality constituted a pool in the reef surface.

The boundary between the reef limestone and the material which filled the depression is very sharp. The base of the depression is somewhat undulous and faintly concave. In the centre it reaches about 25 cm lower than at its margins. The diameter of the depression is about 6 m. The filling of the depression consists mainly of crinoid debris. In the observation points 137A and 137B it can be seen how the crinoid breccia interfingers over a height of about 2 m with the actual reef limestone. In point 137A the average course of the boundary is about vertical. In point 137B in the lower part the crinoid breccia expands towards the northeast over the reef limestone, whereas higher up in the cross-section, the reef limestone in its turn expands again quite rapidly over the debris. The interfingering strongly suggests that the filling of the depression with debris took place mainly in bouts and continued until a level more or less equal to that of the surrounding reef parts was reached.

Close to the boundary between depression and reef limestone in observation point 137C, there is a rather sinuous crack along which later displacement has taken place. This is demonstrated by fossils which are cut off and lack their counterpart. Comparison of the sediment at both sides of the crack suggested that a part of the same fossil now lies about 20 cm lower at the depression side of the crack than at the reef side. Slight horizontal movements, too, are suggested by this exposure. Both may have found their origin in compaction of the depression-filling material.

It is significant that, whereas in points 137A and 137B the depression is almost exclusively filled with crinoid debris, in point 137C stromatoporoid remains are also found, several of which dip rather strongly in one or another direction. There is even a large piece of stromatoporoid limestone, about 40 cm high and 20 cm broad, lying almost upside down. This indicates that the reef of Holmhällar occasionally may have been attacked by storms which were strong enough to tear blocks from the reef edge and toss these onto the reef surface. This phenomenon is reminiscent of the formation of "negro heads" on modern reefs. Point 137C is located at the most seaward part of the depression under discussion.

Debris-filled depressions in other Holmhällar-type reefs are of character and size similar to those in Holmhällar. Thus the largest one in Fågelhammar South has a diameter of about 6 m. It is filled with a distinctly stratified breccia of crinoid fragments and a few small remains of reef builders; it is now found on top of a rauk, well out of the reach of Recent wave action (Fig.84). Some other depressions in this raukar area were found to be filled with stromatoporoid fragments in a matrix of calcareous mud, with crinoid remains in between.

Pools in the reef surface

The term "pool" is used by the present author to denote a part of the reef surface which was situated lower than its surroundings but which nevertheless was covered with living organisms. Exposed reef limestones can be recognized as having been formed in pools in the reef surface, if they differ markedly in faunal composition and/or in matrix percentage from the surrounding reef limestone.

Fig.84. Rauk in Fågelhammar South. Reef limestone with at the top of the rauk a debris-filled depression. Hemse Beds.

A good example of a fossil pool has been found in Holmhällar at the observation points 159–165. An increase in the matrix percentage, of about 25% of the total volume, coincides with a strong decrease in the number and sizes of the stromatoporoids. Underneath the pool and around it, stromatoporoids constitute about 60% of the total reef volume. In the pool they decrease to a volume percentage as low as about 10. Consistent with the decrease in stromatoporoids in the pool, the number of coral and bryozoan colonies rises. Although there is also an increase in the amount of crinoid remains, they do not, by far, reach a volume percentage as high as in several of the debris-filled depressions. In addition to the estimates in observation points 159–165, nine further estimates have been made of the composition of the reef limestone in that part of the Holmhällar raukar field (Fig.224). The results of all sixteen estimates together are given in the graph of Fig.225.

In a number of cases, pools directly overlie reef parts which are very rich in stromatoporoids. These pools have presumably been formed by a local decrease in the velocity of reef growth.

In some of the topographically lower parts in the reef surface, growth came to a complete standstill and debris-filled depressions formed, as discussed before in this chapter. In these parts, debris deposition could continue until the floor of the depression was raised to about the level of the surrounding reef surface. Subsequently the initial depression was overgrown again by reef builders, forming normal stromatoporoid reef limestone. It could also happen, however, that organisms settled on the floor of the depression before it was completely filled, thus converting it into a pool.

An example of this has been seen in Holmhällar in the observation points 131–132. In point 131 the base of a depression lies at a height of about 2 m above present sea level; in point 132 the same base lies at about 1.6 m altitude. The boundary with the underlying stromatoporoid rock is distinct and somewhat undulating. Compared with typical crinoid-filled depressions, crinoid remains in this depression are less numerous, but they are of larger size. Stromatoporoids, corals and bryozoan fragments do not reach high percentages. The amount of matrix mud, mainly red coloured, on the other hand, is greater than in many other depressions. In point 131, higher up in the original depression, a number of coral and bryozoan colonies can be seen that have grown on and between the debris. In point 132, these are already present close to the base of the depression. Further upwards, a few stromatoporoids also appear, but they are generally of rather small size. The more the distinction between pool and surrounding reef limestone fades away, in successive higher parts, the more the stromatoporoids are seen to have regained ground on the other reef builders.

INTERRUPTIONS IN REEF GROWTH

In several places in Holmhällar and related reef-limestone exposures, the raukar show, within their structure, more or less horizontal planes, which may have been further emphasized by erosion. In places they give the reef limestone a somewhat pseudo-stratified character (Fig.85). These planes almost always occur where the reef limestone is very rich in stromatoporoids The reverse is not generally true, as there are also many parts of the reef rich in stromatoporoids, which do not show this kind of wide-apart pseudo-stratification.

Some of these exposed planes are bordered above and below by exactly the same kind of reef limestone, with the same faunal composition. If, moreover, the limestone is of crumbling nature, it is not impossible that these

Fig.85. Heliholm. Raukar of reef limestone very rich in stromatoporoids. Wide apart, sub-horizontal planes which cross the reef limestone give the latter a pseudo-stratified character.

planes were formed only in Quaternary times, e.g., as the result of frost action (see p.200). But a syn-genetic origin is also possible, as will be discussed next for some other planes, and a definite solution of the problem of the origin of all planes cannot be given.

A number of planes suggest a syn-genetic origin. This is especially likely in cases where there are notable differences in the character of the limestone below and above the plane. An illustration of the latter feature has been found at the observation points 272–276. There a plane is exposed which descends from a height of about 3 m above present sea level at point 272 to an altitude of less than 1 m above sea level at point 276. From there, towards the northeast, the plane ascends a little again. At the other side of point 272, the plane can still be traced at point 265. Thus, horizontal extension of this plane is at least 25 m. Immediately below the plane and, still more evident, in the lowermost 0.5 m above it, the limestone is distinctly richer in calcareous matrix than it is in the lower and top parts of the raukar at these points. In the fossil content differences are less well marked, though the reef limestone just above the plane is relatively richer in corals and bryozoans than elsewhere in these raukar. These taxa decrease in number upwards and are replaced by stromatoporoids which, being already numerous slightly above the plane, build the reef limestone almost exclusively in the higher portions, just as they did below the plane. The conclusion from this example is that formation of the plane has been linked to a retardation or interruption in reef growth.

An even more evident interruption in reef growth is found in observation point 139. There a layer, about 5–10 cm thick, of hard, splintery and almost non-fossiliferous limestone is intercalated in the reef limestone, at a height of about 0.75 m above present sea level (Fig.86). The line of outcrop is rather sinuous, but as a whole, the plane shows a distinct southeastward dip. In view of the thickness of the layer and the general lack of fossils in it, it is likely that during deposition of this layer, reef growth was interrupted. There are, however, no apparent differences in the faunal composition of the reef limestone above and below this layer.

Higher up in the same rauk, there is a second layer of such splintery limestone but of smaller extension and thickness, and, therefore, less striking than the lower one. In contrast to the previous case, this second interruption in reef growth additionally appears because of differences in the fossil content of the reef limestone. Going upwards from the first to the second layer, a gradual increase in matrix percentage and in the number of bryozoan remains can be observed, together with a decrease in the volume percentage in stromatoporoids. Above the second limestone layer the number of stromatoporoids is very high again, and the amount of matrix and bryozoans has distinctly decreased. The colour of the matrix is predominantly grey below the second layer and mainly red above it.

A third layer of hard, splintery limestone occurs, at this locality in Holmhällar, at about 2 m above sea level. This one, too, shows a somewhat sinuous line of outcrop. It is the least-well-developed one of the three.

At observation point 140 there is a somewhat sinuous layer of limestone, very rich in crinoid remains but also containing fragments of corals, stromatoporoids, and other organisms. It is situated at a height of about 1.5 m above sea level and shows an average thickness of about 15 cm. Deposition of this layer in all likelihood also indicates a local interruption

Fig.86. Holmhällar, observation point 139. Rauk with two intercalated layers of hard limestone, nearly devoid of fossils.

of reef development. The reef limestone overlying this layer is very rich in stromatoporoids, from immediately above the interruptive layer onwards. This is in contrast to the reef limestone underneath the layer, which is less rich in these fossils and contains more corals, bryozoans, crinoid remains and matrix. There seems to have been a tendency in these reef-building components to gradually replace the stromatoporoids, which at the base of this rauk were the sole reef builders but upwards decrease in importance.

It is likely that this crinoid layer corresponds to the upper layer of splintery limestone, recorded from observation point 139. How the two may have passed into each other unfortunately cannot be established since the rauk is interrupted at this height over a distance of a few metres.

In the examples described above, there is evidence that local interruptions in reef growth have occurred. The same may be true in several other cases, in which the reef limestone is intersected by planes which generally appear as more or less straight horizontal or sub-horizontal lines in the sides of the raukar. More especially if a correlation of the lines in several raukar over not too small an area is possible, the probability that they are due to a temporary break in reef development increases.

However, this argument must be used with caution and additional evidence still needs to be obtained. This is, for instance, illustrated by the vague pseudo-stratification that is found in the observation points 109–111. The two planes exposed in this large rauk can be followed over quite a distance with slight variations in their height above present sea level. But they do not only intersect the reef limestone proper but at the same level also cut through a debris-filled depression and thus, most probably, were not formed at the time of reef growth.

To explain the mode of formation of the actual interruptions in reef growth several possibilities have to be considered, such as (1) the fact that the reef locally reached the sea surface; (2) an increase in the turbidity of the water; (3) a temporary strong increase in sea depth; and (4) the action of waves, more especially storm waves.

(1) It seems highly improbable that reef growth was periodically interrupted because parts of its surface reached sea level. Reef growth in an extremely shallow sea is illustrated, e.g., by the small reefs of the Lower Hamra limestone, which exhibit a much more unorganized character and a higher content of terrigenous material. There is no apparent similarity of these reefs to the reefs of the Holmhällar type.

(2) Where differences in faunal composition are found together with an increase in the matrix percentage, a temporarily greater turbidity of the water may have played a part. Since the matrix is rather pure limestone, however, greater turbidity was not related to an increased supply of terrigenous material. The calcareous mud was probably derived from the reef or its immediate environment and was stirred up by water turbulence. Its deposition and its influence upon reef growth can thus only have been secondary phenomena, which resulted from an alteration in the physical environmental conditions.

(3) A rate of subsidence of the sea bottom in excess of the rate of upward growth of the reef would not have stopped reef growth until the surface of the reef was too deep to receive the required minimum supply of light or until the reef surface came below effective wave base (marl deposition). If such a stronger subsidence actually happened, reef development would have stopped over the entire surface of the reef and not just locally. Moreover, lowering below effective wave base would have permitted the settling of sediment, that was present in suspension. Such sedimentation would have continued until water depth again decreased to depths in which renewed reef growth could take place. All interruptions in reef development thus should then be characterized by layers of sediment of greater extension and thickness than those which are actually found.

(4) Since the remarks made under (2) and (3) indicate that reef development took place above effective wave base, alterations in the effects of wave action remain the most probable cause of the local interruptions in reef growth. During storms, wave action would have greatly increased in intensity. The higher parts of the reef may then have been attacked, and much reef and crinoid debris may have been formed which was deposited in depressions in the reef surface and around the reef. The planes of interruption, thus, should not only indicate a local break in reef development but also partial demolition. The finest debris was kept floating until wave action had decreased. Its deposition may have influenced reef growth, especially in pools, and may even have caused the death of several organisms. In this way, it perhaps caused secondary interruptions in reef development elsewhere on the reef. This explanation clarifies (a) why reef growth was interrupted only locally; (b) why some of the planes do not show indications of the deposition of calcareous mud or reef and crinoid debris, where others do; (c) why some of the interruptions were accompanied by shifts in faunal composition.

FISSURES

The raukar field of Holmhällar is intersected by a number of large, sub-vertical fissures, which cross the reef limestone from one side to the other. Similar fissures occur in the other raukar fields in the southern peninsula of Gotland (Fig.87, 88), as well as in Ljugarn (Fig.89) and Fågelhammar. Generally there are also several smaller cracks linked to the main ones.

The length of the main fissures ranges from about 25 to 75 m, as far as observations are possible; several may have been longer, intersecting also portions of the reef that have been eroded. The total width is generally in the order of some tens of centimetres, but some are much wider. Thus a large fissure on the island of Heliholm, southeast of the lighthouse, reaches a local

A B

Fig.87. Fissures intersecting reef limestone of Holmhällar type. Heliholm. Hamra-Sundre Beds. A. East of the famous rauk "Penningkammare". B. Southeast of the lighthouse; part of the sediment which filled the fissure is in this case still preserved.

Fig.88. Fissure crossing reef limestone of Holmhällar type in the Holmhällar raukar field, near observation point 214. Hamra-Sundre Beds. Comparison of fissured fossils in the right and left limestone parts suggests that in a vertical direction the two parts have been displaced, compared to each other, over a distance of about 8 cm.

maximum width of 94 cm. The widest fissures seen in Ljugarn measured approximately 70 cm.

The fissures usually have a more or less sinuous course. They were filled with fine reef debris, which is generally softer than the true reef limestone. Consequently, Recent erosion has reopened several of the cracks, leaving only some of the original sediment filling locally against the margins of the fissures. Only in a limited number of cases are the fissures still completely filled with a light-coloured, finely-crystalline limestone. In a few of these, the filling limestone appears to be even harder than the actual reef limestone and the fissure is consequently exposed as a ridge. The fissure fillings are usually very rich in small-sized organic remains, such as fragments of branched corals and bryozoans and of other reef-building organisms. Only rarely do they contain larger fossils, such as intact colonies. That the difference between the components of the reef limestone and those of the fissure fillings is mainly a matter of size suggests that the fissures were formed and filled at the time of reef growth.

Fig.89. Ljugarn. Hemse Beds. Fissure in reef limestone.

The fossiliferous sediments in the fissures often show an approximately vertical stratification, which is not found in the surrounding reef limestone. The thickness of individual layers is generally on the order of 1–10 cm. In the large fissure on Heliholm mentioned above, the average thickness of the layers is 1–2 cm, but locally they are much thicker (Fig.87B); in other cracks on that island several layers have been observed of only up to a few millimetres in thickness.

This type of stratification is, in all likelihood, linked with the formation of the fissures. Initially these must have been rather narrow. They then became filled with calcareous mud and small fossil fragments; for larger debris there was no space. This material soon hardened. Thereafter, the fissure widened a little and was filled up again. This may have continued several times and this succession must have led to the vertical layers present in the fissures.

The cause of the formation of these fissures presumably has to be found in compression of the sea bottom underneath the expanding reefs. Small sliding planes and striae that can be observed on the bedding planes of the fissure sediments prove that small, sub-vertical displacements have taken place.

The reefs originated on a very soft limy substratum. At places where reef building took place, this sediment was compressed by the weight placed upon it. With continued upward growth of a reef, the degree of compression decreased. Lateral expansion of the reefs, however, caused the reef flanks to grow on a less consolidated part of the sea bottom. Thus, at a certain stage, compression of the basal sediment under the reef margins exceeded compression under the reef centre. Tension increased until fissures formed and the reef margins moved slightly downwards. With continued reef expansion, this may have been repeated several times. This mode of formation then explains why all fissures are almost perpendicular to the curved longitudinal axis of the reefs (see the map of Holmhällar).

As follows from the above, displacements along these large cracks must have been slight and were directed downwards. Horizontal movements are not to be expected with such a mode of formation as is assumed here. Moreover, these were excluded by the sinuous course of the fissures. Distinct information about the extent of the vertical displacements which have taken place could be gained at the Holmhällar observation points 159–165, where a fissure intersected a pool with a faunal composition different from that of the surrounding reef limestone. The base of the pool at one side was only a few centimetres lower than at the other side. Consequently, the main movement has consisted of a slight sagging of the outer reef part. At the observation points 214–215, there is a fissure which is 12–25 cm wide, where broken fossils suggest that the reef limestone at the west side has presumably moved 8 cm down, compared to the reef limestone at the east side (Fig.88).

There should not be an essential difference in the fissure pattern between cases in which fissures are formed by a downward movement of the margins and others in which there was an upward push in the centre. This is confirmed by data from salt domes and volcanoes.

Beds above rising salt domes often show a radial fault pattern comparable to the fissure pattern of the Holmhällar reef. An example is given in the structure map of the Hawkins Oilfield, shown by De Sitter (1956, p.261). Radial faults are accompanied there by normal longitudinal faults.

If, in a volcano, the supply of magma from the depth exceeds the production of the volcanic vent, the surface is pushed up. This phenomenon has been described from the Merbabu volcano in central Java, Indonesia (Van Bemmelen, 1954, p.81). The vent of the volcano became blocked up, the cone was pushed up by the advancing magma and finally radial faults originated through which the lava appeared at the surface.

Longitudinal fissures, comparable to those that are present above the Hawkins salt dome, are not evident at Holmhällar, but a few have been seen in the neighbouring raukar field on Heliholm. They showed an average dip of approximately 70° inland and may be taken as evidence of the seaward expansion of the reef. Other examples have been observed in Ljugarn and in Fågelhammar.

Since the seaward margin of the reef was in direct contact with the open sea, the reef builders will from there also have grown outward over the forereef in a way similar to their sideward growth over the lateral debris deposits. The forereef rested on deposits whose supporting competency was not yet great and which might be compressed or have flowed from beneath the load. The newer reef parts overlying the forereef caused the latter to subside to a stronger degree than did the sediment underneath the older reef parts, and consequently caused the formation of the fissures perpendicular to the radial ones, in a way comparable to these radial fissures.

In fact, there may even be a direct connection between a radial and a longitudinal fissure. An example of this has been observed in the south of the Ljugarn raukar field where a fissure, of about northwest – southeast direction, ends in a fissure with a more south to southwest direction.

Another type of filled fissure has been found in Holmhällar at about 1.5 m southwest of observation point 145 (Fig.90). This fissure is bounded by two sub-vertical planes which slightly diverge upwards. The width of the

Fig.90. Holmhällar, observation point 145. Fissure, filled with reef debris. The debris is much coarser than in other filled fissures and shows no sub-vertical stratification. This suggests that the fissure was filled in one single stage. Hamra-Sundre Beds.

fissure is about 10 cm at the bottom and more than 40 cm at a height of about 3 m. The difference in relation to the surrounding reef limestone is slightest at the bottom. The fissure filling there consists mainly of stromatoporoid colonies and fragments (all dipping in one or another direction), calcareous mud and some crinoid remains. Upwards the number of crinoid columnals strongly increases, together with the number of bryozoan fragments; some corals are also present there; stromatoporoids still occur, but are relatively less common than in the lower part. This fissure seems to have been filled from bottom to top in one single stage. Indications of this are the presence of much-coarser reef debris and also the complete absence of any sub-vertical stratification, which is so characteristic of the other fissure fillings. Towards the north-northwest of observation point 145 this fissure narrows (Fig.91); towards the southeast, lack of suitable exposures hinders reliable observations.

Fig.91. Holmhällar. The same fissure as that shown in Fig.90, but some metres further north-northeast. The fissure is narrower at this place. It can only be followed upwards until the about horizontal plane in the upper part of the rauk.

The character of the filling material and the increasing width upwards suggest a mode of formation comparable to that of the other fissures but with only one vigorous hinge-like movement instead of several successive stages.

It is interesting to note that north-northeast of observation point 145, where the exposed reef limestone reaches higher, the fissure does not continue until the top of the reef limestone (Fig.91). After the fissure had been filled, the reef builders apparently grew normally over that area.

DEBRIS FLOOR AND TALUS

Neither in Holmhällar nor in the other raukar fields in the southern peninsula of Gotland has anything been preserved of a talus mantle surrounding the reefs. Neither is the debris floor over which the reef flanks are supposed to have expanded exposed.

An impression of the debris underneath the reef flanks can be gained in Fågelhammar South. In the north of this raukar field the raukar get smaller, hardly ever reaching over 1 m in height, whereas the coastline gradually bends inwards. In the reef limestone large stromatoporoids are comparatively less common, flat colonies and also stromatoporoids showing strong vaulting are more common there than in the centre of the reef; corals are anything but rare, crinoid remains are very common. The reef limestone thus shows several indications of representing a peripheral part of the reef. In an elongated rauk-like exposure (see section a–b in Fig.193), this reef limestone passes laterally into a comparable sediment which shows a vague stratification and a reefward dip. At the north side of this exposure, it can be seen how this limestone overlies fossiliferous marly limestone which is very irregularly stratified, and in which fossils occur chaotically. Stromatoporoids predominate, but corals are common. In the matrix of bluish grey marly limestone, irregular flat lenses of hard, brownish grey limestone are also found, usually very rich in crinoid remains. This sediment, too, dips reefwards, that is southeast to south-southeast. Upwards it passes into harder, more resistant sediment, which is much richer in stromatoporoids and crinoids, and gains the character of the stratified, reef-like limestone mentioned before from the south side of this exposure.

Further towards the north a few raukar still follow which are built of reef limestone (Fig.193). This is, however, less characteristic of the Holmhällar type of reef limestone than that in the centre of the reef. Presumably these exposures represent the northern end of the Fågelhammar South reef, in which the reef spread over a base of reef debris. Formation of reef debris must have varied; e.g., more was worked off the reef during heavy storms. The exposure discussed perhaps represents such a stage of increased debris deposition, forming an elevation in the debris floor underneath the reef flank.

The sediments underlying the reef limestone could also be observed in an elongated rauk, about 0.4 km northeast of Sjausterhammar Fiskläge, close to the shoreline (Fig.192). In the southeastern side of this rauk which is about 3 m high, the following succession is present, from top to bottom:

1.30 m + Reef limestone; at the base mainly flattened colonies, upwards soon also rounder specimens.

0.70 m Stratified marly limestone, bluish grey, finely crystalline, layers generally 0.5–5 cm thick; many solitary corals, a few small stromatoporoids, mainly in thin, laminar growth forms, occasionally brachiopods.

0.30 m Reef-like limestone with mainly flattened reef builders, several of which are not in their living position.

0.65 m Calcareous slate, bluish, very thinly cleaving, dense to very finely crystalline; only a few fossils.

Differences in resistance to erosion make the reef limestone protrude over the underlying sediments. All layers dip approximately 5° towards the northeast.

North and south of this exposure, the outcrops do not reach much above sea level. Those north of the rauk mainly show limestone of a reef-detrital character, occasionally some reef limestone or more-normally stratified limestone. The outcrops south of the rauk predominantly show calcareous slate, generally dipping a few degrees seaward, but occasionally up to 20°; the sediment often contains reef debris.

The reef which must have been present in this area, as is indicated by the succession of strata and by the described distribution of the sediments, seems to have been eroded in Recent times.

A third example of the relationship between reef and underlying sediments, is presented by the exposures in the south of the Sjausterhammar peninsula (Fig.191). The outward boundary of the reef limestone in that locality follows a curved course from north-northeast – south-southwest via east – west to about east-southeast – west-northwest, suggesting a crescent-shaped reef with its opening about northwestwards. The reef limestone is underlaid by stratified stromatoporoid limestone, which in its turn overlies calcareous slate. In the centre of the reef, the reef limestone starts close above the calcareous slate; towards the flanks, the intercalated stratified stromatoporoid limestone increases in thickness.

The calcareous slate is similar to that described above from 0.4 km northeast of Sjausterhammar Fiskläge. Its exposed thickness ranges from a few decimetres to a maximum of about 1 m.

The stratified stromatoporoid limestone which underlies the east-northeast side of the reef dips perceptibly about west-southwest. More commonly than underneath the reef, it crops out at the south side of the reef, dipping reefwards. It is found there at a level somewhat lower than that of the adjacent reef limestone. The bedding planes are irregular, due, partly, to the very high content of potential reef-building organisms, many of which are not in their life orientations and may actually have been washed from the reef. In its upper portion the sediment is light grey to greenish light grey, often faintly red-mottled, finely crystalline, very rich in stromatoporoids, but also containing crinoids, corals and bryozoans; the thickness of the layers varies generally between one and a few decimetres, though some may reach about 1 m; several of the layers show an indistinct and irregular subdivision into zones about 5–10 cm thick. The lower portion of the stromatoporoid limestone is light brownish grey, almost dense to finely crystalline and marly; the bedding planes are irregular and knobby; the thickness of the layers is generally 3–10 cm. Part of the stromatoporoids in this lower portion of the stromatoporoid limestone presumably represent reef detritus, but others may have grown on the spot. Also crinoid remains, corals, bryozoans, and some orthoceratids and ostracodes are present. A few thin and rather strongly marly layers, which because of weathering appear as grooves, contain laminar stromatoporoids.

The talus mantle around the reefs can be studied in some of the exposures between Snabben and Sysne-udd, especially in connection with a reef about 1 km south of Snabben. The bulk of this talus is formed by stromatoporoid colonies, which are not in their life orientations and by fragments of stromatoporoids. At the north-northeast side of the reef, about 80% of the sediment consists of stromatoporoid material; corals and crinoid remains are relatively scarce. Although the nature of the rock leaves little doubt that it represents a talus mantle, there is no apparent stratification in this reef talus. Only at a distance of about 3–5 m from the outer boundary of the reef limestone, where the talus zone passes into limestone with reef debris, constituting less than 50% of the total volume, does stratification generally appear, though sometimes inconspicuously. More data about this sediment will be given in Chapter IX.

Sediments with reef detritus which overlie the reef limestone, can also be studied in the above-mentioned coastal area between Snabben and Sysneudd. From the environment of Snabben, two profiles have been described by Hede (1929, p.40). These are of interest because they give an impression of the interrelations between the sediment types found there. Comparable sediments are found in a number of other raukar. The upper boundary of the reef limestone varies between rather undulous and rather smooth, but is generally distinct. The overlying reef debris shows reef builders and fragments of these in a matrix of crinoid breccia. Upwards it becomes less coarse and passes into a crinoid limestone which is more thinly and faintly undulously stratified. Part of the debris may be rounded.

About 0.7 km south of Snabben, it has been estimated that between 0–0.6 m above the reef surface, reef debris constitutes approximately 50–25% of the sediment (for the method of estimation, see Chapter IX). Some layers are richer in debris than others. Upwards, coarse remains decrease in number. At a height of 0.6 m above the reef, the length of the largest reef fossil measured 17 cm; at a height of 0.8 m, only 7 cm. The total volume of reef debris at 0.75–1.50 m above the reef limestone is on the order of 15–7.5%. The cross-section studied was not situated at the place where the reef is presumed to have reached its maximum thickness. Unfortunately the exposures did not permit studying the relationships between the reef debris and its source.

THE FORMATION OF RAUKAR

Although, in fact, a discussion on the formation of raukar does not actually belong in a chapter on Middle Palaeozoic reefs, these erosion forms on the other hand are so characteristic of the reef-limestone exposures, which have been described in the previous pages, that the present author, with Hadding (1941, p.79), feels that a chapter on these reef limestones would not be complete without a few words on raukar formation.

During the Late Quaternary Period, large parts of Gotland were inundated by the Baltic. Fossil beaches dating from the *Ancylus* and *Littorina* time can be found all over the island.

It was during this post-Glacial time that the raukar of Gotland were formed by processes that are still continuing. Among these, wave action is foremost. In Holmhällar, waves and storms have removed not only the stratified sediments around the reef but also the reef margins and large parts of rock with a relatively softer constitution that have been present within the reef. It does not seem to be an exaggeration to estimate the total amount of reef limestone that has already been broken down as 200,000 m^3. Similar amounts may be assumed for the raukar fields on Heliholm and at Hammarshagahällar. Several raukar which now lie in the sea have already been partly undermined, and a number of them appear to have been recently deprived of their pedestals and now lie offshore as loose blocks. Large amounts of broken reef limestone cover the shore in and around the raukar fields.

Some geologists to whom the present author showed pictures of the raukar fields of Gotland, were inclined to explain these as due to karstification. It should be admitted that, at first glance, these may correspond strikingly to some typical karst phenomena. Nevertheless, the data mentioned above suggest that this explanation is untenable. The destruction of the Holmhällar reef took place around sea level. The parts of Holmhällar indicated with blue on the enclosed map, as well as the parts covered with eroded debris, possess a basis of reef limestone, which was eroded down to sea level. Whereas all softer reef parts have disappeared around sea level, some have been preserved in the higher parts of the raukar. Especially the higher parts should have formed suitable points of attack by karstification.

Under the climatic circumstances that prevailed in Gotland, karstification would have taken place slowly. Moreover, an area as Holmhällar has presumably never been covered with a vegetation cover which in true karst areas may largely increase the CO_2-content of the water. In view of the fact that the raukar must have been formed during a rather short interval of time[1], chemical solution can only have been responsible for the removal of a small fraction of the limestone which has actually been demolished.

Extensive limestone plains ("alvars") are found in several places inland in Gotland, such as west and east of Sundre (see the blue portions on the map given by Munthe et al., 1925, tavl. 5, facing p.40), but none of these shows karst phenomena.

Apart from the localities with Holmhällar-type reefs, raukar are also found in the Högklint Beds (Lickershamn), the Slite Beds (Solklint, Lännaberg, Spillingsklint, Bogeklint, Tjelders, Asunden, etc.) and on Stora and Lilla Karlsö. Sometimes an isolated rauk is found, such as Jungfrun, south of Lickershamn (the largest rauk of Gotland), but generally they occur in great numbers together in the well-known raukar fields. All these localities, like those exposing Holmhällar-type reef limestone, were subjected to coastal erosion at one time or another during the post-Glacial period.

Raukar made of stratified limestones are exceptions. An example is Hoburgsgubben (Hoburgen, Hamra-Sundre Beds), included among the raukar by Swedish authors (e.g., Munthe, 1921a,b) though it hardly deserves that name, since it is only a higher portion of the third hillock in the Hoburgen complex. In a few other raukar, stratified limestones may be found together with the reef limestone, such as is described from Sjausterhammar.

In conclusion it can be summarized:

(1) Although raukar are most characteristically developed in the reef limestones of Holmhällar type, they are also found in other reef limestones.

(2) Raukar made of stratified limestones are exceptions.

(3) Since reef limestones are more massively built than stratified limestones and Holmhällar-type reef limestones are generally more massive

[1]During *Littorina* time – that is about 7000 years B.P. – the areas which at present expose the major raukar fields, like Holmhällar, Heliholm, Hammarshagahällar, Ljugarn and Fågelhammar, were still below sea level.

than reef limestones of Hoburgen type, it seems that the most-massive limestones are most suited to raukar formation.

(4) The formation of raukar is explained by Recent marine erosion, not by karstification.

SYNTHESIS

On the basis of what has been said earlier in this chapter, it may be concluded that the reefs of Holmhällar type developed in shallow, though not very shallow, water. The reef limestones present several data which show that the reefs developed above effective wave base. Of these the most important are:

(1) The shape of the reefs has most probably been determined by the dominating wave direction.

(2) Much reef and crinoid debris has been formed which was deposited both around the reefs and in depressions within the reef surface; in the latter, the debris layers may show a certain degree of sorting. Negro-head-like blocks were torn from the reef edge and tossed onto the reef surface.

(3) Calcareous mud, locally affecting the composition of the reef fauna, was presumably mainly derived from the reef or its immediate environment and was stirred up by wave action.

(4) Interruptions in reef growth suggest attacking and demolition of the higher reef portions by storm waves.

Now that it has been shown that Holmhällar-type reefs developed in shallow water, is it possible to estimate the maximum water depth?

In both composition and size, the reefs of Holmhällar type are more closely related to those of the Hoburgen type than to the Upper Visby reefs. The latter may have grown in deeper water than did the reefs of Holmhällar type.

Deposition of fine silt off the New England coast is possible at depths exceeding 60–70 m (Stetson, 1936). It is very likely that this critical depth was situated less deep in the epicontinental Baltic basin of Palaeozoic times. The erosive action of normal waves presumably reaches only a few metres below sea level, but that of storm waves to several dozens of metres. In the English Channel, shells are occasionally injured by the movement of gravel at depths of 70 m, and on the east coast, when ballast foreign to the region is dumped in water 20–35 m deep, the shore after a storm is strewn with these pebbles (Kuenen, 1950, p.228). Since more energy is needed to destroy portions of a growing reef than to displace gravel, development of the Holmhällar reefs in water of less than 40 m depth is most likely.

The important role played by Algae in the construction of the Holmhällar type reefs is also evidence of formation in shallow water. Even if it is taken into account that the reefs were little influenced by deposition of terrigenous debris, hampering light penetration, it is unlikely that water depth exceeded 40 m. Several Algae receive their optimum illumination at about 15 m depth, but have their normal habitat about 15–25 m below low water (Moore, 1958, p.51).

In conclusion, formation of Holmhällar-type reefs in water shalllower than 40 m is most probable.

Was any minimum water depth required for the formation of Holmhällar-type reefs?

Although in their shape the Holmhällar-type reefs show a relation to the contemporaneous shoreline, no reefs of this type have been observed that originated in such extremely shallow water immediately off the coast, as did the Hoburgen-type reefs which developed at the time that the Hamra algal limestone was formed. The exposures in the environment of Snabben suggest that in water which is becoming shallower there is a tendency to pass into reef limestone of Hoburgen type.

In their shape, the larger Holmhällar-type reefs in particular are comparable to the cuspate reefs among the present-day shelf reefs (cf. Maxwell, 1968, p.99). These modern reefs are particularly found along the shelf edge, taking maximum advantage of little polluted, and nutrient- and carbonate-rich oceanic water. Their landward expansion is restricted by the partial insulation of the backreef areas from the open sea. Consequently these reefs generally elongate parallel with the shelf edge. However, when they become longer, and particularly when the ends of adjacent reefs approach one another, the reef ends tend to project backward. These landward arms are narrower than the main, seaward section.

Behind the outer series of scattered cuspate reefs, on modern shelves oval-shaped or elongated platform reefs may develop which generally are smaller than the outer reefs.

Both in the Hemse Beds and in the Hamra-Sundre Beds, reefs of Holmhällar type are found southeast of reefs of Hoburgen type. This suggests that the former developed at some larger distance from the coast than these Hoburgen-type reefs. This would agree with the picture presented by the modern shelf reefs.

It is probable, therefore, that reefs of Holmhällar type for their development required a set of environmental conditions which was not realized in extremely shallow water. It is difficult to specify what the minimum water depth was in which these reefs still could develop. Since several factors were involved (including depth, wave action, currents, degree of water pollution, nutrient supply), this will probably not have been a sharp depth limit. It may, however, be a reasonable guess that the minimum water depth was on the order of perhaps 10–15 m.

A logical final question is whether Holmhällar-type reefs are characteristic for a water depth between 10–40 m or whether other factors also made them what they are.

In the range from relatively deep water in the Lower Visby Beds to very shallow water in the Upper Högklint Beds, no interval occurred in which reef limestone of Holmhällar type was formed. Thus it cannot be concluded that reefs of this type are, as a matter of course, characteristic of a certain depth zone.

It has been noted that reef limestone of Hoburgen type generally formed under changing environmental conditions, more especially during decreasing or increasing water depth (Chapter VII). Available evidence of the most characteristic Holmhällar-type reefs points to a rather even set of conditions. The Holmhällar-type reefs in the Hemse Beds are somewhat older than the majority of Hoburgen-type reefs in that stratigraphical unit. The Lower Hemse Beds, except for the lowermost ones, suggest deposition in a shallow sea of rather constant depth (p.386). The Holmhällar-type reefs in the Hamra-Sundre Beds are presumably somewhat younger than the Hoburgen-type reefs of that stratigraphical unit and may have formed synchronously with the Sundre limestone in the southeast of the southern peninsula of Gotland. The latter deposit also reflects a deposition while water depth remained roughly the same (p.422). It thus appears that in both cases the larger Holmhällar-type reefs may have formed in an interval of time when the epicontinental sea was not affected by epeirogenetic movements. It may have been this latter factor which was in fact of decisive importance in the formation of reefs of Holmhällar type.

The occurrence of, or absence of, epeirogenetic movements seems not only to have determined the character of the reefs which developed, but also – if reefs of more than one type originated – the order in which their growth started. We have seen already, in the previous two chapters, how a continued decrease in water depth in Late Visby and Early Högklint time led to a replacement of Upper Visby-type reefs by Hoburgen-type reefs. As said above, in Early Hemse time Holmhällar-type reefs originated in water of rather stable depth. A transition from these reefs of the Holmhällar type to reefs more of the Hoburgen type occurred when water depth began to decrease (area between Snabben and Sandviken). While the water depth was fluctuating, Hoburgen-type reefs developed between the existing Holmhällar-type reefs and the coastline. In the case of the Hamra-Sundre Beds, a change in water depth first occurred. This was an increase in depth. When the environmental situation stabilized again, Holmhällar-type reefs developed at the seaward side of the existing Hoburgen-type reefs.

Chapter IX

REEF DEBRIS

In the descriptions of the reef types of Gotland in the preceding three chapters, attention was also paid to the sediments around these reefs, and on the reef debris occurring in them. In the present chapter this debris is the subject of some further discussion. Emphasis will be laid on the distribution of reef debris and on the directions of dip that can be measured.

DISTRIBUTION OF REEF DEBRIS

Following the information that reef debris is found in the sediments around the reefs, a logical question is how much of this debris is present, expressed, for instance, as a percentage of the total rock volume, at various distances and directions away from these reefs. Fortunately, the exposures in Gotland are such that at least a partial answer can be given to this question.

Approximation methods

To ascertain the percentage of the total volume of sediment which is made up of reef debris, an approximation method was used, which could be applied in the field. A vertical surface varying between 400 and 2500 cm^2 was mapped as carefully as possible on millimetre paper, at a scale of 1:1 or 1:2.5. All pieces of rock which, macroscopically, could be expected to be reef debris, were drawn in on the map. Crinoid remains were not included among the reef debris. The number of square millimetres which the pieces of debris together occupied was counted. The percentage which this sum

TABLE XIII

Approximation of number of pieces of reef debris per cubic decimetre

Size group	Measured data (400 cm^2)		Calculated data	
	Number of pieces	Average longest dimension (cm)	Represen-tative volume (cm^3)	Number of pieces per dm^3
0.5–1	28	0.7	280	100
1–2	11	1.4	560	20
2–3	4	2.4	960	4
3–4	2	3.4	1360	ca. 1.5
4–5	1	4.6	1840	ca. 0.5
> 5	2	5.8	2320	ca. 1
Total	48			127

constituted of the total surface that was mapped, was considered to be representative of the volume percentage. Before mapping some surfaces needed to be treated with diluted hydrochloric acid. The size of the surface which was mapped was determined by the nature of the exposure and by the coarseness of the exposed debris. The coarser the latter, the larger the surface that was used, if possible.

A more rapid, but less accurate method could be used if it was judged sufficient to have an impression of only the number of pieces of reef debris in a unit volume of sediment (generally a cubic decimetre). The longest dimension of all macroscopically recognizable pieces of reef debris was measured. Generally 5 mm was taken as the lower boundary. The figures obtained were arranged in a number of size groups (see the example in Table XIII). Each size group was considered representative of a rock slice of the same thickness as the average size of that group. By multiplying the surface of the area studied by the average size of the pieces of debris, a volume is found for which the observed number of pieces was more or less representative. On the basis of this, the number of pieces per unit volume can be estimated. By adding up the figures obtained in this way for each size group, an approximation is obtained of the total number of pieces of debris in a certain rock volume. Of course, this method is rather rough, but if used consistently, it gives some impression of the variations in the distribution of macroscopically-recognizable reef debris. It has been used particularly in studying the debris distribution in vertical successions.

Distribution of reef debris in a vertical direction

In order to obtain some impression of the distribution of reef debris in a vertical section, some well-representative exposures had to be studied. As such Slite IV Beds of the Solklint (Slite) and the Bogeklint or Klinteklint (Boge Parish) which are found about 10 m away from the nearest reef, were selected (Fig.92). The rocks exposed there are irregularly stratified crinoid limestones with reef debris, and a matrix of crinoid sand and some marl. The size of the debris varies greatly, from very small fragments to complete colonies of massive corals or stromatoporoids. Debris material of all sizes occurs randomly together in several layers; in some other layers the debris is both coarser and more abundant than is the average for the exposed succession. Examination of thin sections shows that the crinoid material has also been subjected to varying degrees of sorting. Apparently occasional variations in water mobility operated during the development of the reefs. With increasing distance from the reef, the sorting of the material in such layers becomes increasingly less noticeable.

Leaving apart the variations between successive layers, which may be ascribed to varying wave strength, the general trend is that of an increase in the amount of reef debris going upwards, until some distance below the level of the top of the reef, when a decrease sets in. The explanation seems to be obvious. During reef growth the reefs gradually came to extend further over the surrounding sea floor and rose higher. Such larger and higher obstacles were more strongly attacked by water movements. After reef growth ended, the elevation of the reef over the surroundings decreased again and thereby also the effect of the destructive action of moving water on the reef.

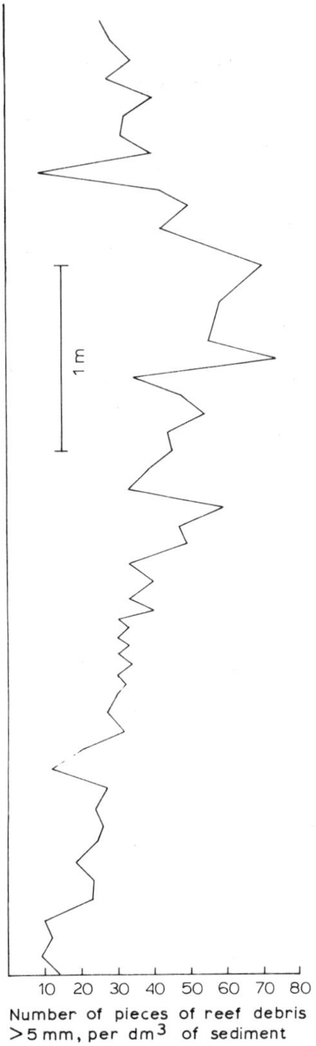

Fig.92. Number of pieces of reef debris (larger than 5 mm) per cubic decimetre of crinoid limestone with reef debris, in a succession of strata. The data were collected in the Solklint (Slite) and Bogeklint (Boge), Slite IV Beds. The graph is representative for a deposit found at the southeast side of the Slite IV reefs, at a distance of about 10 m from a reef. The graph starts about 1–1.5 m above the level of the reef base and ends about 1 m below the level of the top of the reef. On the average there is a gradual increase in the amount of reef debris upwards, until about 2.3 m below the level of the reef top.

In the example of Fig.92, the reversal of debris increase to debris decrease occurs at about 2.3 m below the level of the top of the reefs. These reefs are probably about 7.5 m thick. The vertical distance between the level of quantitative reversal in debris deposition and the level of the top of the nearest reef limestone may be an indication of the height of the reef over its direct surroundings at the end of reef growth. The reefs of the Slite IV Beds developed in water which became shallower (cf. Chapter XI). In reefs of comparable thickness which grew in gradually deepening water, the elevation over the contemporaneous sea floor presumably was somewhat greater (e.g., the Hoburgen reefs of the Hamra-Sundre Beds).

As soon as a young reef extended somewhat over the surrounding sea floor, it was apparently attacked by destructive forces. As mentioned in Chapter VII, the stratified limestone underneath the peripheral parts of the reefs of Hoburgen type, generally shows a distinct increase in the amount of reef debris in upward direction. In the uppermost part of it, the debris may even become the dominant rock constituent, so that reef expansion often took place over a thin layer of reef debris. The debris layer has no sharp lower boundary and, in several instances, also no sharp upper boundary with the overlying reef limestone.

Reef destruction by moving water seems to have continued until an entire reef was buried. Crinoid limestones which overlie reef limestone, e.g., in Hoburgen, generally still contain reef debris. The size of this debris is smaller than in the deposits lateral to the reefs. Within a distance of a few decimetres upwards from the reef limestone, both the size and quantity of the reef debris in the crinoid limestone usually decrease rapidly.

Around reefs of Holmhällar type, the vertical distribution of reef debris is more or less similar. This can be studied, for instance, about 0.3 km north of Sysne-udd, in deposits belonging to the Hemse Beds. The exposures are found along the present coast. Close to the reef limestone and lateral to it, much very coarse debris is present. About 1 m below the level of the top of the reef limestone, debris material up to 70 cm long is present. A few metres away from the reef limestone, most debris is smaller than 20 cm, but coarse fragments up to that size remain rather common in the reef-surrounding sediments up to some tens of metres distance. In an upward direction, there is a decrease in the volume and coarseness of the debris, both close to the reef and at greater distance. The crinoid limestone which overlies the reef limestone still contains much debris, but by far the majority of the pieces is smaller than 2 cm; pieces larger than 10 cm are rare.

Reef limestone covered by crinoid limestone is also exposed between Sysne-udd and Sysne. There it can be observed that in places the uppermost part of the reef limestone itself shows a distinct increase in reef debris and crinoid remains. The covering crinoid limestone also contains reef debris, which decreases rapidly in abundance and size in an upward direction. Together with the decrease in debris volume, the crinoid limestone, in upward direction, shows an increasingly more distinct stratification, which then also becomes smoother.

The sediment found about 0.7 km south of Snabben (cf. Chapter VIII, p. 208) was presumably deposited on top of a lower, peripheral part of a reef. 0.6 m above the reef limestone, the debris still constitutes about 25% of the total volume of rock and the largest piece of debris measured 17 cm. About

1.5 m above the reef limestone, only small fragments are present, but still with a total volume of about 7.5%. The high volume of reef debris in this vertical succession makes it most unlikely that the material is laid down over the highest reef part.

Horizontal distribution of debris around some Hoburgen-type reefs

As appears from the preceding section, care should be taken that the distribution of reef debris in a horizontal direction is studied everywhere at about the same distance above the level of the base of the reef which supplied the debris. In Fig.93, twenty-nine approximate volume percentages of reef debris are given, obtained from a level which generally was slightly above half-way up the reef. Since sufficient data could not be obtained from only one reef, figures are given from the surroundings of a number of

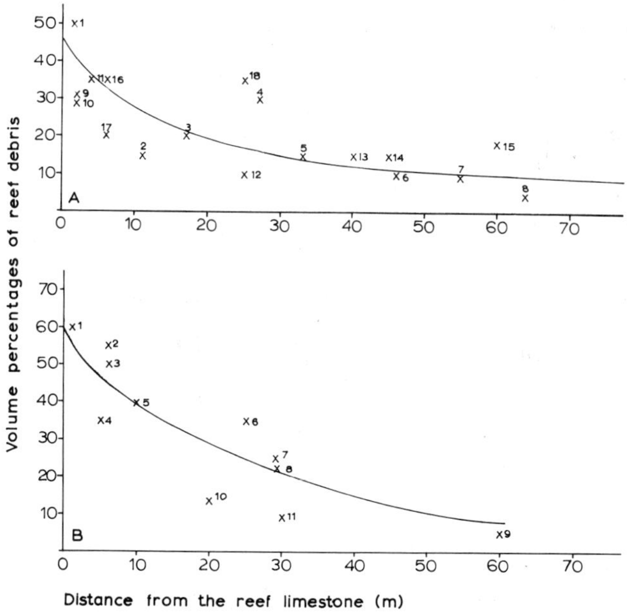

Fig.93. Percentage of reef debris (larger than 5 mm) in crinoid limestone at the southeast (A) and south side (B) of some reefs of Hoburgen type, which developed in comparatively shallow water. At the time of formation of the debris the reefs extended probably 2–3 m above the surrounding sea floor. The approximations of the debris volumes were made in the following localities: A: 1–8. Solklint, Slite; 9–11. Bogeklint, Boge Parish; 12–13. Stora Vede, Follingbo Parish; 14. Endre Backe, Endre Parish; 15. north of Norrbys farm, Follingbo Parish (in all these localities sediments belonging to the Slite Beds are exposed); 16–18. Lindeklint, Linde Parish, Hemse Beds. B: 1. Klinteklint, Gammelgarn Parish, Hemse Beds; 2–9. Lindeklint, Linde Parish, Hemse Beds; 10–11. about 0.7 km southeast of Fole Church, Slite Beds.

different reefs. All of these are assumed to have developed under more or less similar environmental conditions.

From the two graphs of Fig.93, it appears that more debris is present, very close to the reefs, at the south side of the reef than at the southeast side. This means more debris against the shorter lateral sides of the reefs than against the long edge at the seaward side. However, the decrease in the volume of debris is more rapid at the south side. The volume percentages are most strongly influenced by larger pieces of reef debris. The graphs thus mainly illustrate that coarser material was deposited at the places where the water passed along the reef. This appears to be the case even when there is no question of a passage between two neighbouring reefs. However, in such a passage the presence of coarser material is more pronounced (Fig.74).

In the central part of Fig.96, data such as those from the graphs of Fig.93 are generalized to a distribution map. Data for the landward side of the reef are also added. There, close to the reef, the debris volume is higher than anywhere else. The decrease in the amount of reef debris with increasing distance from the reef is very rapid. Apparently much debris was washed off the reef surface and deposited directly at the lee side.

The distribution pattern of reef debris is simplest where reefs are regularly spaced and separated from each other by a distance which is several times their own length. The closer the reefs occur together, the more complicated the picture becomes. This is particularly true if, in addition, the reefs are randomly distributed over a wide zone. The author does not have sufficient data for discussion of such complicated distribution patterns of reef debris.

It is often difficult to determine the exact distance between a vertical section through crinoid limestone with reef debris and the nearest reef limestone. This distance is generally not one that can be measured along an exposed cliff wall, because these walls are usually not perpendicular to the boundary between reef and stratified sediment. The exposures present on the plateau of a cliff may be of some help. However, the boundary was generally not vertical. This implies that the estimated distances from a reef may be in error with one or sometimes even a few metres. The influence of this error on the general distribution pattern as given in the map of Fig.96, however, is negligible.

As mentioned in the general description of the Hoburgen reef type, a number of reefs, particularly in Hoburgen itself, have a distinct mantle of reef talus directly around them. Good examples are found in the third hillock (Fig.67, 68) and in the north of the western cliff of the Storburg of Hoburgen (Fig.59). In both localities, large blocks of tumbled reef rock are included in the talus mantle. These blocks stand at all angles. At the site of such blocks the volume percentage of the reef debris if, of course, 100. Since such extreme values are very local phenomena, they have been excluded from the distribution map in Fig.96. A narrow talus mantle without such blocks is exposed at the northeastern end of the fourth hillock of Hoburgen. Such a talus mantle was found in connection with only one of the reefs which produced the data on which Fig.93 and 96 are based. This is the Klinteklint in Gammelgarn Parish, where it was found at the south side.

Horizontal distribution of debris around some Holmhällar-type reefs

As mentioned in Chapter VIII, some impression of the sediments which were formed around reefs of Holmhällar type can be obtained in the coastal area between Snabben and Sysne-udd (see also Fig.94). Directly against the reef limestone a deposit is present which can best be described as a talus mantle. It may contain as much as 80% reef debris. Over a distance of some metres this passes into a limestone with crinoids, which is still extremely rich in reef debris. Initially this limestone is only vaguely stratified, but with increasing distance from the reef limestone, the stratification becomes clearer. Marl may occur on the bedding planes, with very fine reef debris embedded in it. Generally, at a distance of about 25 m from the reef, the stratified limestone still consists for about 25% of macroscopically recognizable reef debris.

By mapping surfaces of 625–2500 cm^2 in fifteen different places, approximations were made of the percentages which the reef debris takes up of the total volume of reef-surrounding sediment. The rough distribution pattern shown in the central part of Fig.97 is based on the values obtained. At the seaward side of the reef flanks, slightly more reef debris was deposited than at the middle of the seaward side of a reef of this kind. Only two approximations of the volume percentage of reef debris could be made at the concave side, that is the original landward side. These figures suggest that most of the debris found at that side is less coarse than that found at the seaward side and takes up a somewhat lower percentage.

A comparison of Fig.97 with Fig.96, shows that at least at the original seaward side, in the surroundings of the reefs in the Snabben – Sysne-udd area, much more debris was laid down than in the surroundings of the Hoburgen-type reefs in the Slite and Upper Hemse Beds.

It is not easy to give a satisfactory explanation of the significant differences found in the amount of reef debris at the seaward side of the two groups of reefs. A possible cause may be that the Holmhällar-type reefs in the Snabben – Sysne-udd area did not develop under their most favourable conditions (cf. Chapter VIII). There are transitions to the Hoburgen reef type. Moreover, the latter reefs, particularly those found somewhat more northwards along the coast northeast of the harbour of Herrvik and near Kuppen, are surrounded by more reef debris than is normal for Hoburgen-type reefs. Perhaps under the transitional environmental conditions, the reefs were less solidly built, or the destructive agents may have been much stronger.

At any rate, it is clear that different kinds of reefs and different environmental conditions may have led to notable differences in the amounts and distribution of reef debris found around the reefs. No generally valid rule can be produced on this subject.

DIRECTIONS OF DIP IN REEF DEBRIS

Among the reef debris there are generally several flat pieces, particularly fragments of stromatoporoid latilaminae. Some of these flat pieces of reef debris are found in an about horizontal position, whereas others show a certain dip. In a number of exposures, the directions of these dips have been measured.

Two examples are given in the two lefthand diagrams of Fig.95. Both

A

B

C

D

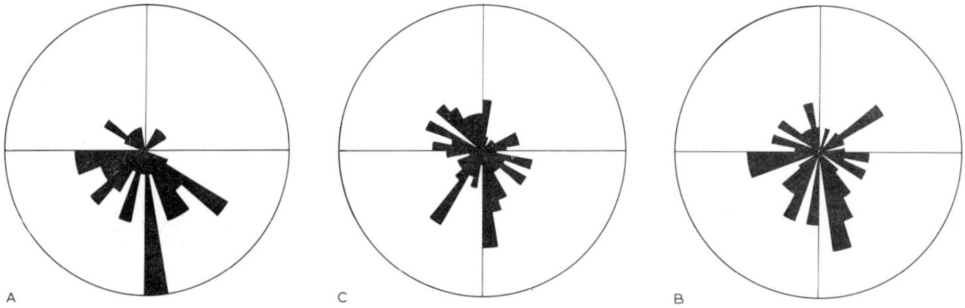

Fig.95. Directions of dip shown by flat pieces of reef debris larger than 1 cm in diameter (A and B) and crinoid columnals longer than 2 cm (C) at the south to somewhat south-southeast side of a reef of Hoburgen type.
A. Bogeklint, Boge Parish, Slite Beds, deposit formed about 6 m away from the nearest reef limestone and about 3.5 m below the level of the top of that reef. B and C. Solklint, Slite, Slite Beds, deposit about 6 m away from and about 3 m below the level of the top of the nearest reef limestone.

show the dip directions found in debris deposited south to somewhat south-southeast of a reef. A fair proportion of the pieces of debris show a direction of dip away from the reef. Dips in other directions also occur, but only few pieces dip in the direction towards the reef limestone. It is understandable that several pieces dip in an about southeastward direction. This is a kind of compromise direction between a dip away from the reef and a dip against the main direction of water movement directly around the reef, which may generally have been away from the middle of the seaward side of the reef and towards and along the peripheral parts. That there also are many pieces which dip more in a southwest to westerly direction is less easy to understand. The positions of these pieces were less stable with respect to the assumed main direction of water movement. Perhaps a rather uneven debris-covered sea floor close to the reef has also played an important part.

The righthand diagram of Fig.95 shows the directions of dip which were found in crinoid stem fragments of 2 cm and longer, at the same side of the reef limestone. The scattering in dip directions is greater here. Particularly there are several fragments which dip in a direction about perpendicular to the dip down from the reef. The greater variation in the direction of dip may have various reasons. Crinoids may have lived both on the flanks of the reef and in the area where they are found. Stem remains of crinoids which grew on the reef flank may rather have rolled down than have slipped down in their length direction. The remains of stems which disintegrated on the spot

Fig.94. Reef limestone of Holmhällar type and surrounding debris deposits. Between Snabben and Sysne-udd, Hemse Beds. A. Horizontal exposure of reef limestone, with large stromatoporoids partly liberated through erosion, diameter of the stromatoporoid elevations about 15–50 cm. B. Vertical section through reef limestone at the southeastern periphery of a reef.
C. Reef debris about 2.5 m away from the southeast side of a reef.
D. Indistinctly stratified limestone with much reef debris, about 25 m away from the southeast side of a reef.

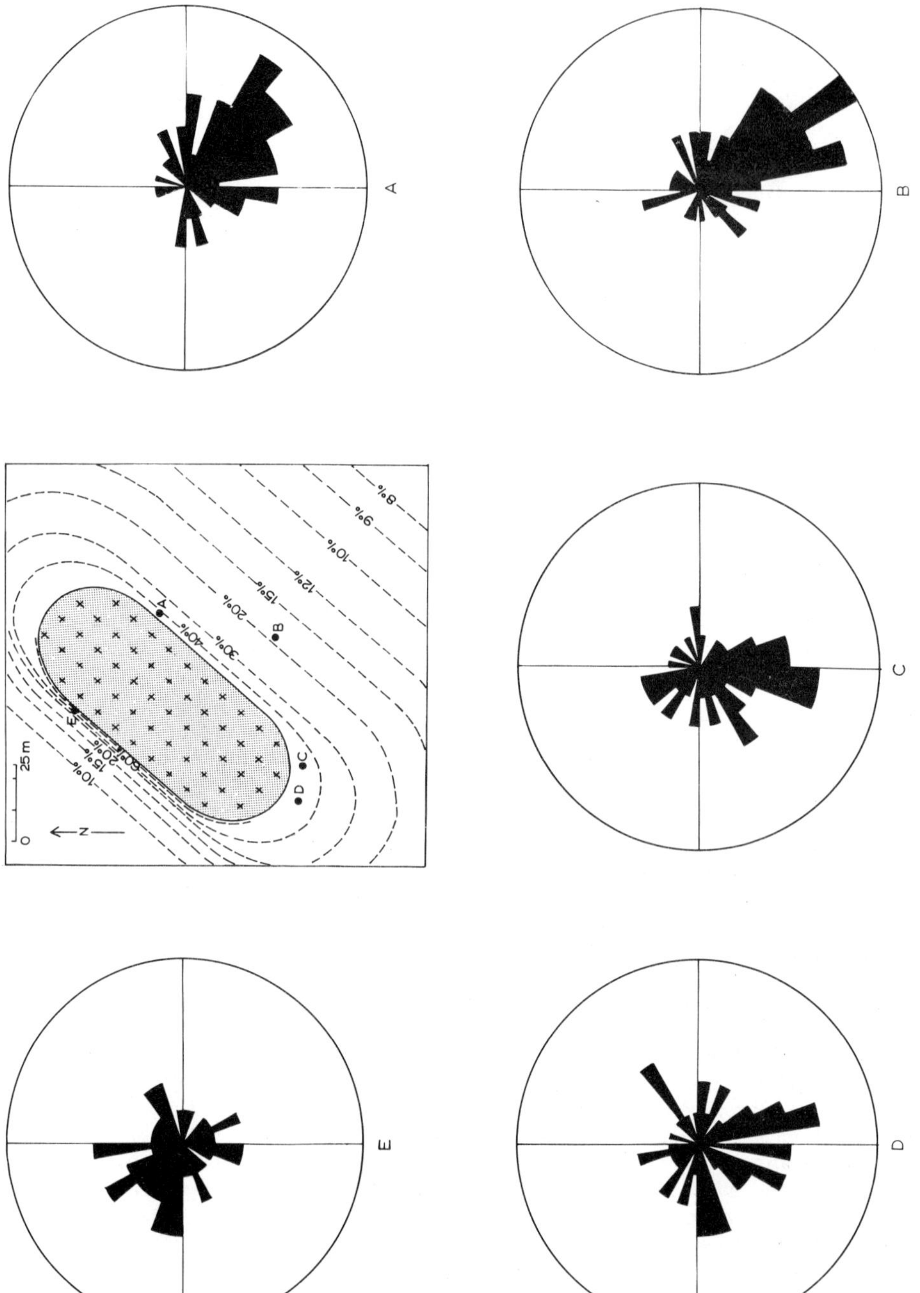

Fig. 96 Legend see p. 223.

will have fallen down in even more random orientations. Finally water movement may have modified the positions also in the case of the crinoid stem fragments. Probably all these causes have played a part, but perhaps rolling down and displacements by moving water have been the two most important ones. At any rate it is clear that the dips found in flat pieces of true reef debris and in crinoid columnals should not be united in one diagram.

In several more places, the directions of dip shown by flat pieces of reef debris were measured. These are combined into the diagrams of Fig.96 and 97. The main direction appears everywhere to be one that is roughly perpendicularly away from the reef margin.

One diagram deserves special attention, that of Fig.97C. This is based on reef debris in a filled depression. The main dips are in a direction towards the convex (original seaward) side of the reef. This may suggest that most of the debris found in this depression came from the southwest. It may be recalled here, that there is an indication that more reef debris was deposited at the convex than at the concave side of the reef. These two indications together may again suggest that during their growth, the reefs of Holmhällar type, as found in the Snabben – Sysne-udd area, had an upper surface which sloped towards the open sea. Perhaps this was the result of some seaward expansion of these reefs over their own debris. In this connection reference is also made to the paragraph on stromatoporoids, early in Chapter VII. It was mentioned there, that in some places in the original-seaward periphery of the Ljugarn reef, tabular stromatoporoids were seen to be dipping moderately reefdownwards (p.183).

For each group of reefs, which belong to the same type and developed under rather similar environmental conditions, a map showing the distribution of reef debris around them would be useful. In the case of sites which expose a reef-debris containing sediment, but whose positions with regard to the nearest reef cannot be located from direct observation, such a map would nevertheless permit a reasonable delimitation - provided that the amount of macroscopically recognizable reef debris and the directions of dip therein are determined at these sites.

Fig.96. Map, showing the abundance of reef debris larger than 5 mm, in percentages of the total volume of sediment, around a reef of Hoburgen type. The reef is schematized. The distribution pattern of the debris is mainly representative for reefs which developed in water which gradually became shallower, such as those found in the Slite IV and Upper Hemse Beds. At the time of formation of the debris the reefs presumably extended 2–3 m above the surrounding sea floor. Five diagrams give an impression of the directions of dip which flat pieces of reef debris display at various places around the reef. The diagrams are based on data from the following localities: A. old quarry in the northeast of the Solklint, Slite, Slite Beds; B. southeastern part of the Lindeklint, Linde Parish, Hemse Beds; C. southern part of the Lindeklint; D. west wall of the Solklint; E. northeastern part of the Klinteklint; Gammelgarn Parish, Hemse Beds.

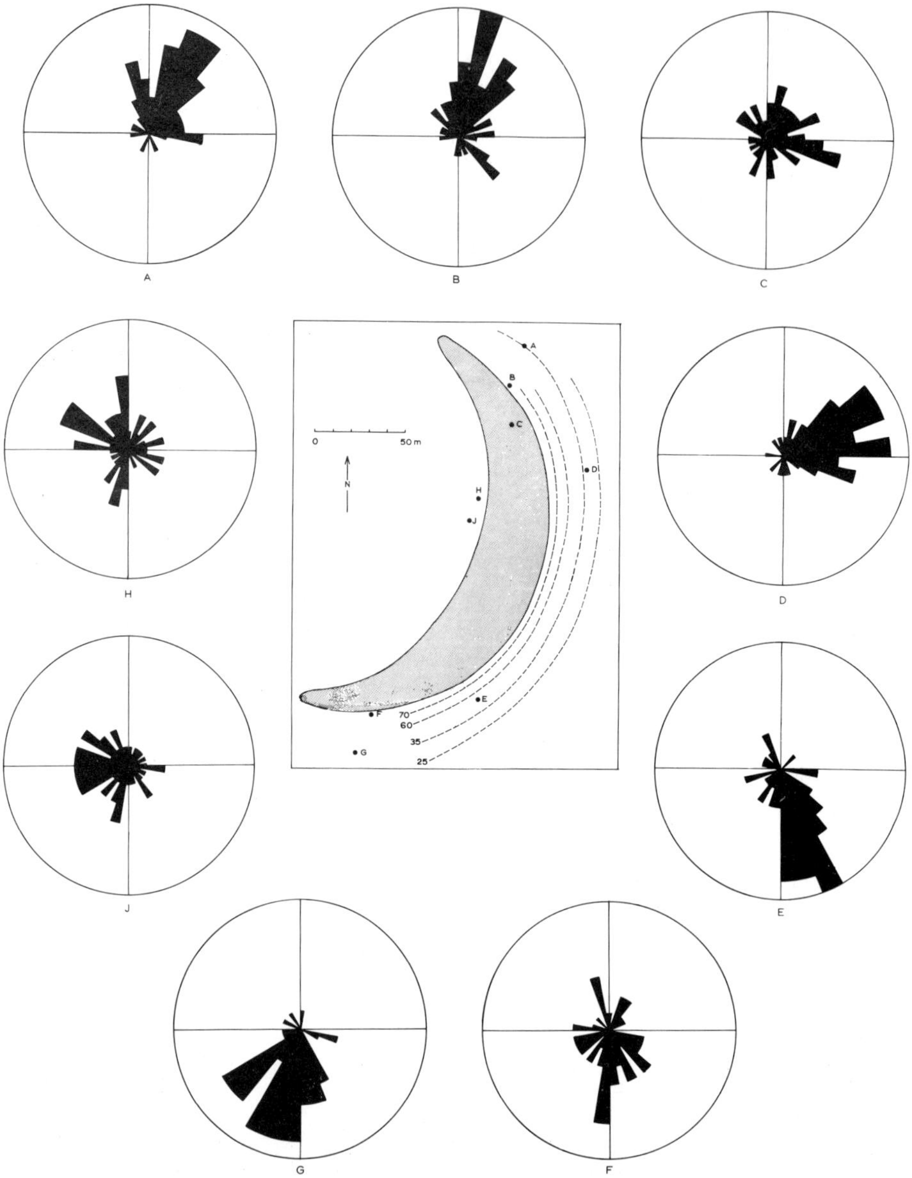

Fig.97. Map, showing the abundance of reef debris larger than 5 mm, in percentages of the total volume of sediment, in sediments at the seaward side of a reef of Holmhällar type. Nine diagrams illustrate the directions of dip found in flat pieces of reef debris in the reef-surrounding sediments and in a filled depression within the reef. All data were collected in the area between Snabben and Sysne-udd, Östergarn Parish, Hemse Beds.

Chapter X

STRATIGRAPHY AND REEFS OF KARLSÖARNA

INTRODUCTION

Karlsöarna (the Carl Islands) comprise two small islands, Stora Karlsö and Lilla Karlsö, 2.5 and 1.39 km^2 large, respectively. The islands are located west of Gotland, in the Baltic, about 16.5 and 11 km southwest of Klintehamn, respectively, with which harbour they have a boat connection during the summer months. Administratively they both belong to Eksta Parish, Gotland. Stora Karlsö is the property of the "Karlsö Jakt- och Djurskydds-förenings Aktiebolag", generally called "Karlsöklubben" for short. Lilla Karlsö is owned by the "Svenska Naturskyddsförening". Both organizations were kind enough to permit me to live and study on their islands for a few weeks.

The reef limestones of Karlsöarna present several problems. These include their mode of formation, their exact position within the stratigraphy of the Silurian, their relationships to the fossil reefs of Gotland, their fauna and environmental significance, and their relation to the general pattern of sedimentation of the region in which they occur.

In the Late Quaternary both islands were partly submerged in the Baltic. Abraded coastal slopes are found from the *Ancylus* and *Littorina* times, with caves formed by the sea at those times. The often vertical cliff walls present the best exposures of both the stratified and reef limestones. Close to the precipices, remains of disappeared reefs stand in the form of "raukar".

STRATIGRAPHY OF STORA KARLSÖ

In the following pages the stratified sediments of Stora Karlsö will be described first, in an attempt to get some impression of the geological nature of the island (cf. Table XV on p.275).

At the base of Stora Karlsö, sediments are exposed which are known as the Lerberg Marlstone. This is overlaid by limestone, which the present author proposes to subdivide into the Spangände Limestone and the Austerberg Limestone.

A list of fossils found in the stratified sediments of both Karlsöarna is given in Table XIV.

The positions of the more important localities, to be mentioned in the following descriptions, are shown in Fig.98.

(Text continues on p. 229)

TABLE XIV

Fossils found in Stora and Lilla Karlsö

Fossils	Stora Karlsö			Lilla Karlsö	
	Lerberg Marlstone	Spangände Limestone	Austerberg Limestone	*Pentamerus gotlandicus* Limestone	Lilla Karlsö Limestone
HYDROZOA					
Clathrodictyon striatellum (d'Orbigny)	+				
Unidentified stromatoporoids	+	+	+	+	+
ANTHOZOA TETRACORALLA					
Acervularia ananas singularis Lang et Smith		+		+	+
Cystiphyllum cylindricum Lonsdale	+	+		+	
Entelophyllum roemeri Wedekind					+
Ketophyllum spp.	+			+	
Rhizophyllum gotlandicum (Roemer)		+			
Schlotheimophyllum sp.					+
Syringaxon cf. *siluriense* (McCoy)		+			
ANTHOZOA TABULATA					
Aulopora sp.	+				
Favosites gothlandicus Lamarck	+	+	+	+	+
Halysites catenularis (L.)	+	+		+	+
Syringopora sp.	+			+	
Thamnopora sp.			+		
ANTHOZOA HELIOLITIDA					
Heliolites interstinctus (L.)	+	+		+	
Heliolites spp.	+		+		+
Plasmopora petalliformis (Lonsdale)	+				
Plasmopora scita Edwards et Haime	+				?
Unidentified corals	+	+	+	+	+
ANNELIDA					
Cornulites sp.			+		+
Scolecodonts	+				
CRINOIDEA					
Unidentified crinoid remains	+	+	+	+	+
BRYOZOA					
Fenestella sp.			+		+
Unidentified bryozoans	+	+	+	+	+
BRACHIOPODA					
Amphistrophia funiculata (McCoy)	+		+		
Anastrophia deflexa (J. de C. Sowerby)	+				
Antirhynchonella linguifera (J. de C. Sowerby)	+	+	+	+	
Atrypa reticularis (L.)	+	+	+	+	+

TABLE XIV (continued)

Fossils	Stora Karlsö			Lilla Karlsö	
	Lerberg Marlstone	Spangände Limestone	Austerberg Limestone	*Pentamerus gotlandicus* Limestone	Lilla Karlsö Limestone
BRACHIOPODA (continued)					
Camarotoechia borealis (Buch)	+	+			+
Chilidiopsis pecten (L.)	+				
Conchidium biloculare (Hisinger)					+[1]
Conchidium sculptum (Walmstedt)		+	+	+	
Cyrtia exporrecta (Wahlenberg)	+				
Delthyris elevata Dalman			+		
Dicaelosia biloba (L.)	+	+		+	
Dolerorthis cf. *rustica* (J. de C. Sowerby)					+[1]
Eospirifer cf. *interlineatus* (J. de C. Sowerby)	+				+[1]
Eospirifer radiatus (J. de C. Sowerby)					+
Glassia obovata (J. de C. Sowerby)	+	+			
Gypidula cf. *galeata* (Dalman)			+		+
Howellella elegans (Muir-Wood)	+				
Leptaena rhomboidalis (Wilckens)	+	+	+	+	+
Meristina obtusa (J. Sowerby)	+				
Nucleospira pisum (J. de C. Sowerby)	+				
Orbiculoidea rugata (J. de C. Sowerby)				+[2]	
Pentamerus gotlandicus Lebedev		+		+	
Plectatrypa imbricata (J. de C. Sowerby)	+	+		+	
Plectatrypa marginalis (Dalman)					+[1]
Plectodonta transversalis lata (Jones)	+	+	+	+	
Ptychopleurella bouchardi (Davidson)	+	+		+	
Resserella elegantula (Dalman)	+				
Resserella sp.					+
Rhynchotreta cuneata (Dalman)		+			+
Skenidioides acuta (Lindström)	+				
Sphaerirhynchia wilsoni cf. *sphaeroidalis* (McCoy)					+
"*Strophomena*" *rugata* Lindström					+
? *Whitfieldella* sp.			+		+
GASTROPODA					
Euomphalopterus alatus (Wahlenberg)	+				
Platyceras cornutum Hisinger	+	+	+	+	+
Platyceras spiratum (Sowerby)			+		
Pleurotomaria limata Lindström					+
Poleumita globosum (Schlotheim)	+				+
Poleumita sculptum (J. de C. Sowerby)	+				
Trochus mollis Lindström					+
CEPHALOPODA					
Dawsonoceras annulatum (J. Sowerby)					+
"*Orthoceras*" sp.	+				+
TRILOBITA					
Bumastus sp.			+		+
Calymene sp.					+
Encrinurus punctatus (Wahlenberg)					+
Proetus delicatus Hedström					+
Proetus sp.					+
Scutellum polyactin (Angelin)		+			
Sphaerexochus scabridus Angelin	+				

TABLE XIV (continued)

Fossils	Stora Karlsö			Lilla Karlsö	
	Lerberg Marlstone	Spangände Limestone	Austerberg Limestone	*Pentamerus gotlandicus* Limestone	Lilla Karlsö Limestone
OSTRACODA					
Craspedobolbina (Mitrobeyrichia) clavata (Kolmodin)					+
Craspedobolbina (Mitrobeyrichia) insulicola Martinsson	+				
Leperditia sp.					+
Unidentified ostracodes	+				+

[1]From the uppermost Lilla Karlsö Limestone
[2]From sandy limestone in the northeast of Lilla Karlsö

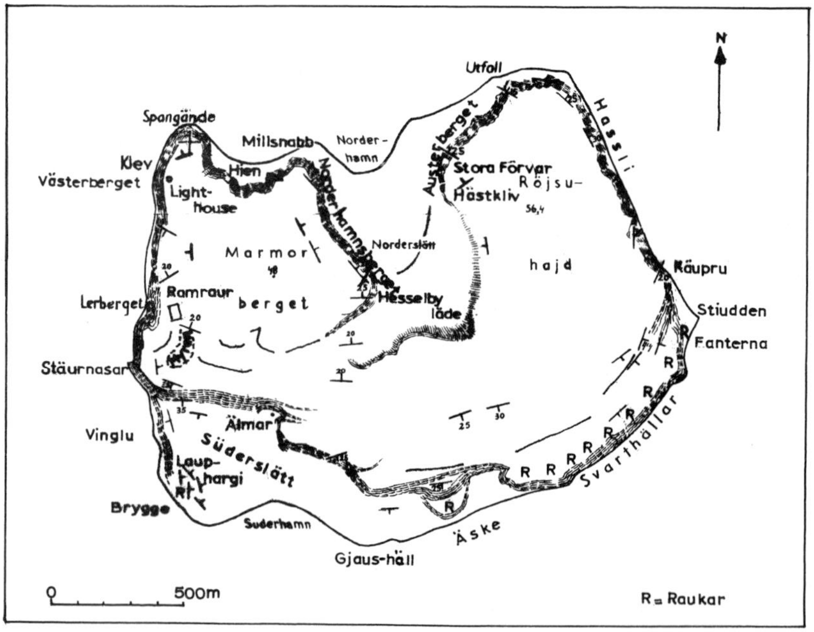

Fig.98. Map of Stora Karlsö. Directions of dip in the stratified limestones are added. RR = raukar.

Lerberg Marlstone

The oldest sediments exposed in Stora Karlsö belong to the Lerberg Marlstone. This is a bluish grey to brownish grey, sometimes greenish coloured marlstone, which is dense, rather soft, and generally very thinly foliating. The marlstone layers alternate with thin layers or lenses of marly limestone, which is harder, finely crystalline and of a light-grey colour.

The Lerberg Marlstone is exposed along the west, northeast and east shores of the island. Along the west coast it forms in the north the so-called "Lerberg" (Swedish: lera = clay), the lower part of the Västerberg, where it reaches a thickness of about 15 m. In this locality it can be followed from Spangände southwards until Klev, generally showing an approximately horizontal stratification. Further south, three consecutive "folds" may be seen, which, though gentle, are easily observed. The extra weight of the overlying reefs presumably caused the sagging of the layers here, as well as some plastic flow of material in lateral directions. Of these "folds" the southern one is the largest. Close to and a little south of Ramraur it reaches a height of about 25 m above sea level; further southwards the upper boundary of the Lerberg Marlstone descends gradually to a lower level and at Stäurnasar it disappears below sea level. South of Stäurnasar the Lerberg Marlstone appears again at the base of the cliff, but there it only reaches a relatively slight elevation. It can then be followed along the shore to directly north of Brygge, sometimes lying horizontally, sometimes faintly undulating. Along the northeastern and eastern shores of the island, the Lerberg Marlstone occurs in the foot of the cliff and along the beach from directly west of Utfall eastwards, where it reaches an exposed thickness of 10–15 m, and further to the south-southeast. In the latter direction the layers in main lines gradually dip away, finally disappearing below sea level at Käupru. Along this shore the layers also manifest a faintly undulating pattern.

As a rule, the Lerberg Marlstone is very rich in fossils, especially in corals, stromatoporoids (tabular and flat lenticular) and brachiopods. Less common are bryozoans, crinoids, gastropods, ostracodes and trilobites. Lamellibranchs are not found.

The occurrence of corals in the Lerberg Marlstone is particularly characterized by the extreme abundance of only one or a few species in a number of special horizons. Among these *Ketophyllum* sp. (in earlier literature *Omphyma subturbinata* and *O. turbinata*) are the most important. These horizons can be partly considered as biostromes. Often the solitary corals found there still occupy the same positions as they did on the Silurian sea bottom. At the foot of Spangände, some coral beds with *Cystiphyllum* sp. are beautifully exposed. The *Ketophyllum* Zone is very striking in the Lerberg, containing a large number of individuals with relatively large dimensions, of up to 33 cm long and 13 cm in diameter. The *Ketophyllum* Zone consists of a number of coralliferous layers lying closely on top of each other and extending to a total thickness of about 1.25 m. The base of the *Ketophyllum* Zone is situated about 1 m above sea level in the Lerberg; at Spangände it is found around sea level. Apart from *Ketophyllum* spp. a number of other corals are also represented (*Halysites catenularius*, *Heliolites* sp.; other solitary forms), but these are much less abundant. Other fossils, such as a few brachiopods and stromatoporoids occur only scarcely. The *Ketophyllum* and *Cystiphyllum* beds apparently represent localized conditions which were so favourable to coral growth that other organisms were more or less excluded. Individual solitary corals are also found in almost all other exposures of the Lerberg Marlstone.

Spangände Limestone

The Spangände Limestone is exposed in several places in Stora Karlsö, generally as a stratified crystalline limestone, surrounding the major reefs of the island and consequently very fossiliferous. Very common are fossils such as crinoid remains, bryozoans, corals and also stromatoporoids; brachiopods are also common, and actually more so in these flanking beds than in the reef limestone. The colour of the rock is generally bluish grey to red-mottled. To some extent there is a direct relationship between the thickness of the layers and the purity of the limestone of which they consist.

At Spangände the lower layers are thin and separated by thin layers of marlstone. As a rule the layers are about horizontal. Higher up in the section the layers are thicker and the limestone somewhat purer. There they show a strong dip, directed roughly towards the north-northwest. Both the layers at the bottom of the cliff and those exposed higher (as is shown by large blocks that have come down) are very fossiliferous. The stronger dip of the higher layers and the reef-detrital nature of the rock suggest deposition close to a large reef (cf. Fig.98).

Sediments as those exposed at Spangände can be followed from there southwards, in the higher parts of the Västerberg. About 150 m south of the lighthouse, the dip is 15–25° to the southwest, but further south the dip is directed more to the west. Downward from the plateau at Ramraur the dip is maximally about 25° but decreases away from the Marmorberg reef (Fig.99).

Fig.99. Ramraur, Stora Karlsö. Stratified sediments deposited around the Marmorberg reef show a dip which gradually decreases with increasing distance from the reef.

Fig.100. Detail of Fanterna, Stora Karlsö, seen from the south. In the foreground, thin-layered marly Spangände Limestone. Behind it, reef limestone of Fanterna type.

The limestone there is greyish white to red-mottled and middle crystalline.

In the steep cliff which extends from slightly south of Ramraur to south of Stäurnasar, the limestone at the bottom is predominantly thin-layered, finely crystalline, partly very marly, grey or bluish grey to red-mottled in colour, and enormously rich in crinoid fragments, corals, bryozoans and stromatoporoids. Higher in the wall the rock has thicker layers and a lighter, yellowish grey to red-mottled colour, and is coarser crystalline and less marly than in the lower parts; it remains highly fossiliferous.

At Stäurnasar the stratified limestone alternates with unstratified reef limestone. Small reefs also occur at Vinglu. At the time of deposition of the lower Spangände Limestone the sea bottom must have been particularly ideal for the existence of corals, especially *Halysites* , of which many large colonies are found.

From Stäurnasar southwards, the stratified limestone described above shows directions of dip which differ quite a bit from one place to another. In the southwest it passes into thick-layered limestone, which is finely to middle crystalline, and of a light-grey colour. Of this limestone the beach slope at Brygge is built up. This limestone is especially characterized by the occurrence of the brachiopod *Pentamerus gotlandicus*. In the north at Brygge, the layers show a strike and dip of 90°/5–10°; 50 m further south these are about 295°/5°; and another 50 m further south 110°/5–10°; thereafter the layers are mainly horizontal.

Laup-hargi, the raukar field near Brygge, is mainly built up of grey, dense to finely crystalline, stratified limestone. In its lower parts this limestone is marly and the thin layers alternate with thin layers which are strongly marly. In an upward direction the layers are thicker, the limestone less marly, and finally even rather pure. In its uppermost part it is a little sparry. The lower layers are rather rich in fossils. Among these fossils crinoids, bryozoans and stromatoporoids should especially be mentioned, as well as the corals *Favosites* sp.; *Heliolites interstinctus* and *Syringaxon* cf. *siluriense*, and the brachiopods *Atrypa reticularis*, *Camarotoechia borealis*, *Dicaelosia biloba*, *Leptaena rhomboidalis* and *Plectodonta transversalis lata*. Higher up in the section the number of species is not as plentiful, although *Pentamerus gotlandicus* is manifest in a great wealth of individuals.

In between the stratified limestone, a remnant is seen here and there of the Laup-hargi reef limestone, which may be considered as being about synchronous with the Spangände Limestone. The dips which are found in the stratified limestone of the raukar field are most probably connected with the reef-limestone occurrence. The dips are mainly directed landwards. Deviations from this direction may be related to unevennesses in the topography of the reef and its surrounding talus. Moreover, it is also quite possible that some of the larger rauk-like rock masses are loose blocks rather than solid-rock exposures.

Stratified limestone is also found between Laup-hargi and Äske. In the middle of a range of exposures in Suderslätt, an exposure slightly east of Älmar shows an alternation of layers of a hard and generally rather strongly recrystallized marly limestone with thinner layers of a softer and much more marly sediment. The bedding planes are strongly rugged. The limestone is bluish grey or grey to red-mottled in colour and locally contains several crystals of calcite. The sediments are very rich in fossils. Solitary and social corals, crinoids, bryozoans (generally small), stromatoporoids and a number of brachiopods can be found. The sediments show a strongly reef-detrital character. Also in this exposure the marliness of the rock decreases upwards. Strike and dip of the layers are about 60°/10°, some tens of metres further east-southeast even dips of 12° and 17° are found in about the same direction; west-northwest of the described exposure the dip varies somewhat, but is on the average 5–10° to the south-southeast or east. Occasionally, e.g., directly west of Älmar, the stratified limestone is overlaid by a small occurrence of reef limestone.

Between Spangände and Norderslätt, including the localities Millsnabb and the Norderhamnsberg, similar stratified limestone is found as at Spangände.

The lower part of Hassli, along the east shore of Stora Karlsö, is built up of stratified, dense to finely crystalline limestone which is grey to red-mottled, but partly rather marly and then more bluish in colour. In the lowest part of the steep cliff near Buckkliv *Pentamerus gotlandicus* and *Conchidium sculptum* are rather common, but such fossils as crinoids, bryozoans, corals and - to a lesser extent - stromatoporoids are also abundant. At Käupru a thin-layered finely crystalline limestone is found, alternating with marly layers. *Pentamerus gotlandicus* occurs there as well.

The dip in the Spangände Limestone in Hassli is small; the overlying Austerberg Limestone shows greater dips. Presumably this is due to the further growth of the Röjsuhajd reef. The margins of this reef at the end of the time of deposition of the Spangände Limestone were closer to the place of the present cliff than before and its height over the surrounding sea bottom may then also have been greater. The marly facies of the Spangände Limestone shows a dip of up to 10° to the south-southeast.

At Fanterna the Spangände Limestone is overlaid by the Svarthällar reef limestone (Fig.100). The Austerberg Limestone, which a little further north is found at the same level as the Svarthällar reef limestone, is completely replaced by this reef limestone at Fanterna.

The thickness of the Spangände limestone varies and may reach 30–35 m. Generally it is much thinner. These differences in thickness probably are mainly related to the distance and direction away from the large reefs and consequently to the amount of reef-debris deposition.

A breccia and an unconformity

In the north of the beach horn of Käupru, at the bottom of the cliff, a breccia is exposed, containing *Pentamerus gotlandicus*. In a marly matrix fragments occur of layers of marly limestone which originally alternated with the marl; also embedded are stromatoporoids and corals and fragments of these, brachiopods and other fossils. The whole is a completely disorderly mass, but the disturbance is greatest at the top of the breccia. At its bottom, limestone slabs may still have an approximately horizontal position and often

Fig.101. Breccia of Spangände Limestone near Käupru, Stora Karlsö. At the base still indistinct remains of the original stratification. On top of the breccia stratified limestone in undisturbed position. The breccia is probably caused by a down-slipping flank reef.

Fig.102. Unconformity in Spangände Limestone between Käupru and Stiudden, Stora Karlsö.

occur next to each other, reminding one slightly of the original layers (Fig.101).

Underlying the breccia is a badly exposed alternation of layers of marlstone and of marly limestone. They show a dip of about 8–10° east-southeastwards. The thickness of the breccia is 1 m. It is overlaid by an up to 1 cm thick layer of marl, followed by stratified limestones. The latter contain some material which is presumably reef debris. They show slightly rugged bedding planes which are covered with some marl. These stratified limestones have a dip of about 8° towards the east-southeast.

At Fanterna a similar, but thinner, breccia is found, which is overlaid by reef limestone. The origin of the breccia will be explained in the discussion at the end of this chapter.

Between Käupru and Stiudden, low in the cliff, an obvious unconformity is found (Fig.102). At the bottom there is a stratified, highly marly sediment. This rock is very fossiliferous. Its layers dip, on the average, about 7° towards the northwest. Overlying it, but dipping about 8° southeastwards, is an alternation of harder and more calcareous, and softer and more marly layers. These sediments are also very fossiliferous. Except that some fossils are larger, there is no difference in fossil content with the sediments underlying the unconformity: corals, coral fragments, bryozoan remains, brachiopods, ostracodes, and only a few smaller stromatoporoids can be found. The origin of the unconformity will be discussed at the end of this chapter, in conjunction with that of the *Pentamerus gotlandicus* breccia.

Austerberg Limestone

The steep cliff at the northwest of the Austerberg is mainly built up of distinctly stratified limestone. This sediment is finely crystalline, sometimes nearly dense, yellowish white to white grey, locally somewhat marly and there bluishly coloured, and generally very rich in bryozoans. In the northeast of the Austerberg, especially near Utfall, the rock is enormously rich in bryozoan fragments. Corals and, to a lesser extent, stromatoporoids are also present, but it is highly probable that most of these are not on their place of growth in the Austerberg Limestone, but belong to reef debris. At Hästkliv, the sediment is somewhat more marly. There is a rather strong dip, which is about 15° towards the northeast between Stora Förvar and Utfall.

At the top of Hassli a similar stratified limestone is found as in the Austerberg; it overlies the Spangände Limestone.

The reef limestone, exposed along the southeast shore from Fanterna to west of Äske, is thinner at both ends than in the centre. At Äske, slightly more inland, stratified limestone still occurs above the raukar field. In view of the height at which is is found this limestone has probably, in a similar development, partly overlaid the reef limestone whereas other layers will have abuted against the reef limestone. The stratified rock is a finely crystalline limestone, grey to greyish white, sometimes yellowish grey, rather thin layered in layers of about 1–10 cm thick, with very rugged bedding planes. As fossils, brachiopods, corals, bryozoans and an occasional crinoid were observed in it. The direction of dip varies; sometimes the layers are almost horizontal, sometimes they dip south; locally even a slight northern dip will occur.

Thus, at Fanterna the whole thickness of the reef limestone is synchronous with the Austerberg Limestone. Also in the southwest of Svarthällar this stratified limestone abuts against the reef limestone (Fig.103). At some

Fig.103. Svarthällar, Stora Karlsö, seen from the west. At the right, reef limestone of Fanterna type; at the left side, layers of stratified limestone are seen, which abut against the reef limestone.

distance from both ends of Svarthällar, however, the thickness of the reef limestone increases. In some localities, in vertical section stratified limestone alternates with reef limestone. A short distance from the shore more inland the reefs are locally covered by finely crystalline limestone which is grey coloured and usually very thin layered. This rock is strongly recrystallized but in some exposures an abundance of small crinoid fragments can still be observed in it. The dip varies in degree and direction, but is generally a few degrees to the southeast. It also happens that in an exposure this stratified limestone is overlaid by reef limestone. For the major part, the Svarthällar reefs can in all likelihood be considered as having been formed synchronously with the Austerberg Limestone. In view of the great thickness in the centre of Svarthällar, the higher parts of the complex of Svarthällar reef limestones and the stratified limestones alternating with them are presumably younger than the Austerberg Limestone occurring further north in Stora Karlsö.

The raukar and rauk-like exposures at Gjaus-häll are built up of a hard limestone, finely to middle crystalline, greyish-white to grey and thin layered, which presumably corresponds in age to the Austerberg Limestone. The bedding planes are rugged. Due to a strong recrystallization and a recent overgrowth of weathered surfaces by lichens, the fossil content of the rock is not always easy to determine. However, in a few less-overgrown parts, as, e.g., the undersides of overhanging layers, a wealth of fossils can be observed. These include coral colonies (*Favosites*, *Heliolites*, and others), solitary corals, bryozoan fragments, brachiopods (*Atrypa reticularis*, *Conchidium sculptum*, *Leptaena rhomboidalis*, *Plectodonta transversalis lata*, and others), crinoid remains and an occasional stromatoporoid. Crystals of calcite are very common. The distance from the large reefs to Gjaus-häll is rather great, and presumably for that reason the layers lie fairly horizontally. There are no overlying smaller reefs of the Fanterna type (cf. p.243).

STRATIGRAPHY OF LILLA KARLSÖ

The oldest sediment exposed in Lilla Karlsö, is called *Pentamerus gotlandicus* Limestone, according to Hede (1927, pp.43–50, *Pentamerus gotlandicus*-förande kalksten). This local stratigraphic unit is overlaid by a sedimentary complex, with a maximum thickness of about 55 m, which the present author will refer to as Lilla Karlsö Limestone. To further subdivide this Lilla Karlsö Limestone would be unrealistic. Not only is it difficult to study the rock in sufficient detail, since it is almost entirely exposed in vertical cliff walls, but it was also laid down as a detrital reef-surrounding deposit of a strictly local character (Fig.113).

The present author, in accordance with Rutten (1958, pp.380–381), assumes that the interior of the island consists of the Central Lilla Karlsö reef limestone. On its mantle deposits flank reefs developed.

In the following sections of this chapter, the *Pentamerus gotlandicus* and Lilla Karlsö Limestones will be described. After the reef limestones have also been discussed in later sections, such problems will be touched on, at the end of the chapter, as the correlation with Stora Karlsö and with Gotland, and the environment of deposition of the Lilla Karlsö sediments.

The position of localities in Lilla Karlsö is shown in Fig.104.

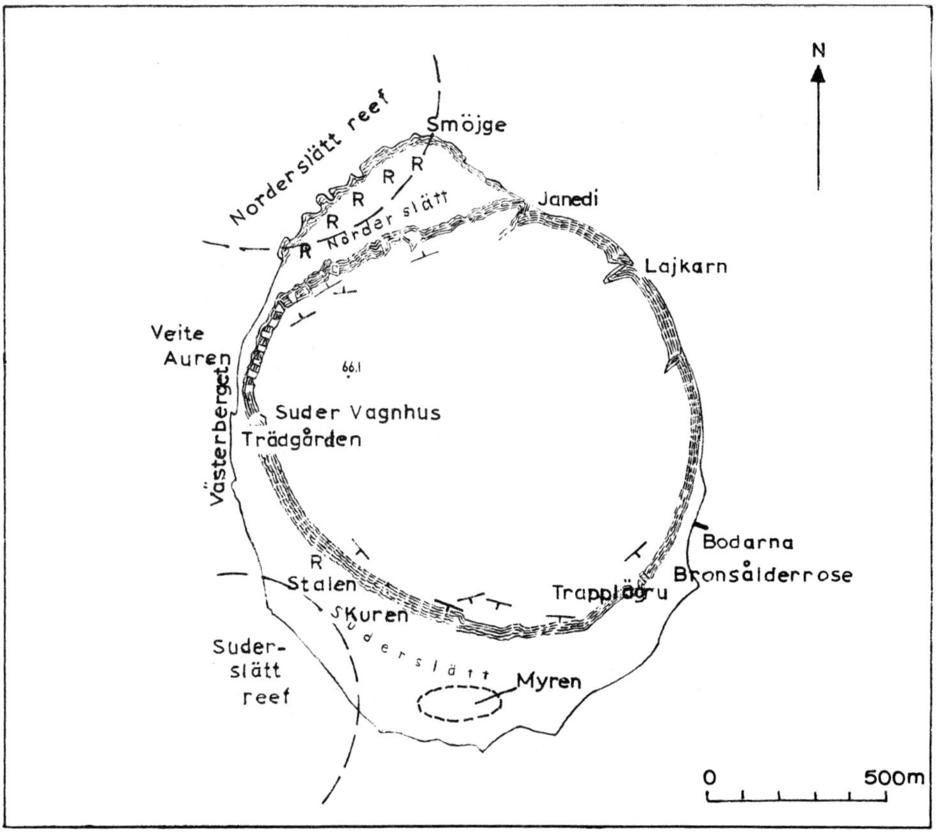

Fig.104. Map of Lilla Karlsö. Directions of dip in the stratified sediments around the Central Lilla Karlsö reef are shown. The assumed positions of the Norderslätt and Suderslätt reefs are also indicated. RR = Raukar.

Pentamerus gotlandicus Limestone

The *Pentamerus gotlandicus* Limestone is a clearly stratified marly limestone, generally with layers of 2–10 cm thickness, which sometimes, however, approach up to 20 or 30 cm. These limestone layers are separated by thin layers of a bluish grey, dense and soft marlstone, usually less than 2 cm thick. The limestone is light grey to bluish or brownish grey, normally finely crystalline but sometimes middle crystalline. The individual beds often show great variations in thickness over short distances, and can even completely thin out (Fig.105). Usually the rock is enormously rich in fossils, especially crinoids, bryozoans and corals; also brachiopods and some tabular stromatoporoids are present. No lamellibranchs were found. Among the fossils observed, *Pentamerus gotlandicus* is the most important for stratigraphical correlation.

The best outcrops of this limestone are found in the west of the island. There it is found along the beach from southwest of Stalen to Trädgården and in the lower beach cliff at Veite Auren and northwards, to a little to the north-northeast of the cave Norder Vagnhus. Some small exposures also occur in the northwest and northeast of the island. The thickness above sea level differs greatly from one place to another; in the south of Veite Auren, it reaches up to about 10 m above sea level, but generally its upper boundary is lower. This phenomenon is connected with the fact that in several places and at varying heights the stratified *Pentamerus gotlandicus* Limestone is overlaid by a *Pentamerus gotlandicus* breccia of varying thickness, which in its turn generally forms the basis of flank reefs of Fanterna type (Fig.106, 107,108). Where no reefs occur, as in several places of the lower part of the high wall between Suder and Norder Vagnhus, the alternation of marly limestone layers with thinner marlstone layers continues until about 10–12 m above sea level. Generally this succession shows an increase in the thickness of the limestone layers upwards and a corresponding decrease in the thickness of the marlstone layers. Sometimes the limestone layers remain thin, generally 2–10 cm, but then they are more numerous because of the thinner marlstone layers. In such cases the boundary with the overlying Lilla Karlsö Limestone is rather blurred. This is the more true as it is likely that the name-giving fossil, *Pentamerus gotlandicus*, is not abundant and possibly does not even occur in the sediments higher than about 4 m above present sea level. Because of the generally very steep walls, however, this cannot be proven.

Fig.105. *Pentamerus gotlandicus* Limestone, Västerberget between Norder and Suder Vagnhus, Lilla Karlsö. Limestone layers alternating with thinner marlstone layers. Some limestone layers thin out; the bedding planes are irregular and rugged.

Fig.106. The coast of Lilla Karlsö near Norder Vagnhus, showing how at varying level the stratified *Pentamerus gotlandicus* Limestone (foreground) is overlaid by reef limestone.

About 75 m north of Trädgården, along the beach, a discordance of bedding is found in the *Pentamerus gotlandicus* Limestone (Fig.109). The alternation of limestone and marlstone, there partly abuts against a bluish grey to grey crinoid limestone, which is thicker bedded, hard, sparry and rather poor in fossils other than crinoid remains. A short distance further north the crinoid limestone again disappears from the exposure. It apparently represents the result of local conditions so very favourable to crinoid development, that their remains built a slight elevation on the sea floor. The lowermost layers of the overlying normal limestone arch over it. A somewhat similar exposure is found at the bottom of the exposure in the south of Trädgården.

Pentamerus gotlandicus breccia

Directly underlying most of the flank reefs of the Central Lilla Karlsö reef, is found a breccia of *Pentamerus gotlandicus* Limestone (Fig.107,108). This breccia consists of a marly matrix in which fragments are embedded of the marly-limestone layers, which originally alternated with the marlstone layers. The limestone slabs show a great variety in orientation. Locally the breccia also contains some fragments of reef limestone.

This *Pentamerus gotlandicus* breccia shows great differences in its thickness underneath the various reefs. It does not always occur at the same height in the stratigraphical column of Lilla Karlsö.

At some other places the *Pentamerus gotlandicus* Limestone shows bucklings, which are caused by lateral compression (Fig.110).

A little north of Suder Vagnhus is found a flank reef with stratified sediment dipping upwards against it at its present landward side, which is the side directed towards the Central reef. This stratified sediment is a marly limestone, grey to greyish brown in colour, with rugged bedding planes; the rock is very fossiliferous and also contains reef debris. Hardly any bedding is left close to the reef. There, a completely unorganized mixture of fossils, fossil fragments, marly limestone and a marl matrix is found.

Lilla Karlsö Limestone

The northwest cliff of the island is almost entirely built up of stratified limestone, which is light brownish grey, dense to finely crystalline, partly finely oolitic, with layers generally varying in thickness from a few centimetres to about 15 cm. The fossil content of the sediment varies; most common are corals, bryozoans and stromatoporoids, which presumably belong in part to the outwash of the Central Lilla Karlsö reef, brachiopods (e.g., *Leptaena rhomboidalis*, *Atrypa reticularis*), ostracodes (*Leperditia* sp. and others) and trilobites (*Proetus* sp., *Encrinurus punctatus*).

Comparable sediments are found in the high and steep cliffs in the southwest, west, northeast and east of the island. The bedding planes are

Fig.107. Lilla Karlsö, about 50 m south of Trädgården. At the base, stratified *Pentamerus gotlandicus* Limestone; on the top, reef limestone; in between both, an unstratified breccia of *Pentamerus gotlandicus* containing sediment.

Fig.108. Lilla Karlsö, about 30 m south of Trädgården. The section of Fig.107 can also partly be seen in the right of this photograph. It shows how the thickness of the *Pentamerus gotlandicus* breccia under one reef differs from that under another reef.

---

generally rugged and often covered by a film of marl. As a result of weathering the rock often falls to pieces easily, building high screes at the foot of the walls. This is especially well demonstrated, e.g., between Suder and Norder Vagnhus. The amount of reef debris which the sediment contains varies. Exposures of flank reefs, which developed on the slopes of the Central reef (Fig.121), are intercalated. In more marly parts of the stratified limestone, caves have developed. None of these is excavated in reef limestone. The best-known caves are Suder Vagnhus, over 20 m deep, over 10 m high and close to 20 m broad, and Norder Vagnhus, 31 m deep, 10–1 m high and generally 4–6 m broad. Both are formed by the *Ancylus* lake, as are at least sixteen smaller caves in the west of the island. A few other small caves date from the Baltic Ice Sea.

Because of the steep walls, it is difficult to gain access to the Lilla Karlsö Limestone for detailed observations. It can be well studied at Trapplägru (staircase), about 0.2 km southwest of Bodarna, the place where boats dock in Lilla Karlsö. There the deposit is developed as a marly limestone, dense to finely crystalline, with a bluish grey, sometimes red-mottled colour, which alternates with thin layers of bluish grey marlstone. The limestone is rich in crinoids, bryozoans, stromatoporoids and corals. Its layers have a south-southwestward dip varying from 25–32° at the plateau side to about 20° at the seaside. The occurrence of marlstone in this section is notable, since it does not seem to be as common in other sections through the Lilla Karlsö Limestone, except locally in the uppermost parts.

REEF LIMESTONES OF STORA KARLSÖ

Several reefs are assumed to have contributed towards the shaping of Stora Karlsö. In the west are the Marmorberg reef, the reef limestone of Stäurnasar, and some remains of reef limestone at Laup-hargi. In the main they are probably of the same age as the Spangände Limestone.

The Röjsuhajd reef in the east is also mainly synchronous with the Spangände Limestone, although in its uppermost part it is perhaps more so with the Austerberg Limestone. The latter is suggested by the high content of the surrounding Austerberg Limestone in bryozoan fragments and other reef debris, and by the rather strong dips in the lowermost Austerberg Limestone.

The Svarthällar reef limestones in the southeast may be considered to be, in the main, of the same age as the Austerberg Limestone, although in their centre, where they reach a rather great thickness, the upper parts are probably still younger. Some other, smaller reefs, which in their appearance are comparable to the Svarthällar reef limestones, are most likely older.

The descriptive names for the various reefs are partly introduced here for the first time, but Rutten (1958) has already pointed out that the interior of Marmorberg and Röjsuhajd should consist of reef limestone.

The reef limestones of Stora Karlsö are not well exposed in all cases. Poor exposure may occur either because the reef limestones are still mainly enveloped by stratified sediments or because they have mainly been eroded.

Fig.109. Discordance of bedding in *Pentamerus gotlandicus* Limestone, about 75 m north of Trädgården, Lilla Karlsö. The discordance is caused by a local deposition of crinoid limestone.

Fig.110. Buckling in *Pentamerus gotlandicus* Limestone, about 90 m south of Trädgården, Lilla Karlsö.

As will be shown, there are some significant differences with the reefs of Gotland proper. Therefore, the present author proposes to classify them into two new reef types, to be known as the Stäurnasar and the Fanterna types.

The reefs of the Stäurnasar type are relatively large. They are in their generally not very well exposed lower parts presumably built up mainly by huge coral colonies; in their higher parts by stromatoporoids and corals together. The ratio between greatest thickness and greatest diameter will be in the order of 1/10; at the time of greatest extension, they may have covered an area of more than 100.000 m^2.

The Fanterna reef type is named after Fanterna, in the east of the island (Fig.100). The reefs are built up mainly by bryozoans and corals in a nearly pure calcareous matrix. Branched bryozoans occur in intact colonies measuring to over 1 m in all directions, but especially in an abundance of small fragments. The fauna of the Fanterna reefs is poorly known in detail. Because of strong recrystallization, the fossils are more altered and difficult to collect from the generally hard and splintery limestone. A weathered surface of Fanterna-type reef limestone is massive and often strongly brecciating. A number of the reefs are not very large. Presumably they developed as flank reefs, but some of the younger ones managed to survive the main reef and became strongly extended.

Western reef limestones

In the western half of Stora Karlsö exposures of what are at least presumably three large reefs are found, which all developed about synchronously with the Spangände Limestone and which are grouped here together as Western reef limestones.

Stäurnasar reef limestone

At Stäurnasar, in the west of Stora Karlsö, the steep cliff along the shore shows a very beautiful exposure of reef limestone, up to a thickness of about 20 m locally. The reef limestone rests almost directly over the Lerberg Marlstone. The lower part of the reef is built up almost exclusively by huge colonies of *Halysites catenularius*. These colonies are on the average, about 70 cm in diameter and 100 cm thick. The colonies practically occur "shoulder to shoulder" over a thickness of several metres. In between the corals a few tabular stromatoporoids are found. The matrix is formed by a bluish grey to brownish grey marly limestone, which also fills nearly all the *Halysites* colonies. At some places there is hardly any space left in between the colonies for a matrix. Evidently the corals were able to multiply rapidly and to colonize some areas to the nearly complete exclusion of other organisms. Succeeding generations used the skeletons of their predecessors as a substratum. Most coral colonies are wonderfully well preserved and little seems to have happened to the reefs after the death of its builders.

In the upper part of that same Stäurnasar exposure, the predominance of *Halysites* in the reef limestone is considerably less; colonies of other coral genera are also present there. Moreover, several more stromatoporoids occur. These generally formed tabular colonies which are rather thin and show undulating surfaces. The matrix of the reef limestone is less marly and constitutes an important portion of the total rock volume. Over the full height of the section the weathered surface of the reef limestone is rather massive.

Some tens of metres south of the above exposure, the reef limestone shows a distinct subdivision into zones. At the bottom, 1.70 m of the reef limestone consists almost entirely of *Halysites catenularius*. This zone is succeeded, without a perceptible transition, by reef limestone built up by tabular and flat-lenticular stromatoporoids, in between which only an occasional coral colony can be found. Then again, a sudden change-over to a second *Halysites* limestone zone takes place. This zone can be followed upwards over several metres. In the top of the reef the number of tabular and flat-lenticular stromatoporoids increases again, now gradually replacing the corals.

The sudden end of coral growth at the bottom of this section was probably due to increased sedimentation, causing the stifling of the *Halysites* colonies. These colonies stand very close against each other and more mud is found in the colonies (in the tubes and between the walls of tubes) than in between them. With an increase in sedimentation there was little opportunity to remove the extra amount of sediment coming down on the corals to places where it could do little harm. The lowest centimetres of the overlying stromatoporoid zone are very marly and the stromatoporoids there are extremely thin. In the whole stromatoporoid part, the matrix constitutes an important part in the volume of the rock, but at the bottom and locally at the top of the zone, the matrix volume is highest. Thus, the transition to the second *Halysites* part may probably also be explained by stifling, in that case of the stromatoporoids, making place for a new period of colonization by the corals. The fact that over the full height of the section those stromatoporoids which occur remained very thin suggests that life conditions for them were marginal at all times.

Since this subdivision into zones is only a local phenomenon, its causes must also have been of a local rather than a general nature. This excludes, for instance, a temporary increase in water depth.

In the second *Halysites* zone in this section a saucer-like depression was found to have developed between *Halysites* reef limestone found left and right of it. At the bottom of this depression a layer of marly limestone is found, containing tabular stromatoporoids and a few small *Halysites* colonies. The latter generally lie upside down. The depression was probably formed through an occasionally greater influx of mud than could be bypassed, which overwhelmed a local part of the reef, causing the death of the corals there. Upwards, *Halysites* colonies again replace the stromatoporoids, which are, in their turn, higher in the section over the entire reef and are replaced anew by stromatoporoids. There the depression fades away in the reef. It is also likely that the greater zonal division described above from this part of the cliff represents a reef portion which developed in a depression in the reef surface. This part then acted as a mud trap and, moreover, did not grow upwards at the same rate as the surrounding reef surface. These two factors together may have resulted in the higher matrix volume. Temporarily more favourable conditions following the last greater influx of mud gave the corals a chance to build up the depression to the level of the surrounding reef portions.

The Stäurnasar reef limestone as a whole shows an orderly nature. Notably, the great predominance of corals especially in the lower part of the reef and the increase in stromatoporoids upwards in the reef make it differ from all three reef types found in Gotland and give reason to consider it a separate type of reef formation.

In the south, stratified limestone with a varying content of reef debris buckles down under the Stäurnasar reef.

Marmorberg reef limestone

It is most likely that the heart of the Marmorberg, in the northwest of Stora Karlsö, consists of reef limestone. Indications in favour of this supposition are:

(1) Locally both in the surrounding cliffs and on top of the plateau of the Marmorberg, reef limestone is exposed.

(2) The rather strong dips shown by the stratified sediments exposed in the Marmorberg (Fig.99,111). All around the hill the dip is generally directed away from its centre. With increasing distance from the centre of the Marmorberg, the dip decreases. The Middle Palaeozoic strata of Gotland and Karlsöarna were never subjected to a general folding or to other powerful tectonical disturbances. Glacial pressures during the Pleistocene Ice Ages can not have brought about this phenomenon either.

(3) The reef-detrital character of most of the surrounding sediments.

(4) The occurrence of crinoid limestone around and on top of the Marmorberg ("Karlsö marble", the Spangände Limestone is also often

Fig.111. Stora Karlsö; the most southwestern part of the Marmorberg, seen from the southeast. Flank deposit of the Marmorberg reef, with a decreasing dip of the layers away from the reef.

red-mottled). Everywhere on Gotland crinoid remains are extremely abundant in the immediate environment of fossil reefs (cf. the "Hoburg marble" covering the reefs at Hoburgen, as mentioned in Chapter VII).

Exposures of Marmorberg reef limestone are rather scarce. They are found on top of the Marmorberg, in the higher parts of the Västerberg cliff, high in the section exposed at Spangände, in the Norderhamnsberg and in the southeast of the Marmorberg. The exposure found at Stäurnasar and described above most likely belongs to another reef and not to that of the Marmorberg. On the plateau of the island it can be seen that Stäurnasar forms a hill, separated by a valley from the Marmorberg proper. The highest part of the Stäurnasar hill is situated at the edge of the cliff. The great sag in the underlying Lerberg Marlstone, which dips from a height of about 25 m at Ramraur to below sea level under Stäurnasar also indicates a separate reef.

Except from the steep cliff, information about the exposures in the Västerberg can also be derived from loose blocks lying on the beach at the foot of the cliff. These blocks are especially common northwest of the Västerberg. Three types can be distinguished, viz. blocks of reef limestone with stromatoporoids and corals, blocks of bluish grey limestone very rich in crinoid-stem fragments which are generally red coloured, and blocks which consist partly of this crinoid limestone and partly of limestone very rich in reef builders. The latter blocks probably originate from the very outside of the reef.

The stromatoporoids in the reef limestone are mainly tabular forms with a thickness of less than 1–1.5 cm in many instances. The thickest stromatoporoid was a bullet-shaped colony of 25 cm thick and 10 cm in diameter, found in crinoid limestone. The corals generally form larger colonies. A few observed colonies of *Halysites catenularius* had a diameter of 75 cm and a thickness of 30 cm, and an exceptional giant measured not less than 190 cm in diameter and was 80 cm thick. The other *Halysites* colonies seen were smaller. This applies still more to the massive, lens-shaped coral colonies.

Generally stromatoporoids are more abundant in the reef limestone than corals. However, the latter also contributed essentially to reef building and some portions of the reef, up to a few cubic metres, consist almost exclusively of *Halysites* colonies. Bryozoans did not play an important part. The recrystallization of calcium carbonate in the stromatoporoids and in most of the lens-shaped corals was generally so strong that it blotted out the finer microstructure. This obstructs not only specific or generic identification of the colonies but even the distinction between the two groups. The matrix of the reef is a marly limestone which usually forms a rather important percentage of the total volume of reef limestone. The weathered surface of reef limestone is rather smooth, not conglomeratic or brecciated to such an extent as in many of the reef-limestone exposures in Gotland. For this reason it is very difficult, when standing on the narrow beach, to distinguish between reef and stratified limestone in the upper part of the cliff wall.

Some of the blocks found at the foot of the Västerberg are very rich in solitary corals; others show a transition from reef limestone to the reef-detrital Spangände Limestone through its disorderly nature, as well as through the presence of many fragments of reef-forming organisms and more crinoid remains.

The small and weathered outcrops of reef limestone in the Norderhamnsberg and the southeast of the Marmorberg do not present any new information. Tabular and flat-lenticular stromatoporoids and some coral colonies are recognizable in this rock. The general nature of the reef limestone is similar to that found in some of the blocks on the beach at Västerberg. The reef limestone in the southeast is surrounded by a light-grey crinoid limestone, which is red-mottled. It contains, in addition to the many usually small crinoid remains, several stromatoporoids and corals, many of which are not in their position of growth. The rock is comparable to the crinoid limestone at Spangände.

The reef character of the limestone which is found in a few small outcrops on top of the Marmorberg is vague. The many crinoids in it suggest that this reef limestone belongs to the very top of the reef.

At Spangände, the reef limestone in the cliff is inaccessible; on the plateau it is overlaid by crinoid limestone.

If one realizes that all exposures of the Marmorberg reef limestone belong to the higher parts of the reef, it becomes probable that the Marmorberg reef may be of similar type to the Stäurnasar reef. The latter shows a comparable composition at its top. Through its more orderly nature, more and larger coral colonies, and smoother surfaces, this kind of reef limestone as found in the higher parts of the Marmorberg and Stäurnasar reefs differs from that of the general Hoburgen reef type of Gotland. As has been demonstrated before, the difference between the lower parts of the Stäurnasar reef and the Hoburgen-type reefs is very pronounced.

In some places on top of the Marmorberg, such as east of Myren and southeastwards from there, a red-mottled, grey to white-grey crinoid limestone is found, which is very rich in predominantly small crinoid fragments. It also contains some small remains of corals, brachiopods and stromatoporoids. This rock is known as "Karlsö marble". The limestone is rather strongly recrystallized, finely to middle crystalline and usually very thickly, sometimes also irregularly bedded. The strike and dip vary; east of Myren the dip is towards the northwest; north and northeast of Suderslätt it dips in a south-southeast or southeast direction. The plateau of the Marmorberg shows distinct fluctuations in height. The dips mentioned above, are connected with one of the elevations of this plateau, situated southeast of Myren and are similar to the topographical slopes. The same situation is found around other elevations on the plateau. This situation suggests that the upper surface of the Marmorberg reef limestone is rather strongly uneven.

Laup-hargi reef limestone

In the raukar field Laup-hargi, in the southwest of the island, a small remnant of reef limestone is preserved between the Spangände Limestone. In all likelihood the very greatest part of the reef was situated further seawards and has been removed by erosion (Fig.112). Indications of this are the large depression, faintly saucer-like, which is found in the stratified sediments exposed around sea level off the shore, and the dips in the reef-surrounding stratified limestones in this region, which are mainly directed inland. The size of the basal depression and the strong dips suggest a rather large reef, even if the dips in the overlying rocks are somewhat exaggerated by differential compaction of the sediments. In all likelihood the reef

Fig.112. Laup-hargi, Stora Karlsö, seen from the north-northwest. Stratified sediments, dipping seaward, suggest that the centre of the Laup-hargi reef has been completely removed by erosion and only some parts of the very periphery have remained.

Fig.113. Laup-hargi, Stora Karlsö. Rauk, showing reef talus with an abundance of fossils, fossil fragments and also with parts of layers of crinoid limestone; the whole is embedded in a matrix of marly limestone. At the top, crinoid limestone.

---

was synchronous with the Spangände Limestone.

The reef limestone which is preserved, contains corals (*Favosites*, *Heliolites*, *Acervularia*, *Cystiphyllum cylindricum*, and others), stromatoporoids, bryozoans, crinoid remains (generally small) and brachiopods (*Pentamerus gotlandicus*, *Atrypa reticularis*, *Camarotoechia borealis*, and others). Brachiopods are rather common, although nowhere abundant. Corals, occurring in both solitary forms and colonies, outnumber the stromatoporoids. The latter almost never reach large measurements. The matrix of the reef is a marly limestone, which forms an important element of the total rock. Fragments of fossils and colonies that are not in their growth positions are very common. It is likely that at least part of the exposed reef sediment is reef talus rather than true reef limestone. This holds true especially for those parts in which the matrix contains an abundance of crinoid remains together with a large amount of relatively small bryozoan fragments.

In these talus-like parts a vague and very lumpy stratification can be observed locally. This kind of deposit also contains slabs of hard crinoid limestone layers, with a length of up to 25 cm and with an average thickness of 3 cm, showing a strong variation in orientation. The dip of fourteen of these slabs in one of the raukar, close to the southeast end of the raukar field (Fig.113), showed an average direction to the north-northeast.

The number of exposures in Laup-hargi is too few and the exposed reef limestone too vague to reliably classify the reef in one of the generalized reef types. There is a resemblance to the reefs of the Hoburgen type of Gotland but the dominance of corals over stromatoporoids is a characteristic which it has in common with the upper part of the Stäurnasar reef.

Röjsuhajd reef limestone

The same reasoning as applied to the Marmorberg can be used to argue that the heart of Röjsuhajd also consists of reef limestone. Exposures of this limestone, however, are exceedingly scarce. Some small and weathered outcrops are found on the plateau of Röjsuhajd. The weathering makes it difficult to distinguish these outcrops from those of stratified limestone. The only reliable statement which can be made about this Röjsuhajd reef limestone is that it is comparable to that exposed in the Marmorberg and the higher parts at Stäurnasar, but seems to contain more bryozoans.

Since exposures of reef limestone are found in the plateaus of both the Marmorberg and Röjsuhajd, it seems likely that the Röjsuhajd reef is thicker than its western equivalent, because Röjsuhajd reaches about 8 m higher.

There is a remarkable crack, about 25 cm wide, which is found on Röjsuhajd some tens of metres southeast of Linnés Ask. It has an orientation about east-northeast – west-southwest, and is partly filled up with some vertical layers, consisting of a grey to white-grey limestone, which is hard and finely crystalline. Locally a fossil can be observed in this limestone, or it may be red-mottled as a result of small crinoid fragments. The surrounding rock is a hard, grey to red-mottled limestone with crinoids, and intercalated small exposures of reef limestone. The crevice has a somewhat sinuous course and can be followed over some tens of metres before it disappears under vegetation. Presumably this crack formed during compaction of the sediments; its formation would have been favoured by the dips in the stratified sediments arching over the Röjsuhajd reef. Similar cracks of smaller measurements are found in the southeast of Norderslätt, in a light grey to brownish grey, dense limestone with stromatoporoids and crinoid remains.

General discussion on the older reef formations

The Western reefs and presumably also the Röjsuhajd reef started to grow from a muddy bottom in the upper part of the Lerberg Marlstone or at the latest, at about the transition from Lerberg Marlstone to Spangände Limestone. Although studies of Recent reefs have shown that most of them started growing on hard bottoms, a consolidated platform is not essential for a reef foundation. Observations of present-day reefs have proven that a relatively small number of them definitely rose from unconsolidated bottoms. The large reefs of the two Karlsöarna presumably present fossil examples of this. Nowhere is there any evidence that a hard substrate for reef growth has been present within this region. As has been shown, a marlstone or perhaps a marly limestone acted as the foundation for reef building. Reef-forming organisms, nevertheless, require a solid surface for attachment, and, in the absence of a suitable flatform especially hard skeletal remains as, e.g., coral colonies or solitary corals, may have served as objects for attachment of the pioneer populations. The extra weight of the reefs upon the underlying sediment caused a plastic deformation, as the bucklings in the Lerberg Marlstone indicate.

When the reefs were built up to slightly above the surrounding sea bottom, formation of reef debris began, which became embedded in the sediment deposited directly around it. Since during their growth the water presumably became shallower (cf. the discussion at the end of this chapter) and because the reefs became higher as their growth rate exceeded that of sediment accumulation in their environment, gradually more reef debris was produced. The growing obstacles on the sea bottom caused increased water turbulence. Debris, formed in the first place directly around the reef, was reworked and redeposited on the surrounding bottom. In part the skeletons of the reef builders and reef-surrounding fauna disintegrated post mortem into skeletal sands. All possible intermediates from fine reef sand to complete colonies can be observed in the flank deposits. The percentage of still recognizable reef-derived debris in the flank beds generally ranges from a common figure of less than 5–10% to a rare maximum of about 40%.

As described above, the sediments deposited marginally to the reefs incline from the reefs with angles of about 25° close to the reef, gradually decreasing with increasing distance from the reef, to an approximately horizontal stratification at some hundreds of metres from its centre (Fig.99). The dips are about the same on all sides of the reef mass, indicating that the main form was a dome, situated at some distance from the shore. There is no field evidence from which this distance can be reliably deduced, but 5 km or more might be a reasonable guess.

As the top of the Marmorberg reef suggests, in the course of reef development a number of smaller growth centres on the reef could lead to a rough surface. All the elevations probably had the form of smaller domes. They, too, were enveloped by stratified sediments, dipping away from the centre of the dome, although the angles are smaller than with the main reef.

Originally the dips in the flank deposits may have been smaller than they are at present. Later compaction of the reef-surrounding deposits may have exaggerated the difference in level between a given reef surface and its contemporaneously accumulating flank sediments. The structure developed by the building of the framework-forming organisms clearly had great rigidity and did not undergo any important change in volume under the weight of any overburden placed upon it. Lime muds, deposited in connection with the reefs, on the other hand, will have been filled with water at the time of deposition. As this water was expelled, compaction took place and the original dips increased. After reef building came to a close, sediments were deposited over the reefs. These arch over them in the form of domes. On all margins they have dips approximately equal to the underlying sediments that drape around the reefs. These additional deposits produced a further compaction of the bordering sediments and a consequent intensifying of the inclination of the layers.

Even if the occurrence of a large compaction factor is assumed to compensate for the present exaggerated difference in levels, an ultimate height of these reef masses of about 10 m, or perhaps 15 m, above the contemporaneous topography of the normal sea bottom may be posited. After the death of the reef it became slowly buried under younger sediments, but still projected upwards through several metres of these deposits, as can be observed on the plateau of Röjsuhajd.

Svarthällar reef limestones

In the south of Stora Karlsö, at Äske, an elongated raukar field begins, which can be followed over a distance of about 1.3 km east-northeastwards till Fanterna. The raukar are built up by a hard and splintery reef limestone, which contrasts sharply with the stratified sediment in its surroundings. In these reefs vertically extended reef-forming organisms, mainly bryozoans, but also a few branched corals, as well as more horizontally expended reef builders, especially lenticular and tabular coral colonies, occur. There are practically no stromatoporoids. Among the reef dwellers brachiopods are most common. The reef builders occur in such an alternation with each other and with other material that the reefs as a whole present a massive structure without any distinct building pattern. The stratified limestones sag under the reef limestone.

Fig.114. Reef, in the south of Svarthällar, Stora Karlsö. At the base, an alternation of hard limestone layers with layers of a more marly composition; this is overlaid by about 0.5 m of reef debris and on top of that a reef of Fanterna type, about 6 m thick.

As this type of reef limestone cannot be directly compared with any of the three types distinguished in Gotland, the author proposes to give it a name of its own, the Fanterna reef type, after the locality in the east of Stora Karlsö.

Both at Äske and in the west of Svarthällar, in the direct environment of the reefs, stratified limestones still occur. These partly come to a dead end on the reefs. Earlier perhaps they have also covered the reef limestone to some extent. The reef limestone here has a maximum thickness of about 9 m; further northeast, however, the thickness increases considerably. Perhaps there the top of the reef limestone is younger than any stratified sediment in the island, but in the opinion of the present author there is no reason to consider the whole complex of Svarthällar reef limestones as younger, as Hede (1927b, p.42) supposed.

The usually very crumbly weathering sometimes leads to a pseudo-stratification of the reef limestone. A real, but vague stratification is found at some places at the top of raukar at the edge of the raukar field. Parts of reef limestone, located further inward in the reef-limestone occurrences, which are very rich in bryozoans, sometimes show a kind of weathering reminiscent of an industrial slag. Due to the very strong recrystallization, the organic constituents of the reefs are often barely recognizable in the exposed solid rocks. However, the reef builders can be very well studied in some loose blocks on the beach that are selectively eroded by the waves. In most instances these show a dominance of bryozoans, directly followed by coral colonies. Solitary corals are also very common; brachiopods are less numerous. After some searching a few small stromatoporoids were also found. The bryozoans occur in large colonies, which are up to 0.8 m in diameter and 1.15 m thick. Most bryozoans, however, have apparently fallen apart into an abundance of small pieces under the disintegrating action of contemporary waves and currents on the somewhat elevated reefs. These pieces play a dominant part in the reef matrix. This matrix is probably composed to a very high degree of autochthonous material, since it is remarkably pure for a reef limestone and leaves practically no residue after solution in cold dilute acid. It is dense to finely crystalline. Calcite crystals, up to 1 cm large are very common. In comparison to other reefs, the matrix does not occupy a high part of the total reef-limestone volume, generally less than 20%.

Just before the southwest end of Svarthällar, the rocks underlying the Svarthällar reef limestone are also exposed (Fig.114). These consist of an alternation of 1–5 cm thick layers of hard limestone of bluish-grey, grey to brownish-grey colour, with thicker layers, generally 2–13 cm, of a more marly limestone, greyish blue to bluish grey in colour and splitting into very thin flakes. Both types of sediment are enormously rich in fossils, such as solitary and social corals, bryozoans, brachiopods, crinoids, and an occasional cephalopod.

At the top of the sequence, a number of layers of the hard limestone occur, with a combined average thickness of about 0.5 m, in between which the flaky, marly limestone is nearly missing. The rock consists mainly of reef debris, particularly bryozoan fragments with corals in between. Sometimes a certain sorting to size of the reef debris can be observed; there are layers with only bryozoan fragments, layers with bryozoan remains, fragments of coral colonies and some solitary corals, and layers which also contain complete coral colonies. This sorting is presumably brought about by wave action on the reefs and was also dependent upon the distance from the reef to the place of deposition. Upwards the coarseness of the reef debris generally increases somewhat. The boundary between this debris-rich limestone and the overlying reef limestone is not sharp. Lowermost in the reef limestone, debris is still very common in between corals which are found in their positions of growth. Over a short distance upwards, the number of these corals increases rapidly, whereas at the same time the percentage of debris decreases. In the lowermost part of the reef a vague stratification can be seen locally.

A weathered surface of the reef limestone is massive in appearance, in part crumbling. Locally, where the rock is less recrystallized, bryozoans and corals show a different reaction to weathering. Parts rich in bryozoans weather massively or like industrial slag; coral colonies are partly liberated between them as lenses. Where many lenticular corals occur together, they sometimes cause a kind of stratifiation. Colonies of *Favosites* are often filled with limestones only in their outermost few millimetres; for the entire remainder the polyparia are empty. This gives the rock a porous nature locally.

The reefs of Svarthällar and rather similar reefs found in Lilla Karlsö are the strongest and most uniformly recrystallized of all reefs found in Gotland and both Karlsöarna. This may be a consequence of the rather pure nature of the matrix, which is a dense calcite cement with hardly any inorganic clastic material. Probably recrystallization has also partly been stimulated by the occurrence of the above-mentioned corals which escaped internal cementation and thus retained their natural porous, cellular structure. In this way they permitted access to invading fluids and rendered the rock liable to diagenetic action. Comparable observations were made by

Fig.115. Detail of an intercalation of crinoid limestone within a reef. Älmar, Stora Karlsö.

Henson (1950) in some rudist reefs of northern Iraq. There, patchy alteration generally followed the distribution of the porous rudist colonies; elsewhere in the same outcrop, where these organisms were tightly cemented inside and outside, the whole rock was unaltered.

The reefs of Holmhällar type, too, are rather pure, but nevertheless they are less strongly recrystallized than the reefs of Svarthällar. This may have been caused by the fact that between the stromatoporoid colonies, which are their main reef builders, Algae generally occur. These assisted in the building of the framework that holds the reef together, are much finer in texture, are less liable to attack, and as a result are generally relatively less altered.

Still much less altered are reefs of other types in Gotland, which have too high a percentage of clastic terrigenous material in their matrix, which acted against uniform recrystallization.

In addition to the reefs in the Svarthällar area some other, isolated and smaller reefs occur elsewhere in Stora Karlsö. These will be briefly described below. They probably belong to the same type of reef.

Directly west of Älmar, in the southwest of Stora Karlsö, the Spangände Limestone is overlaid by a small reef, comparable to the Svarthällar reef limestones. The reef limestone is a hard and splintery limestone in which many bryozoans and compound and solitary corals are recognizable; also an occasional stromatoporoid is present, generally small and very flat. The reef reaches a thickness of up to about 5 m. It apparently was too small to show faunal differentiation on its various sides. With the other reefs of Fanterna type it shows a remarkably uniform fauna over the whole reef, a uniformity which in this reef type can also be noted in widely separated reefs.

In the Älmar reef an 1.20 m long intercalation of crinoid limestone is well exposed, with a strongly concave lower boundary and a faintly convex upper boundary; in the centre the thickness is about 70 cm, at the margins about 30 cm. The whole presumably represents a small depression in the reef surface, which at a later stage was again overgrown by the reef builders. At the bottom of the depression, some relatively large fragments of reef limestone were observed, 10–30 cm in diameter. The rest of the filling material is of smaller size. Crinoid-stem fragments are strongly dominant. These are, on the average, 1.5 cm in diameter and about 3.5 cm long, with the longest remains measuring about 11 cm (Fig.115). Small crinoid remains are rare. In between the crinoid material are solitary corals and remains of reef-frame builders. The whole is embedded in a matrix of faintly marly limestone.

The stratified limestone directly underlying this reef is very rich in small, red-coloured remains of crinoids, which give the rock as a whole a red-mottled appearance. In the reef limestone itself crinoid fragments are less numerous, but of greater measurements, with stem fragments of up to 2 cm in diameter and 5 cm long.

Also on top of the reef some strongly recrystallized crinoid limestone occurs, in lumpy layers of 1–3 cm thick. In addition to the many crinoid-stem fragments, which often fall apart into crystals of calcite, some corals and brachiopods are found.

Some tens of metres northwest of the Älmar reef described above, another reef of the same type, but slightly larger, is exposed (Fig.116). No stratified limestone is seen actually overlying and underlying the reef. However, both northeast and southwest of the reef, stratified sediments incline to disappear under the reef. The greyish white reef limestone reaches a thickness of about 6 m. Locally, especially at the top, it is red-mottled by crinoid fragments. Together with the vague stratification, which locally begins to occur higher in this reef, this may be an indication that the reef has not been much thicker than is now exposed. The reef limestone is very strongly recrystallized, both the fossils and the matrix, which is a rather pure limestone; the weathering is crumbling and breccia-like. Crystals of calcite, generally small, are common. Among the reef builders some stromatoporoids are also found, a few of which are rather large;

Fig.116. Reef of Fanterna type, exposed some tens of metres northwest of Älmar, Stora Karlsö.

Fig.117. Vinglu, Stora Karlsö, seen from the south. Reef of Fanterna type, exposed just above the boundary between Lerberg Marlstone and Spangände Limestone. At the right-hand margin of this photograph, a second reef is found.

Fig.118. Rauk gate, named Hesselby Läde, Stora Karlsö. The gate has been excavated by the *Ancylus* lake; the height of its opening is about 9 m.

---

the largest one observed was a "tower" of about 50 cm high and at its base 20 cm in diameter, built up of very strongly convex latilaminae.

At Vinglu, a small reef is exposed with stratified sediments sagging under it (Fig.117). The lowest part of these enveloping sediments belongs to the Lerberg Marlstone, the higher parts to the Spangände Limestone. Near to the reef, crinoid limestone is found, rich in thick and long crinoid-stem fragments. The reef limestone is of about the same nature as that at Svarthällar. A few, thin stromatoporoids were observed in it.

A short distance southeast of this reef, another similar reef is located. It is separated from the stratified sediments by a breccia, 2–1 m thick, which was formed from the stratified rocks. Thus, in all likelihood, the reef has slipped down from an originally higher position. The reef is now found close to the boundary between Lerberg Marlstone and Spangände Limestone. It is, therefore, more likely that it developed synchronously with the Spangände Limestone than with the Austerberg Limestone.

It appears from the discussion of both the larger and the smaller reefs of Stora Karlsö that although stromatoporoids are present in most of them, and in some parts of them may even be dominant, they generally play a subordinate part. Actually, only in the higher parts of the Stäurnasar reef type, are they generally the main reef builders. It is, therefore, difficult to understand Hadding (1941, p.30), when he writes: "the reefs on Stora Karlsö are built up essentially in the same manner as other Gotland reefs of corresponding type. The stromatoporoids form the bulk of the reef-building organisms. Their form is as a rule tabular. The reef structure is similar to that found in other places where stromatoporoids of this type dominate." In reality there are quite distinct differences with the reefs of Gotland, as pointed out in the previous pages, and the present author considers it justified to classify the reefs of Stora Karlsö in two special types.

REEF LIMESTONES OF LILLA KARLSÖ

Also for Lilla Karlsö it is assumed that the nucleus of the island consists of reef limestone of relatively great thickness. Compared to this central reef mass the other reefs in the island are smaller. Remains of reefs of rather large extension are found in Norderslätt and Suderslätt, but it is very unlikely that they had a thickness comparable to the Central Lilla Karlsö reef. Because the reef limestone is generally more resistant to weathering and erosion than the overlying and surrounding stratified sediments, fragments of the reefs may still be preserved when the other sediments have

Fig.119. Raukar southwest of Smöjge, Lilla Karlsö. Remnants of the Norderslätt reef.

already been removed. This applies especially to the Norderslätt reef, of which only a great number of erosional remnants (raukar, stone giants) are still present, especially along the shore between Norder Vagnhus and Janedi (Fig.119). Also almost all of the reef limestone of the Suderslätt reef has already been eroded.

The centres of the three major reefs of Lilla Karlsö form the angular points of a triangle. The plan of the Central Lilla Karlsö reef is about circular. This suggests that the three reefs probably formed too far offshore to bear any special relationship to a shore line.

In the steep cliffs of Lilla Karlsö small reefs crop out, which must have developed on the flank deposits of the Central Lilla Karlsö reef. In some parts of the island, such as between Suder and Norder Vagnhus, the flank reefs are exposed in sections crossing the reefs in various directions. A good idea of the characteristics of the flank reefs can be obtained by combining the data from these various outcrops.

Central Lilla Karlsö reef limestone

About the same arguments as were used to show the existence of the Marmorberg reef in Stora Karlsö can be advanced for the assumption that in the centre of Lilla Karlsö a comparatively large reef is present. This Central reef, however, is only scarcely exposed. It is not impossible that some small outcrops are to be found on the plateau of the island, but they are very indistinct and do not permit any statement about the type of reef to which they belong. Corals, bryozoans and stromatoporoids were presumably the chief reef builders. The reef limestone on the plateau is surrounded by a grey to white-grey, locally red-brown crinoid limestone. The surface of the plateau shows a slight accidentation.

Already in the *Pentamerus gotlandicus* Limestone, indications are found which suggest that the Central Lilla Karlsö reef was growing strongly. These indications include local small dips, the occurrence of fossil remains which may be reef debris and the presence of an abundance of crinoids. The dips in the Lilla Karlsö Limestone are, as a rule, much steeper than those in the *Pentamerus gotlandicus* Limestone. This greater dip might have been caused partly by settling of these deposits, but partly by deposition at a lesser distance from a higher reef (Fig.120). It is difficult to establish when growth of the Central reef came to an end. If indeed the small exposures on top of the island belong to this reef, it may have continued growing during almost the entire time in which sedimentation of the stratified Lilla Karlsö Limestone took place.

The sediments surrounding the Central reef show a reef-detrital character. They contain bryozoan fragments, corals and stromatoporoids, brachiopods, trilobites and other fossils. Remains of crinoids are the dominant constituents of the rocks. Many corals and also several stromatoporoids appear to have been tumbled over and redeposited. Together with most of the bryozoan colonies from which the fragments are found, they presumably once belonged to the reef community. Only a small percentage of corals and stromatoporoids is definitely found in their positions of growth. Yet the percentage of reef builders in the sediments surrounding the Central reef is rather insignificant; the remains of bryozoans, corals and stromatoporoids usually make up less than 5% of the total rock volume.

Fig.120. Lilla Karlsö Limestone, dipping away from the centre of the island, as a result of its being the mantling deposit of the large Central Lilla Karlsö reef. Photograph taken from the south of the plateau of Lilla Karlsö. In the background, the plateau of Stora Karlsö, inclining southward.

The detrital limestones almost everywhere dip away from the centre of the island. In all likelihood they were mainly correlative with the growing reef, as is the case with many present-day reefs. The original topography of the reef sides and the direct environment of the reef became gradually buried and to a certain degree levelled off by the debris accumulation. Locally on this mantling deposit, reef-forming animals seem to have been able to establish themselves, in some of the cases giving rise to smaller reef bodies. Hard skeletal remains probably served as objects for attachment of the pioneer populations. In these detrital limestones several individual coral colonies, especially *Favosites*, were also observed to be attached to older colonies, to solitary corals or even to fragments of crinoid stems.

During the formation of these smaller flank reefs, the mantle deposit of the Central reef presumably was rather unconsolidated, as can be seen from the position of the upper of the two reefs shown in Fig.126 and 127. On the other hand, the well-developed bedding planes show that the deposit was not very pappy and there were enough hard remains to serve as a substratum. Under such conditions an unstable bottom was evidently not a hindrance to luxurious reef growth, as the number of flank reefs and the sizes of the individual reefs indicate.

Norderslätt reef limestone

The raukar field in the northwest of Lilla Karlsö is presumably the remnant of a large reef, comparable in its original extension to the Central Lilla Karlsö reef. In the centre, reef limestone occurs down to below present sea level. Southwest of the raukar field, stratified sediments are exposed up to 5–6 m above sea level. These sediments were presumably deposited close to the reef and have also been overlaid by reef limestone. The stratified sediments show an alternation of layers of marlstone with layers of harder marly limestone. The latter increase upwards in thickness and number and gradually pass into crinoid limestone which occurs in layers of generally 2–15 cm thick, separated by thin layers or films of marl. Apparently the reef produced little debris, for these stratified sediments rather close to the reef usually contain only a relatively small amount of such debris. This may suggest development in rather calm water.

In the north the reef limestone of the last raukar shows a tendency to stratification. It was probably formed close to the lateral boundary between reef and surrounding sediments. The southeast boundary of the Norderslätt reef may be sought at over 0.4 km further southwest. This distance is a peripheral section through the reef and its diameter at the time of maximum expansion may, therefore, be supposed to have measured 0.5 km or more. In a northwest – southeast direction, reef limestone is exposed over a distance of up to 0.15 km. All the reef limestone farther off the coast has been demolished by the destructive action of the present Baltic Sea.

In the northeast the highest raukar reach about 10 m above present sea level; in the southwest of the Norderslätt reef, the reef limestone occurs up to about 17 m high. In this area, in the top of the rauk-like formations, local interruptions in reef growth are represented by rather horizontal planes which sometimes give the reef limestone a thick to very thick stratified appearance.

Since the reef limestone at present thus reaches from below sea level at one place to about 17 m above sea level elsewhere, the thickness of the reef in its centre has presumably been more than 20 m.

The reef limestone consists substantially of coral colonies. Massive colonies are about five times as abundant as branched colonies. However, because the latter are often much larger (often more than 0.5 m in diameter and thickness) they are more conspicuous. Some colonies of *Acervularia ananas* showed a diameter of about 1.5 m. Also many bryozoan colonies are present, often intact and of large diameter. Some reached a great thickness, others remained thin; thus an almost massive colony of *Ptilodictya lanceolata* was seen with a diameter of about 1.40 m, but a thickness of not more than 12 cm; the branches grew so closely next to each other that hardly any space was left for the reef matrix.

The matrix in the reef is practically alike in all exposures and consists of a fine-grained or dense limestone which is only faintly marly. In this matrix fragments of bryozoan branches are abundant. The relative pureness of the matrix is notable, since marl is a common component of the *Pentamerus gotlandicus* Limestone. The fossils are strongly recrystallized; this often deleted the finer structural features and thus makes a study of the fossils more difficult.

The many large and intact colonies of the delicate bryozoans and branched corals suggest that the Norderslätt reef has not been subjected to much disturbance by contemporary wave action or currents. In modern reefs (e.g., the Great Barrier reef, Maxwell, 1968) there is often an inverse relationship between the abundance of branched bryozoans and the strength of water movement. Out of 240 coral colonies in the Norderslätt reef, 76% were found to be in their growth orientation, about 12% were distinctly not in their growth orientation and the remaining 12% were presumably tilted during compaction under the influence of differences in pressure caused by the overburden. Thus it seems likely that the position of not less than 88% of the coral colonies was not seriously affected by water movement. As will be further discussed in Chapter XII, this is an indication of formation in relatively deep water. The great scarcity, and in part of the exposures even complete absence of stromatoporoids may further support this conclusion. In summary, it may be said that the composition of the Norderslätt reef limestone is indicative of reef development in rather calm, clear and not very shallow water.

Suderslätt reef limestone

South and southeast of Stalen, exposures of solid rock are found at about present sea level, both along the shore and in the sea, up to about 75 m offshore. In the foot of these exposures, thin-bedded limestone is found, containing crinoid remains. Between the limestone layers thin layers of marlstone are present. At the surface of sections through this deposit the marlstone layers have been excavated by erosion for some depth. This deposit is overlaid by a thin complex (up to 0.5 m) of true crinoid limestone with reef debris; this rock is hardly stratified. On top of this crinoid limestone, remains of the lowermost part of a reef are found.

The preserved fragments of this Suderslätt reef consist mainly of coral colonies, with a contribution by branched bryozoan colonies. Among the coral colonies, branched forms are most conspicuous because of their large sizes of up to 1 m in diameter and thickness; massive forms are more abundant but generally smaller. The underlying stratified sediments show dips in varying directions, with the result that in one place stratified limestone is found at present sea level, in another vaguely stratified crinoid limestone, and in a third reef limestone; all within a short distance of each other. The three rock types together can be found over a length of about 0.3 km along the shore. West of Stalen, the reef limestone is exposed offshore to a height of just over 3 m above sea level. West-southwest of Stalen in the coastal area a few raukar-like exposures of reef limestone are also found. The rock of these exposures has perhaps also formed part of the Suderslätt reef. Slightly west of Myren, five raukar-like exposures present reef limestone up to 5 m above sea level; this almost certainly belonged to this major reef.

The great distance over which traces of the Suderslätt reef can still be followed in a northwest – southeast direction makes it clear that this reef must have had a relatively great horizontal extension. Nothing can be said about its original thickness, except that it must have been more than 5 m. Almost all of the reef has been destroyed by erosion. The present beach around its remaining fragments is a typical rubble beach as is also found in the raukar fields of Gotland proper.

Flank reefs

It has already been mentioned that the alternation of marlstone and limestone, which forms the lower part of the exposed stratified sediments of Lilla Karlsö, shows a gradual increase upwards in the thickness of the limestone layers and a corresponding decrease in those of the marlstone. Debris from the Central reef may have contributed to this increase in limestone deposition in the reef environment. The presence of well-defined stratification planes indicates that these reef-surrounding sediments were not very pappy. The same can be said of other similar deposits in the area of Gotland. However, in general contrast to these, on the mantle of stratified material around the Central Lilla Karlsö reef, reef builders settled in several places, giving rise to smaller reefs. With continuing growth of the Central reef the amounts of debris produced by it became greater and consequently the slope of the surrounding deposit increased. This increased slope is reflected in the shape of the younger flank reefs, which reached their greatest thickness at their seaward side, whereas towards the side closest to the Central reef a gradual thinning can be observed (Fig.124). At this inner side the flank reef sometimes interfingers with the stratified sediments. Further within the reef local lenses or layers of stratified sediment are only rarely found, indicating that once such a flank reef was growing well, it was not easily overwhelmed by mud and by the debris of the major reef. The lower and upper boundaries of the flank reefs are generally sharp, whereas the lateral boundaries are sometimes sharp and in other cases rather vague.

The length and width of the flank reefs generally do not surpass 25 m and many are smaller. Along the east coast between Bodarna and Janedi proportionally large reefs occur. Access to them is difficult and they are most easily visible from a boat on sea. The stratified sediments neighbouring them show dips in varying directions (Fig.121). Relatively large flank reefs also crop out between Stalen and Suder Vagnhus in the southern half of the west cliff.

Weathered surfaces of the flank reefs are usually massive in structure, in several instances irregularly crumbling. The unstratified masses generally contrast strongly with the surrounding stratified sediments.

The flank reefs which occur lowest in the stratigraphical column are faunistically characterized by a great profusion of corals, especially tabulate corals, and bryozoans. Stromatoporoids are rather rare and small. Higher up in the succession of deposits they are more common, but never did they contribute to reef formation to such an extent as they did in the reefs of Hoburgen type in Gotland. The occurrence of several intact colonies of branched bryozoans and corals in the oldest flank reefs, especially in the west, indicates that these reefs did not suffer very much damage by wave action or currents. In the higher reefs, a notable part of the branched reef builders is replaced by massive forms. Studies of recent corals (e.g., Vaughan and Wells, 1943) have shown that branching forms retreat in favour of massive forms when the water becomes rougher. While not all ecological data for recent organisms can positively be applied to related extinct taxa, it is likely that the reactions to rough water would be similar even in unrelated groups of similar forms. Thus the faunal composition of the flank reefs suggests that in the course of time the water became shallower. This deduction is further supported by the fact that the higher reefs are more disorderly in appearance.

Fig.121. Cliff wall north of Bodarna, Lilla Karlsö, with flank reefs exposed between the stratified limestone.

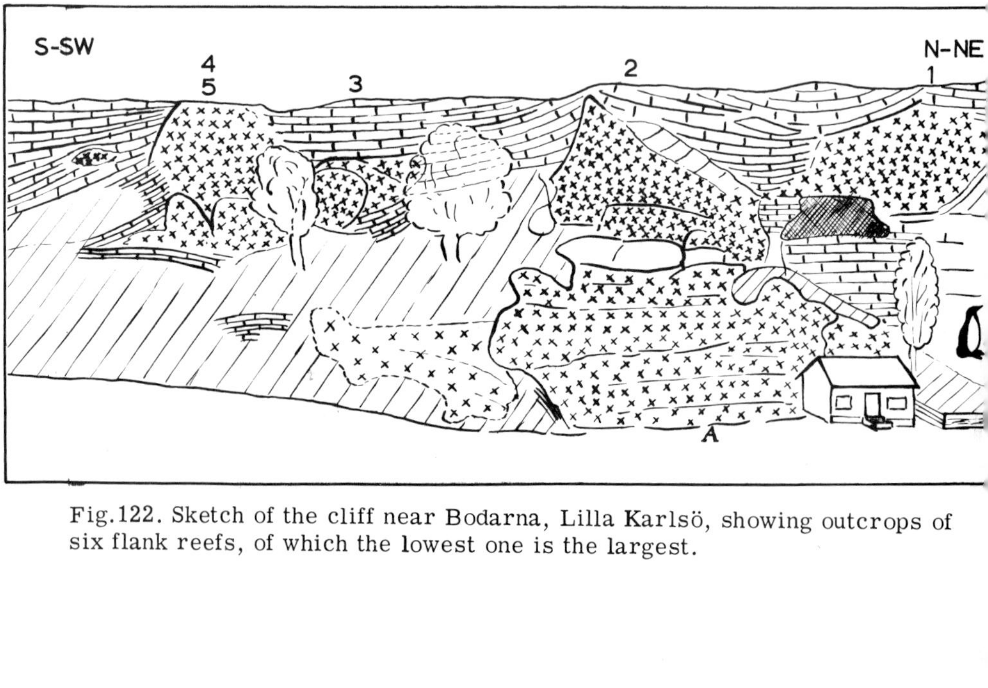

Fig.122. Sketch of the cliff near Bodarna, Lilla Karlsö, showing outcrops of six flank reefs, of which the lowest one is the largest.

Fig.123. Coastal cliff of Lilla Karlsö near Bodarna. Compare Fig.122.

Weathering results in that some bryozoan-rich parts, especially in the reefs in the west, show a structure reminiscent of industrial slag.

As can be seen directly on arrival in Lilla Karlsö at Bodarna the younger flank reefs are generally smaller than the older ones (Fig.122).

Compared with the flank reefs elsewhere in Lilla Karlsö, the reefs in the southern cliff are usually smaller; they are also less abundant than in the east and west. In the southwest the reefs get larger. A short distance east of Kuren a comparatively large one occurs, some tens of metres broad and about 8 m high. Southeast of the rauk Stalen the reef of Fig.128 is found. Stalen itself is also mainly built up of reef limestone; only at the very top is some stratified limestone present. Between Stalen and Suder Vagnhus the cliff shows mainly reef limestone. The flank reefs there are on the average distinctly larger than those at the southeast and south side of the Central Lilla Karlsö reef. This applies to both the older and the younger flank reefs. Between Suder and Norder Vagnhus outcrops of stratified sediments and flank reefs alternate, both in a horizontal and vertical direction. In the cliff wall from Norder Vagnhus to slightly west of Janedi only stratified limestones are exposed. Along the east coast, between Janedi and Bodarna, relatively large reefs crop out, especially lower in the wall. These latter reef-limestone occurrences do not form part of the Central Lilla Karlsö reef, as Rutten (1958, p.381) supposed, but belong to flank reefs, surrounded by stratified sediments, as can be seen in indentations in the cliff wall.

The stratified sediments that are found between Norder Vagnhus and Janedi are sparry, grey to light-brownish coloured limestones, very rich in fossils, especially bryozoans, corals, stromatoporoids and brachiopods. Part of these almost certainly came from the Central reef. Perhaps the narrow passage between the Central Lilla Karlsö reef and the Norderslätt reef made the environment less suitable for the development of flank reefs. The passage between the Central and the Suderslätt reef was not so narrow, and there some reefs developed on the flanks of the Central reef, but they are more disorderly than the flank reefs which are found elsewhere in Lilla Karlsö.

It is notable that the largest flank reefs are found in the east and west, at the latter side especially in the area about 350 m north of the Suderslätt reef, whereas the reefs in the south and southeast are smaller and less in number. It may be that as a result of the water becoming more shallow, a relationship to a shore line began gradually to make itself felt and that water movement at the seaward (southern) side increasingly worked against settlement of reef builders. In this way the occurrence of the highest number of reefs, their larger size and the large branched colonies in the oldest of these, as found in the southern half of the western side, can be explained from development at the calm-water side of the Suderslätt reef.

Several of the flank reefs, especially in the west, overlie *Pentamerus gotlandicus* breccia and seem to have slipped down on the dipping mantle of the Central Lilla Karlsö reef. This phenomenon will be further discussed later in this chapter (pp.273–275).

In the following pages a few outcrops of flank reefs will be discussed in more detail.

Fig.122 and 123 are a sketch and a photograph of the reefs which crop out near Bodarna. Reef A is exposed lowest in the wall. It is built up by bryozoans and corals, embedded in a matrix which is generally rather pure but locally faintly marly. The reef shows some vague stratification, is strongly recrystallized and weathered surfaces crumble irregularly.

At a level directly above that of reef A, several smaller reefs are found. Of these the one overlying reef A, reef 2, is the most important. At its right side, stratified limestone is found, very rich in reef debris, and dipping north-northeast. Close to the reef, the layers show a kind of kink; the transition from reef to debris-rich stratified sediment is very gradual there. In contrast to this, the boundary at the left side is very sharp; there the stratified limestone dips south-southwest. Between reefs 1 and 2 a depression is found, containing stratified limestone rich in reef debris. Below the depression both reefs touch each other. Reefs 3, 4 and 5 are smaller than the other two. The south-southwest boundary of reef 4 with the stratified deposits is also rather sharp. Almost everywhere in the environment of these reefs too, the stratified sediments are rich in reef debris; they are somewhat marly. The bedding planes are generally irregular and rugged and often covered by a film of marl. Behind reef A, between A and 1, a very disorderly mass of extremely marly limestone is found, exceedingly rich in crinoid remains and with a varying amount of reef debris, including an occasional stromatoporoid. Reef 1 has a more marly matrix than reef A. This also holds for the other higher-exposed reefs. These reefs contain a number of stromatoporoids, of small size but not rarely of rather round shape. Weathered surfaces generally are irregularly crumbling, sometimes rather smoothly massive.

The stratified limestone exposed near Bronsålderrose, northeast of Myren, is light brown, sparry, contains some reef debris, shows marl films on its bedding planes and dips south. Intercalated in this sediment is the flank reef of Fig.124 and 125, and its surrounding crinoid limestone. The most common reef builders are compound corals (*Halysites*, *Favosites*, and others); bryozoans are numerous but usually less conspicuous. Some stromatoporoids are also present, generally tabular forms of rather great extension and up to a few centimetres thick. Often these thin stromatoporoids are secondarily broken. Some stromatoporoids are lens-shaped. The reef matrix is marly and constitutes a high portion of the total reef volume. Embedded in the matrix are several crinoid stem fragments. The reef as a whole is of a rather disorderly appearance. It is highly recrystallized, but because of the marly matrix, not homogenously so; crystals of calcite, up to a few centimetres large, are not rare. The weathered surface of the reef is more brecciated than it is conglomeratic. Locally the reef shows a vague stratification; the nature of these planes suggests that this

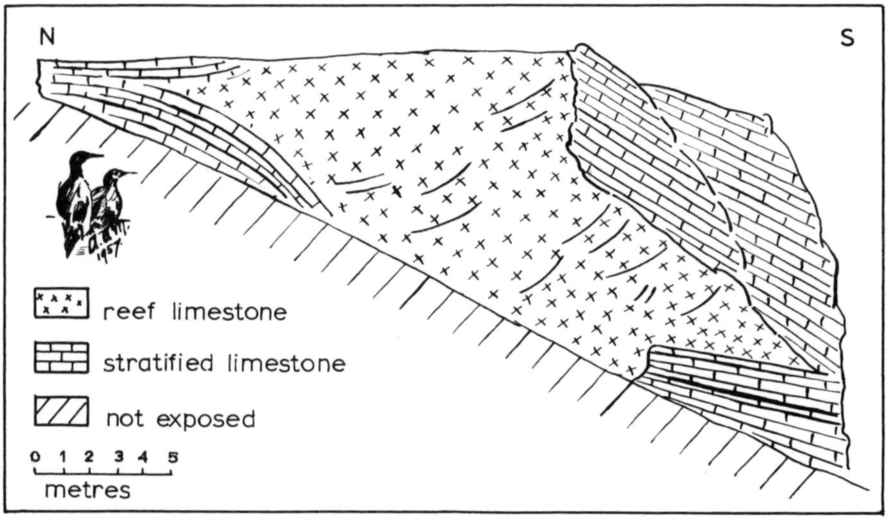

Fig.124. Sketch of a flank reef, exposed in the southern cliff of Lilla Karlsö near Bronsålderrose.

Fig.125. Flank reef, exposed in the southern cliff of Lilla Karlsö near Bronsålderrose. Compare the sketch in Fig.124.

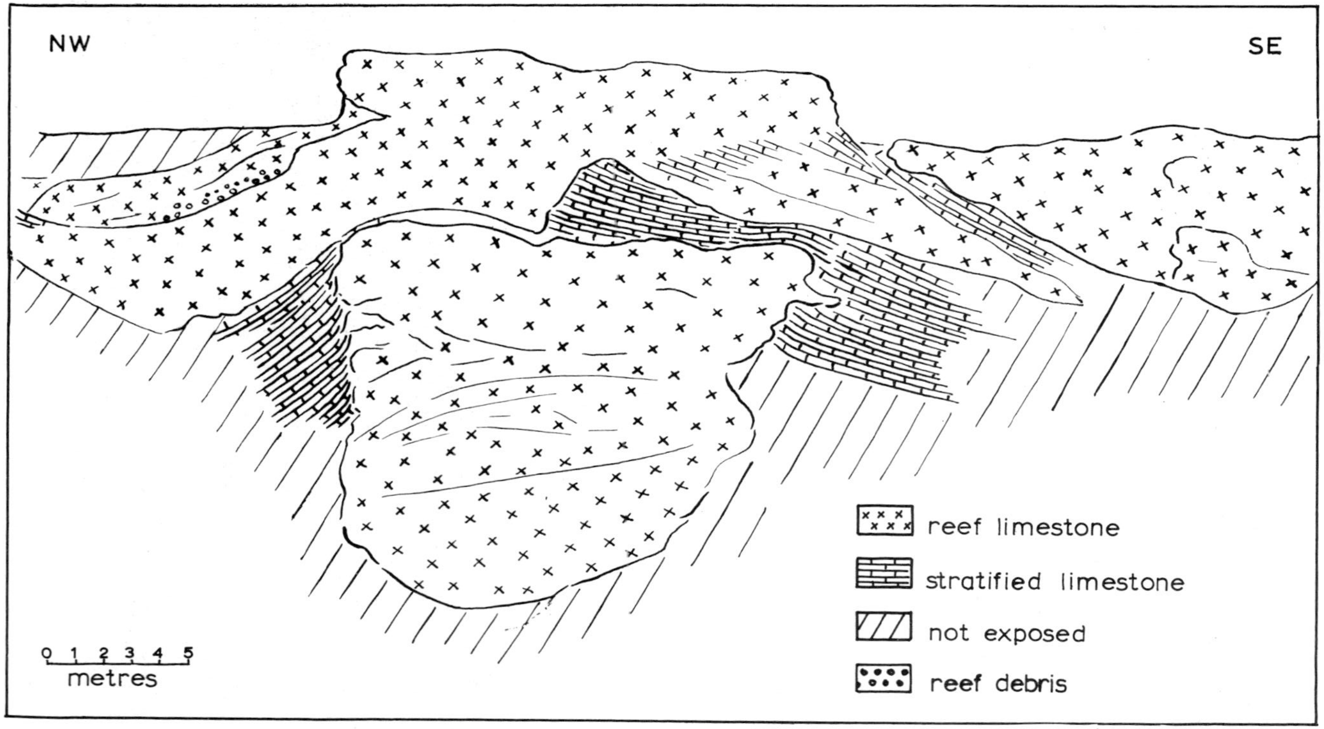

Fig.126. Lilla Karlsö, southern cliff about 0.1 km west of Bronsålderrose. Two flank reefs are sketched. Note how settling of the stratified sediments caused a subsidence of the lateral parts of the top reef.

Fig.127. Photograph of the two reefs sketched in Fig.126. Lilla Karlsö, about 0.1 km west of Bronsålderrose.

originated due to lateral compression during an attempt to slip down. Also the dip in the stratified limestones at the top at the south side may be intensified by this. The reef at this south side wedges out between crinoid limestone. The stratified limestone at the north side inclines down under the reef; it is rather rich in *Halysites* colonies and solitary corals.

The crinoid limestone in the environs of this flank reef is thin bedded with marl films on the bedding planes. The latter are generally rugged. The rock is bluish grey, grey to brownish grey in colour. Through weathering it usually first becomes brownish, later turning more greyish. The sediment is very fossiliferous, containing some trilobites, brachiopods (*Howellella elegans, Camarotoechia borealis, Dicaelosia biloba, Rhynchotreta cuneata, Ptychopleurella bouchardi, Chonetus striatellus,* and others), some gastropods, stromatoporoids, *Fenestella* sp. and other bryozoans, an enormous amount of crinoid fragments including some bases of crinoid calyces, and solitary corals. The crinoid remains are partly red, partly grey-white in colour. Typical for this deposit are colonies of *Favosites* sp., attached to older colonies of the same coral (both may or may not be in their growth position), to solitary corals, and even to crinoid remains. On the coral colonies *Fenestella* sp. has also often settled.

About 0.1 km west of Bronsålderrose, two reefs occur above each other (Fig.126, 127). The lower of these shows a vague stratification. It is built of a light-greyish reef limestone with bryozoans, corals and a few stromatoporoids. In between both reefs there is a very marly and fossiliferous rock, or an alternation of limy and thin marly layers, both with irregular bedding planes, especially at the northwestern side. The upper reef has a somewhat more marly matrix and the weathered surface is more crumbling. Corals (*Favosites, Halysites,* and others) and bryozoans are the main reef builders. There are some more stromatoporoids than in the lower reef, but they still play a subordinate part. In the upper reef is a rather large intercalation of debris in a strongly marly matrix; fossils and fossil fragments occur in all possible orientations.

In the northwest the upper reef as it were comes down from the lower one. It apparently started growth on a rigid foundation that hardly changed volume under the weight placed upon it. Through lateral spreading it came to rest on a soft limestone bottom. The weight of the reef caused a compaction of that sediment and with it the reef sank down. The stratified limestone is thin bedded with marl films on the rugged bedding planes.

The large block of reef limestone at the foot of the cliff about 0.15 km southeast of Stalen probably is not in its original position.

Near the rauk Stalen, Lilla Karlsö–from the beach inland–shows the following picture: a low cliff, then a gradually mounting topography with some raukar, the main cliff of the island, and finally its upper plateau. In the main cliff directly southeast of Stalen, the reef of Fig.128 is exposed.

The reef limestone is rather uniform, light grey in colour, a little marly, as a whole recrystallized, finely crystalline, with a flaky or splintery weathering surface.

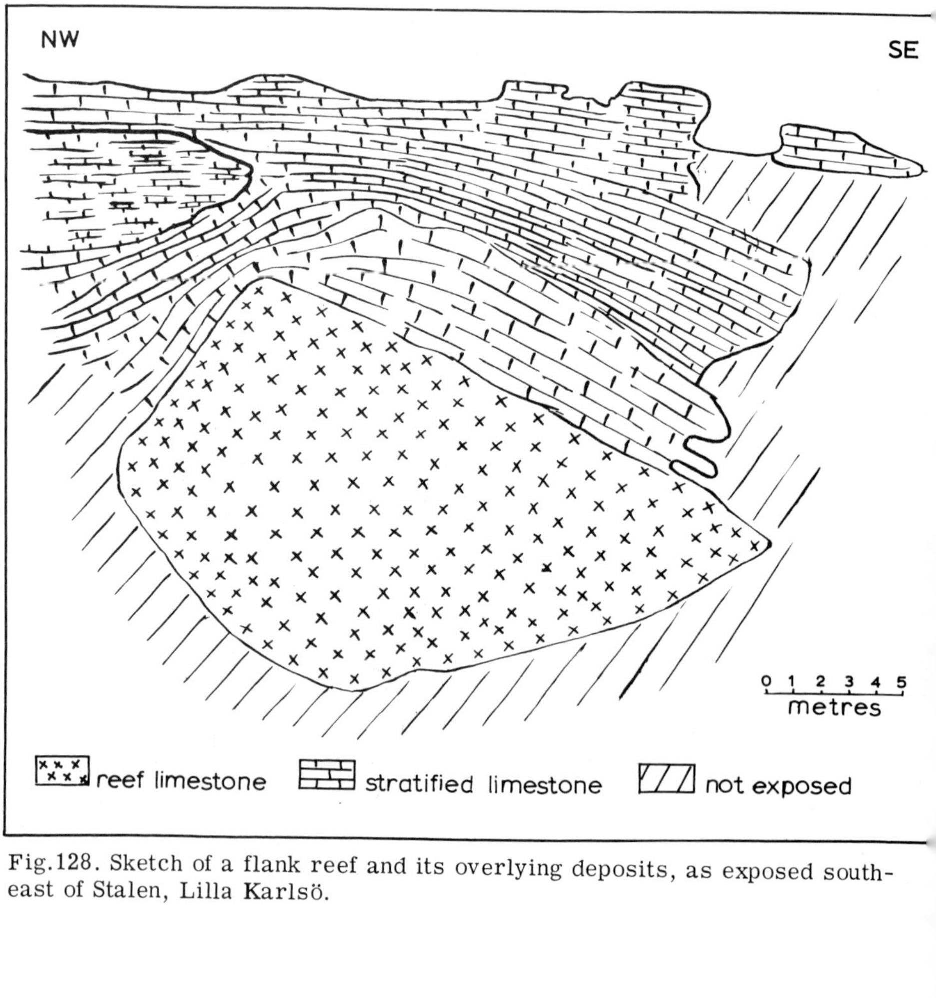

Fig.128. Sketch of a flank reef and its overlying deposits, as exposed southeast of Stalen, Lilla Karlsö.

Fig.129. Cliff wall south of Suder Vagnhus, Lilla Karlsö. Stratified Lilla Karlsö Limestone and two outcrops of flank reefs.

The fauna consists of corals, bryozoans, an occasional stromatoporoid, some brachiopods and cephalopods and crinoid remains. Locally there is a pocket of a somewhat more marly sediment.

The reef is overlaid by stratified limestone rich in reef debris. Some layers are rather strongly marly. The bedding planes are often rugged. Fossils include large coral colonies (*Acervularia ananas*, and others), some stromatoporoids and many brachiopods. Part of the stromatoporoids are tabular forms of generally less than 2.5 cm thick, but often some tens of centimetres in horizontal extension; most of these are found in their position of growth. Another part of the stromatoporoids is flatly lens-shaped; these colonies generally lie oblique. The layers of this debris-containing deposit dip about 5° towards the centre of the island. Together with the crinoid limestone overlying this deposit this is an indication that probably not the central reef but the flank reef itself has supplied most of the debris.

Stratified limestones dipping up against a flank reef can also be seen in the cave Suder Vagnhus. The transition to the crinoid limestone, also very fossiliferous, is gradual. In the northwest (top left in Fig.128) a rather massively-built part is included in this deposit.

The reef limestone of the rauk Stalen itself occurs at the same level as the reef limestone described above and is of the same general nature.

In the main cliff south of Suder Vagnhus, a reef crops out which is shown in Fig.129. It consists of dense, light-grey reef limestone with bryozoans, corals (*Favosites*, and others), and some trilobites and orthoceratids, in a slightly marly matrix. At the south side of this reef the boundary with the stratified limestone is not very sharp. This stratified rock is a light-grey, grey to greyish-brown limestone with a little marl on its bedding planes. This limestone contains crinoid-stem fragments and some stromatoporoids, a few of which show rather round forms. Higher up the sediment becomes more sparry and contains comparatively more stromatoporoids. At the bottom the layers dip towards the reef after having arched over a small reef at the south side of the larger flank reef. This small reef contains corals, most of which are small and partly also rather thin, and also bryozoans, crinoids, brachiopods and an occasional stromatoporoid. At the north side of the larger reef, the stratification of the adjacent rock is in part difficult to perceive, especially at close quarters. This sediment is a brownish, marly limestone, sparry, with many corals (*Favosites, Halysites, Heliolites,* and others). On top of the reef is crinoid limestone.

DISCUSSION

Correlation between Stora and Lilla Karlsö

The Lerberg Marlstone, exposed in the west, northeast and east of Stora Karlsö has no exposed equivalent in Lilla Karlsö.

The Spangände Limestone of Stora Karlsö is characterized, especially in its lower part, by the brachiopod *Pentamerus gotlandicus*. This fossil gives its name to the *Pentamerus gotlandicus* Limestone of Lilla Karlsö. These two deposits may, therefore, be correlated.

It is not certain whether the boundary between Spangände and Austerberg Limestone and between the *Pentamerus gotlandicus* and Lilla Karlsö Limestone correlate; probably the latter is older than the Stora Karlsö boundary.

The Lilla Karlsö Limestone then, correlates with the uppermost Spangände Limestone and the Austerberg Limestone. However, the greater height of Lilla Karlsö makes it likely that the upper part of the Lilla Karlsö Limestone is younger than any of the Stora Karlsö deposits.

A tentative sketched correlation is presented in Table XV.

Environment of formation of the various sediments

The sequence of Lerberg Marlstone and Spangände Limestone in Stora Karlsö probably represents a period of slowly decreasing water depth. In the shallower water, wave action at the bottom was stronger and, therefore, terrigenous clastic material had more difficulty in settling and was kept floating to a higher degree. This, together with an increasing production of limestone debris from the developing reefs, led to a shift from marlstone deposition to formation of marly limestone. The faunistic succession in the Western reef limestones, especially the Spangände reef, also suggests a change in environment during the time of growth, most likely a shallowing of the water. The remains of the Norderslätt and Suderslätt reefs contain several large and intact colonies of branched reef builders, suggesting slight disturbance by wave action. A significantly high percentage of reef builders in these latter two reefs are found in their growth orientations; this is indicative of relatively deep water (cf. Chapter XII, p.437). The higher parts of the major reefs in Stora Karlsö are closer in appearance to the Hoburgen-type reefs of Gotland, thus pointing to shallower water.

Because of a combination of the shallowing of the water and the upward growth of the reefs, their tops approached the water surface. This did not take place at precisely the same time for the Western and Röjsuhajd reefs of Stora Karlsö and the Norderslätt and Suderslätt reefs of Lilla Karlsö, but the time intervals were probably not very long. It is possible that the decrease in water depth was temporarily stronger.

Of the major reefs of Karlsöarna only the Central Lilla Karlsö reef seems to have survived the shallow-water stage suggested above. Unfortunately there are no exposures in this reef which can be studied for indications of its subsequent stages of development. The fact that it could continue growing suggests that the very-shallow-water period must have been of only short duration. The strong development of the Svarthällar reefs which took place mainly thereafter indicates that the water soon became rather deep again, for in their composition the Svarthällar reefs are much more comparable to the Norderslätt and Suderslätt reefs of Lilla Karlsö than to the common Hoburgen-type reefs of Gotland.

Downward-slipping phenomena

The occurrence of the *Pentamerus gotlandicus* breccia in Lilla Karlsö underneath reefs suggests a relationship between the two, such that the breccia was formed when reefs slipped downwards over the stratified sediments, consisting of an alternation of marlstone and marly limestone layers. Also the bucklings found in the *Pentamerus gotlandicus* Limestone, caused by lateral compression, could well be caused by such a moving flank reef.

The breccia and the upward dipping sediments which are found against the reef north of Suder Vagnhus suggest that blocks of stratified sediments have also probably moved down over a marly surface in the sedimentary sequence dipping down from the Central reef.

Close to Suder Vagnhus another reef has apparently come down together with about 1.5 m of its underlying stratified sediments. Underneath this combined mass, *Pentamerus gotlandicus* breccia is found, thinning out seawards between it and the stratified sediments exposed at the base. The latter are assumed to still be in their original place. The displaced stratified rocks show a distinctly greater dip in seaward direction than the basal sediments.

The question now is whether these downward slippings occurred during the Silurian or later.

Slightly southwest of Trädgården, in Lilla Karlsö, there is a displaced reef-limestone mass, overlying *Pentamerus gotlandicus* Limestone, which is exposed to a thickness of 4 m. Of these, the lower 3.5 m are normally stratified, but the upper about 0.5 m is strongly brecciated. The reef limestone has apparently come down from a higher and more central position in the island. The high cliff facing the displaced reef limestone shows a distinct indentation. This is strongly indicative of a relatively recent displacement of the reef limestone.

Another argument for this is provided by the large reef exposed at Trädgården. A short distance north-northeast of Trädgården the reef limestone has its base about 6 m above present sea level. A vague stratification can be observed, which is approximately horizontal. In a huge block, northeast of Trädgården (D in Fig.130B), this vague stratification dips distinctly seawards. The reef limestone together with an amount of stratified sediment annexed to it discordantly overlies the normal alternation of *Pentamerus gotlandicus* marlstone and marly-limestone layers. In between the block and its underlying sediments a breccia has developed. In the south the lower

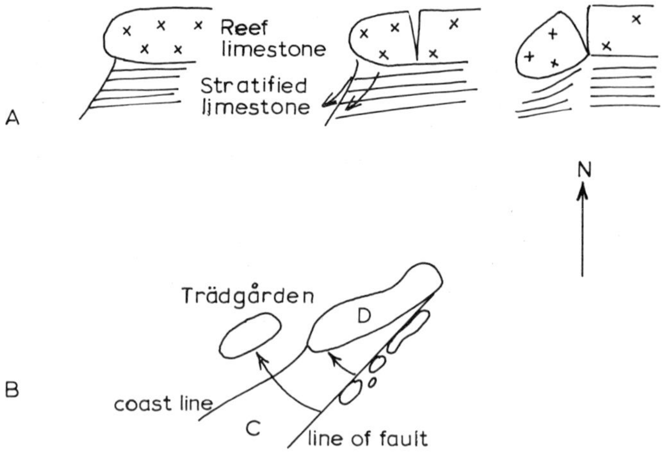

Fig.130. Sketch illustrating the origin of Trädgården, in the west of Lilla Karlsö. The heavy reef limestone at the top of the sedimentary complex caused the wall to sag laterally. Stress made the reef limestone fault and tilt. At a certain moment, the south of the detached reef limestone slipped down and a catapult effect caused Trädgården to be displaced furthest from its original position.

---

boundary of the reef limestone has disappeared below sea level. It seems likely that Trädgården itself, a large reef-limestone block, now forming a small island just off the shore, was once connected with the large onshore block. The two together were again connected with the reef limestone as is exposed in raukar at the landward side. These raukar mark with their west sides an approximately straight line, which may be assumed to represent the line of fault.

It can now be explained how the situation at Trädgården has developed (Fig.130). In the wall of Lilla Karlsö, heavy reef limestone overlaid stratified limestone. The weight of the reef limestone caused lateral displacement of the underlying sediments. The movements in the lower part of the wall caused stresses in the reef limestone and at a certain moment the reef-limestone mass was faulted open, with the seaward part of the reef limestone starting to tilt downwards. At a certain stage gravity led to a décollement of the southern part of the block and a catapult effect caused the Trädgården block to become detached from block D and to be moved furthest away from its original position. In the north of block D only the fault was opened further.

A similar phenomenon as explained above for west Lilla Karlsö can still be seen in development on a smaller scale at Hoburgen, southern Gotland, where a large block of the plateau of Storburg is in the process of being separated from the remaining top of the hillock along a widening fissure, caused by a gradual lateral sagging of the underlying sediments.

In Stora and Lilla Karlsö the dip in the sediments mantling the large reefs and the occurrence of marl in the lower part of the sedimentary sequence have certainly facilitated a downward slipping of rock masses. The heavy reef limestones of the flank reefs were especially prone to slip down, but as has been said before, masses of stratified sediment could go that way too. This explains the unconformity between Käupru and Stiudden in Stora Karlsö (Fig.102). The breccia north of Käupru (Fig.101) may be caused by either type of sediment moving down, but not improbably by stratified limestones as are now found overlying it.

The example of Hoburgen teaches that downward slipping can still occur today. The slippings in Stora and Lilla Karlsö may also be of a very recent date. However, it can be imagined that especially suitable conditions for these phenomena existed during and following the Pleistocene Ice Ages. Plucking of the southwards moving land ice may have caused rock displacements. Further a suitable situation for such phenomena occurred at the time of melting of the ice of a glaciation. During an Ice Age, the high walls of the island were enveloped and supported by the land ice, as is demonstrated by the presence of erratic blocks on top of the island. When the ice melted, the walls lost their support, great amounts of water became available, the rocks unfroze and became wet throughout and the soft marlstones at the base of the sedimentary complex presented ideal lubrication layers.

Correlation with Gotland

Pentamerus gotlandicus Lebedev is in Gotland characteristic of the Slite IV Beds. This suggests that the Spangände Limestone and *Pentamerus gotlandicus* Limestone correlate with the Slite IV Beds and the Lerberg Marlstone probably with the Slite III Beds.

Whereas in the east of northern Gotland, Slite limestone is found to extend southwards over the Slite marlstone, in the west of central Gotland, between Djupvik Fiskläge, Fröjels Fiskläge and Väte, the Slite marlstone is overlaid by thick sandy limestone to limy sandstone, of a maximum thickness of 3 m. Moreover, the lowermost part of the Halla limestone in that area contains rounded sandstone pebbles and ripple marks. This suggests that an increasingly stronger shallowing of the water occurred at the time of their formation when going westwards.

TABLE XV

Schematical representation of the supposed correlation between the rocks of Stora Karlsö, Lille Karlsö and Gotland

Stora Karlsö		Lilla Karlsö		Gotland
Stratified sediments	Reef limestones	Stratified sediments	Reef limestones	
		Lilla Karlsö Limestone		Klinteberg Beds
Austerberg Limestone	Svarthällar reef limestones	Lilla Karlsö Limestone	Flank reefs / Central Lilla Karlsö reef limestone	Halla-Mulde Beds
Spangände Limestone	Western reef limestones / Röjsuhajd reef limestone	*Pentamerus gotlandicus* Limestone	Norderslätt and Süderslätt reef limestones / Central Lilla Karlsö reef limestone	Slite Beds
Lerberg Marlstone				Slite Beds

This period of shallowing of the water may have been the same as the one which probably contributed to the end of the growth of almost all the major reefs of Karlsöarna. This phenomenon should have taken place then in Karlsöarna at about the time of the transition from deposition of the Slite Beds to formation of the Halla-Mulde Beds.

The shallowing of the water did not lead to a break in the deposition of stratified sediments. The discordances which are mentioned from both Karlsöarna by Lindström (1882a) and Hede (1925a, p.20) are only local phenomena, identical with the phenomena which the present author described above as being due to the downward slipping of flank reefs.

As will be pointed out in the discussion of the Halla-Mulde Beds in Chapter XI, there are strong indications that during Halla-Mulde time, water depth soon increased again in western Gotland, even though in northeastern Gotland a decrease in water depth continued. This must have been caused by a change in the direction of the hinge line of epeirogenetic movement of the basin floor, which resulted in a relatively strong increase in water depth in the area of Karlsöarna.

The Central Lilla Karlsö reef was probably not yet elevated high enough above the sea floor to have its growth terminated by the Late Slite – Early Halla-Mulde shallowing of the water and, therefore, managed to survive, growing again more rapidly when the water became deeper again. The Svarthällar reefs benefited even more from the increasing water depth. Being still small reefs at the more unfavourable time, growing at the foot or on the flanks of large reefs, they thereafter developed into an extensive reef-limestone complex. In their composition they bear witness to formation in relatively deep water in the later period.

During Late Halla-Mulde time the water began to become shallow again. This is evidenced by the sediments in Gotland and also by the characteristics of the highest parts of the Lilla Karlsö Limestone (more marl, more stromatoporoids, less bryozoans).

The presence in the uppermost Lilla Karlsö Limestone of such fossils as *Conchidium biloculare*, *Dolerorthis* cf. *rustica*, *Eospirifer* cf. *interlineatus* and *Plectatrypa marginalis* might suggest that these youngest sediments of Karlsöarna correlate with the lowermost Klinteberg Beds (cf. Hede, 1927b, p.50).

Chapter XI

STRATIGRAPHY OF THE SILURIAN OF GOTLAND

INTRODUCTION

In the following pages, the stratigraphy of Gotland will be discussed. The system of Hede (1921 and later), with some modifications proposed by the present author, will be used. A short survey of the stratified sediments of each unit will first be given. Thereafter, the more important exposures of reef limestone and directly related sediments will be briefly described, as far as this has not been done in earlier chapters. These reviews of reef localities may be helpful to future students of the reefs of Gotland, in preparing excursions or research programmes. Concluding the treatment of each stratigraphical unit will be a section discussing the environment of deposition of the various sediments, the stratigraphical implications of the study of the reefs and related deposits, and similar subjects.

For a correlation between the stratigraphical units distinguished by Hede and those of earlier authors, the reader is referred to Table IV and, for more detailed information, to Munthe et al. (1925).

VISBY BEDS

Stratified sediments

At the base of the exposed Silurian of Gotland, Hede distinguished two units. Both consisting predominantly of marlstone, Hede called these two units the Lower Visby Marlstone and the Upper Visby Marlstone. Both are relatively thin units, if compared to such units as the Högklint, Slite, Klinteberg or Hemse Beds. There is no distinct lithological boundary between the two. A gradual decrease in the amount of marl upwards in the profile can be noted, which, however, continues throughout the entire Upper Visby Beds and even in the lowermost Högklint Beds. Except particularly some ostracodes, only a few fossils seem to be restricted to the Lower Visby Beds, and still fewer are known to occur only in the Upper Visby Beds (Table XVI). A greater number seems to be restricted to the two Visby units together. In all these cases, it is likely that part of the fossils are facies, rather than index fossils. For instance, on the basis of the literature, *Holophragma calceoloides* (Lindström) should be a fossil from only the Visby Beds, but the present author also found it in the Hemse marlstone.

The author prefers to consider the Lower and Upper Visby Beds together as the Visby Beds with a lower and upper subunit. This is more in agreement with the subdivision applied to some of the other stratigraphical

TABLE XVI

Fossils which in Gotland are found only in the Visby Beds

Phylum, classis, or subclassis	Species	Lower Visby Beds	Upper Visby Beds
Anthozoa	*Catenipora escharoides* Lamark	+	+
	Clisiophyllum (Dinophyllum) involutum Edwards et Haime	+	+
	Goniophyllum pyramidale (Hisinger)	+	+
	Lykophyllum tabulatum Wedekind	+	+
	Phaulactis angusta (Lonsdale)		+
	Pholidophyllum tabulatum Schlotheim	+	+
	Planalveolites fougti (Edwards et Haime)	+	+
	Thecia hisingeri (Jones)	+	+
	Porpites (Palaeocyclus). porpita (Linnaeus)	+	
Bryozoa	*Ceramopora lindströmi* Hisinger		+
	Crepipora lunariata Hisinger		+
	Mesotrypa suprasilurica Hisinger		+
Brachiopoda	*Brachyprion walmstedti* (Lindström)	+	
	Eospirifer marklini De Verneuil	+	
	Hesperorthis (Orthis) davidsoni (De Verneuil)	+	
	Liljevallia gotlandica Hedström		+
	Plectodonta (Sowerbyella) transversalis (Dalman)	+	
	Resserella visbyensis (Lindström)	+	+
	Rhynchonella exigua Lindström	+	+
	Stricklandia lirata J. de C. Sowerby	+	
Gastropoda	*Callonema obesum* Lindström	+	+
	Callonema scalariforme Lindström	+	+
	Cyclonema delicatulum Lindström	+	+
	Cyclonema giganteum Lindström	+	+
	Poleumita roemeri (Lindström)	+	+
Cephalopoda	*Phragmoceras convolutum* Hedström	+	
	Phragmoceras costatum Hedström	+	
Trilobita	*Encrinurus laevis* (Angelin)	+	+
Ostracoda	*Apatobolbina simplicidorsata* Martinsson	+	
	Apatobolbina tricuspidata Martinsson		+
	Barymetopon infantile Martinsson	+	
	Beyrichia hirsuta Martinsson	+	
	Craspedobolbina juguligera Martinsson	+	+
	Leperditia hisingeri Schmidt	+	
	Leptobolbina hypnodes Martinsson	+	
	Noviportia simpliciuscula Martinsson	+	

units of Gotland. It should be emphasized, however, that this is purely a formal procedure. The distinction made by Hede remains essentially unaltered.

Lower Visby Beds

The Lower Visby Marlstone is the oldest Silurian sediment exposed in Gotland. It is found along the northwest coast from Stavsklint (Tofta Parish) in the south to Hallshuk in the north. Its maximal thickness above sea level is about 10 m, but usually less. It generally forms the lowermost part of the cliffs which are found along this coast.

The rock consists of thin bedded, rather soft marlstone of bluish grey colour, which alternates with harder marly limestone which is dense to finely crystalline, light grey in colour, and which often contains crystals of pyrite. This limestone occurs partly as thin layers of restricted horizontal extension and partly as knolls or thin lenses. If occurring as knolls or lenses, these often also lie in special horizons. In such horizons the limestone knolls or lenses occur either closely together or up to some decimetres apart. Their thickness varies generally from 2 to 4 cm, but in exceptional cases may reach 15 cm. The limestone layers are 1–4 cm thick, but with a majority in the group of 1–2 cm thickness. The marl layers are 2–10 cm thick.

The fossil content of the Upper Visby Beds is high, especially in corals (many solitary corals, *Halysites*, heliolitids, favositids) and brachiopods. Lamellibranchs, ostracodes (some Beyrichiidae and large *Leperditia* sp.), trilobites, bryozoans, crinoids and stromatoporoids are rather scarcely represented. The occurrence of the brachiopod *Stricklandia lirata* (J. de C. Sowerby) is notable; it is not common in several horizons, but is particularly abundant in others. A thin layer very rich in this fossil also approximately marks the boundary between the Lower and Upper Visby Beds in several places.

Upper Visby Beds

The Upper Visby Marlstone can be followed along the northwest coast from Gnisvärds Fiskläge (Tofta Parish) in the south to again Hallshuk in the north. It is generally found in the lower part of the coastal cliffs. Seen on a large scale, the layers show a faintly wavy course, especially in the northern part of their area. At least in part, this was caused by the differential compression exerted by the overlying rocks. The thickness of the Upper Visby Beds varies somewhat. It is on the average about 10 m, but may locally reach up to 15 m. This variation in thickness may also be caused by the overburden, more particularly the Högklint Beds. The Upper Visby Beds are often thinner underneath Högklint reef limestone masses than under stratified Högklint sediments. Hadding (1956, p.3) stated: "There is reason to believe that the marly mud originally had a comparatively even surface which was

Fig.131. *Halysites* biostrome with some *Omphyma* sp. Upper Visby Beds, south side of Ihrevik.

deformed later on. The compact reef bodies must have pressed the marly mud downwards and outwards-upwards, at the same time locally elevating the limestone sediment deposited on the marl." Moreover, the fact that the boundary with the Högklint Beds is not very easy to determine and probably is not always laid at exactly the same stratigraphical height, should also be recognized. This problem will be discussed further when treating the Högklint Beds (pp.282, 311).

The Upper Visby Beds consist of thin layers of rather soft bluish grey marlstone which alternate with harder marly limestone. The latter is dense to finely crystalline and light grey in colour. Lithologically, therefore, there is little difference between the Upper Visby Beds and the underlying Lower Visby sediments. The limestone occurs partly as small elongated knolls or thin lenses and partly as thin layers which thin out in the marl after varying distances. The lenses and knolls are found particularly in the lower part of the Upper Visby Beds. The number of limestone layers increases upwards, as well as their thickness, which increases from about 2 cm to sometimes 15 cm at the top of the unit. Thicker than normal limestone layers are found in the environment of several Upper Visby reefs. The bedding planes of the limestone layers are often rugged. Together with the upward increase in thickness of the limestone layers in the profile, the marlstone layers decrease in thickness. In the marly limestone, pyrite is found locally in isolated crystals or in groups of crystals. The pyrite cubes are generally of the order of a millimetre of less, but locally the crystals have edges of more than half a centimetre.

In the higher parts of the Upper Visby Beds, small reefs also occur. Over 140 of these have been observed in the mentioned coastal cliff. Details about these reefs have been given in Chapter VI.

Except for the reefs, the Upper Visby Beds almost everywhere present a similar overall lithological and palaeontological picture. The rocks are more fossiliferous than those of the Lower Visby Beds. Bedding planes often present real "zoos" of fossils. Only very few fossils have been observed that were embedded in their position of growth. This might indicate that sedimentation did not take place at a high rate. Neither in the marlstone, nor in the limestone have tracks or burrows been found. Corals (many solitary species, favositids, *Halysites*, heliolitids, and occasionally *Syringopora* sp.) and brachiopods (many *Dicaelosia* sp., *Leptaena* spp., *Atrypa* spp., Rhynchonellida and occasionally also spiriferids) play an important part among the fossil content, and bryozoans and crinoids are rather common. Stromatoporoids (e.g., *Stromatopora discoidea* (Lonsdale)) are not particularly abundant, but increase in number upwards in the profile. Trilobites and graptolites are scarcely represented, and Algae are still absent. The primitive thick-shelled brachiopod *Dinobolus davidsoni* (Salter) can be rock forming in some limestone layers. The coral *Halysites* is also occasionally found in extreme abundance, in some cases forming biostromes (Fig.131). Often a cap of curious stromatoporoids is found on the shell of the common gastropod *Poleumita roemeri* (Lindström).

Discussion

The Visby Beds present a number of characteristics which indicate a gradual decrease in sea depth during the time of their deposition, such as the upward decrease in the amount of marl, the increase in the amount of limestone in that same direction, the fact that stromatoporoids become more abundant in the higher parts of the Upper Visby Beds and the occurrence there of the oldest, though small, reefs of Gotland.

During Early Visby time (deposition of the Lower Visby Beds), the sea floor likely was generally uninfluenced by wave action; a depth of more than 50 m at that time is very likely (Hadding, 1941, p.66).

When the lower Upper Visby Beds were laid down, wave action presumably began to periodically influence the sea floor and this was permanent in middle and late Late Visby time when the reefs developed. The fauna of these reefs needed current water refreshment, whereas water movements also strongly contributed to the formation of the limestone mantles around part of the reefs. Hadding (1941) also explained the formation of the lenses and layers of marly limestone through wave action which removed the surface layer of the sea floor and washed out the fine continental debris, but left behind the coarser fossils and fossil fragments. Hadding (1941, p.67) stated: "These are found in the series as thin limestone beds filled with mostly well preserved, slightly worn calcareous brachiopods, corals, bryozoan branches, crinoidal fragments, and the like. In places where these limestones are abundantly present within the series the sedimentation took place at so slight depths that strong waves or occasional currents could influence the bottom deposits. The depth was, however, still so great that clayey mud was deposited under normal conditions." The present author does not consider this theory of Hadding to be fully satisfactory. Among others, the rather random alternation of marlstone and marly limestone in both horizontal and vertical directions, is not easily understood in the reasoning of Hadding. The problem of the mode of formation of the Visby Beds is worth to be studied in detail using modern geological and geochemical methods, including detailed analyses of the insoluble fractions.

HÖGKLINT BEDS

The Högklint Beds derive their name from the well-known hill and cliff of Högklint about 6.5 km southwest of Visby. The beds concordantly overlie the Upper Visby Beds. The Tofta limestone is considered by the present author to be a facies of the Högklint Beds.

The Högklint Beds are found in the northwest of the island. Exposures occur mainly in the cliff coasts formed by the *Ancylus* lake, the *Littorina* sea and the present Baltic. Of these, the latter two are by far the most important. The dominant exposures in the present coastal cliff are those at Högklint and along the long coastal stretch between Nyhamn and Hallshuk. Unfortunately, most of the walls are so steep that in general only the lowermost parts can be directly studied; the uppermost parts, moreover, are usually strongly weathered.

The *Littorina* sea, which reached heights above the present sea level around Gotland of about 14 m in the south and 27 m in the north, has also produced important exposures. These are found especially at Snäckgärdsbaden, but also at Brissund, Kinner (southeast of Nyhamn), Lickershamn (with raukar) and Hallshuk (Hjännklint). The *Littorina* cliffs often show a scree of huge blocks at their feet, which is much larger than usually found at the foot of the present coastal cliffs.

The fresh water of the *Ancylus* lake reached to a height of maximally 19 (southern Gotland)–45 m (northern Gotland) over the present sea level. At that time, Gotland was only about 1,300 km^2 large, compared to its size of about 1,900 km^2 in *Littorina* time and about 3,100 km^2 now. The cliffs left by the *Ancylus* lake are usually of little importance. They are the oldest outcrops in Gotland and the rock is generally very strongly weathered. Among others, some walls of the Galgberg (north of Visby) owe their origin to the *Ancylus* lake.

Earlier, quarries also presented a great number of good exposures. The once frequent lime kilns, however, have disappeared and the walls of their many quarries are strongly weathered, overgrown, or (especially in the Visby area) levelled. Nowadays, only a few quarries in Gotland are occasionally worked for other purposes and the total number of valuable exposures of this kind is but a fraction of what it used to be.

Stratified sediments

The author proposes a subdivision of the Högklint Beds into two subunits, the Lower and the Upper Högklint Beds. The sediments discussed in the descriptions to the geological maps of Gotland are assigned to these subunits as follows:

Lower Högklint Beds: Hede (1933), Högklint-kalksten, ledet a, b; Hede (1940), Högklint-kalksten, pp.20–24.

Upper Högklint Beds: Hede (1933), Högklint-kalksten, ledet c, Tofta-kalksten; Hede (1936), Högklint-kalksten; Hede (1940), Högklint-kalksten, pp.24 (De översta delarna. . . .)–27 (. . . . 1 à 1,5 km från kusten), Tofta-kalksten.

The Högklint Beds originated in a period of continuous sedimentation, during which reef growth also took place. Consequently, the boundary between the Lower and Upper Högklint Beds is not very distinct. This is even true for the boundary between the Visby and Högklint Beds. The very great majority of fossils continues over such an artificial boundary. Local conditions, especially the development of reefs and their influence on their direct environment, played an important part in obscuring stratigraphical boundaries. Even during Late Visby time, certain places of the sea bottom apparently were more favourable than others for the growth of a great many organisms. In these places, more limestone was laid down, wedging out laterally in a marlier environment. Such a limestone lens might have been overlaid by a marlstone layer, but if favourable conditions continued, two or even several limestone lenses formed on top of each other or reef building began. This situation is realized around the transition from Visby to Högklint Beds. The latter are often regarded to begin at the first thicker limestone layers that can be followed over distances of at least some hundreds of metres. But this is not always at exactly the same level; differences of some decimetres to sometimes a metre or more occur. Reefs which began their growth in late Late Visby time, when the uppermost Upper Visby Beds were laid down, partly continued their development during Högklint time. In this way, these reefs and even more the crinoid limestones enveloping them obscure the boundary between the two stratigraphical units.

This is already the situation at the boundary between these two units, and it applies even more to the subdivision of the Högklint Beds, a unit so rich in reefs. Therefore, when speaking about Lower and Upper Högklint Beds, one may not expect to find a sharp boundary. Nevertheless, the terms are of use to help describe some variations in the character of the rocks belonging to the Högklint Beds and in the conditions under which they were deposited.

Because of the general occurrence of bioherms in the Högklint Beds, they show a great variety of sediments, to a total thickness of maximally about 35 m, including the Tofta limestone. The stratified rocks are in the majority.

The stratified sediments of both subunits will now be shortly described.

Lower Högklint Beds

The Lower Högklint Beds are exposed particularly along the northwest coast of Gotland, to a thickness of 15–20 m. Due to the general slight southeastward dip of the Silurian strata of Gotland, only a few exposures of Lower Högklint sediments occur in the inland direction; they are not at all exposed in Fårö.

More or less marly limestone is dominant in the Lower Högklint Beds. This sediment is dense to finely crystalline or occasionally somewhat fine sandy; the colour of the rock is grey to bluish or brownish grey and the thickness of the layers generally varies between 2–20 cm. The individual limestone layers thin out after varying distances and, therefore, the layers from one exposure generally cannot be directly correlated with those in another locality. Thin layers or films of bluish grey to dark brown marl are often interbedded between the limestone layers; these marl layers are thickest (up to maximally 3 cm) in the lowermost Högklint Beds. Locally, lenses of a stronger marly limestone are also found there.

Especially the lower part of the Lower Högklint is very fossiliferous; bedding planes are often irregular and rugged. Bryozoans, corals, stromatoporoids and brachiopods (many *Leptaena* spp., rhynchonellids and spiriferids) are common; Algae occasionally occur in special layers, but are rarely abundant. The generally small coral colonies which are found embedded in the stratified limestone in part of the localities are not restricted to particular horizons. They are often surrounded by a few limestone layers which either bend under and over them or abut against them. A coral colony is commonly directly overlaid by a thin layer of marlstone. Generally the colonies are in their positions of growth and show no indications of distortion. Crinoid remains are usually abundant only in the environment of reefs.

In the higher parts of the Lower Högklint Beds, the fossil content varies in abundance, whereas the rock is sometimes finely oolitic. Especially in the south of the area where the Lower Högklint Beds are exposed, the rock is often thin, sometimes thick bedded, dense to finely crystalline, finely oolitic in many places, and apparently relatively poor in fossils. Locally there are more fossiliferous parts and small, indistinctly bedded reef-like developments generally occur, together with some real reefs. The true, large Högklint reefs generally began their growth already during the first part of Early Högklint time, but usually continued their expansion in the later part of the Early Högklint and often also during part of Late Högklint time.

Further north, exposures without reefs often show an alternation of finely crystalline limestone and somewhat coarser limestone. Little regularity can be found in this alternation. The coarser limestone is dominantly light grey in colour. Due to strong recrystallization of most of the fossils, only "shadows" are visible; among these, tetracorals and brachiopods are most common. Locally calcareous Algae occur, but only rarely in great numbers. Fossils are relatively best distinguishable on bedding planes, where they sometimes are in part liberated from the enveloping sediment. This coarser limestone is irregularly bedded; on the bedding planes, there is often a film of marl. The weathered rock often falls apart in irregular pieces. The finely crystalline limestone generally falls apart through weathering in more regular, rectangular pieces. This limestone is more yellowish brown in colour. The layers, on the average a few centimetres thick, are usually thinner than those of the coarser limestone, whereas internally these layers often also show a millimetre-thin lamination. The rock is poor in fossils; locally some brachiopods are found. Locally (e.g., in the southeast of Hallshukklint and in the Hjännklint, not far from there) the bedding planes of the finely crystalline limestone layers are rugged, with hummocks up to 5 cm high. These hummocks may have originated through wave action. On the rugged bedding planes, a film of marl is often found. Another indication of wave action is cross-bedding in some of the coarser layers. In the northwestern part of Hallshukklint, it was found that an apparently homogeneous coarse layer really consisted of a cross-bedded alternation of layers of coarser and finer limestone.

Upper Högklint Beds

The Upper Högklint Beds are very commonly exposed. The Tofta limestone in the south, which forms an independent stratigraphical unit in Hede's stratigraphy, is also assigned to the Upper Högklint by the present author. In the north, the Tofta limestone is absent and the Högklint limestone, in Hede's definition, is directly overlaid by Slite sediments. New reefs occur only sporadically in the Upper Högklint, and those which are present are generally of small size. Well developed wave-ripple marks were described by Hede (1936, p.14) from 0.9, 1.1 and 1.2 km north of Lauterhorn (about in the centre of the west coast of Fårö), with a direction of strike of the crests of respectively 60°, about 45° and about 90°. The thickness of the Upper Högklint is about 20–25 m.

The Upper Högklint is built of generally thin-bedded, but sometimes thick-bedded, light-grey limestone, which is almost dense to finely crystalline, and sometimes finely oolitic. As a rule, the rock is very fossiliferous, with especially calcareous Algae playing an important part. The Upper Högklint Beds differ from the Lower Högklint Beds especially in this way. Also, bryozoans, stromatoporoids, crinoids and corals occur. Locally stromatoporoids are abundant and may have given rise to indistinctly bedded reef-like developments. Stylolites are not rare.

Locally in the lower part of the Upper Högklint, less fossiliferous parts also occur. In several places films or thin layers of greenish marl are found on the bedding planes. Locally the limestone is hard and breaks shell-like. In certain limestone layers, the fossils are rounded and strongly worn. This often gives the rock a conglomeratic appearance. One of the best developed conglomeratic layers (thickness generally 2–10 cm, but locally up to 30 cm) is the one which Hede took as his stratigraphical boundary between Högklint and Tofta limestone. It shows a great many fossils, especially calcareous

Algae, crinoids and bryozoans, and to a lesser degree also corals, the majority of which are rounded and worn. In this specific layer the fossils are generally embedded in greenish grey marlstone in stead of in limestone.

The upper part of the Upper Högklint Beds consists of generally well-bedded, often somewhat marly limestone, which is dense to finely crystalline, and sometimes finely oolitic. The content in calcareous Algae varies from place to place, but is generally high. Stromatoporoids occur especially in the southeast, and occasionally in vaguely bedded reef-like developments. Hede ascribed these sediments in the southeast to the Tofta limestone, which he gave the rank of an independent stratigraphical unit. The present author considers the Tofta limestone to be only a facies. A still accessible exposure of this limestone is in the quarry behind the Galgberg (north of Visby), where the mentioned layer of marlstone with strongly rounded fossils can also be observed, although it is thin there. The Tofta limestone is often cross-bedded. The stromatoporoid colonies are generally not large and were sometimes tilted during growth, presumably by wave action. The tilting appears from the fact that the latilaminae are all much thicker at one side than at the other. Generally, the larger the stromatoporoid colony, the less rounded it is. The rock is very rich in tuberiform colonies of calcareous Algae of various sizes, which may give it an oolitic, pisolitic or conglomeratic appearance. Brachiopods are notably poorly represented. As a rule, the difference between the Tofta limestone and the other sediments of the Upper Högklint Beds is small.

In a few places, such as at Gutevägen (Visby), faults occur in the Upper Högklint Beds, with vertical displacements from a few centimetres up to a few metres. Their strike varies from 55 to 85°. Their origin is presumably linked with differential compaction of the various underlying sediments, and especially with a difference in settling between stratified and unstratified sediments. The fact that epeirogenetic movements perhaps made the Högklint sediments rise above sea level a few times, during Middle Silurian times, may have further promoted compaction and thus have made the differences in volume decrease between the various sediments more pronounced.

Reef limestones and related sediments

A survey will now be given of the main occurrences of reef limestone of Högklint age.

The hillock Högklint (Fig.2), from which the name of the stratigraphical unit has been derived, is from the point of view of reef study not the most valuable of the exposures showing rocks of that age. The best exposure is the cliff at the seaward side, which is, however, too steep and high to be accessible for detailed studies. The reef limestone is of the common Hoburgen type, grey to greyish white. Stromatoporoids were the main reef builders, but also a great variety of other fossils is present; among these solitary and social corals, bryozoans, crinoids and brachiopods are the most apparent. In between the reef limestone, parts are present of a distinctly to vaguely stratified nature, measuring from a few decimetres to several metres. At the very top of the exposed reef limestone, this shows locally a tendency to stratification or it is covered by crinoid limestone, suggesting that the reef has not reached much higher there than the present topography. Underneath the Högklint reef, in the Upper Visby Beds, some small reefs are found and it seems fairly likely that the Högklint reef itself began its growth already at the end of Upper Visby time. At that time, conditions for reef growth were favourable and several began development. About at the place where the coastal cliff turns from a south–north to a west–east direction, a small reef is found in the lowermost Högklint Beds. It is only a few metres large, and covered by some stratified limestone, over which the main Högklint reef expanded.

Some parts of the stratified Högklint Beds close to the reef limestone are extremely rich in crinoid remains of generally reddish colour; these parts remind one of the well-known "Hoburg marble" of southernmost Gotland.

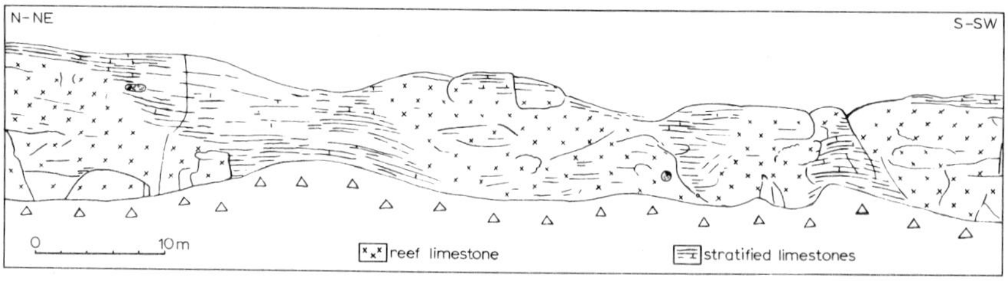

Fig.132. Part of the *Ancylus* cliff between Galgberget and Snäckgärdsbaden Hotel; taken about 0.5 km south of the hotel. Four parts of reef limestone are shown, separated by stratified sediment, indicating how closely reefs locally followed after another in a north – south direction. Further south in the cliff the distance between successive reefs is generally larger.

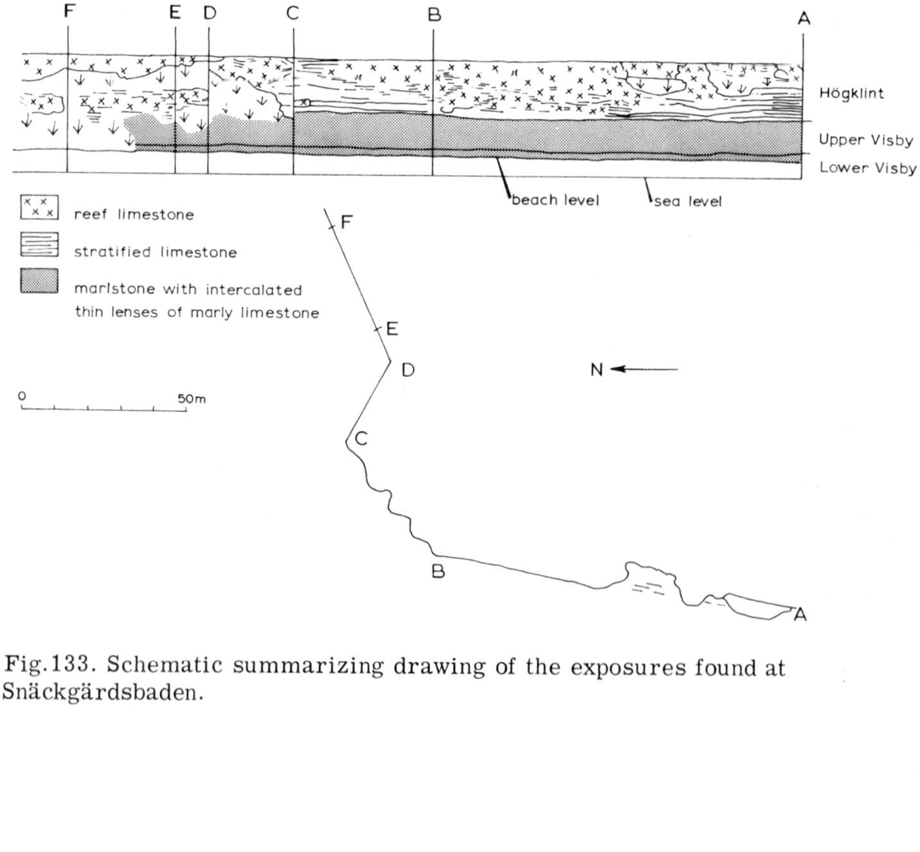

Fig.133. Schematic summarizing drawing of the exposures found at Snäckgärdsbaden.

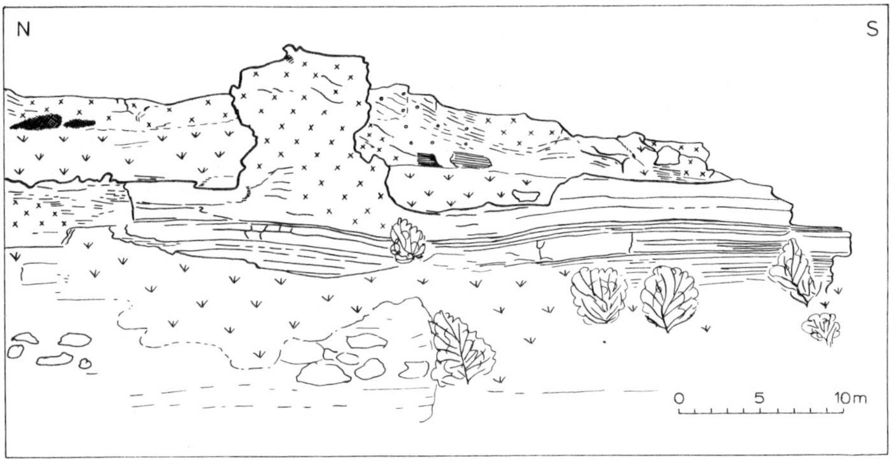

Fig.134. Southern half of the cliff wall AB of Fig.133. The protruding rock mass just left of the middle of the drawing is known as "Predigstolen" (the pulpit). It consists of reef limestone overlying stratified limestone; the boundary between the two is not very distinct because of the only gradually disappearing stratification. The part of the wall south of "Predigstolen" is shown in more detail in Fig.135.

Fig.135. Snäckgärdsbaden, Högklint Beds. Detailed drawing of the southern part of Fig.134, from A (Fig.133) at the right to "Predigstolen" at the left. South of the rock gate at the right some reef limestone. Underneath and especially left of this gate stratified crinoid limestone with increasing crinoid contents upwards; also the number of colonies of calcareous Algae increases upwards. Northwards the crinoid limestone is followed by an unstratified to very vaguely stratified body of reef limestone with many stromatoporoids, solitary corals and crinoid remains. Underneath the reef limestone a small cave with slightly folded layers of stratified limestone. They belong to the limestone occurrence which separates the above-named reef body from the reef at the left, of which it apparently forms the flank deposit. It contains remains of reef builders but also lumps of reef limestone of identical nature to that of the reef north of it.

Nothing in particular needs to be described from the Högklint reef limestone exposed in the Korpklint, southwest of Högklint.

Also in the Stavsklint the Upper Visby Beds are overlaid by Högklint Beds, both stratified and reef deposits. At the southwest side, two layers of the adjacent stratified limestone penetrate into the reef limestone over a horizontal distance of about 5 and 8 m, respectively. The latter, which occurs highest in the reef, shows a distinct rise reef-inwards. In addition to stromatoporoids and corals, the reef limestone, especially locally, is also rich in bryozoans. Several reefs appear to have contributed to the reef limestone mass of Stavsklint. It reaches a thickness of up to about 8 m and is in many places overlaid by stratified crinoid limestone.

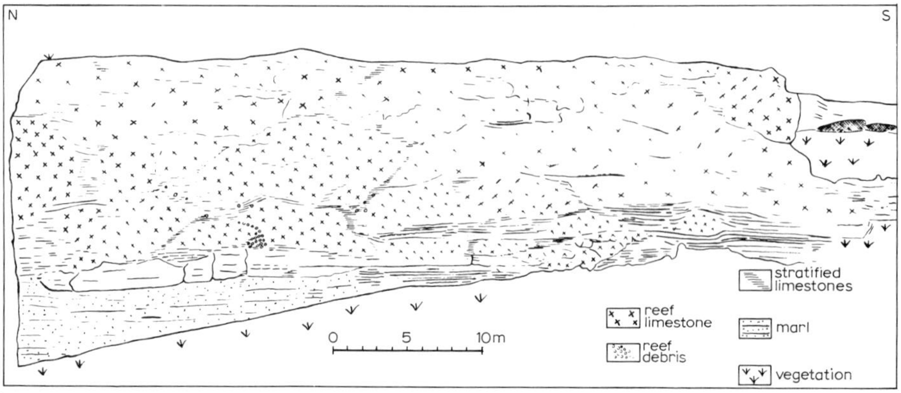

Fig.136. Northern half of the cliff wall AB of Fig.133. At the base some small reef-limestone bodies. The southern one of these consists of hard unstratified limestone in which as fossils mainly some light grey stromatoporoid colonies are recognizable. It rests over hard sparry limestone rich in heliolitids and favositids, which generally occur in their attitudes of growth, solitary corals, stromatoporoids and many crinoid remains. Downwards the sediment is more marly. Some of the layers sag under the overlying weight of the reef limestone. North of this small southern reef body a larger reef-limestone unit. There is a characteristic difference in the positions of the coral and stromatoporoid colonies in the reef and stratified limestone. In the stratified sediment under the reef they occur all with their largest diameter parallel to the stratification, in the reef limestone they are found in all possible orientations. The boundary between stratified and unstratified limestone follows an irregular course, when seen in closer detail, but is distinct. The reef limestone contains pockets of marly sediment of which the layers have been in part deformed by differential compression within the rock complex. The overlying stratified sediment is less thick than over the lower reef south of it. Further north stratified limestone also occurs as mantles around parts of the higher reef limestone, thus leading to a complex reef-limestone massif as also found in several other places in the Högklint Beds. The individual units are roughly lens-shaped. The stratified mantles become upwards increasingly indistinct. Stromatoporoid development in the higher reef portions was more vigorous.
Snäckgärdsbaden. Högklint Beds.

Fig.137. Snäckgärdsbaden. Photograph of cliff wall BC of Fig.133. Reef limestone overlies stratified limestone. On top of the reef limestone locally again some covering stratified limestone. Högklint Beds. The locality of the reef of Fig.138 is at the left in this photograph.

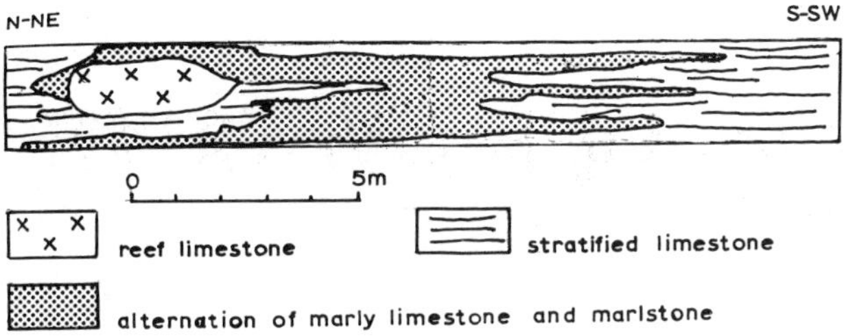

Fig.138. Small body of reef limestone in the lowest of the Högklint Beds in section BC (Fig.133). The body is found just southwest of C, and is built of hard limestone rich in stromatoporoids and coral colonies. In its normal development the basal Högklint bed consists of hard stratified limestone with crinoids. However, about 28 m northeast of B it passes into an alternation of layers of marly limestone and marlstone, characteristic more for the underlying Upper Visby Beds. This deposit also covers the reef-limestone body, but at its base and in its lateral environs it is replaced again by hard stratified limestone.

At Axelsro, reef limestone is exposed over quite a large horizontal surface, but the thickness in the cliff wall does nowhere exceed 6 m. The rock is of the common grey Hoburgen type. Underlying the reef limestone is crinoid limestone with an increasing content of crinoid remains and reef debris in an upward direction. In all, about 6–7 m stratified Högklint limestone is found in between the Upper Visby Beds and the Högklint reef limestone at Axelsro and it should be noted that the Visby–Högklint boundary continues underneath the reef mass at the same level as in the environs of Axelsro.

The Högklint reef limestone at Kneippbyn is exposed over an area of about 200 m long and about 50 m broad and in the coastal cliff to a thickness up to about 25 m. The longest axis of the reef-limestone area is about northeast – southwest. Reef growth began very early in Högklint time or probably even earlier. In the wall some small reefs are also found that grew at about the Upper Visby–Högklint transition.

The boundary between the Upper Visby and Högklint Beds, which is generally laid at the first thicker limestone layers which can be followed over a reasonable long distance, occurs distinctly lower underneath the Kneippbyn reef. Locally it is as much as about 5 m lower than at some distance north or south of Kneippbyn. In addition to the causes mentioned while discussing the stratified sediments (p.282), the wealth of crinoid growth here in the close environments of the reef and the deposition of reef debris also will have contributed to an early development of thicker limestone layers. Further differential compression of the Upper Visby Beds may have been of importance. The enormous weight of the reef mass made it subside into the comparatively soft sea floor, together with the limestone deposit directly around the reef, causing the marly sediment underneath to compress and to move outwards-upwards, in the latter way also leading to some elevation of the limestone layers covering the Visby Beds further away, around the reef. The dips in the Upper Visby Beds in the Kneippbyn area support this thesis of central subsidence and peripheral elevation of the stratigraphical boundary.

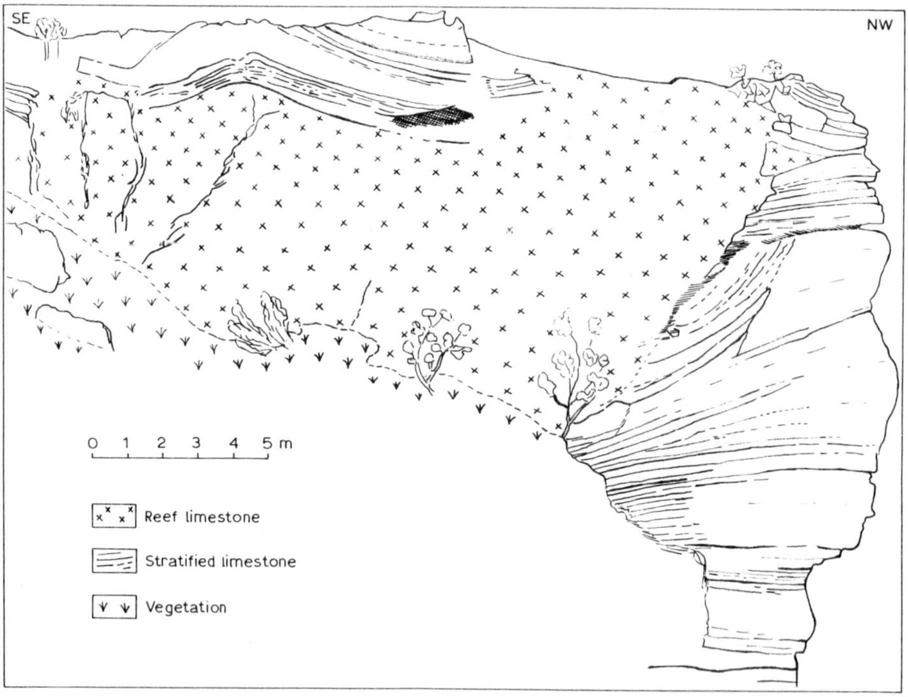

Fig.139. Part of Korpklint, Snäckgärdsbaden (section DC of Fig.133), with reef limestone and stratified limestone, belonging to the Högklint Beds. In the stratified limestone at the right upthrust faulting occurs possibly caused by gravitational sliding in the reef-limestone mass towards the main (southwest – northeast) cliff at the right. (After Manten, 1962, fig.18.)

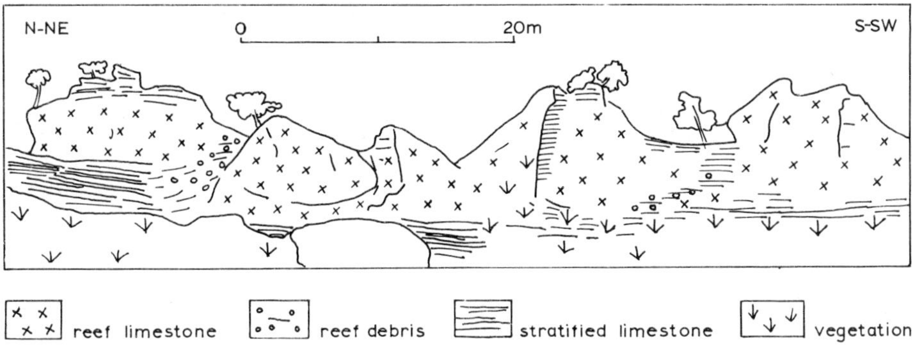

Fig.140. The cliff south of Brissund. A is a reef, belonging to the Högklint Beds, with some Upper Visby marlstone at its base. This marlstone is also found at B, where it shows cross-bedding. C is reef-like limestone. D is well-stratified marly limestone, rich in crinoid fragments, but also containing some reef debris. It is overlaid by reef limestone (E) with a brecciated structure, which is locally rich in calcareous Algae. F is stratified limestone, which is richer in reef debris than D; going upwards the stratification becomes less distinct. G is strongly-weathered reef limestone, which is rooted at H in the underlying marlstone (Fig.141). J is again reef limestone, dense, with stromatoporoids and calcareous Algae. The rock shows several stylolites, with an amplitude of up to 5 cm and a length of a few metres, which also intersect the Algae and stromatoporoids. The reef limestone rests over crinoid limestone with reef debris. At the northwest side of the reef limestone, debris (K) has been deposited between this reef-growth centre and its northern neighbour. This detrital deposit contains remains of stromatoporoids, favositids and calcareous Algae. Intact crinoid calices suggest a rather quiet environment. The northern reef (L) overlies stratified crinoid limestone (M) and is overlaid by variably-well-stratified limestone with crinoids. Where not otherwise stated the sediments belong to the Högklint Beds.

The first exposures of Silurian rocks north of Visby are some old and mainly levelled quarries behind the ruins of St. Göran Church. In fact only parts of their northern walls are still there, but these are all strongly weathered. They consist of stromatoporoid reef limestone. These northern quarry walls occur in the southern slope of the Galgberg, a hillock on which in earlier times public death sentences were carried out. Around this hillock solid rock can be seen at a number of places, comparatively well in the south of the west side and on the north side. In the first locality, the boundary between Upper Visby and Högklint Beds is exposed. The Högklint Beds are represented by both stratified crinoid limestone and unstratified reef limestone. Along the staircase in the north it can be observed how the reef limestone mass is composed of smaller reef units. On the average, the reef limestone higher upwards shows larger and rounder reef builders. Notable are the, in part, very large stromatoporoid knolls, up to 2 m in diameter; in between them occur head-size acervularians and favositids. No differences were observed in the species composition of the fauna in the reefs and the stratified sediments. Apparently the reefs were too small to lead to any significant faunal differentiation. At the top of the hillock mainly crinoid limestone is present. None of the exposures is particularly valuable as far as reef study is concerned.

The west wall of Galgberget forms part of the *Ancylus* cliff. This cliff can be followed north of Galgberget, over a distance of about 2 km until the Snäckgärdsbaden Hotel. This cliff will here be called "Galgberg Extension". It shows cross-sections through several reefs, which generally are not of great length. The average is about 25–60 m; only a small number of reef-limestone exposures is longer. Northwards the average distance between successive reefs decreases and in the last kilometre, there is, in several cases, only a triangle of stratified material between adjacent reefs, which then are in direct contact in the upper part of the exposure. The sides of the

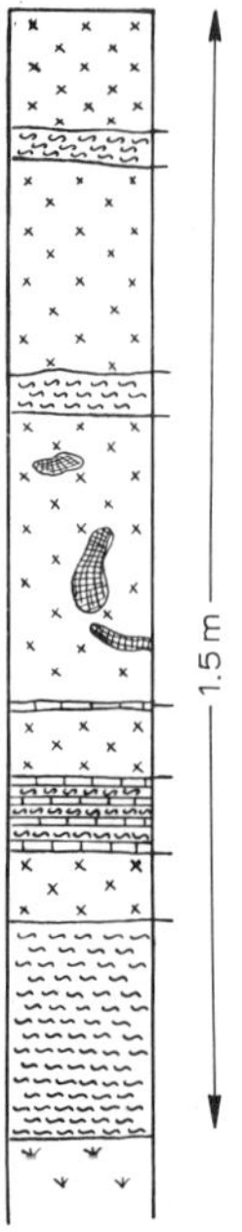

Fig.141. Stratigraphical succession at H in Fig.140. From base to top: Upper Visby marlstone; massive limestone; interstratification of marl and limestone layers; massive limestone; a limestone layer; massive limestone with favositids and several stromatoporoids; marl; massive limestone; marl; massive limestone (G in Fig.140). Apparently the beginning of reef growth was rather difficult, so that horizontal expansion of the reef was four times interrupted by periods in which the reef builders were repelled to a smaller territory. Brissund.

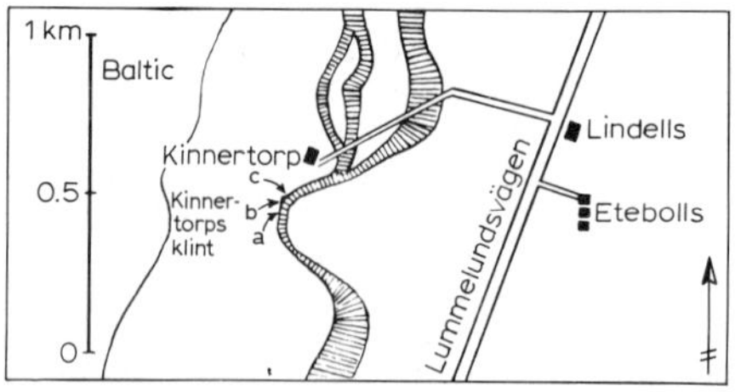

Fig.142. Map showing the location of the Kinnertorpklint. a, b and c indicate the positions of the sections shown in Fig.143, 144, and 145, respectively.

triangle make an angle with the base varying between 20–60°. In other instances, a broader zone of stratified sediment occurs between two reefs. Generally, fossiliferous marly limestone is present, in layers of 1–15 cm, with interbedded marl films, which may sometimes, however, become real marl layers of up to 1–2 cm thick. It is evident that not all reefs are sectioned alike; in some reefs, the centre is still hidden in the solid rock east of the wall, in others the main part must have been present somewhat further westwards and has been removed by erosion. A small part of the wall is shown in Fig.132.

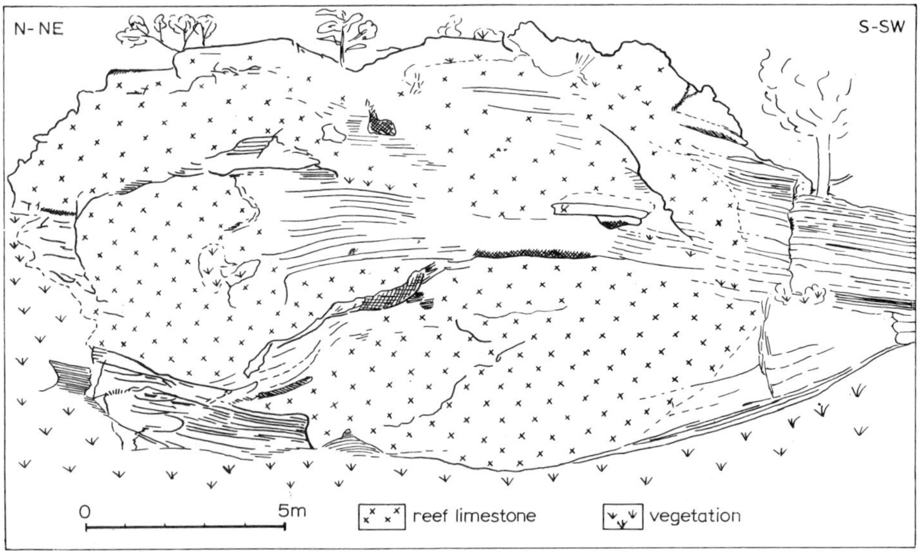

Fig.143. Reef exposed in the Kinnertorpklint, about 45–65 m south of the northern end of this cliff (a in Fig.142), Högklint Beds. At the right and the left crinoid limestone. At the very left, below, the crinoid limestone is replaced by very fossiliferous marlstone. Within the reef limestone some large parts of more or less well-stratified crinoid limestone.

One of the classical exposures of reef limestone in the Högklint Beds is formed by the cliff walls at Snäckgärdsbaden, a well-known bathing locality about 3 km north-northeast of Visby. A summary view of that locality is given in Fig.133. Some parts of the exposure are shown in more detail in Fig.134–139.

In section BC (Fig.137), it can be seen how the boundary between Upper Visby and Högklint Beds east of the reef-limestone mass (which is the same as is to be seen in the left of section AB) moves upwards in a flexure-like way. A phenomenon as shown in Fig.138 further complicates the drawing of an exact stratigraphical boundary.

The Högklint reef limestone section BC (Fig.137) shows no subdivision into smaller reef-limestone units, but this can be observed in the northern part of AB (Fig.136), disappearing again southwards in that section.

The most interesting part of the cliff south of Brissund, approximately 9 km north-northeast of Visby, has already been described earlier (Manten, 1962, fig.9–11). A summarizing drawing of this cliff is given in Fig.140. Fig.141

(Text continues on p.295)

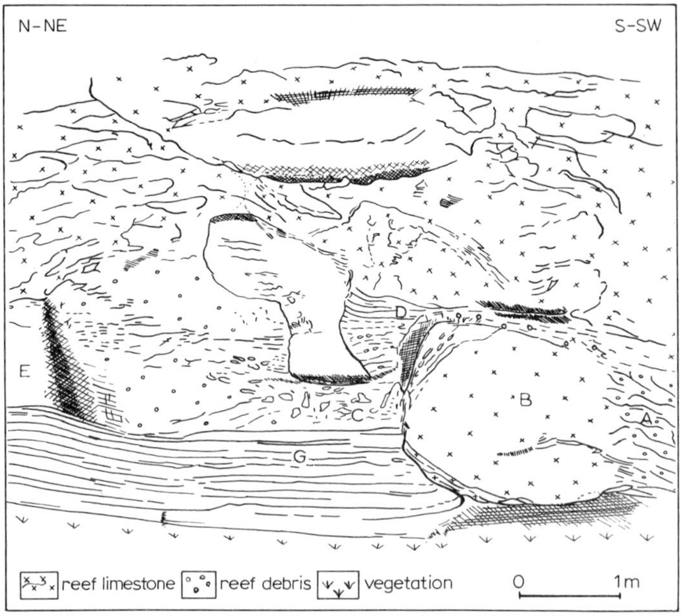

Fig.144. Detail of the exposure in the north-northeast trending cliff wall in the Kinnertorpklint, located 7–14 m before the northern end of this cliff (b in Fig.142). In the south at the top reef limestone, underlaid by a more debris-like deposit (A). This contains many crinoids, some brachiopods, thin stromatoporoids and halysitids. Towards the base it is increasingly more marly. Northwards it ends against a lump of massive limestone (B), which is very rich in crinoid fragments, but also contains the remains of several reef builders. At the top this lump is surrounded by thin fossil colonies and detrital layers, which partly follow the surrounding of the massive lump. At the base of it and behind it a small cave, which shows somewhat cross-bedded marlstone layers with interstratified more calcareous layers. The marlstone layers are rich in heliolitids and favositids. Close to the lump B the layers are strongly plicated. The whole makes it likely that the lump was in its entirety detached from a reef and moved downwards. At the top left and north of the lump (C) an agglutinated mass of coarse remains of stromatoporoids and corals, with crinoids and smaller solitary corals in between. On top of it a more regular accumulation of fossils in a calcareous matrix, which is vaguely stratified (D). This stratification lacks in the overlying reef limestone. Towards the north the reef debris in the detrital deposit C becomes less coarse. At the left in the drawing it is replaced by a thick layer of crinoid limestone (E) with some favositids, stromatoporoids and thick solitary corals. At the base of it an interstratification of marlstone and more calcareous layers (G) is found, which is similar to the one that has been mentioned from underneath and behind B. The fact that this deposit is overlaid by limestone with reef debris shows that the beginning of reef growth was synchronous with the deposition of the marly sediment. The crinoid limestone layer E has apparently been deposited in rather quiet water. It was found to contain a flat fan of solitary corals, descending of one ancestor, of which the successive generations descended through budding. The longest chain comprised seven successive generations.

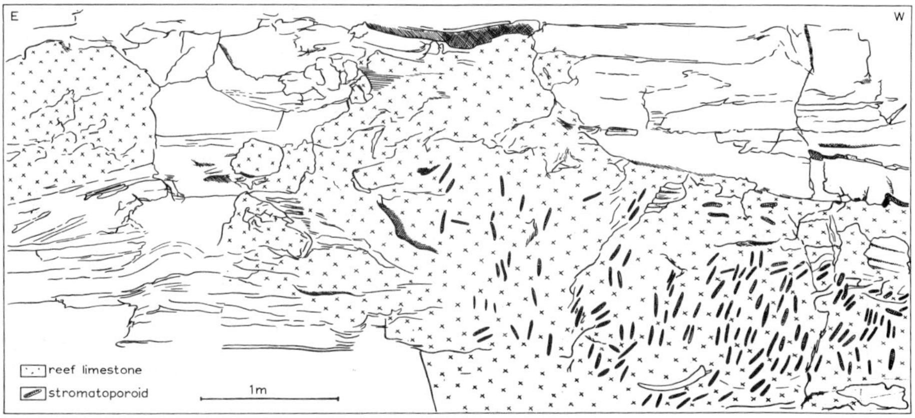

Fig.145. Section of the north side of the Kinnertorpklint (c in Fig.142). Högklint Beds. The reef limestone at the right shows a great number of stromatoporoid colonies which are not in their growth orientations. Presumably this rock has been formed at the northwestern margin of the reef. Seen the fact that in some parts the orientation of the colonies is remarkably alike it seems possible that these parts represent blocks of reef rock that were detached from the reef by storm waves and deposited in their entirety at the lee side. Also a few pockets with crinoid limestone are found. Upwards the number of marl pockets increases. The reef limestone is overlaid by rather thick layers of coarse-spathic crinoid limestone. Towards the east the reef limestone is less detrital in character. Over a short distance in the exposed cliff the reef limestone is replaced by stratified limestone, which at the base is a coarse crinoid limestone, but upwards becomes more marly and then contains longer crinoid stem fragments. In the left the crinoid limestone is again overlaid by reef limestone, which is rich in crinoid remains. Other recognizable fossils are stromatoporoids, halysitids and small algal balls. At the top a layer of crinoid limestone. East of the drawn section the cliff is hidden behind a Recent scree slope.

---

shows how the beginning of reef formation during Högklint time was not always very easy. Also elsewhere in the Brissund cliff, the reef limestone shows, in its lower parts, portions with intercalated stratified sediment. Sagging of the stratified sediments underlying the reefs and bucklings of the stratified material intercalated in the reefs are witnesses of differential compressions during later stages of reef development. Main reef builders were the stromatoporoids, but also corals of various kinds have contributed, as well as calcareous Algae. The cliff north of Brissund is exposed over a length of about 0.6 km, but presents no new data on Högklint reef formation. In the south, mainly stratified limestone is exposed, after a short distance being replaced by reef limestone. The latter weathers to various remarkable forms, including niches and dome-shaped roofs. Water seepage occurs at the boundary of stratified and reef limestone at the foot of the cliff in several places.

The next most important exposure, northwards, is the cliff wall at Kinnertorp (Kinner's croft). Its location is shown in Fig.142. Three parts of the wall are pictured and described in Fig.143–145.

From Nyhamn Fiskläge north-northeastwards, a great many good exposures are found in the present coastal cliff. Of these, several have already been described in Chapter VII (in geographical order: Fig.51, 48, 44, 52, 49, 54, 55, 56, 37). A few others are shown in Fig.146 and 147.

The reef limestone in the cliff about 0.5 km west of Stuguklint reaches a thickness of up to about 30 m. Intercalated over generally the full exposed length, however, are approximately horizontal zones of stratified limestone, of about one to a few metres thick. These stratified intercalations seem to suggest alternating expansion and retreat of the growing reef surface. The stratified limestone which bounds the reef at the northeast side shows sliding phenomena.

Sliding within the reef limestone has taken place in the cliff a few hundred metres west of Stuguklint. On its west side, the reef limestone is underlaid by a brecciated zone of one to a few decimetres thick, which has been formed through shattering of thin-layered marly limestone. The brecciated zone is distinctly discordant with the normal stratification. The cause of these phenomena will have to be sought in the action of gravity upon the rock masses which are no longer supported at the seaward side.

Just north of the largest reef-limestone occurrence in the last-mentioned cliff, stratified limestone is found in which many longer crinoid-stem remains are present, up to 12 cm in length.

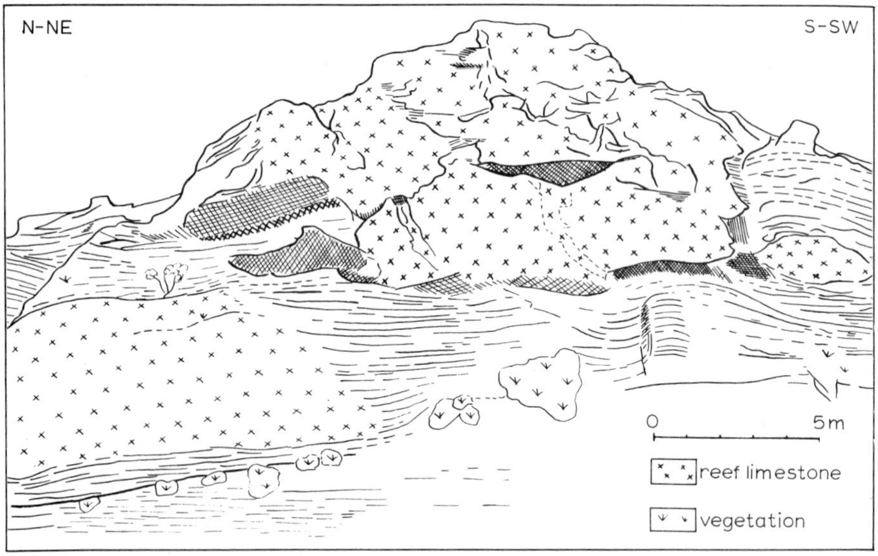

Fig.146. Högklint reefs, approx. 0.45 km north-northeast of Lundsklint. At the base of the section the uppermost Upper Visby Beds are exposed. The oldest reef, at the left in the drawing, is already surrounded by stratified Högklint limestone. At the southwest side of the main reef a small one, which probably could not compete in vigour with its strongly expanding neighbour.

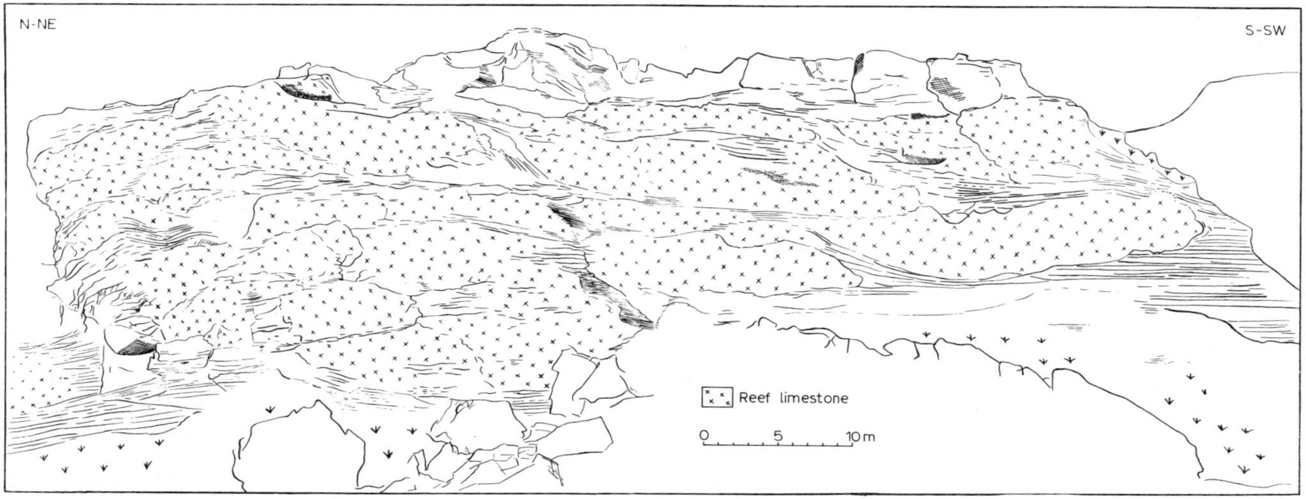

Fig.147. Southern part of a reef limestone complex, about 0.18 km long, approximately 0.7–0.9 km north of Lundsklint. Högklint Beds. At the foot of the cliff in which it is exposed an impressive amount of loose blocks, caused by cliff fall, is found. At the left is a small Upper Visby reef. The boundary between the Upper Visby and Högklint is about 1 m above the top of this reef. This shows that the lowest parts of the Högklint reef mass, to the right of the Upper Visby reef must have originated during Late Visby time. They rest over layers of hard, marly limestone. The Högklint reef limestone is subdivided into smaller units, due to intercalated stratified limestone of varying thickness. (After Manten, 1962, fig.13.)

Fig. 148

Fig. 149

Fig. 150

Stuguklint is the last large exposure of Högklint reef limestone before the rauk Jungfrun is reached. At the southeast side, the main reef shows a rather steep and interfingering contact with the stratified limestone. A small dome of detrital stratified limestone separates this reef from a smaller one west-northwest of it.

Jungfrun, in the west of the southern shore of the Bay of Lickershamn, is the largest rauk (sea stack or stone giant) of Gotland, 11.5 m high (Manten, 1962, fig.14). It consists of reef limestone. Lenticular stromatoporoids were the main reef builders, but social and solitary corals and bryozoans have also contributed.

South of Lickershamn, an extensive, elevated raukar field is found with raukar of fairly large size (Fig.148). They are built of Hoburgen-type reef limestone similar to the reef limestone occurring elsewhere in the Högklint Beds, but in contrast to the raukar of reef limestone in the Hemse and Hamra-Sundre Beds. In the Lickershamn raukar there are several intercalations of stratified limestone (Fig.149). Depressions in the reef were filled mainly with crinoid remains; in between these, there are reef builders and fragments of reef builders (Fig.150). The crinoid limestone is usually thin and irregularly bedded. Generally the depressions once filled became again covered by reef limestone.

Three of the exposures east-northeast of Lickershamn are drawn in Fig.46, 151 and 152.

At several places the reefs in the Högklint Beds do not occur in a single belt parallel to the Silurian coast line of that time, but in a distinctly wider belt, with also in a direction perpendicular to the coast, the one reef having followed after the other. They influenced each other in their development and thus created a rather complicated environmental pattern. This pattern is often difficult to reconstruct only from direct field observations, since generally only the cross-sections of the coastal cliff present good exposures. For the Lickershamn area, the direction parallel to the coast line of Högklint time was presumably around 45° (cf. p.311) and thus the north–south inland row of exposures east-northeast of Lickershamn is particularly valuable since it presents some of the few cross-sections at a distinct angle to the trend of the Högklint reef zone.

Fig.152 gives an illustration of how reefs were competing. The three reefs exposed at the base of the section are all overgrown by reefs which developed southeast of them. The reefs closest to the open sea presumably were most vigorous; the others grew at their lee side, and moreover had

(Text continues on second half of p.301)

Fig.148. Raukar of Högklint reef limestone at Lickershamn.

Fig.149. Lickershamn. Högklint Beds. Part of a rauk with some stratified limestone with crinoid remains and reef debris intercalated between the reef limestone. The intercalated rock is less compact than the reef limestone and stronger attacked by weathering.

Fig.150. Depression within reef limestone, filled with a crinoid breccia in which reef debris is inbedded. The rock is somewhat more liable to weathering and erosion. Raukar field at Lickershamn, Högklint Beds.

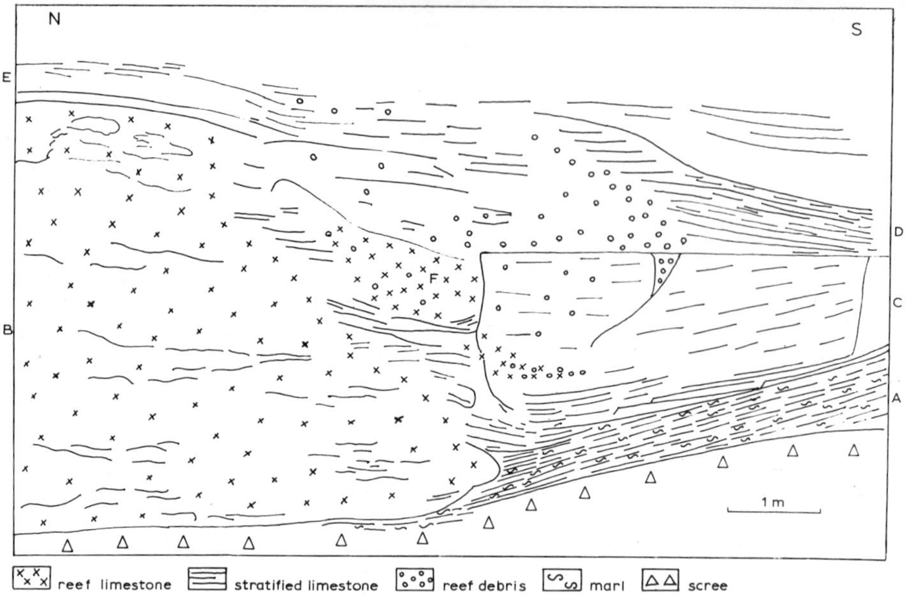

Fig.151. Section about 2 km east-northeast of Jungfrun, at the northeast side of the sand and gravel plain of Lickershamn, about at the place where the cliff wall, which further northeastward lies parallel to the present coast line, has bent inland.

A = Alternation of 1–4 cm thick layers of marly limestone and 0.2–2 cm thick marlstone layers, characteristic of the Upper Visby Beds. The rock is very fossiliferous. Directly south of the drawn part of the exposure the stratification is about horizontal, but soon it dips again down under another occurrence of reef limestone.

B = Light brownish to bluish grey marly reef limestone with stromatoporoids, corals and some bryozoans; most characteristic is the very high number of solitary corals. In between the reef builders also some crinoid remains and brachiopods are found. The reef builders are generally rather flat; about 25% of them are not in their orientations of growth; in the lower half meter the number of colonies is not high and those present are on the average only about 5 cm in diameter and 1.5 cm thick.

C = Crinoid limestone, overlying the sediments described under A. At the base some 5–10 cm thick layers, covered by a thicker one, which in the south of the drawn section is about 1 m, increasing northwards to about 1.5 m; internally this layer shows a somewhat vague bedding, further fading northwards, towards the reef limestone. In addition to the abundant, especially small, crinoid remains there are some solitary and social corals and also some reef debris is present. Because of its rather homogeneous nature the rock distinctly contrasts to the marlstone and the reef limestone.

D = Crinoid limestone of a less homogeneous nature, more distinctly stratified, richer in reef debris, which partly is rather coarse.

E = Crinoid limestone, distinctly stratified, overlying the reef limestone. The rock is extremely rich in crinoid remains; some stem fragments reach to 8 cm in length.

F = A small outcrop of reef limestone in between B and C, separated from reef B by a narrow zone of irregularly and partly vaguely stratified rock, in which much reef debris and a number of mainly small potential reef building colonies in their growth positions are observed. Probably a small flank reef formed at the south to southwest side of reef B. In its lowermost part in the south the reef limestone F passes without distinct boundary into a reef debris deposit, which thins out in the crinoid limestone C.

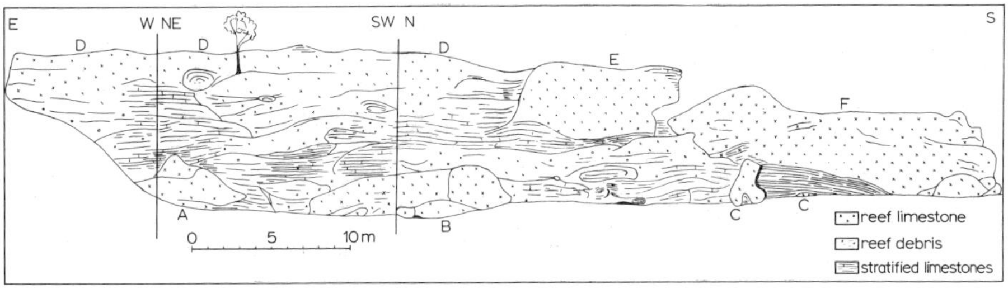

Fig.152. Section about 2 km east-northeast of Jungfrun, at the northeast side of the sand and gravel plain of Lickershamn, about where the cliff, which further northeastwards lies parallel to the present coast line, has bent inland. The section shown is that of the fourth exposure, going inland. Remains of three reefs (A, B, C) are shown, which from the southeast were overgrown by three other reefs (D, E, F). In between a varying amount of stratified limestone with reef debris. The reef D did presumably not extend much further northwestwards than the place of the present cliff. Especially in its lower part, but locally also at the top, it shows clear transitions to stratified sediment; that much marl was deposited there in between the reef builders is reflected by several fissures and small caves which recent erosion has caused. Most of the reef builders are relatively small and flat and quite a number of them are not in their positions of growth.

The exposed parts of the older reefs are also marly and full of marl nests. The stratified limestone varies between almost pure crinoid limestone to a rock very rich in reef debris, which in some places is rather coarse. The sediment exposed underneath reef **F** is an Upper-Visby-like alternation of layers of marlstone and marly limestone; except for the uppermost 20–25 cm below the reef almost as much marlstone is present as marly limestone, which is more than is normal for the uppermost Upper Visby Beds. The sediment has presumably been laid down at the northwest side of the oldest part of reef **F**; it is rich in fragments of coral colonies (especially *Halysites* sp.), crinoid stem fragments and solitary corals.

---

more to suffer from the deposition of continental debris (much marl in the left of the section underneath reef **F**).

Consequently the number of growing reefs decreased in the course of Early Högklint; but the distribution of the exposures of reef limestone shows that also in the later situation, when less but larger reefs were present, the general pattern of a zone of reefs, rather than a single belt was maintained.

Northwards from the place where the cliff is again parallel to the present coast line, in the lower part of the exposed wall Upper Visby Beds occur in outcrops. The upper boundary is distinctly lower where Högklint reef limestone overlies the Visby Beds; the difference in level may be up to 2 m or more. Slightly north of the curve in the direction of the cliff, a small reef is present high in the Upper Visby Beds; it is covered by stratified limestone. Solitary corals in the latter, which are cut by thin calcite seams, indicate that vertical displacements in the order of magnitude of 0.5 cm have taken place also after consolidation of the sediment, presumably because of the weight of the overlying Högklint reef mass. In the lowermost part of that reef limestone, many of the reef builders are not in their growth orientations; the great majority of them is relatively small and flat, the only larger coral colonies were of *Halysites catenulatus* (Martini). There are many solitary corals and brachiopods, both in the matrix and in the several small marl nests (5–30 cm wide, 2–10 cm high). In the stratified limestone enveloping the Högklint reef, the amounts of reef debris are comparable to those found around other reefs of the Hoburgen type (see Chapter IX).

Further northwards, the Högklint Beds present the usual lateral alternation of stratified and reef limestones. Where no reefs are present, the wall is not so steep and usually less well exposed. The reefs are generally not longer than 75 m with a maximum thickness of 20 m, but as a rule not more than 15 m. There are several intercalations of stratified sediment and also horizontal planes (exposed as lines) of one or a few metres in length which may reflect local interruptions in reef growth.

Of several extensive patch reefs found along the shore from Ihrevik northwards, two have been presented in Fig.47. Fig.153 is a detail taken from one of these reefs, showing vaguely stratified reef limestone on top of massive, unstratified reef limestone. A notable exposure, somewhat further northwards in this area, is the one represented in Fig.155, where stratified sediment and reef debris occur in the lowermost part of a reef, surrounded and overlaid by reef limestone. A drawing of the entire reef-limestone exposure is given in Fig.154.

Several details of the Hjännklint, southwest of Halls Fiskläge, have already been given in Chapter VII (Fig.35, 62, 66). Another detailed section is that of Fig.156, in which a notable contact between stratified limestone and overlying reef debris and reef limestone is shown. About 12 m south-southwest of the locality of Fig.156, within the predominantly stromatoporoid limestone, part of that rock consists almost entirely of branched coral colonies. This part is almost 6 m high and 4 m broad. Directly south of it,

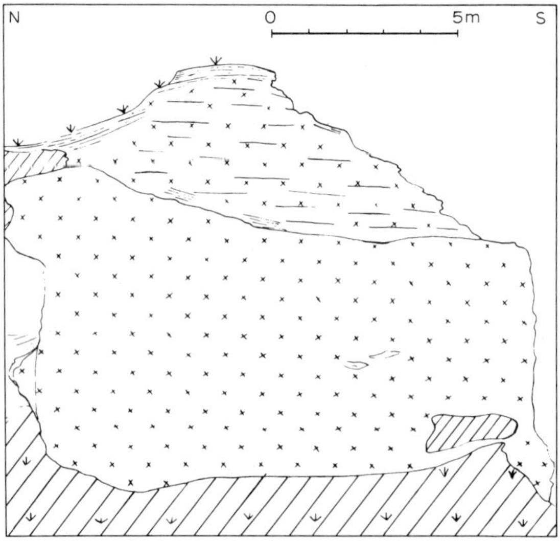

Fig.153. Detailed section through reef limestone, about 0.9 km south of Sigsarvebodar, in the east of Ihrevik. The lower part consists of massive reef limestone, mainly built by stromatoporoids in a matrix of marly limestone. The marl content is highest in the north. In marl pockets crinoids, small algal balls, brachiopods, solitary corals and an occasional gastropod are found. In the upper part of the section reef limestone is found, which is vaguely stratified. The matrix volume and the marl content of this matrix are higher. More tabular reef builders are present.

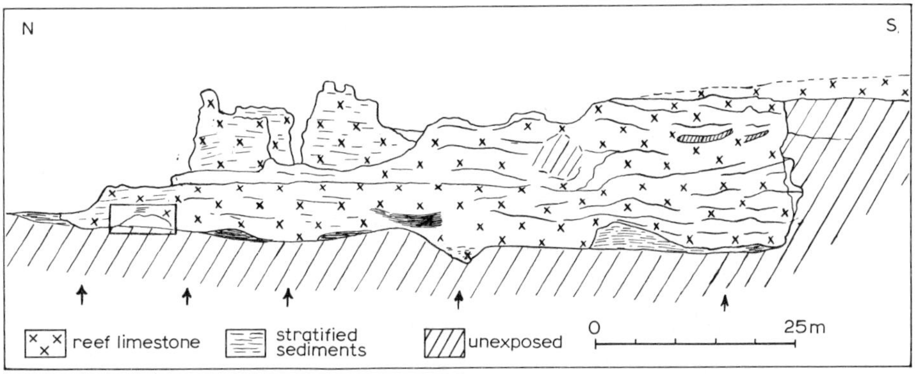

Fig.154. Högklint reef limestone south of Sigsarvebodar. The rectangle at the left marks the position of Fig.155. Over its entire thickness the reef limestone shows a vague stratification, which is, however, strongest at the top, where it may show dips up to 20°, e.g., in the northern of the raukar. The boundary with the underlying crinoid limestone is undulous and suggests that the reef had several roots (indicated by arrows). At the right part of a reef limestone occurrence found about 50 m landinward is drawn. Also this reef limestone is vaguely stratified; moreover it is stronger marly and contains reef debris, all suggesting formation at the periphery of a reef; this interpretation is further strengthened by the presence of a block, about 1 x 1 m, in which all tabular stromatoporoids are in about vertical positions, and which block has likely as a whole been loosened elsewhere from the reef.

---

corals are also common, but their number decreases further southwards. At the base of the coral limestone is reef limestone with tabular stromatoporoids, an occasional large stromatoporoid and colonies of *Halysites*. North-northwest of the locality of Fig.156, the reef limestone is strongly dominated by stromatoporoids, several of which are large; a size of 90 cm in horizontal and 60 cm in vertical diameter is not exceptional. Locally, they are roughly arranged in layers. Since in some places layers of fragment limestone also are intercalated, the reef limestone has in parts a vaguely stratified appearance.

Two parts of the Hallshukklint are drawn in Fig.63 and Fig.70.

The Norsklint, about 2 km southwest of Hallshuk shows mainly reef limestone of the general type. Only in the south is it seen to be underlaid by stratified limestone, which over some distance is gradually replaced. Locally, however, it appears again at the base of the exposures, thus suggesting that the Norsklint reef began its development in a number of growth centres, which in the course of their expansion united to form a patch reef, that reached a thickness of about 8–10 m. In general stromatoporoids, but locally corals were the main reef builders. Intercalations of a stratified nature are not rare, especially in the lower parts of the reef limestone.

A survey of the fossils which occur in the Högklint reef limestones and surrounding crinoid limestones is given in Table XVII.

The stratigraphical position of the Tofta limestone

Some remarks are needed here with regard to the stratigraphical position of the Tofta limestone.

When Hede (1921) published his stratigraphy of Gotland, he included the Tofta limestone as an independent stratigraphical unit. Later investigators have questioned the correctness of this.

Hadding (1956, p.18) especially, was of the opinion that the Tofta limestone may be regarded as "a local, extreme-shallow-water facies of the uppermost part of the Högklint group and the lowest part of the Slite group".

Also Martinsson (1962a, 1967) doubted the independency of the Tofta limestone as a stratigraphical unit. Because the Upper Högklint Beds are poor in marlstone, Martinsson (1962a) could sample these only very badly in his study on the beyrichiid ostracodes. Nevertheless, he found three species (*Craspedobolbina uncilifera* Martinsson, *Beyrichia halliana* Martinsson, and *B. hystricoides* Martinsson) in both the Högklint Beds sensu Hede and the Tofta limestone. *Beyrichia bicuspis* Kiesow occurs in Högklint, Tofta and Slite Beds. Two species are shared by the Tofta limestone and the

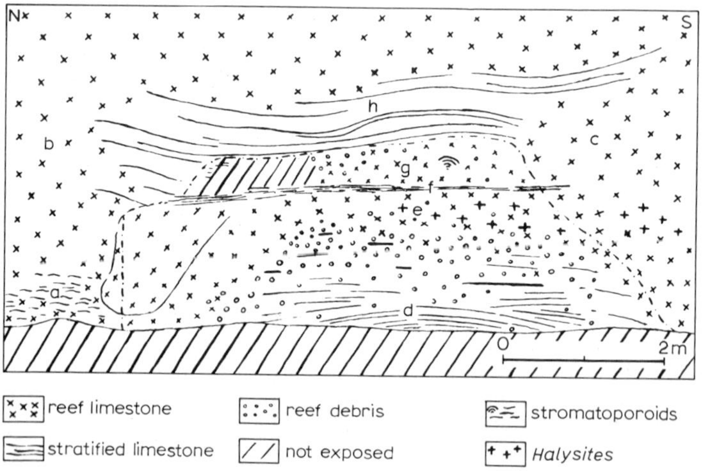

Fig.155. Drawing, showing an intercalation in reef limestone of stratified sediment and reef debris. The intercalation occurs at the base of a reef. South of Sigsarvebodar. Högklint Beds (cf. Fig.154). The dash-dot line marks the boundary of a shallow cave. At a, reef limestone is present mainly built by stromatoporoid covers in a matrix of middle-coarse limestone with crinoids; b and c indicate unorganized reef limestone with stromatoporoids of various form; d is stratified marly crinoid limestone, with upwards an increasing amount of reef debris, at e passing into reef limestone rich in solitary corals and colonies of *Halysites*; f marks a few thin (on the average 1.5 cm) layers of crinoid limestone in which some small stromatoporoids are found in their positions of growth; g is stromatoporoid reef limestone with intermixed still some reef debris; h is stratified crinoid limestone exposed directly above the shallow cave, which deposit contains relatively long crinoid stem fragments and also some calices.

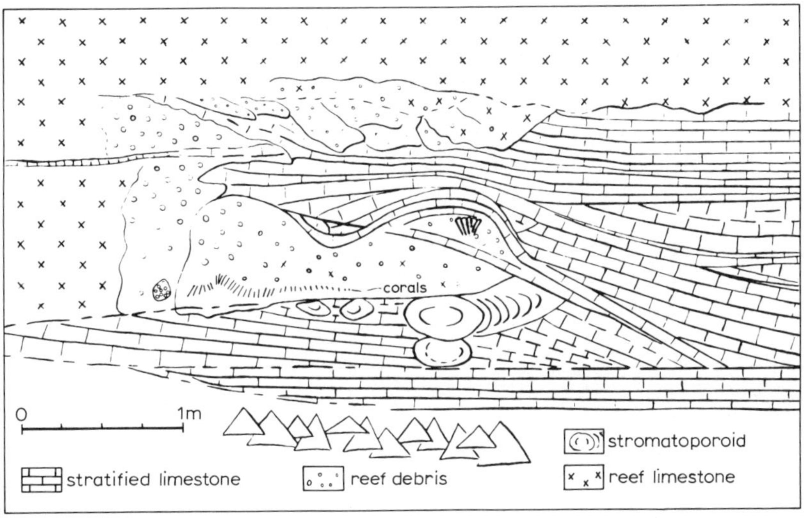

Fig.156. Hjännklint, detailed section, about 1.4 km southwest of Halls Fiskläge. Högklint Beds. Contact between stratified limestone, reef limestone and reef debris. The stratified limestone is cross-bedded. The small body of reef debris in the centre is at all sides surrounded by the stratified sediment. It contains an intact coral colony in growth orientation. At the base within the larger intervening debris body a number of coral colonies, which are generally not in their attitudes of growth. The lines within the larger debris parts and between the debris and the reef limestone are grooves, brought out by weathering. At the left a layer with parallel arranged remains of branched corals and bryozoans, crossing both the reef limestone and the debris deposit.

Slite Beds (*Craspedobolbina mucronulata* Martinsson, and *Beyrichia ponderosa* Martinsson), plus one (*Craspedobolbina percurrens* Martinsson), which occurs in the Tofta limestone, the Slite Beds and the Halla-Mulde and Hemse Beds. Three species, finally are known only from the Tofta limestone (*Clintiella bingeriana* Martinsson, *Bingeria zygophora* Martinsson, and *B. cyamoides* Martinsson); their presence is presumably palaeoecologically determined. There are, thus, connections of the Tofta limestone with both the Högklint and the Slite Beds. Martinsson (1967) felt inclined to consider the ties with the lower parts of the Slite Beds (his Slite unit 1, or Slite I–III of the present author, see p.312) strongest. Limestones containing the *Craspedobolbina mucronulata* fauna of the Tofta limestone were found by Martinsson also northeast of the pinch-out of the Tofta limestone as supposed by Hede, and these in addition contain some of what Martinsson considered as transient Slite faunistic elements.

Because of the difference in expressed opinions and the importance of the Tofta limestone in the interpretation of the stratigraphical succession and the palaeogeographical and palaeoecological history of the Gotland Silurian, the position of the Tofta limestone with respect to the Högklint Beds and the Slite Beds needs to be discussed also at this place.

(Text continues on p.309)

TABLE XVII

Fossils found in the reef limestones and surrounding crinoid limestones of the Högklint Beds of Gotland

Fossils \ Localities \ Lithology	Reef limestone							Crinoid limestone				
	Högklint	Snäckgärdsbaden	Galgberget	Lickershamn	Sigsarvebodar	Hallshukklint	Other localities	Högklint	Snäckgärdsbaden	Sigsarvebodar	Hallshukklint	Other localities
ALGAE												
Rothpletzella sp.	+			+			+					+
Solenopora gotlandica Rothpletz	+	+	+	+		+	+		+	+		+
Unidentified calcareous Algae						+	+					
HYDROZOA												
Actinostroma astroites (Rosen)		+										
Actinostroma sp.				+			+					
Clathrodictyon cf. *variolare* Rosen				+								
Clathrodictyon cf. *vesiculosum* Nicholson et Murie				+								
Labechia conferta (Lonsdale)	+	+	+			+	+					+
Stromatopora discoidea (Lonsdale)	+		+		+							
Unidentified stromatoporoids	+	+	+	+	+	+	+	+	+	+	+	+
ANTHOZOA TETRACORALLA												
Acervularia ananas (L.)	+		+			+	+					+
Acervularia breviseptata Weissermel						+	+				+	+
Acervularia sp.	+		I									
Calostylis denticulata (Kjerulf)									+			
Cystiphyllum cylindricum Lonsdale							+					
Diploëpora grayi (Edwards et Haime)				+								
Dokophyllum högbomi Wedekind	+											
Hedstroemophyllum articulatum Wedekind				+				+				
Hedstroemophyllum sp.	+											
Kodonophyllum truncatum (L.)			+			+	+				+	
Lykophyllum hisingeri Wedekind							+					
Omphyma sp.						+						
Polyorophe glabra Lindström			+	+	+				+			
Polyorophe lindströmi Wedekind								+				
Rhabdophyllum striatum Wedekind				+				+				
Rhegmaphyllum conulus (Lindström)	+			+							+	
Schlotheimophyllum patellatum (Schlotheim)					+							
Syringaxon dalmani (Edwards et Haime)	+	+	+	+	+		+		+			
Zelophyllum intermedium Wedekind	+											
Zelophyllum spinosum Wedekind	+											
Zelophyllum högklinti Wedekind	+											
ANTHOZOA TABULATA												
Aulopora sp.	+		+		+	+						
Favosites asper d'Orbigny			+									
Favosites gothlandicus Lamarck	+	+	+	+		+					+	
Favosites sp.	+	+	+	+	+	+	+	+	+	+	+	+
Halysites catenularius (L.)	+											
Halysites catenulatus (Martini)	+		+		+	+	+		+	+		
Planalveolites fougti (Edwards et Haime)			+		+							
Roemeria kunthiana Lindström						+						
Striatopora halli Lindström	+	+	+	+	+	+			+			
Striatopora stellulata Lindström					+							
Thamnopora lamellicornis (Lindström)	+				+			+				

TABLE XVII (continued)

Fossils \ Localities \ Lithology	Reef limestone							Crinoid limestone				
	Högklint	Snäckgärdsbaden	Galgberget	Lickershamn	Sigsarvebodar	Hallshukklint	Other localities	Högklint	Snäckgärdsbaden	Sigsarvebodar	Hallshukklint	Other localities
ANTHOZOA HELIOLITIDA												
Heliolites interstinctus (L.)	+	+	+	+	+	+	+		+	+		
Heliolites parvistella Ferd. Roemer	+	+		+								
Heliolites sp.	+		+						+			
Propora tubulata (Lonsdale)			+		+				+			
Thecia swinderniana (Goldfuss)	+		+			+	+		+		+	+
Unidentified corals	+	+	+	+	+	+	+	+	+	+	+	+
ANNELIDA												
Conchicolites sp.					+	+						
Cornulites sp.						+						
Spirorbis sp.			+		+	+						
CRINOIDEA												
Cyathocrinus sp.									+			+
Eucalyptocrinus granulatus Lewis								+				+
Euspirocrinus spiralis Angelin												+
Gissocrinus sp.												+
Hypanthocrinus sp.												+
Polypeltes sp.	+											
Unidentified crinoid remains	+	+	+	+	+	+	+	+	+	+	+	+
BRYOZOA												
Fenestella mobergi Hennig			+			+						
Fenestella reticulata (Hisinger)	+			+	+	+	+		+	+		+
Fenestella sp.	+	+	+				+			+		+
Fistulipora sp.			+			+						
Helopora lindströmi Ulrich									+	+		
Phaenopora lindströmi Ulrich					+							
Ptilodictya lanceolata (Goldfuss)			+				+	+	+	+		+
Ptilodictya triangularis Hisinger												+
Unidentified bryozoan remains	+	+	+	+	+	+	+	+	+	+	+	+
BRACHIOPODA												
Atrypa reticularis (L.)	+	+	+	+	+	+	+	+	+	+	+	+
Atrypina angelini (Lindström)									+	+		
Camarotoechia borealis (Buch)		+	+	+	+	+	+	+	+	+	+	
Camarotoechia nucula (J. de C. Sowerby)							+			+	+	+
Chilidiopsis pecten (L.)									+	+	+	
Chonetes sp.					+							
Cliftonia lindströmi Ulrich et Cooper			+						+			
Delthyris elevata Dalman			+			+					+	+
Dicaelosia verneuilana (Lindström)	+	+	+		+		+		+	+		
Dictyonella capewelli (Davidson)									+			
Eospirifer radiatus (J. de C. Sowerby)			+		+				+	+		
Howellella elegans (Muir-Wood)		+			+	+	+			+		

TABLE XVII (continued)

Lithology / Localities / Fossils	Reef limestone							Crinoid limestone				
	Högklint	Snäckgärdsbaden	Galgberget	Lickershamn	Sigsarvebodar	Hallshukklint	Other localities	Högklint	Snäckgärdsbaden	Sigsarvebodar	Hallshukklint	Other localities
BRACHIOPODA (continued)												
Leptaena lovéni De Verneuil			+		+							
Leptaena rhomboidalis (Wilckens)	+	+	+	+	+	+	+	+	+	+	+	+
Linoporella punctata (De Verneuil)						+			+		+	+
Platystrophia sp.	+	+	+		+	+	+		+	+		+
Plectatrypa imbricata (J. de C. Sowerby)			+		+				+	+		
Plectodonta transversalis lata (Jones)									+	+		
Protoathyris sp.									+			
Resserella basalis (Dalman)			+		+				+			
Resserella elegantula (Dalman)											+	+
Resserella sp.						+						
Rhipidomella hybrida (J. de C. Sowerby)			+	+	+				+	+		+
Rhynchotreta cuneata (Dalman)	+	+	+	+	+	+	+	+	+	+	+	+
Sphaerirhynchia wilsoni (J. Sowerby)					+							+
Stropheodonta semiglobosa (Davidson)					+		+	+		+	+	+
"*Strophomena*" *testudo* Lindström in museo							+					+
Trimerella sp.			+									
Unidentified brachiopods	+	+	+	+	+	+	+	+	+		+	+
LAMELLIBRANCHIATA												
Conocardium sp.						+	+					+
Cypricardinia sp.						+				+		+
Rhombopteria sp.					+							
Unidentified lamellibranchs	+		+									
GASTROPODA												
Cyclonema turritum Lindström			+				+					
Euomphalopterus alatus (Wahlenberg)	+								+	+		
Holopea applanata Lindström							+					
Oriostoma alatum Lindström							+					
Oriostoma angulatum (Wahlenberg)							+					
Oriostoma contrarium Lindström							+					
Platyceras cornutum Hisinger	+		+	+	+	+	+	+	+		+	+
Pleurotomaria aequilatera Wahlenberg							+					
Pleurotomaria claustrata Lindström							+		+			
Pleurotomaria limata Lindström						+						
Poleumita discors (J. Sowerby)						+	+				+	+
Poleumita globosum (Schlotheim)						+	+				+	+
Pilina unguis (Lindström)						+	+					+
Subulites ventricosus Hall (according to Hedström, 1923)							+					
Trochus gotlandicus Lindström						+						
Trochus incisus Lindström							+					
Trochus visbyensis Lindström							+				+	
Trochus sp.							+					
Tryblidium reticulatum Lindström						+	+				+	+
Unidentified gastropods	+	+	+	+	+		+	+	+			
TENTACULITIDA												
Tentaculites sp.									+	+	+	+

TABLE XVII (continued)

Lithology / Localities / Fossils	Reef limestone							Crinoid limestone				
	Högklint	Snäckgärdsbaden	Galgberget	Lickershamn	Sigsarvebodar	Hallshukklint	Other localities	Högklint	Snäckgärdsbaden	Sigsarvebodar	Hallshukklint	Other localities
CEPHALOPODA												
Dawsonoceras annulatum (J. Sowerby)						+	+					
Orthoceras sp.	+						+					+
Unidentified cephalopods			+		+		+					
TRILOBITA												
Arctinurus ornatus (Angelin)							+				+	
Arctinurus sp.			+									
Bumastus sp.		+	+		+	+	+		+	+		+
Calymene tuberculata (Brünn)	+					+	+			+	+	+
Calymene sp.			+									
Encrinurus punctatus (Wahlenberg)						+	+	+	+		+	+
Proetus sp.						+	+					+
Warburgella rugulosa (Lindström)												+
Unidentified trilobites	+		+								+	
OSTRACODA												
Beyrichia sp.											+	+
Leperditia sp.						+						
Unidentified ostracodes	+	+	+	+		+	+		+		+	+

Hede (1940) assumed two stratigraphical hiatuses in the Visby area, the one between the Högklint and Tofta limestone, the other between the Tofta and Slite limestones. Towards the north of Gotland, the Tofta limestone gradually wedges out and the two hiatuses finally should unite to one large hiatus. Hadding (1956), on the other hand, thinks it likely that the hiatuses in the Visby area are a local phenomenon, connected with deposition of the sediments in shallower water. In north Gotland the stratigraphical sequence was thought to be complete.

On the basis of data as shown below, the present author defends a more than local importance of the upper hiatus. This implies that the Tofta limestone should as a whole be included in the Högklint Beds.

In those places in the Visby area where the contact between Tofta limestone and Slite Beds is or has been observable, indications for a hiatus are usually found, e.g., about 1.2 km north-northeast of Stora Hästnäs (Visby Parish), south-southwest of Bingerskvarn (Visby Parish) and especially about 1 km southwest-westsouthwest of Suderbys (Västerhejde Parish). The Tofta limestone there is formed by a dense, marly limestone, rich in Algae (*Spongiostroma holmi* Rothpletz), together with corals, crinoids, etc. The upper side generally is irregular, often with elevations where fossils are embedded. In the softer parts, up to 4–5 cm deep holes and rills are present, filled with Slite sediment. The uppermost 0.5 mm or less of the Tofta limestone is coloured rusty brown by limonite.

In these exposures in the Visby area, the Tofta limestone is overlaid by limestone belonging to the first stage of the Slite Beds. This Slite I gradually thins out towards the northwest and seems to be completely lacking northwest of the line Snäckgärdsbaden–Kappelshamn–Fårö. With some doubt with regard to the stratigraphical position, Hede (1940, p.41) mentions two small exposures about 3 km east-northeast of Lummelunds-bruk. He felt himself compelled to ascribe the layer, only a few centimetres thick, of finely crystalline to finely oolitic limestone, which is exposed there, to the base of the Slite Beds because it directly overlies the Tofta limestone, rich in Algae, with a sharp boundary. For faunistic reasons, however, he states that resemblance to a zone, higher in the profile – and included by the present author in the Slite II Beds – is much greater than to that forming the base of the Slite Beds further southwards. A hiatus, also present there, may be the solution to the problem. Moreover, the fauna of the lowermost Slite sediment at Lummelunda is partly rather fragmentary and possibly deposited there secondary.

Strong indications of an absence of the Slite I Beds is also provided by the exposure about 0.5 km east-southeast of Vialms, on the coast of Fårösund, as described by Hede (1933, p.34). There, Högklint reef limestone is exposed, showing a very rugged upper side (erosion?), filled with thinly stratified, light-grey, crystalline limestone of the Slite II Beds. That this limestone actually belongs to the Slite II is indicated by the fact that, according to Hede (1933), at a level about 4–5 m higher, the base is found of the *Ilionia prisca–Megalomus* Zone, which according to the experience of both the present author and others (cf. Hadding, 1956, p.12) is a good key bed, shortly below the top of the Slite II. Unfortunatcly the present author could not personally visit this exposure, since it is at the present time situated on military ground.

Thus there are distinct indications of the presence of a stratigraphical hiatus between the Högklint Beds (including the Tofta limestone) and the Slite Beds. The Tofta limestone is generally found there, where the Högklint Beds, otherwise to about 35 m thick (Hede, 1940, p.19), show a smaller thickness, such as near Visby only 20 m (Hadding, 1956, p.15, fig.12). Moreover, in the north, e.g., on Fårö, where the Tofta limestone is missing, the Upper Högklint shows a limestone, rich in Algae, that is deposited in somewhat deeper water, and that stratigraphically corresponds to the Tofta limestone.

Discussion

Such facts as less marlstone deposition, the occurrence of larger reefs and the presence of more and larger stromatoporoids and of Algae, suggest that the Lower Högklint Beds, highly presumably, were deposited during a period of gradually shallowing water. This decrease in depth was a continuation of the trend notable during Late Visby time. At the end of Upper Visby time and in very early Högklint time, several small reefs developed on the sea floor, in a broad zone with a general direction which presumably was about parallel to the contemporaneous coast line. With further expansion, they became heavy competitors. Those closest to the open sea had better chances of survival and may have caused the deaths of their direct landward neighbours. They often overgrew them, or fused with them. Fusion could also take place with reefs occurring near both ends of their longest axis, leading to the formation of elongated patch reefs. Thus only part of the reefs present at the beginning of Högklint time reached large size in the course of that period; but the pattern of a broad zone of reefs rather than a single belt remained. During deposition of the Upper Högklint Beds the sea was probably always very shallow. Small shifts of sea level occurred; during a few of these the sea floor at the place where the Högklint Beds are now exposed probably temporarily emerged dry. During the Late Högklint, there was little reef growth. Almost no new reefs began development, but some already in existence continued growth.

For the first time in the Silurian of Gotland in the Högklint Beds stromatoporoids and Algae became abundant. Both appear to have had a distinct preference for shallow water with not too much pollution by continental debris. Consequently they are often found together. In the Lower Högklint they are more common in the reefs than in the stratified deposits, but in the Tofta facies, as well as elsewhere in the uppermost Högklint, they are abundant also in the normal sediments.

It was shown that the fixing of the boundary between the Visby and Högklint Beds at the first longer and thicker limestone layers is not very exact, since favourable conditions for increased limestone formation did not begin simultaneously at all places (p.282). Nevertheless differences in the level of the boundary at various localities are not great under normal conditions. Great differences in level are restricted to places where close to the boundary the stratified sediments are overlaid by large Högklint reefs (Högklint, Kneippbyn, north of Lickershamn); crinoid and debris deposits caused extra limestone deposition there and the heavy reefs led to extra compaction and outwards-upwards movement of the underlying Visby sediments (pp.156, 290).

If we draw a line along the points where a stratigraphical boundary (Visby – Högklint or Högklint – Slite) is found at the same topographical height, a somewhat curved line is obtained with a direction of 40° in the south (Visby, Lummelunda) to 55° in the north (Fårö). The course of this line is probably more or less parallel to the coast line of Högklint time.

In the lower parts of the Lower Högklint Beds a dip of about 0°.30', or slightly more, in a southeastward direction, can be deduced in the areas of Lickershamn and Halls Fiskläge. During or more likely preceding the shallow water phase of the Upper Högklint, a slight epeirogenetic movement seems to have taken place, since the boundary Högklint – Slite only dips about 0°.20'–0°.25'.

SLITE BEDS

The Slite Beds constitute the stratigraphical unit which has the greatest geographical distribution in northern Gotland. It occurs in a strip, often about 15 km wide, from the east coast (between the northeastern point of Fårö and a place about 8 km south of Slite) to the west coast (between Tofta and Klintehamn) (Fig.11). In the southwestern part of this strip, Slite marlstone is present, whereas the northeastern part shows mainly limestones. The topographical height of the limestone area varies between 20 and 70 m, with an average around 40 m; the marlstone area is much lower, generally less than 15 m above sea level.

Most exposures of the Slite Beds are found inland. There are a number of large, rather bare plains (e.g., File Haidar, Hejnum Hällar). Most of the exposed cross-sections through parts of the Slite Beds were formed by the *Littorina* sea (Bogeklint, Graunsklint, Gisslauseklint, Spillingsklint, Barabacke), but an occasional one was made by the *Ancylus* lake (Patvalds).

The name of the Slite Beds is derived from the village of Slite, on the east coast, in Othem Parish. Initially Slite was just the harbour of Othem, but now it has expanded and overtaken the original village, mainly because of its good seaport and cement industry.

The total thickness of the Slite Beds is about 100 m.

Stratified sediments

To treat the thick complex of Slite Beds as one unit would imply that no detailed picture could be obtained of the evolution of Gotland during the time that these beds were laid down. On the other hand a reliable stratigraphical subdivision of the Slite Beds is difficult to make, since there is rather a great variety in the rocks constituting these beds and the age relations between the sediments found in the various outcrops are still far from clear. Most important for our purpose is an insight into the stratigraphical succession of the limestone deposits. With some reservations, a subdivision into four subunits is presented here. The rocks described by Hede (1927a, 1928, 1933, 1936, 1940) presumably should be divided into these four stages as follows:

Slite I Beds: Hede (1927a), pp.23–24 (Lagrets äldsta del . . . .); Hede (1940), pp.37–41 (ledet a), pp.42–43 (contact Tofta–Slite), pp.43–45 (ledet b).

Slite II Beds: Hede (1927a), p.25 (Tomtmyr); Hede (1928), p.14 (kalksten, karakterisere av bl. a, *Ilionia prisca* och *Megalomus*); Hede (1933), pp.31–37, Hede (1936), pp.19–20, 21–22 (tunnlagrad . . . . finkristallinisk eller finoolitisk kalksten)?; Hede (1940), p.41 (på tvenne ställen i Lummelunda), pp.45–47 (ledet c), pp.47–49 (ledet d).

Slite III Beds: Hede (1927a), pp.26–28 (en serie kalkstenar, vilka . . . .); Hede (1928), p.15 (kalksten, i regeln synnerligen rik på stromatoporider), p.16 (kalksten, ställvis rätt rik på leperditior; en av lagrad kristallinisk kalksten och revkalksten bestående skiktserie), pp.17–20, pp.20–24 (Slite märgelsten); Hede (1933), pp.38–39, pp.46–53 (Slite-märgelsten); Hede (1936), pp.22–23 (en upp till omkring 5 m mäktig skiktserie; Kalbjerga-kalksten), p.24 (tunnlagrad . . . . tämligen starkt märglig kalksten), pp.25–28 (en intill c:a 2 m mäktig skiktserie), pp.29–33 (Slite-märgelsten); Hede (1940), pp.49–50 (ledet e).

Slite IV Beds: Hede (1927a), pp.31–32 (sandkalksten); Hede (1928), pp.24–43 (kristallinisk kalksten och revkalksten); Hede (1933), pp.40–45, p.53 (contact Slite marlstone/Slite IV Beds), p.54; Hede (1936), p.32 (Ryssnäs-kalksten), pp.34–36 (lagrade kalkstenar och dem ekvivalerande revkalksten), pp.36–39 (tunnlagrad . . . . finkristallinisk, ställvis finoolitisk kalksten); Hede (1940), pp.51–58 (lagrade kristalliniska kalkstenar och dem ekvivalerande revkalksten).

In each of the four stages, a gradual change in the character of the sediments can be observed in the direction from about southeast to northwest.

These changes are, as a rule, most apparent in the east of Gotland (geological map sheets Slite and Kappelshamn).

Martinsson (1962a), made a provisional subdivision of the Slite Beds into three units. His unit (1) comprises the Lower Slite Beds, up to and including the beds with *Conchidium tenuistriatum* (now also called *Rhipidium tenuistriatum*), and is thus equivalent to the Slite I–III proposed here, plus the *Conchidium tenuistriatum* Zone, which is taken by the present author as the base of the Slite IV Beds. Martinsson's unit (2) comprises the marlstone belt and his unit (3) corresponds to the Slite IV Beds except for the *Conchidium tenuistriatum* Zone.

Slite I Beds

Sediments formed during the first phase of Slite time only occur in part of the area where Slite Beds are found, viz., southeast of the line Korpklint (Snäckgärdsbaden)–Kappelshamn–Fårö. Northwest of this line, sediments of this age seem to be lacking.

The northwesternmost site where rocks that definitely belong to the Slite I Beds, are exposed, is located about 1.4 km east-northeast of Stora Hästnäs (in the northeast of Visby Parish). As mentioned while discussing the Tofta limestone, the present author considers the rocks of two small exposures about 3 km east-northeast of Lummelunds-bruk (Hede, 1940, p.41) to belong to the Slite II Beds.

The basis of the Slite I Beds is exposed in several, generally small exposures, especially in the parishes of Västerhejde, Tofta, Träkumla and Stenkumla. It consists of generally thin bedded, more or less marly limestone, dense to finely crystalline, concordantly overlying the Tofta limestone. The lower part of the Slite I Beds (zone a of Hede, 1940; maximal thickness about 3.5 m) is fossiliferous and contains some small reef-like parts (e.g., near Bjärs in Västerhejde Parish, near Martille and Forse in Stenkumla Parish, about 0.25 km northwest of Liksarve in Tofta Parish), which do not, however, generally constitute true reefs. The upper part of the Slite I Beds (zone b of Hede, 1940; maximal thickness about 3 m) is poorer in fossils. Occasionally the rock is finely oolitic (e.g., at Liksarve), but as a rule, it is rather marly. In the north, at Stora Hästnäs (Visby Parish), the sediment is, in part, more fossiliferous and somewhat bituminous. In the south of the Slite limestone area, the marl content of the sediment is distinctly higher (e.g., Norrgårde in Tofta Parish), until this finally passes into the Slite marlstone (Vallve, north of Paviken, in Eskelhem Parish).

The thickness of the Slite I Beds decreases northwestwards.

Slite II Beds

North of Tomtmyr, in Stenkumla Parish, a limestone complex, up to 10 m thick, occurs, which is considered to constitute the Slite II Beds. Stratified limestone alternates with reef limestone. The stratified sediment is brownish grey to greyish white, finely crystalline to dense, often finely oolitic and, as a rule, rather hard. The fossil content is moderate, with mainly bryozoans, stromatoporoids and Algae. The layers are usually thin. The reefs alternating with these layers are small.

Slightly south of the above locality, the thickness of the limestone complex decreases, the marl content increases, and the whole is, within a short distance, replaced by marlstone and strongly marly limestone.

Northwards, no other reefs of the Slite II Beds are known. The stratified limestones there, will be described in short in the following paragraphs.

In the south and southeast of the Slite limestone area, the Slite II Beds rest, presumably everywhere, on Slite I Beds. Exposures in these Slite II are found, among others, in the area of Hallbro Slott, near Martille (Stenkumla Parish), east of Visby, between Österby, Katrinelund and Stora Hästnäs.

Further northwestwards, the Slite II Beds rest directly over the Högklint (incl. Tofta) Beds. The latter is the case, among others, in exposures 0.9 km north of Nors in Fleringe Parish, and east of Vialms, also in Fleringe Parish. The boundary between the Högklint and Slite II Beds is generally very sharp, the Tofta facies generally being absent there. A most interesting exposure is the one about 0.5 km east-eastsoutheast of Vialms, along the coast of Fårösund, described by Hede (1933, p.34). There, Högklint reef limestone is exposed, which before deposition of the Slite II Beds, had been subjected to severe erosion, against which only the most solid parts of these Högklint reefs have resisted. As a result, a very irregular surface developed. On top of these reef remains and in between these, later Slite II Beds were deposited, in the form of thin-bedded, light-grey, crystalline limestones, which generally are rather loosely granular and rich in fossils.

That this limestone belongs to the Slite II Beds is apparent from the fact that in this area, at a level only about 4–5 m higher, the basis is found of an *Ilionia prisca* – *Megalomus* Zone, which constitutes a good marker zone slightly below the top of the Slite II Beds. Similar exposures are reported by Hede (1936, p.19) from 0.6 km east-southeast of Lansa, 0.9 km southeast of Lautur, 1.2 km north of the northern farm of Broskogs, directly west of Kalbjerga, and 1 km northwest of the northwestern end of Aiketräsk; all these localities are in Fårö.

In general the lower part of the Slite II Beds consists of a complex, up to 8 m thick, of thin-bedded, grey limestone. In the southwest, this sediment is dominantly finely oolitic, sometimes almost dense or finely crystalline. In the north, the basal 3 m are often dense or almost dense and rather strongly marly; further upwards, the rock becomes somewhat less marly, and is there, as in the south, dominantly finely oolitic, sometimes dense or very finely crystalline. The rock is almost always rich in fossils. Crinoids and bryozoans are especially abundant. Upwards, an increasing number of stromatoporoids is added, whereas calcareous Algae and several other fossils are also found. The bedding planes are usually uneven and rugged.

The above complex is overlaid by 2–4 m of limestone which is dominantly dense to finely crystalline; in the south it is only rarely, but in the north more frequently, though always locally, finely oolitic. The bedding planes are usually uneven also there; the thickness of the layers varies generally between 3 and 10 cm. The limestone is more or less marly and generally rich in fossils, especially in the more marly parts. Especially characteristic among the fossils are *Ilionia prisca* (Hisinger), which seems to be absent in the lower 8 m of the Slite II Beds, and "*Megalomus*" sp. Of these two fossils, *Ilionia* is most common in the marlier parts, whereas "*Megalomus*" seems to be found mostly in less marly limestone.

In the area of the geological map sheet Slite, this uppermost Slite II zone is the oldest sediment, found exposed, of the Slite Beds. Only some small outcrops are known in the environs of the line between Lokrume Church and Tingstäde Church.

Slite III Beds

Sediments ascribed to the Slite III Beds are found mainly in a strip, running approximately southwest–northeast, from Västerhejde to Vialms in the northeast of Fleringe Parish, and from there continuing over part of western Fårö. This strip is broader in the middle than in the southwest and northeast. Exposures occur, among others, near Västerhejde, south to southwest of Gällungs in Väskinde Parish, west of Suderbys in Bro Parish, in an extensive area in the parishes of Fole, Hejnum, Tingstäde and Othem, near Uppgårds in Lärbro Parish, near Mölnars in Fleringe Parish, south of Trälgar and Vialms in Fleringe Parish.

The lowermost part of the Slite III Beds consists generally of stratified limestone of light grey to brownish light grey colour, usually very rich in stromatoporoids. Exposures have been found, among others, in the area between Vidmyr and Tingstäde Träsk, in the environs of the country road northwest of Tingstäde, at the base of a quarry about 1 km southeast of Trägårds in Tingstäde Parish and of a quarry 0.45 km south-southwest of Lauks in Lokrume Parish, west of Stora Ryftes in Fole Parish and about 0.6 km east-northeast of File in Othem Parish. The rock is generally finely crystalline and locally finely oolitic. The bedding planes are generally rather irregular, sometimes smoother. Fossils are abundant: stromatoporoids, corals, crinoids, bryozoans, Algae; also present are lamellibranchs, brachiopods, cephalopods, gastropods and some ostracodes. In several localities cross-bedding has been observed. Hede (1928, p.15) reported faint ripple marks from a locality 1 km west-northwest of Tingstäde Church; the ridges of the marks strike about 170°. The maximum thickness observed for this zone is just over 4 m.

In some places in the Tingstäde–Lokrume area, it can be seen how the stromatoporoid-rich limestone, described above, is covered by a thin layer of thin bedded, dense limestone, rich in *Leperditia* sp. The rock is greenish or brownish light grey and also contains small stromatoporoids, some lamellibranchs and gastropods and some other fossils.

The upper part of the Slite III Beds is exposed primarily in the area between the churches of Tingstäde, Fole, Bäl and Othem. It consists predominantly of thin bedded, sometimes thick bedded, limestone of light grey to faintly greenish or brownish light grey colour, sometimes red-mottled through red crinoid remains. The limestone is, as a rule, rather pure and finely crystalline, sometimes finely oolitic. In the north, it is locally somewhat bituminous (e.g., in the large quarry of Storugns in Lärbro Parish). Interbedded between the limestone layers often are films or thin layers of marl. Generally the lower limestone layers are thicker than those higher in this zone. This decrease in the thickness of the limestone layers and consequent increase in the number of marl layers may be connected with a northwestward shift of the sedimentation belts in Slite III time, due to a temporary somewhat greater water depth in this part of the basin. Deeper water provided more chances for terrigenous debris to be laid down. The northwestward shift also concerned the deposition area of the Slite marlstone, the northwestern boundary of which was, at the end of Slite III time, in the southwest about 6 km further northwest than during Slite I time, in the east and northeast perhaps even somewhat more. (See also pp.329, 334).

The bedding planes of the Upper Slite III Beds are often rugged or knobby. This may be partly due to the fossils embedded in the layers, but partly it must have another reason, possibly related to water depth. Among the fossils present, brachiopods, stromatoporoids and crinoids dominate, but also corals, bryozoans, Algae and lamellibranchs (of which especially *"Megalomus"* sp. is characteristic) are anything but rare, and gastropods, cephalopods, ostracodes and trilobites are represented as well. The maximum thickness of this zone is about 6 m; the total thickness of the Slite III Beds is about 10 m.

In some places, reef limestone is found to occur in the Slite III Beds, such as southeast, east and northeast of Fole, southeast and east of Graute in Hejnum Parish, and southwest of Othem. In the surroundings of reefs, the stratified limestone can be rich in calcite veins, up to 7 cm wide, and in stylolites, e.g., in the large quarry in File Haidar.

The marl zone, which during Slite III time expanded from the southeast over the limestone area, consists generally of an alternation of usually thin layers of marlstone and thin layers or lenses of strongly marly limestone. Presumably close to the northwesternmost line reached by the Slite marl are the exposures directly north of Bunge Church and near Utbunge, about 3.2 km east to slightly east-southeast of Bunge. There the marl zone consists of a dense or quite dense, marly limestone, which, in part, is rather strongly sandy (sandy limestone). The marl zone there is not very thick. Somewhat further southeast, for example along the coast near Enenäs (Grundudden), about 3.5 km southeast of Bunge, thin marl layers also occur, although there are, as well, some layers which are of a somewhat fine sandy nature.

Slite IV Beds

During Slite IV time, the stratified limestone expanded in its turn over the marlstone. This can be seen, among other places, in the Bogeklint and the klint of Tjeldersholm, about 4.5 and 8 km, respectively, south of Slite; both show a basis of marlstone with interbedded layers of marly limestone. From the base of the Slite IV Beds upwards a decrease in marl content can be established, e.g., in the Bogeklint and in Kvarnbacken or Lotsbacken (south of Slite harbour).

In the southwest, from a few fundred metres north of Sion (2.4 km west-northwest of Träkumla Church) south-southwestwards, the basis of the Slite IV Beds is formed by a 2–6 cm thick layer of light grey, finely crystalline limestone which is rich in fossils; in particular *Conchidium tenuistriatum* (Walmstedt) is very abundant.

The most common stratified sediment in the Slite IV Beds is light grey, very finely to middle crystalline limestone in layers of varying thickness. In the lower part of this unit, between the limestone layers, films of bluish or greenish marl often occur. In the higher parts, the rock is purer; oolitic members also occur there. The limestone is very rich in fossils. *"Megalomus"* sp. occurs regularly, but especially characteristic of the Slite IV Beds is *Pentamerus gotlandicus* Lebedev. Crinoid remains often give the limestone a red-mottled appearance. The thickness of the Slite IV Beds reaches more than 30 m.

From two localities in Fårö, which expose sediments from the lower part of the Slite IV Beds, Hede (1936, p.33, 36) reported well-developed ripple marks. These are 1.8 km west-southwest of the Holmudden Lighthouse (strike of the ridges 20° and 60°) and 0.9 km north-northeast of Ryssnäs (strike of the ridges 50°).

Because of the frequent occurrence of reefs of various size, there are many local facies differences in the Slite IV Beds. Consequently, one should be very careful in drawing stratigraphical conclusions from the occurrence of a similar rock type in different localities. The fossil-rich crinoid limestone of the Tjeldersholmklint and the fossil-rich limestone of Ytterholmen are not so definitely synchronous as Hede (1928, p.41) presumed, despite the fact that both contain abundantly *Atrypa reticularis* (L.). Both kinds of rock were formed in the vicinity of reefs and they owe their similar development in the first place to this situation.

Slite marlstone

The southeastern half of the area where Slite sediments occur at the surface, shows mainly marlstone. Close to the limestone area, this marlstone varies somewhat lithologically, but south of the line which roughly runs from Västergarn to Slite, the marlstone is rather uniformly developed, bluish grey to brown grey, dense, distinctly stratified, very fossiliferous, with interbedded thin layers or lenses of harder, finely crystalline, grey marly limestone. The marlstone is beautifully exposed, to a great thickness, in the large quarry of the Skånska Cement A.B. in Slite (Fig.157). The thickness of the marlstone complex is at least 68 m. In the Slite quarry, this thickness of marlstone is overlaid by stratified limestone (Fig.158) with intercalated reef limestone.

The mineralogical composition of the upper 30 m of Slite marlstone in the Slite area is as follows: 60–70% calcite, 13–17% dolomite, 7–8% quartz, 5–6% feldspars, 2–3% mica and illite, 2–3% kaolinite, and traces of iron hydrate and oxide. Carbonate-free, the figures are: 38–44% quartz, 28–33% feldspars, 11–17% mica and illite and 11–17% kaolinite.

The chemical composition of glowed rock of these upper 30 m is: 62–70% CaO, 18–23% SiO_2, 4.4–5.8% Al_2O_3, 3.5–4.5% MgO, 2.3–3.3% Fe_2O_3, 1.2–1.4% K_2O, 0.7–1.0% S and 0.1% or less Na_2O. The $CaCO_3$ content of the unglowed rock is 76–83%. Large samples were taken, comprising both marlstone layers and interbedded marly limestone. The $CaCO_3$ content of the marlstone alone is about 70–78%.

An index zone is found in the Slite marlstone complex at a depth of about 43–51 m below the top of the marlstone, which in the Slite area is about 35–43 m below sea level. It consists of a zone with only 43–50% CaO and as much as 30–34% SiO_2 at a depth of 49–51 m, overlaid by a zone which is much richer in CaO (65–73%) and poorer in SiO_2 (16–20%) at a depth of 43–48 m. Below this double index zone, the average CaO content is 50–58%, directly above it 53–60%. Above this, for a thickness of about 10 m, the values develop, with fluctuations, in the direction of the percentages given before for the upper 30 m.

Perhaps the index zone corresponds with the lower part of the Slite II Beds in the limestone area.

Reef limestones and related sediments

In this section, a review will be given of the more important-reef limestone occurrences within the Slite Beds. Observations of general value will be shortly described. Most of the reef limestone is of the Hoburgen type.

(Text continues on p.320)

Fig.157. Quarry of Skånska Cement A.B. in Slite, showing Slite marlstone of great thickness, overlaid by Slite limestone.

Fig.158. Detail of the Skånska Cement A.B. quarry in Slite, showing Slite limestone.

Fig.159. Raukar of reef limestone, belonging to the Slite Beds. Solklint, Slite.

Fig.160. Raukar, consisting of reef limestone belonging to the Slite Beds, on the Lännaberg in Slite. The planes crossing parts of the reef limestone form the basis of shallow local depressions which one occurred on the surface of the reef. At these levels growth probably was temporarily interrupted. Such depressions may have been present in about the same part of the reef during various stages in its development.

Unless otherwise stated, the reef limestone exposed in the localities mentioned in the next paragraphs, belongs to the Slite IV Beds.

Bogeklint or Klinteklint (Boge), about 2 km south-southeast of Boge Church is one of the most interesting localities of Slite reef limestone. Although the highest local point of this hillock is approximately in the centre, the area with the greatest average height part is in the north of the klint. In the northeast are some raukar, consisting of reef limestone. The same reef is also exposed in the northern part of the east wall of the klint, in the north wall and in the northern part of the west wall. The reef limestone is grey and weathers mainly in a brecciated way. Stromatoporoids dominate. As in most Slite reefs, coral colonies are commonly found and occasionally reach great sizes. Some local stratified intercalations occur with dips of up to 25° in various directions. Marly pockets of a few square centimetres to a few square decimetres are common and usually leave holes after weathering. The marl contains crinoid remains, solitary corals, brachiopods and a few smaller coral colonies.

This northern reef presumably has not been much higher than is exposed. In a number of places on top of the northern part of the klint, crinoid limestone is found, the reef limestone present in the highest part often showing a tendency towards stratification. Also a zone of at least 5 m length, containing several pockets of crinoid limestones embedded in the

Fig.161. Rauk on the Lännaberg in Slite, showing reef limestone of a brecciaceous nature. Slite Beds.

Fig.162. Exposure at the exit of the quarry of the Skånska Cement A.B. in Slite, south of the Lännaberg. At the top a rauk of reef limestone. This rock has its base at the level of the land as shown at the left of the photograph. The reef limestone overlies stratified limestone. At the base of the section marlstone is exposed.

---

reef limestone, about 1.5 m below its present top, suggests decreasing vigour in reef development. Nowhere is the base of this reef limestone exposed. The maximum thickness of this reef can thus be estimated to be at least 12 m. The original shape of the reef cannot be established since parts of it, especially in the northeast and north, have been removed by erosion. A northeast-southwest major axis seems likely.

Southeast of this reef, there is another occurrence of reef limestone, but of lesser thickness, and overlying crinoid limestone with reef debris which probably belongs to the northern reef. Southwards, the base of the southeastern reef limestone is on an increasingly higher level, suggesting southeastward expansion of reef growth.

Observations on the plateau of the klint and in the corresponding west wall, indicate the possibility that the reef limestone of the middle part of Bogeklint belongs to a number of smaller and thinner reefs.

In the southern and lowest part of Bogeklint, only stratified limestone is present, showing decreasing marliness upwards. At the bottom, marl pockets and thin marly layers are not exceptional.

No faunistic differences between the smaller reefs and the northern reef were found.

Solklint, west-northwest of Slite, is a hillock, which is about 250 m long in a north–south direction, and about 175 m broad in an east–west direction. In the northeast, a number of raukar are found (Fig.159); from thereon raukar also occur on top of the Solklint in a southwest direction. On the west to north side of the hillock, reef limestone is exposed, the east to southwest side show stratified limestone with reef debris. With increasing distance from the reef, the sizes and amount of reef debris decrease and are least in the southeast. Since most of the reef limestone is found in an about northeast–southwest direction and the hillock in this direction is higher than elsewhere, it seems very likely that this is the orientation of the original major axis of the reef. Its present length is about 150 m and its breadth about 100 m; the observed thickness of the reef limestone is 7 m. In view of the occurrence of several raukar in the northeast, which are, on the average, only 3 m high, it may be assumed that the northeast part of the reef has been eroded and that the reef length has been more than the present 150 m.

Lännaberg is a hillock of reef limestone north of Slite. On the north and east side of it, a group of large raukar is found (Fig.160, 161). The reef limestone is a mass of stromatoporoids and corals, crinoid-stem fragments and some bryozoans in a matrix of marly limestone. Most reef builders are rather small. Other fossils, such as brachiopods and lamellibranchs are relatively rare. The general character of the rock in these raukar is rather different from that in the raukar of Holmhällar or Ljugarn, where the

Fig.163. Detail of Kvarnbacken, Slite, showing reef limestone, enclosing a marl pocket which has been eroded to some depth.

average reef builders are much larger, the matrix purer, the number of coral colonies smaller.

Dips, up to about 20°, but only perceptible for a short distance (usually less than a metre), can be seen in some places in the reef limestone. These are caused by the positions of the reef builders. Upwards, within a thickness of a few decimetres, the colonies may gradually return to a horizontal position. This suggests that in these places the reef builders have straightened out unevennesses in the topography of the growing reef surface.

It seems likely that the Lännaberg and its raukar are the remains of one reef, slightly oval in plan, approximately 250 m long and with its longest axis about northeast–southwest. The thickness has been 10 m or more. The lowermost parts of the deepest exposed raukar show a more marly and vaguely stratified matrix, which may be an indication that these parts were formed in an early stage of reef growth.

At the south side of the hillock and reef, some reef debris is exposed, which was deposited rather close to the reef. It is notable that there are few large reef builders nor larger fragments of these. Crinoid remains and small fragments (up to a few centimetres) of stromatoporoids, bryozoans and corals are dominant. Bedding is irregular, becoming more distinct as the distance from the reef increases.

In between the Lännaberg in the north and the cement factory in the south, a raukar field is present, which will be called the *cement-factory raukar field.* On the south side, in an exit of the cement-factory quarry, the base of the reef limestone is exposed in the form of 5.5 m stratified limestone, underlaid by Slite marlstone (Fig.162). This stratified limestone is, in its lower parts, a normal kind of limestone with some crinoid remains, but passes upwards into crinoid limestone with reef debris.

About 60 m north of this exposure, the base of the reef limestone is situated about 7 m lower. Underlying it is crinoid limestone of unknown thickness. 75 m further north, the reef base is another metre lower. This shows that the contact of Slite marlstone and Slite limestone is not a horizontal surface. There are two possibilities: (a) reef formation began immediately after deposition of the Slite marlstone was replaced by limestone sedimentation and the non-horizontal contact is caused by differential compression of the marl; (b) reef formation began already during the time in which in the close environment still the uppermost Slite marlstone was formed. The exposures present no conclusive evidence but to judge from their general nature possibility (b) seems the most likely.

As far as can be established, all reef limestone belongs to one reef. The exposed thickness reaches about 8 m; for the centre of the reef a thickness of 10–12 m may be assumed.

Kvarnbacken (Lotsbacken). This hill is situated directly south of the harbour of Slite. On its east side, along the beach, a section is exposed. This section shows three successive zones.

At the base is a zone with many irregular limestone lenses in a matrix of bluish grey marl. The lenses are 1–4 cm thick. The limestone is dense to finely crystalline and marly; locally it contains some pyrite. The marl bands between the limestone lenses are generally less than 1 cm thick; only occasionally a specific marl band can be followed for a distance of more than 10 m. Most wedge out after only a short distance. This zone is moderately fossiliferous. It is exposed around and directly above present sea level, up to a visible thickness of between 1 m and a little more than 2 m. The sediments of this zone are less resistant to erosion. In consequence, the zone retreats underneath the two overlying ones.

The second zone, 0.1–0.8 m thick, consists of marly limestone with marly intercalations. The limestone is finely to middle crystalline, light grey to faintly bluish grey, weathers to brownish grey to yellowish brown and is more fossiliferous. It occurs in lenses, 1–12 cm thick and one or a few metres long; some are of greater extension and have the character of true layers. Marl is mainly concentrated in pockets. Marl bands also occur, generally less than 1 cm thick and only extending slightly.

The upper part of the section shows crinoid limestone, which is middle crystalline, light grey but through weathering bluish grey, highly fossiliferous (crinoids, bryozoans, corals); very small crystals of pyrite occur. The layers vary in thickness between 1 cm and several decimetres. Locally thin marl bands are interstratified. The thicker limestone layers often contain larger fossils than the thinner ones.

The third zone of these three is most resistant against erosion. All three show dips in varying directions and degrees (up to a maximum of about 20°).

Further towards the north, marly reef limestone is exposed (Fig.163) for a distance of more than 100 m. Stromatoporoids predominate, several of which have grown strongly in an upward direction. Coral colonies are common. South of the harbour, sediments, comparable to those of the lower zone in the east of the hill, are exposed.

Around the mill (Swedish: kvarn) marly reef limestone is exposed, belong to a reef other than that along the east coast. This central Kvarnbacken reef is larger and presumably also a little younger.

Slottsbacken. This hill, about 10 m high, is situated south of Slite, on an isthmus between the Baltic and Bogeviken (Boge Bay). It consists almost entirely of grey reef limestone, which is best exposed in a number of raukar in the east. The reef limestone is strongly weathered. Locally it shows parts with a more or less vague stratification, which parts are up to 2 dm high and a few metres long, and show dips of 0–20°. At the very top of the hill, there are a few very small outcrops of crinoid limestone.

Along the western shore of Hydeviken, in the south, sediment is exposed, up to 2 m above sea level, which belongs to the uppermost Slite marlstone. It is similar to the rock at the base of Kvarnbacken and high in the quarry of the Slite cement factory and consists of lenses or thin local layers of limestone within a softer marlstone. At a height of 5 m above sea level, stratified, crystalline, marly limestone is exposed. A little further north, it can be seen that the limestone has its base about 3 m above sea level. It overlies a transition zone of about 0.5 m thick, consisting of somewhat irregular layers of marly limestone, 0.5–5 cm thick, alternating with softer marlstone layers, which upwards decrease in number and thickness. Some marlstone layers can also be noted in the lowermost part of the overlying limestone. Both the marlstone and the limestone of the transition zone are very fossiliferous.

Further northwards, about 0.6 km south of the north side of the bay, another exposure is found. In the highest part of the marlstone, a local, large limestone lens is found, up to 50 cm thick. The marlstone is also rich in colonies of *Halysites catenulatus* (Martini), up to 120 cm long and 40 cm thick. The gradual transition zone from marlstone to limestone varies in thickness from 20 cm to about 2 m. A number of raukar occur here, partly consisting of stratified limestone with crinoids and partly of weathered reef limestone. The base of the exposed reef limestone descends from south to north, from about 7 m to about 3 m above sea level, which is only about 1.5 m or less above the top of the marlstone. The exposed reef limestone is not thicker than 3.5 m. However, the original thickness should have been at least 7.5 m, judging from a comparison between the raukar situated at the highest and lowest levels in this area.

On and around *Graunsklint*, northeast of Lärbro, only little reef limestone is exposed. Most of the exposures show normal stratified limestones or stratified limestones with crinoids and reef debris. The distribution of the sediment types and the topography of the surface of the hillock make it likely that a number of small reefs are present. Nowhere is reef limestone exposed thicker than 5 m, but the true thickness of the reefs may have been somewhat more.

The hillock of *Patvalds*, about 3.5 km southeast of Lärbro, shows grey reef limestone and some stratified crinoid limestone with reef debris, exposed in some parts of a low cliff around the hillock and in several places on its surface. Nowhere is the observed thickness of the reef limestone more than 3 m. It seems very likely that a number of reefs played a part here.

Gisslauseklint, about 3 km east of Othem, shows exposures in a low and interrupted wall and scattered on top of the klint. Reef limestone, reef debris and stratified limestone with reef material crop out. The reef limestone is traversed by vertical and subvertical cracks (see p.152). About 160 m northeast of the road Othem–Hellvi, on the east side of the hillock, a contact of reef limestone and the surrounding sediments is exposed. At the contact, a number of coral colonies are found, up to 25 cm in diameter, apparently in their positions of growth. Some of the stromatoporoids, most of which are lenticular in shape, are not in their growth positions. Very close to the reef, there are several large washed-out reef builders and fragments of these. At a distance of 1.5 m away from the reef, the debris is already rather fine, with only a few elements of 5–10 cm, embedded in stratified limestone with crinoids.

Spillingsklint (Othem Parish) is a rather extensive hillock with exposures in a low and interrupted cliff wall and in some raukar. In the north and east, reef limestone is dominating. It seems likely that reef growth began in several places, but that in expansion, part of the reefs united. In the south of the southeast wall, thick bedded limestone with reef debris is found. Since the angle between the southeast boundary of the reef limestone and the position of the exposed wall is very small, the sediment in the wall, southwestwards, shows clearly, how the amount and size of the reef debris decreases with gradually increasing distance from the reef.

The plain of *File Haidar* shows badly exposed Slite III reef limestone in between scattered vegetation. There is only one quarry, made in stratified limestone. When visited in 1959, it showed a vaguely bedded intercalation, about 4 m thick and 6 m long, consisting of crinoid remains with some stromatoporoids and a number of coral colonies, most of which were not in a position of growth. The stratified limestones abuted against it and deposition apparently took place very close to a reef.

The reef limestone of *Hejnum Hällar*, too, is less suitable for detailed studies since it is only exposed in horizontal outcrops in between the vegetation. It belongs to the Slite III Beds. Normal grey reef limestone, occasionally red-mottled, dense, strongly recrystallized, occurs. Stromatoporoids are the main reef builders, but coral colonies are common. About one third of the latter are not in their position of growth (see also Chapter XII, p.436). The reef limestone is traversed by vertical fissures (see p.152).

Spinnbjersbacke. This is a very low hill, situated north of Tjelders, or 6 km south of Boge Church. Three raukar areas are found there. Of these, the northern one, possesses only two raukar, a northern rauk of about 5 m and a southern rauk of about 2 m high. The central raukar field is the largest one, with raukar up to about 4 m high.

The three fields represent the remains of at least three fossil reefs. All raukar consist of stromatoporoid reef limestone, which is greenish grey and dense. In their lower parts tower-shaped colonies of stromatoporoids occur, up to 25 cm high, and consist of a pile of inverted basin-shaped latilaminae, 5–10 cm in diameter. Most of these are not in their positions of growth. Coral colonies, too, are common in the lower parts. Higher in the raukar, dome-shaped stromatoporoids grew close together and united into covers with dome-shaped elevations. Also thin colonies of large horizontal extension occur. Pockets with marl or crinoid limestone are found locally. Horizontal or concave lines of limited extension, which suggest interruptions or retardations in reef growth, are also seen in some raukar.

The sediment underlying the reef limestone is exposed in the north of the hill, not far away from the northern raukar field. This is a stratified marly limestone, finely crystalline, light grey, with layers of on the average 1–4 cm thick and bedding planes which are often irregular and knobby. Some marl pockets occur, which as a result of weathering, often leave holes. The maximum height over which this sediment is exposed is 2.5 m. At the base, the fossil content is moderate, but this increases upwards, with crinoid and bryozoan fragments, solitary and social corals, stromatoporoids, brachiopods. This increase in the number of fossils suggests reef development not far away. Also the slight topographical elevation of Spinnbjersbacke is indicative of the presence of reef limestone. The top of the 5 m high rauk in the southern field contains some crinoid limestone, which might indicate that the former reef has not been much higher. Altogether, an original thickness of about 7–8 m seems likely for the reefs at Spinnbjersbacke.

Tjelders. In the forests west, southwest and south of Tjelders, some slight elevations are found, with small exposures of reef limestone. 0.25 km south of Tjelders an isolated rauk is found and 0.6 km west of Tjelders, a few raukar occur together. All show weathered reef limestone of Hoburgen type. Another few, very small outcrops show crinoid limestone with a varying content of reef debris.

Tjeldersholm. This is a promontory, about 0.3 km long and 0.5 km broad, east-northeast of Tjelders. In the northeast, a section is exposed from sea level up to a height of 6–7 m at maximum. This shows predominantly marly crinoid limestone, which through weathering has become yellowish brown to brownish grey in colour. The layers are generally 0.5–5 cm thick; bedding planes are irregular and rugged. Dips of a few degrees, in various directions, are common. The sediment is very fossiliferous, with crinoid remains, solitary and social corals, stromatoporoids, bryozoans, brachiopods, gastropods, trilobites. Part of the fossils are certainly in secondary positions and presumably are reef debris. Bryozoans, corals and stromatoporoids are also represented by several fragments, part of these being somewhat rounded. In general, the sediment is most marly in its lowest 2.5–3 m. About 4 m above present sea level bedding becomes somewhat more regular. The content and coarseness of recognizable reef debris decreases. The section is cut by several joints, most of which are filled with calcite.

North of the two best exposures of crinoid limestone in the Tjeldersholm area, two outcrops of reef limestone are found, the largest of which measures about 3.5 m in hight and 6 m in length. Since this section is cut through the reef periphery, and the reef limestone descends below the present sea level, the original dimensions of the reef must have been larger. That much more reef limestone has been present in Tjeldersholm than is found there now, is also indicated by the amount of crinoid limestone and the dips within this sediment, and by the large amount of reef-limestone fragments on the shore.

The lower part of the exposed solid reef limestone is very marly. Next to the stromatoporoids, many coral colonies are found, several of which are relatively thin.

Barabacke is a low, elongated hill, west of Bara Ödekyrka, which shows reef limestone for a distance of more than 0.5 km and a visible height of about 7 m maximum. The reef limestone is very weathered. The reef grew at the end of Slite time and possibly also still at the very beginning of Halla–Mulde time. In the south, the reef limestone is surrounded by the Bara oolite, which forms part of the Halla limestone.

An interesting small hill is found about *1.4 km north-northeast of Bara Ödekyrka*[1]. In the north, this hill is cut by a road. Exposures are up to 2.5 m high. North of the road, stratified limestone is exposed with layers of on the average 2–15 cm thick. The sediment is finely to partly middle crystalline, light-grey coloured, through weathering light brownish grey. At the base, it is extremely rich in small fossils and fossil fragments (bryozoans, crinoids, brachiopods, corals, stromatoporoids). Higher up, the fossil material is much coarser, whereas loose blocks show that in the highest part, some reef limestone has been present. Reef limestone also forms the central part of the exposure at the south side of the road. The dimensions of the hill and the distribution of sediments suggest that the reef has not been longer than approximately 60 m and in breadth was still somewhat smaller. Weathered outcrops in the east and south of the hill show reef limestone with a maximum thickness of about 1 m, underlaid by stratified limestone. In its centre, the reef may have been thicker, but presumably did not exceed 4 m even there. The stratigraphical position of the reef is about at the top of the Slite Beds.

Slight differences in the amount of reef debris present in the stratified limestone around the reef suggest that most debris was deposited east and south of the reef. On the southeast side, the amount of debris is slightly less and the average size of the fragments slightly smaller. North of the reef, there is again somewhat less and smaller debris.

In the environment of this hill, a number of other similar elevations occur, which also owe their origin to the presence of fossil reefs, which are, however, badly exposed.

The farm *Simunde,* 0.7 km northeast of Bara Ödekyrka, is situated on a low hill, which rises about 4–5 m above its surroundings. The hill consists predominantly of reef limestone, which is, however, badly exposed. On the south side, some stratified limestone is present.

Directly southwest of the farm *Hommunds,* 0.25 km northeast of Simunde, a similar hill is found. It contains a reef, which in a northeast–southwest direction measures about 45 m and in a direction perpendicular to it about 30 m. The maximum visible thickness of the reef limestone is 3 m; in the centre of the reef, the thickness may be expected to be about 5 m. In two old quarries in the south of the hill, stratified crinoid limestone is exposed, which also contains some reef debris. The amount and coarseness of this debris decrease from the top to the bottom of the sections and also in horizontal directions away from the reef.

1.6 km south-southwest of (former) Simunde Station, which is 2.25 km west-southwest of Bara Ödekyrka, a 4–5 m high elevation is present, whose origin is also connected with a fossil reef. A section exposed in the west of this hill is interesting; it shows stratified sediments that have been deposited at the northwest, i.e., leeside, of the reef. At the base, finely crystalline limestone is found, which is thin bedded and strongly marly. It alternates with thin bands of bluish grey marl. Within the limestone small pockets of marl occur. The limestone is very fossiliferous, especially in small fragments of crinoids and bryozoans. Upwards the limestone is

[1]The Swedish word Ödekyrka denotes a church which is not longer used and, therefore, has fallen into disrepair.

less marly, the marl bands disappear and the amount and coarseness of reef debris increase. This reef debris also includes now colonies of corals and stromatoporoids. The general picture indicates deposition at a gradually decreasing distance from the reef. Reef limestone, finally forms the top of the section. Reef limestone is exposed for a thickness of about 2 m. In the centre, the reef is presumably 4–5 m thick.

Near *Lillfole,* 1.7 km southwest of Fole, a reef limestone area is present, which appears as a topographical elevation of little height. No important exposures are found here.

Southwest, east and northeast of Fole Church, reef limestone is present, which belongs to the Slite III Beds. Only small and weathered horizontal outcrops are found. Stratified sediments alternate, horizontally, with reef limestone, suggesting the presence of several reefs in this area. The reef limestone is rich in stromatoporoids, most of which are dome shaped and seem to be in their positions of growth. Coral colonies are rare. The reef limestone is cut by a number of fissures, with a general direction of 110° (E 20° S).

Close to the cross-roads, about 0.7 km southeast of Fole, an old quarry shows stratified limestone, containing reef debris. At the top of the low section, which is exposed, the recognizable reef debris constitutes approximately 13.5% of the total rock volume. This sediment seems to have been deposited at the south-southeast side of a reef, at a distance of about 20 m. The size of the reef could not be established. In a few places, cross-bedding is found, directed away from the reef. The rock about 1.5 m lower in the section, has presumably been deposited at a somewhat greater distance from the reef and contains about 7.5–10% macroscopically recognizable reef debris.

Between Stora Hellvigs and Lilla Fjells, about 2 km east-northeast of Endre, a reef limestone area is found, about 0.4 x 1.2 km in size, which is orientated northeast–southwest. The reef limestone is badly exposed and of the general type. A few fissures have been observed, directed 140° (E 50° S). An exceptional fissure is perpendicular to this direction.

The hill of Endre, *Endre Backe,* consists in its centre and north almost entirely of reef limestone. In the south, stratified limestone with reef debris is present. No particularly valuable observations were made.

In between the farm Norrbys and the sanatorium, *along the road Stora Vede–Follingbo,* is a wall which shows stratified limestone with reef debris. Close to Stora Vede, a former railroad crosses underneath the road. It is mainly cut in reef limestone. Neither locality presents anything of special interest.

A list of fossils found in the Slite reef limestones and the crinoid limestones adjacent to these, is given in Table XVIII. The reef limestone in column 1 is from the Slite I Beds; the locality is 1.2 km north-northwest of Tofta Church. The fossils mentioned in column 2 are from reef limestone in the Slite I Beds, occurring southwest of Suderbys in Västerhejde Parish. To the Slite III Beds belongs the reef limestone from west of Gardrungs, Stenkumla Parish, of which the fossil content is given in the 3rd column. Columns 4–11 give fossils from Slite IV reef limestone; the locality given as Simunde is 1.6 km south-southwest of the former railway station of Simunde (or 3.2 km south of Källunge Church).

The first crinoid limestone column lists fossils found in the Slite I Beds 0.25 km northwest of Liksarve in Tofta Parish, the second column shows which fossils have been noted from Slite III crinoid limestone south-southwest of Gardrungs (or about 2 km south-southwest of Stenkumla Church).

The further crinoid limestone sites are in the Slite IV Beds; the one listed as Bara is 1.4 km north-northeast of Bara Ödekyrka, the one listed as Simunde being the same as that in which the reef fossils of column 10 were observed.

Discussion

From the data in the preceding sections on the Slite sediments, one can now attempt to compose a rough picture of what has happened in the course of Slite time.

Deposition of the Slite I Beds followed the strong decrease in water depth at the end of Högklint–Tofta time. The beds are missing, apparently, in the north of the Slite limestone area. During Slite I time, water depth presumably fluctuated somewhat, but was shallow, with during the first part perhaps some increase in depth and during the second some slight decrease. The initial increase in water depth is suggested by the fact that after the deposition of the Tofta limestone in very shallow water there is again a beginning of reef growth. The next decrease in depth is thought likely from the observations that the just started reef growth ends again, whereas there was locally deposition of oolitic limestone and elsewhere of somewhat bituminous marly limestone.

During Slite II time, the area of sediment deposition expanded northwestwards and Slite II Beds are found over the entire Slite limestone area; water depth increased, probably with fluctuations.

The increase in water depth presumably continued during Slite III time, in which a north and northwestward extension of the area of marl deposition took place.

The Slite IV Beds reflect a retreat of the area of marl sedimentation southwestwards. Presumably a new epeirogenetic movement had occurred, which initially influenced the water depth and the distribution of the sedimentation belts rather significantly. Thereafter, a more gradual shallowing of the water is assumed. Intercalated in these beds are most of the Slite reefs and also the most important ones.

During the entire Slite Period, marl deposition took place southeast of the area where the limestones were laid down; that is, at larger distance from the coast. Marl sedimentation was faster than limestone formation; about 70 m of marlstone corresponds to about 30 m of limestone (Slite I–III Beds). Marl deposition undoubtedly took place in deeper water than that in which the limestones were laid down. In comparison to the Slite limestones, the marlstone lacks cross-bedding, ripple marks or the presence of Algae. Within the Slite Beds fossils such as *Cyrtia exporrecta* (Wahlenberg) and *Dicaelosia biloba* (L.) were only found in the marlstone, and *Plectodonta transversalis* (Dalman) was found more commonly in this deposit than in the limestones. Ziegler (1965), working on the Silurian of Wales, considered these three fossils as belonging to the *Clorinda* community, the deepest-water community of the five which he could distinguish.

Slight variations in water depth caused shifts of the area of marl sedimentation, suggesting that water depth during Slite III and the beginning of Slite IV was about at effective wave base; in Fårö, ripple marks are found in limestone directly overlying marl.

It is difficult to give absolute figures for the variations that took place

(Text continues on p. 334)

TABLE XVIII

Fossils found in the reef limestones and surrounding crinoid limestones of the Slite Beds of Gotland

Lithology / Localities / Fossils	Reef limestone											Crinoid limestone										
	Tofta	Västerhejde	Stenkumla	Bogeklint	Solklint	Kvarnbacken	Spillingsklint	Tjeldersholm	Barabacke	Simunde	Other localities	Tofta	Stenkumla	Bogeklint	Solklint	Kvarnbacken	Spillingsklint	Spinnbjersbacke	Tjeldersholm	Bara	Simunde	Other localities
ALGAE																						
Rothpletzella sp.				+	+		+															
Solenopora sp.					+		+				+											
Unidentified calcareous Algae				+				+	+		+											
HYDROZOA																						
Actinostroma sp.				+																		
Clathrodictyon striatellum (d'Orbigny)								+												+		
Unidentified stromatoporoids	+	+	+	+	+	+	+	+	+	+	+	+	+	+	+	+	+	+	+	+	+	+
ANTHOZOA TETRACORALLA																						
Acervularia ananas (L.)	+	+									+	+	+									
Acervularia sp.				+	+						+									+		
Cystiphyllum cylindricum Lonsdale																			+			
Diploëpora grayi (Edwards et Haime)					+		+	+	+		+											
Entelophyllum fasciculatum Wedekind																			+			
Hedstroemophyllum articulatum Wedekind																			+			
Lykophyllum hisingeri Wedekind																			+			
Omphyma sp.											+		+						+			
Rhegmaphyllum conulus (Lindström)					+	+					+											
Stauria favosa (L.)											+			+								
ANTHOZOA TABULATA																						
Aulopora sp.				+				+	+		+					+			+			
Favosites asper d'Orbigny		+																				
Favosites gothlandicus Lamarck	+			+		+					+	+		+		+		+	+	+		+
Favosites sp.		+		+	+	+	+	+	+		+		+	+					+	+	+	+
Halysites catenularius (L.)													+			+						
Halysites catenulatus (Martini)		+									+											
Halysites sp.				+	+	+	+	+	+		+			+					+	+	+	+
Planalveolites fougti (Edwards et Haime)														+								
Roemeria sp.				+																+		
Syringopora sp.				+	+	+	+		+		+			+						+		+
ANTHOZOA HELIOLITIDA																						
Heliolites barrandei Penecke								+	+		+								+			
Heliolites interstinctus (L.)				+	+	+	+	+	+		+			+		+	+	+	+	+	+	+
Heliolites parvistella Ferd. Roemer						+	+	+	+	+	+			+					+	+	+	+
Heliolites sp.					+						+		+									
Plasmopora foroensis Lindström						+	+	+	+		+								+		+	
Plasmopora petalliformis (Lonsdale)																			+		+	
Plasmopora rudis Lindström				+			+															
Plasmopora sp.				+																		
Propora speciosa Billings									+													
Thecia swinderniana (Goldfuss)		+		+			+		+		+								+	+		
Unidentified corals	+	+	+	+	+	+	+	+	+	+	+	+	+	+	+	+	+	+	+	+	+	+

TABLE XVIII (continued)

Fossils \ Lithology / Localities	Reef limestone: Tofta	Reef limestone: Västerhejde	Reef limestone: Stenkumla	Reef limestone: Bogeklint	Reef limestone: Solklint	Reef limestone: Kvarnbacken	Reef limestone: Spillingsklint	Reef limestone: Tjeldersholm	Reef limestone: Barabacke	Reef limestone: Simunde	Reef limestone: Other localities	Crinoid limestone: Tofta	Crinoid limestone: Stenkumla	Crinoid limestone: Bogeklint	Crinoid limestone: Solklint	Crinoid limestone: Kvarnbacken	Crinoid limestone: Spillingsklint	Crinoid limestone: Spinnbjersbacke	Crinoid limestone: Tjeldersholm	Crinoid limestone: Bara	Crinoid limestone: Simunde	Crinoid limestone: Other localities
ANNELIDA																						
Conchicolites sp.				+									+									
Cornulites scalariformis Vine											+											
Cornulites serpularius Schlotheim													+									
Cornulites sp.											+					+			+	+		
Spirorbis sp.							+		+		+		+									
Unidentified annelid remains																				+		
CRINOIDEA																						
Barrandeocrinus sceptrum Angelin																						+
Botryocrinus sp.							+															+
Calceocrinus sp.																						+
Cyathocrinus sp.																+						+
Euspirocrinus spiralis Angelin																						+
Gissocrinus sp.									+													+
Herpetocrinus sp.																						+
Pisocrinus sp.																						+
Promelocrinus sp.																						+
Streptocrinus crotalurus (Angelin)																						+
Unidentified crinoid remains	+	+	+	+	+	+	+	+	+	+	+	+	+	+	+	+	+	+	+	+	+	+
BRYOZOA																						
Coenites repens (Wahlenberg)				+															+	+		
Coenites sp.				+										+					+	+		
Fenestella mobergi Hennig		+																				
Fenestella reticulata (Hisinger)				+				+	+		+								+		+	
Fenestella sp.				+							+		+	+					+	+		
Fistulipora sp.				+	+						+											
Ptilodictya lanceolata (Goldfuss)								+			+		+	+				+	+	+	+	+
Unidentified bryozoan remains		+	+	+	+	+	+	+	+	+	+	+	+	+	+	+	+	+	+	+	+	+
BRACHIOPODA																						
Amphistrophia funiculata (McCoy)																		+	+			
Atrypa reticularis (L.)	+	+	+	+	+	+	+	+	+	+	+	+	+	+	+	+	+	+	+	+	+	+
Camarotoechia borealis (Buch)	+	+		+	+	+	+	+	+	+	+		+	+	+	+	+	+	+	+	+	+
Camarotoechia diodonta (Dalman)	+	+		+			+		+	+	+	+		+						+		+
Camarotoechia nucula (J. de C. Sowerby)		+																				
Camarotoechia sp.											+											+
Chilidiopsis pecten (L.)								+					+					+	+	+	+	+
Chonetes sp.																		+	+		+	+
Chonetoidea grayi (Davidson)				+																		
Conchidium sculptum (Walmstedt)																				+		
Conchidium tenuistriatum (Walmstedt)													+									
Cyrtia exporrecta (Wahlenberg)			+					+						+			+		+	+	+	
Delthyris elevata Dalman		+				+					+									+		
Dictyonella sp.							+															
Dolerorthis rustica (J. de C. Sowerby)											+											

TABLE XVIII (continued)

Lithology	Reef limestone											Crinoid limestone										
Fossils / Localities	Tofta	Västerhejde	Stenkumla	Bogeklint	Solklint	Kvarnbacken	Spillingsklint	Tjeldersholm	Barabacke	Simunde	Other localities	Tofta	Stenkumla	Bogeklint	Solklint	Kvarnbacken	Spillingsklint	Spinnbjersbacke	Tjeldersholm	Bara	Simunde	Other localities
BRACHIOPODA (continued)																						
Eospirifer globosus (Salter)											+		+									
Eospirifer grandis (Hedström)							+				+											
Eospirifer interlineatus (Hedström, non J. de C. Sowerby)		+																				
Eospirifer sinuosus (Hedström)			+				+				+											+
Eospirifer sp.				+				+		+	+			+				+	+	+	+	+
Gypidula galeata (Dalman)					+		+											+	+			+
Leptaena rhomboidalis (Wilckens)	+	+	+	+		+	+		+	+	+	+	+		+		+	+	+	+	+	+
Linoporella punctata (De Verneuil)												+										
Meristina obtusa (J. Sowerby)														+					+			+
Orbiculoidea sp.											+											
Pentamerus gotlandicus Lebedev														+					+	+		
Platystrophia biforata (Schlotheim)				+	+	+				+		+	+									
Platystrophia sp.		+									+											
Plectatrypa imbricata (J. de C. Sowerby)								+		+			+	+		+		+	+			+
Plectatrypa lamellosa (Lovén)		+									+					+						
Ptychopleurella bouchardi (Davidson)				+	+	+		+		+	+								+	+	+	+
Resserella elegantula (Dalman)					+	+		+					+	+		+			+	+	+	+
Resserella sp.				+	+				+		+			+	+		+	+	+	+	+	+
Rhipidomella hybrida (J. de C. Sowerby)				+															+			
Rhynchotreta cuneata (Dalman)	+			+	+		+	+	+	+	+	+	+	+	+		+			+		+
Sphaerirhynchia wilsoni (J. Sowerby)				+	+																	
Stropheodonta semiglobosa (Davidson)			+				+			+	+		+	+	+	+	+	+	+	+		+
"*Strophomena*" sp.						+										+		+				
Trimerella lindströmi (Dall)											+											
Trimerella sp.					+		+				+											+
Unidentified brachiopods		+		+	+	+	+	+	+	+	+			+	+				+			+
LAMELLIBRANCHIATA																						
Conocardium sp.				+			+	+			+					+			+		+	+
Cypricardinia sp.		+																				
"*Megalomus*" sp.											+											
Mytilarca acuta Lindström in museo									+													
Pterinea sp.											+											
Unidentified lamellibranchs											+											
PTEROPODA																						
Conularia laevis Lindström									+													
GASTROPODA																						
Bellerophon sp.								+														
Craspedostoma elegantulum Lindström				+			+				+											

TABLE XVIII (continued)

Fossils \ Lithology / Localities	Reef limestone											Crinoid linestone										
	Tofta	Västerhejde	Stenkumla	Bogeklint	Solklint	Kvarnbacken	Spillingsklint	Tjeldersholm	Barabacke	Simunde	Other localities	Tofta	Stenkumla	Bogeklint	Solklint	Kvarnbacken	Spillingsklint	Spinnbjersbacke	Tjeldersholm	Bara	Simunde	Other localities
GASTROPODA (continued)																						
Craspedostoma sp.											+											
Cyclonema sp.											+											
Euomphalopterus alatus (Wahlenberg)											+		+			+				+		+
Hormotoma sp.											+					+			+			
Lophospira bicincta (Hall.)											+											
Murchisonia imbricata Lindström				+		+		+	+													
Oriostoma acutum Lindström											+											
Oriostoma angulatum (Wahlenberg)							+		+		+											+
Platyceras cornutum Hisinger				+	+	+	+				+			+						+		+
Platyceras cyathinum Lindström																						+
Platyceras enorme Lindström											+											
Pleurotomaria limata Lindström											+											
Poleumita discors (J. Sowerby)											+											
Poleumita globosum (Schlotheim)											+											
Poleumita sp.				+							+					+			+			
Trochus incisus Lindström											+											
Trochus sp.											+											
Unidentified gastropods					+	+						+		+	+				+			
TENTACULITIDA																						
Tentaculites multiannulatus Vine																					+	+
Tentaculites sp.											+		+		+	+			+			+
CEPHALOPODA																						
Ascoceras fistula Lindström							+															
Ascoceras lagena Lindström							+															
Ascoceras sp.											+											
Choanoceras mutabile Lindström							+				+											
Ophidioceras reticulatum Angelin							+				+											
Ophidioceras sp.											+											
Orthoceras sp.				+							+							+	+			+
Phragmoceras inflexum Hedström											+											
Phragmoceras sp.											+											
Unidentified cephalopods					+		+				+											
TRILOBITA																						
Arctinurus ornatus (Angelin)								+														
Bumastus sulcatus Lindström											+											
Bumastus sp.		+		+	+		+		+		+											
Calymene tuberculata (Brünn)				+			+				+											
Calymene sp.			+		+			+			+		+	+		+	+	+	+		+	+
Dalmanites imbricatulus (Angelin)																+						
Encrinurus punctatus (Wahlenberg)	+		+	+				+			+		+	+		+	+	+	+		+	+
Proetus verrucosus Lindström								+								+						+
Proetus sp.				+		+					+							+				+
Sphaerexochus sp.			+																			
Unidentified trilobites				+				+			+							+	+			
OSTRACODA																						
Beyrichia sp.											+			+		+		+				+
Craspedobolbina clavata (Kolmodin)				+							+					+		+				+
Leperditia baltica (Hisinger)						+		+		+	+			+	+	+	+	+	+	+	+	+
Unidentified ostracodes								+			+			+	+	+	+	+	+			+

in water depth. An attempt, however, can be made for the depth increase during Slite III, on the basis of the distance over which the northwestern boundary of the area of marl deposition moved and the assumption that the slope of the sea floor amounted to slightly less than half a degree. An increase in water depth in the order of 25 m seems likely then. Estimated in a similar way, the decrease in water depth during Slite IV time would have been about 30 m. The reefs then developed in water deeper than that in which the Slite I and II Beds were laid down, about equally deep as that in which the southeastern Slite III Beds formed and less deep than the water in which marl sedimentated.

If we connect the locations of the northwesternmost outcrops where Slite marlstone is found, a line is obtained which in all likelihood was a depth contour of that time. This line runs from Liksarve (1.7 km north-northeast of Tofta) in the southwest, via Hejdeby and Othem, to Bunge in the northeast, with a strike of about 55°. This might indicate that the coast-line direction at that time was about northeast-eastnortheast–southwest-westsouthwest.

HALLA-MULDE BEDS

Halla limestone

The Halla limestone is best developed in the east of Gotland, where it reaches a thickness of approximately 15 m. There, its lower part consists partly of oolite, partly of dense to finely crystalline limestone. The oolite is, among others, exposed immediately south of Barabacke (west of Bara Ödekyrka), and from there the name Bara oolite is derived. It is a light-grey sediment, weathering brownish to yellowish grey, with only a moderate fossil content. At its base, the limestone is generally finely oolitic. Upwards the oolite structure becomes coarser and there ooids occur of a millimetre or more in diameter. Interbedded are layers which are only finely oolitic or in which the oolite structure is completely lacking. The thickness of the oolite may amount to about 5 m. It is partly overlaid by, partly equivalent to, an other stratified limestone, which generally is dense, sometimes finely crystalline or finely oolitic, moderately to faintly marly and locally rather bituminous.

Towards the southwest, the lower Halla limestone thins out. In the west, it is only about a few decimetres thick. No true Bara oolite occurs there, but mainly the dense to finely crystalline or oolitic limestone variety. From the environment of Klintebys, Munthe (1915b) has described rounded sandstone pebbles and ripple marks, occurring in the lowermost part of the Halla limestone, close above the calcareous sandstone of the Slite Beds.

The upper part of the Halla limestone is built in the east by a stratified, grey limestone, which is dense to sometimes finely crystalline and more or less marly. Sometimes there is an alternation of limestone layers with thin layers of marlstone. The limestone is rich in Algae and stromatoporoids; the latter are predominantly small. The marl content decreases towards the top. This top, in the east of Gotland, is developed as a distinct erosion level, e.g., about 1.5 km east-southeast of Bryggans Fiskläge. This erosion level normally shows a smooth surface, in which fossils are cut off straight, to the level of the surrounding rock, suggesting that the sediment had already undergone a certain consolidation before this crosion took place. The uppermost few millimetres of limestone in these eroded areas are generally more or less strongly oxydized and red or red-brown in colour.

The upper Halla limestone may reach a thickness of about 10 m. This thickness decreases towards the southwest, where the upper Halla is generally developed as dense, occasionally finely crystalline, marly limestone, locally rich in Algae and small stromatoporoids.

Mulde marlstone

The Mulde marlstone, distinguished as an individual stratigraphical unit by Hede (1921, and later, under the name of Mulde märgelsten), is built up by an alternation of marlstone layers with layers of marly limestone. The marlstone is a bluish grey to brownish grey, soft and dense sediment. The limestone is harder, finely crystalline and has a bluish grey colour which weathers brownish grey. In general, the Mulde sediments are very fossiliferous. The name has been derived from a comparatively large and well-known quarry, Mulde Tegelbruk (Fröjel Parish) from which many fossils have been collected and described or listed. This quarry is now abandoned and dilapidated and at present the Mulde sediments are not clearly exposed.

The Mulde marlstone does not contain reefs.

In west Gotland, the Mulde marlstone reaches a maximum thickness of 20–25 m. It thins out towards the northeast, and disappears in Hejde Parish, still in the western half of central Gotland.

Reef limestones and related sediments

The reef of Barabacke, which grew during Late Slite time, is enveloped, especially at its southern side, by the Bara oolite of the lowermost Halla Beds. This gives reason to presume that the reef continued growth during the very beginning of Halla time, for it is less likely that the reef had grown so high over its surroundings that all of the Bara oolite should be regarded as a subsequent deposit.

Also in the northeast of the Halla Beds, some small reefs occur in the Bara oolite and the limestones which are partly synchronous with, partly overlie this oolite. Growth of these reefs soon came to an end and most of the Halla succession in the east is devoid of reef limestone. Thus, the decrease in reef size, notable in the upper part of the Slite IV Beds, goes on in the Halla Beds and in this respect, too, the Halla Beds in the northeast are a normal continuation of the Slite IV Beds.

Further southwest in the Halla Beds, reefs are larger. They are found especially in the area of Väte and Viklau (cf. Hede, 1927a, p.34) but the exposures are poor. East of the road Hejde – Väte – Atlingbo, hardly any reef limestone is exposed in the surroundings of Väte, but on the west side some marly reef limestone with corals and stromatoporoids occurs in outcrops near Rovalds farm in the region known as Kvie Gräne. It is possible that most or all of the hill on which the mill is built, consists of such reef limestone, but there are too few outcrops to establish this with certainty.

If this presumption is correct, the reef should have reached a length of about 150 m. Some reef limestone is found in two other localities between Väte and Kvie, which may represent two similar reefs, but the outcrops here are even scarcer and poorer.

In the surroundings of the northern two Vikare farms, about 2 km north of Viklau, exposures are also few and poor. Generally crinoid limestone with reef debris is found, with occasionally some marly reef limestone. In this area, Hede (1927a) noted a greatest reef-limestone thickness of about 6.5 m.

A list of fossils found in the Halla reef limestone and directly surrounding sediments is included in Table IX (pp.60–67).

Discussion

In the interpretation of the Halla and Mulde Beds, the following points need to be taken into account:

(1) The oolite at the base of the Halla Beds is less well-developed in the southwest than in the northeast. Oolite formation in the southwest also took place during a shorter time.

(2) Erosion levels at the top of the Halla Beds are found only in the northeast.

(3) The Upper Halla Beds in the southwest are marlier than those in the northeast.

(4) Whereas reefs in the northeastern part of the Halla Beds area only occur in the lowest part, and are small, they occur through most of the profile in the southwest and there are also larger, have a higher marl content and a greater contribution from corals.

(5) The Halla Beds decrease in thickness southwestwards, whereas the Mulde Beds increase in thickness in the same direction.

(6) The Lower Klinteberg Beds, which follow over the Mulde Beds, shows in part very marly limestone in the southwest.

(7) The Austerberg and Lilla Karlsö Limestones of Karlsöarna, which are mainly synchronous with the Halla-Mulde Beds of Gotland, show characteristics of deposition in relatively deep water.

Summarizing, there are strong indications that during deposition of the Halla-Mulde Beds, water depth in the northeast of central Gotland decreased, whereas simultaneously water depth in the southwest increased. This may have been caused by a change in the direction of the coast line from about southwest–northeast to rather west–east. This will again have been due to a change in the direction of the hinge line of epeirogenetic movement of the basin floor. In view of the shallowness of the basin and the flatness of the bordering land at its northwest, relatively small movements are already sufficient to account for this. Palaeogeographically, in eastern Gotland the coast line thus moved southwards, in western Gotland northwards. As was generally the case in the Silurian Baltic basin, deeper water allowed the deposition of marl and, therefore, in the southwest of central Gotland marl was laid down (Mulde marlstone, marlstone in the Lower Klinteberg Beds) whereas in the northeast, limestone formation went on (Halla limestone). As a result, there is reason to assume that Halla limestone and Mulde marlstone are contemporaneous deposits. Thus, they should be regarded as belonging to one chronostratigraphical unit, the Halla-Mulde Beds.

KLINTEBERG BEDS

The Klinteberg Beds occur, in central Gotland, in a wide strip which narrows in the west. The name is derived from the Klinteberg, a hillock of up to 52.5 m high, situated in Klinte Parish. This hillock is one of the best exposures of rocks belonging to this unit.

Stratified sediments

In the Klinteberg, it can be seen how limestone belonging to the lowermost Klinteberg Beds conformably overlie the Mulde marlstone. The limestone is marly, predominantly thin bedded, dense to finely crystalline, partly finely oolitic, light grey in colour and when weathered, brownish to yellowish light grey. The rock alternates with reef limestone and detrital crinoid limestone related to it. Of the fossils which are present, *Conchidium conchidium* (L.) needs especially to be mentioned; it is considered by Hede (1925a, p.22) to be quite characteristic of the Klinteberg Beds.

In Hejde Parish, in a somewhat younger part of the Klinteberg Beds, thin bedded, finely to middle crystalline, somewhat marly limestone which is rich in *Conchidium conchidium* (L.), alternates with thin bedded, dense to finely crystalline, rather hard limestone, in which *Spongiostroma holmi* Rothpletz and stromatoporoids are abundant. The latter, *Spongiostroma*-rich rock can be followed from there further northeastwards, in a zone which is about 5 m thick and is exposed, among others, northeast of Guldrupe Church, east-northeast of Viklau, northwest of Sjonhem, about 2.5 km northeast of Sjonhem and northeast of Ganthem. Locally, in between the layers of this limestone, thin layers are interbedded of greenish grey, calcareous marlstone. In addition to *Spongiostroma*, *Rothpletzella* is rather common.

The next zone in the stratigraphical succession, is especially characterized by the presence of the lamellibranch *Ilionia prisca* (Hisinger) and the almost complete absence of the Algae, which were common in the preceding zone. The limestone of the *Ilionia prisca* Zone is thin bedded, predominantly dense, but locally very finely oolitic and sometimes finely crystalline, varying marly and often containing pyrite. The bedding planes are often irregular and rugged. This zone is about 10 m thick. The rock is by far the best developed in the east of the area where the Klinteberg Beds are found. There, it constitutes the solid rock of large parts of the parishes of Sjonhem, Ganthem and Norrlanda. It is rather closely comparable to the limestone of the overlying zone, which is only a few metres thick, and does no longer contain *Ilionia prisca*. The rock of that zone is thin bedded, finely oolitic or finely crystalline and less marly. It is especially to be found in the north of Norrlanda Parish.

The Upper Klinteberg Beds, finally, consist of a succession of generally thin, sometimes thick limestone layers of up to about 40 m thick. The sediment is as a rule dense, but sometimes finely crystalline or finely oolitic; it is usually rather hard. The colour is light grey, on a weathered surface often brownish light grey. Locally scattered small crystals of pyrite are present. Compared to other deposits in Gotland, the fossil content of the limestone is rather moderate; stromatoporoids are relatively the most common.

Adding up, the presumable total thickness of the Klinteberg Beds is approximately 65 m.

Reef limestones and related sediments

Although Klinteberg reefs are best exposed in the west of Gotland, in the Lower Klinteberg Beds, e.g., in the Klinteberg and at Klintebys, they do occur through most of the Klinteberg Beds. Reef limestone is for instance found in limestone rich in *Spongiostroma* in a canal about 3 km west of Bjerges Station (a former railway station between Viklau and Vänge), with the stratified *Spongiostroma* limestone doming over (Hede, 1927a, p.44). It is also found about 3.7 km southwest of Vänge. Elsewhere in this *Spongiostron* limestone the rock is locally rather marly and also usually rich in crinoid fragments and a varying amount of fossil remains which could be reef debris; the bedding planes are irregular and somewhat rugged; all this may be taken as an indication of reef development that will have taken place close to where these deposits were laid down.

In the *Ilionia prisca* limestone, indications of reef development are found in Bjers' Hällar, southeast of Guldrupe, whereas also the crinoid

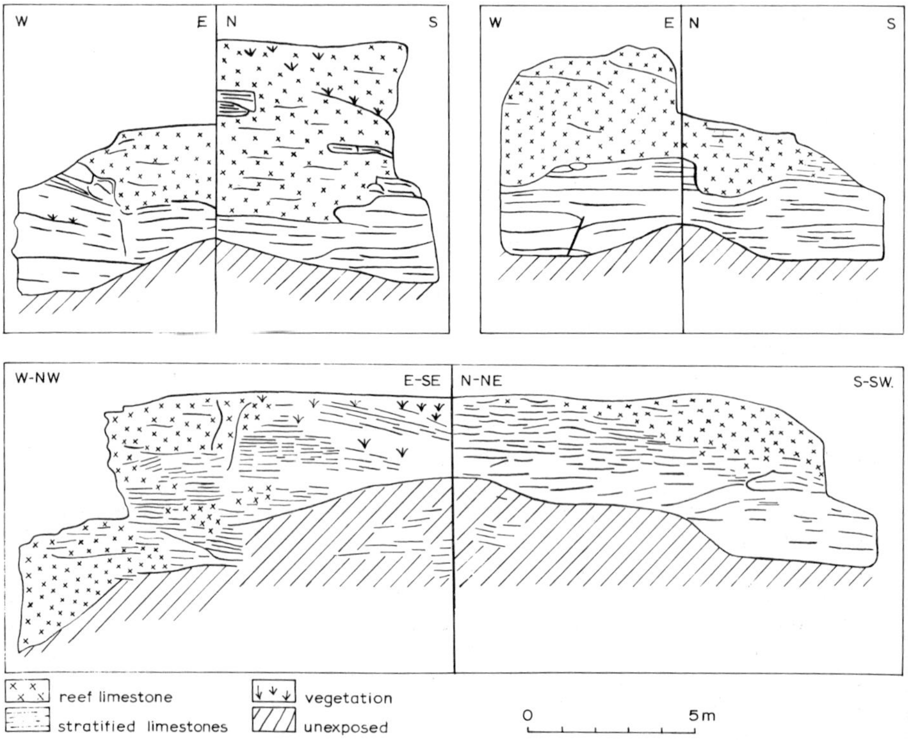

Fig.164. Reef-limestone exposures in the south of the Klinteberg, found close to each other and presenting in the two above sections and the right half of the lower section (which is the northern of the three) reef limestone belonging to the same reef-limestone body. The lower boundary of the reef limestone shows fluctuations, which leads to the presumption that reef development probably began independently in a few close by places, with a fusion taken place through further expansion.

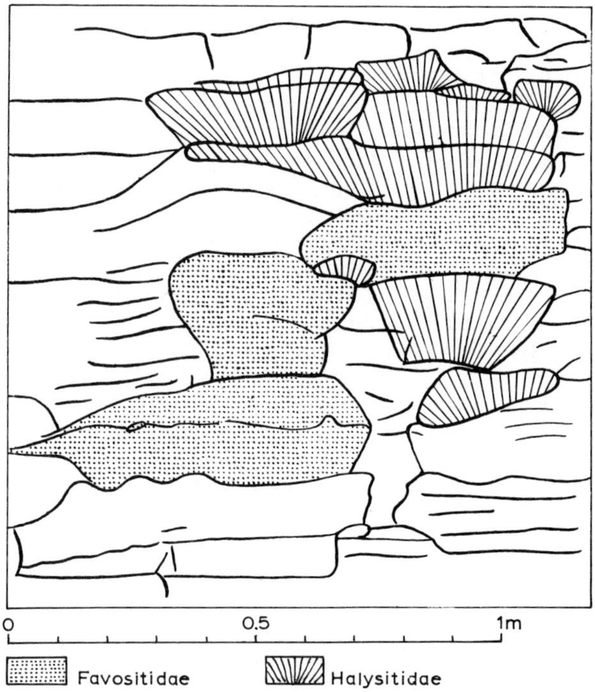

Fig.165. Detail of the exposed wall in the central part of the west wall of the Klinteberg. In the crinoid limestone many coral colonies are found close to each other and generally in their positions of growth, but without building a reef.

limestone which is found at Fjäle in Anga Parish (Hede, 1929, p.21) in close connection with *Ilionia*-containing rock, can be seen as an indication that in that environment reef limestone should be present.

In the Upper Klinteberg Beds, reef limestone is found about 2 km southeast of Guldrupe, east-northeast of Kräklingbo and in the hill about 0.7 km north-northwest of Kräklingbo. In the latter locality, the reef limestone is covered partly by stratified sediments, belonging to the lowermost Hemse Beds, which arch at all sides over it.

A list of fossils found in the Klinteberg reef limestones and directly surrounding sediments is included in Table IX (pp.60–67). The great majority of fossils identified from this stratigraphical unit came from the Klinteberg itself.

The more important exposures of reef limestones and related sediments of the Klinteberg Beds are described briefly in the following pages.

Describing the *Klinteberg* from south to north, one first encounters the reefs described in Fig.40–43. A few reefs which are found north of these consist of similar marly reef limestone with mainly flat reef builders. At their north or northwest side, they are bordered by a very marly limestone which is thin and often irregularly bedded with some marl on the bedding planes. The high fossil content comprises crinoids, bryozoans, brachiopods and also some solitary and social corals. The southern boundary of the reefs is generally steeper than that in the north and northwest; the stratified sediment there is less marly and thicker bedded.

Next follows the reef drawn in Fig.164. Possibly the uneven lower boundary of the reef limestone is caused by the beginning of reef growth in this area simultaneously in a few places; rapid horizontal expansion may soon have led to a fusion into one larger reef. There are several intercalations of stratified crinoid limestone. These, together with the marly matrix and flat reef builders, give the reef limestone, locally, a vaguely stratified appearance. The crinoid limestone at the north-northeast side of the reef is generally thin, occasionally also cross-bedded. The direction of dip of the layers shows small variations from one place to the other, which presumably are also mainly synsedimentary. Part of the rock is rather rich in reef debris.

Only a short distance north of the reef of Fig.164, another one is found, of which the southernmost part can still be seen in the left half of the lower section in Fig.164. This reef is exposed over a distance of about 150 m and to a height of 3–9 m. It is likely that this reef limestone body, which is very

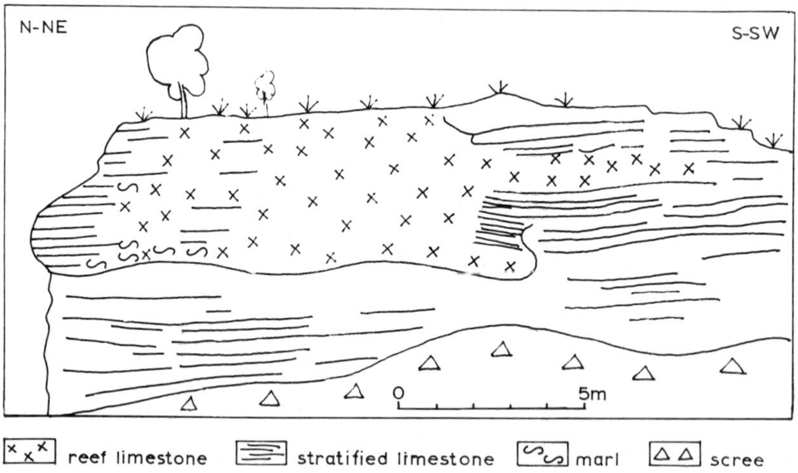

Fig.166. Sketch of a reef-limestone exposure in the central part of the west wall of the Klinteberg, a couple of metres north of the place where the coral colonies of Fig.165 are found. About 7 m north-northeast of the right margin of the drawing the following sedimentary succession is found:

1.1 m stratified brownish grey crinoid limestone
0.9 m unstratified reef limestone with stromatoporoid and *Halysites* colonies in a marly matrix; the weathered rock shows a brecciaceous structure
0.6 m distinctly but irregularly bedded crinoid limestone, grey to brownish grey in colour
0.75 m indistinctly bedded marly limestone with crinoid remains, brownish coloured
0.7 m very vaguely stratified limestone, less marly, rich in very strongly recrystallized stromatoporoids
1.5 m brownish grey crinoid limestone

Fig.167. Klinteberg, stratified sediments.

large for the Klinteberg Beds, also originated through the fusion of a number of neighbouring centres of reef growth. In parts of this large reef, the reef limestone nature of the rock is vague. At its north side, crinoid limestone is exposed again.

About 50 m north of the large reef within the irregularly bedded crinoid limestone, there is a part which is very rich in coral colonies, almost all of which seem to be in their growth orientations. A detail of this is shown in Fig.165. North of this locality is the reef of Fig.166. The reef limestone reaches its lowest position in the wall near its southwestern side, suggesting that reef expansion during development took place northeastwards, whereas a southwestward expansion occurred only during one stage of the life time of the reef. The layers of the crinoid limestone southwest of the reef decrease upwards in thickness.

In a northwest – southeast orientated wall north of the above mentioned reef, the crinoid limestone is again overlaid by vaguely stratified reef limestone over a distance of 7 m.

The next 250 m are badly exposed, except for an old quarry in stratified limestone. Then there are about 40 m of crinoid limestone, thick bedded at the base, upwards thinner bedded and in that direction containing more marl and less crinoid remains (Fig.167).

The sections of Fig.168 and 169 are about perpendicular, but do not connect,

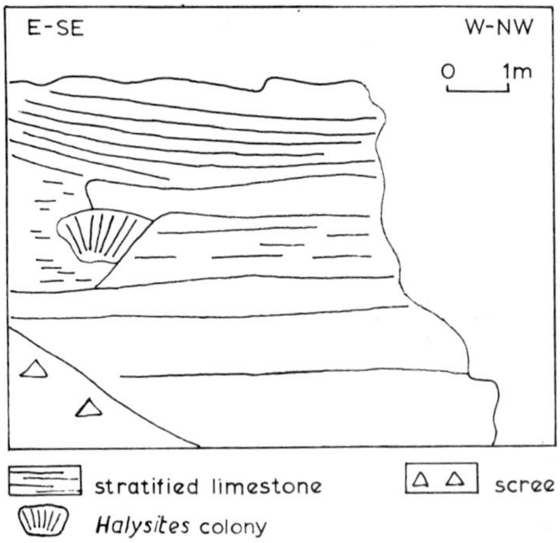

Fig.168. Sketch of an exposure in the north of the east wall of the Klinteberg, showing in the west a succession of thick and thinner bedded limestone, eastward abuting against or arching over a body of indistinctly bedded, very marly limestone with many corals, especially *Halysites*, and flat-lenticular stromatoporoids. This body appears to form part of a reef, which is found to be better developed in the exposure shown in Fig.169.

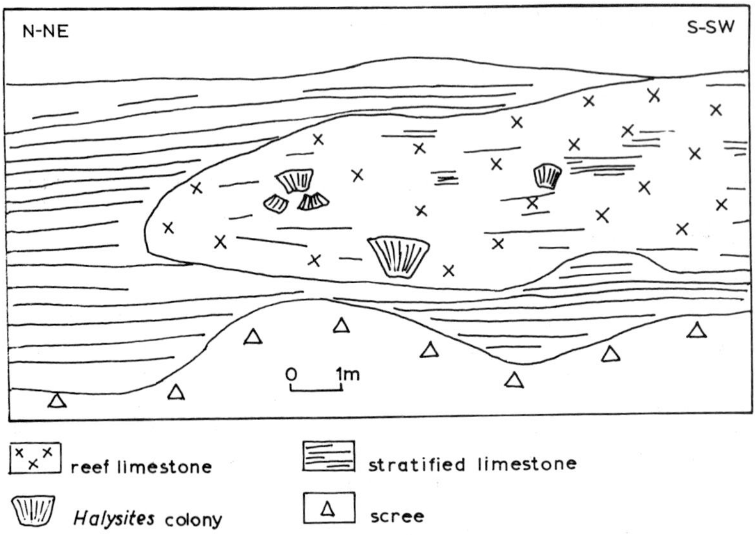

Fig.169. Exposure of very marly and vaguely stratified reef limestone in the Klinteberg, shortly north and about perpendicular to the exposure of Fig.168.

since in the angle between the two there is a high cone of rubble. The reef which they show was certainly not an example of vigorous growth. For the part drawn in Fig.168 it is even difficult to recognize it as belonging to a reef. Fig.169 shows how the stratification in the surrounding sediments abuts against the reef, whose vague stratification does not correspond with it. In the reef, several thin layers of marl occur and also areas, up to 0.5 m^2 in cross-section, of crinoid limestone. The matrix is marl, reef builders are corals and stromatoporoids, the latter generally flat-lenticular in shape and in a smaller number that is usual for Hoburgen-type reefs.

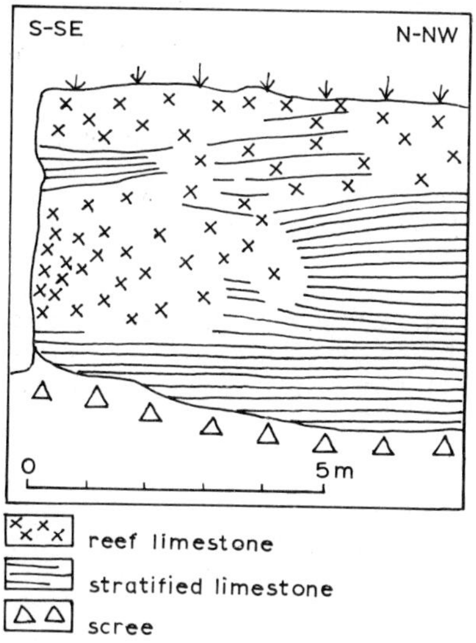

Fig.170. Section in an old quarry in the north of the east Klinteberg. Two reefs, in the northwest only vaguely separated by a vaguely bedded transition form of reef limestone and crinoid limestone.

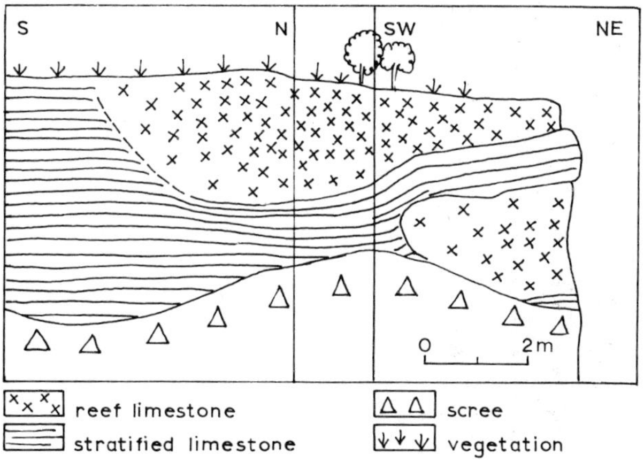

Fig.171. The same two reefs as drawn in Fig.170. They are well-separated here. The upper reef shows a strong expansion northwards.

Coral colonies, dominantly *Halysites* sp., can be large, up to 1.3 m with a height of 0.9 m, and seem to be partly in their orientations of growth. Underneath, above and northeast of the reef, marly limestone occurs with marl films on its rugged bedding planes; the beds are generally 2–10 cm thick. Northeastwards, this rock is replaced, starting both at the base and top of the wall, by thick and smoothly bedded marly limestone in which small corals, brachiopods and crinoid fragments are found, often surrounded by an algal crust. This deposit can be followed northwards for about 80 m, with a decrease in Algae in that direction.

The exposure in the north of the Klinteberg has already been dealt with in Chapter VII, Fig.39.

On the east side of the Klinteberg is a quarry of a former lime kiln. In this the sections of Fig.170 and 171 are drawn. Two reefs are found above each other, separated over most of their extension by some layers of crinoid limestone, which is also exposed underneath and lateral to the reefs. In the north, the boundary between the two reefs fades. The crinoid limestone shows cross-bedding (Fig.172).

Just south of the wall of Fig.172, the section of Fig.173 is found, which illustrates, once again, how small the difference between stratified sediments and reef limestone in the Klinteberg Beds may be.

The *road Klinte – Hemse* crosses the extension of the Klinteberg close to its beginning in Klinte. On the east side of the road, a wall about 75 m long and with a maximum height of 6 m is found, mainly exposing very marly reef

Fig.172. Crinoid limestone in an old quarry in the northern part of the east Klinteberg, showing cross-bedding.

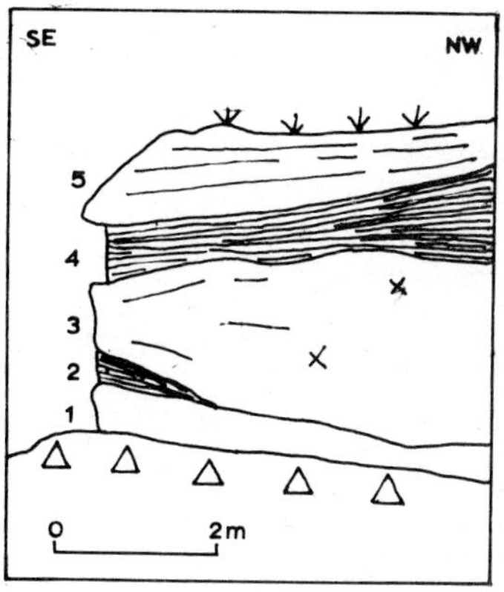

Fig.173. Sketch of the succession of sediments only a short distance south of Fig.172.

1 = thick-bedded crinoid limestone, dipping slightly northwestwards;

2 = thin-bedded marly limestone with interbedded very thin layers of marlstone or clayish marlstone; the whole thinning out northwestwards;

3 = in the southeast crinoid limestone which passes northwestwards, especially in its lower and upper part, into a reef-like limestone with coral colonies (*Halysites* sp., *Acervularia ananas*), some stromatoporoids, bryozoans, brachiopods and some marl lenses with crinoid fragments; this zone strongly increases in thickness northwestwards;

4 = rather thin bedded crinoid limestone with thin marlstone layers interbedded; together arching over 3;

5 = thick bedded, somewhat brownish grey limestone with crinoid remains.

---

limestone rich in flat-lenticular stromatoporoids, but also containing corals, brachiopods, crinoid remains and orthoceratids. All fossils are strongly recrystallized. The colour of the rock is light brownish grey to bluish grey. There are a few intercalations of crinoid limestone in the reef limestone. The whole reef is traversed by several calcite veins and also nests of calcite crystals are common. The reef is overlaid by marly crinoid limestone, with a rather irregularly wavy boundary between the two. The crinoid rock is brownish grey, irregularly bedded and very rich in reef debris.

The *hill at Klintebys* represents an extension of the Klinteberg. Outcrops of stratified sediments and reef limestone occur there, of the same character as in the Klinteberg itself, but the exposures are not so good. The stratified deposits are predominantly crinoid limestone or even crinoid breccia, generally containing reef debris and are often cross-bedded (Fig.174). The sediments exposed in Klintebys quarry, on the west side of the hill, are shown in Fig.175.

(Text continues on p.348)

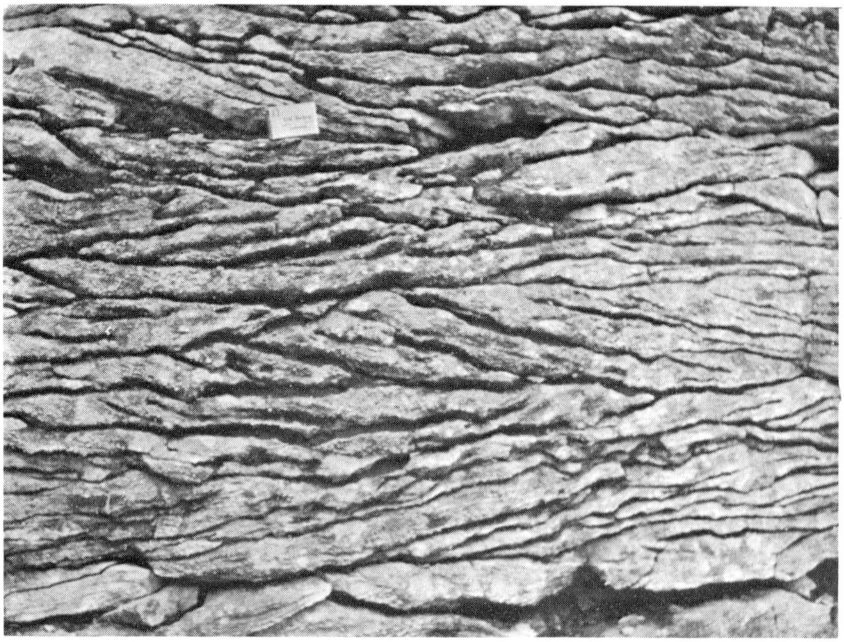

Fig.174.

Fig.175.

Fig.176. Fröjelklint, Klinteberg Beds. Two large stromatoporoid colonies, the one overlying the other, with on top of these still a third, but smaller colony.

Fig.174. Cross-stratified crinoid limestone with reef debris, Klinteberg Beds, Klintebys.

Fig.175. Reef in the Klintebys quarry, Klinteberg Beds. Note how the boundary between reef limestone and crinoid limestone at the north side is much steeper than at the south side. The amount of reef debris in the crinoid limestone is distinctly higher at the north side. Attached to the reef debris several roots of crinoids were found. The crinoid limestone, in places rather a crinoid breccia, is locally cross-bedded. In the reef limestone stromatoporoids and corals are the main organogenic constituents.

Prästklint, 0.8 km north of Fröjel Church, is only exposed over its uppermost few metres. At about the highest part of the exposed wall, the following sediments are found:

1.70 m + Reef limestone of Hoburgen type.
1.65 m Dense to finely crystalline stromatoporoid limestone, marly, light brownish grey, in layers of generally 1–6 cm thick and with very irregularly rugged bedding planes.
0.30 m (average thickness) Marly limestone very rich in small crinoid fragments, brownish grey, bedding planes less rugged than in the stromatoporoid limestone.
1.10 m + Stromatoporoid limestone, as above.

The upper stromatoporoid limestone is thicker locally, reaching a maximum of 2.05 m, and is then covered by grey to brownish grey crinoid limestone with a varying content of reef debris and in layers of generally 5–15 cm thick. In the environment of reef limestone, this crinoid limestone shows faint dips.

The *Fröjelklint* shows predominantly thinly bedded marly crinoid limestone, which in several places is so rich in crinoid fragments that it should rather be called a crinoid breccia or crinoid coquina. The bedding planes are irregularly rugged. The deposit often contains some debris, increasing in amount and size towards the reefs, making the bedding planes even more uneven. Locally some cross-bedding is found. Here and there a stromatoporoid or coral colony is found embedded in the crinoid limestone, probably in the position where it grew. Some are large. One of the stromatoporoids observed, measured 80 cm long and 50 cm high. A few such large colonies together may form a reef-like development (Fig.176) with the stratified sediment layers sagging under it, abuting against it or arching over it, just as with a normal reef. Close to the true reefs the sediment layers may show dips of 1–10°. In one place the stratified limestone underneath a reef was very rich in tabular stromatoporoids.

The reefs exposed in the Fröjelklint are generally small, and nowhere are they thicker than 4 m. They are rather unorganized and in some cases partly vaguely stratified. Small, flat stromatoporoids are in the majority, but larger, rounded colonies are also common. The matrix is marly.

Bjers' Hällar is a low hill, with at several places at its edge and in some places on the plateau, small exposures of one to a few square metres. Most of these show stratified limestone, generally dense to finely crystalline, thin, but somewhat irregularly bedded, light brownish grey, rich in crinoid remains and locally also with a high volume of reef debris. Some marly stromatoporoid reef limestone also occurs. This rock presumably constitutes the nucleus of the hill.

A topographic elevation *about 2 km southwest of Hejde Church* possesses in the north a small quarry in stratified crinoid limestone with some reef debris, some parts of which show signs of rounding. Similar sediment is also found in the wall, a few metres high, which bounds the hill on its north side. The crinoid limestone there is partly marlier, more irregularly bedded and often with more and coarser reef debris. Some marly stromatoporoid limestone is exposed, it is light brownish to bluish grey in colour, brecciates rather irregularly upon weathering, is unorganized in composition and is laterally bounded by crinoid limestone rich in reef debris. Similar rock is also found in pockets between the reef builders.

Tyrvalds Bakke, 2.7–4.4 km northeast of Klinte Church exposes in an old quarry and at its north and northwest sides in a 2–5 m high wall, over quite a distance, crinoid limestone with reef debris, the latter increasing upwards in amount and size. The sediment is marly, generally brownish grey in colour but occasionally bluish grey to white grey. Locally there is some cross-bedding.

Discussion

If one ignores the small area with marlstone in the very southwest of the territory where the Klinteberg Beds are found, then there are no differences of any great importance between the stratified sediments which together build this stratigraphical unit. In almost the entire succession Algae are common, the only exception being the *Ilionia prisca* Zone. Stromatoporoids are common throughout. Cross-bedding is frequently observed. Around reefs there is much reef debris, part of which is rounded. All data available suggest that during Klinteberg time, no important variations in sea depth occurred. The entire limestone succession seems to have been laid down at slight depth, presumably in somewhat shallower water than that in which most of the other limestone complexes of Gotland were deposited. This opinion was also expressed by Hadding (1941, p.71). Also the generally small and thin reefs suggest formation in relatively very shallow water. The marlstone in the extreme southwest is believed to belong to the Upper Klinteberg Beds and forms a natural transition to the overlying Hemse marlstone. Its deposition may reflect the beginning of a new period with somewhat stronger epeirogenetic movements; this will be discussed further when dealing with the Hemse Beds.

HEMSE BEDS

Stratified sediments

The Hemse Beds, which like several others of the stratigraphical units in Gotland, owe their name to the parish in which they are most extensively exposed, appear either at the surface or underneath a cover of Quaternary sediments in quite a substantial part of southern Gotland. The petrological and palaeontological characteristics of the deposits belonging to this unit vary considerably. As appears from the maps (cf. Fig.11), in the northeast mainly limestones are found, whereas in the south and west of the area of the Hemse Beds, marlstone is present at the surface. The limestone area is wide in the east, but gradually narrows westwards, to disappear completely in western Gotland. The bipartition is also apparent in the topography. The limestone area lies higher (on the average 50 m above sea level) and shows hillocks and klintar. The marlstone area generally presents a flat or very faintly undulating surface with only a slight (less than 15 m) height above sea level.

It is difficult to draw a sharp geographical boundary between the two areas. This is partly due to the often thick cover of Quaternary material in the boundary area, which prevents detailed mapping. But in addition to this practical difficulty, there is the penetration of both main rock types in each other's territories and the existence of gradual transitions. Roughly the line between the marlstone and limestone areas runs as follows: Pejnarve in Levede Parish – south of Lindeklint – Allmungs in Stånga Parish – directly south of Rotarve in Lye Parish – Bönde in Lau Parish – Ekmyr – the northern end of Lauviken.

Limestones

In order to have some impression of the sedimentary succession in which the Hemse reefs are intercalated, a survey of the stratified limestones of the Hemse Beds will be given in some detail.

In the northernmost part of the area where Hemse sediments are found in east Gotland, the oldest deposit present is a limestone characterized by the lamellibranch *"Megalomus" gotlandicus* Lindström. The zone reaches a thickness of only a few metres. The rock is a finely crystalline, partly finely oolitic, pure limestone, light grey to brownish light grey in colour, sometimes almost white. The fossil content consists of stromatoporoids, crinoids, bryozoans, corals, brachiopods and some cephalopods and ostracodes, as well as the alga *Solenopora.* The index fossil for the zone is only locally common; elsewhere, especially in the parts rich in stromatoporoids, it is only scarcely represented. Some of the stromatoporoid-rich parts are reef-like developed or indistinctly stratified; elsewhere the rock shows distinct layers of 2–15 cm thick.

In the area of Kräklingbo – Östergarn, this limestone is overlaid by a generally dense, occasionally very finely crystalline, more or less marly limestone, which is thinly stratified and light grey to faint-brownish grey or bluish light grey in colour. Often it is found alternating with thin layers of dense, bluish grey marlstone. The deposit is very fossiliferous; brachiopods and ostracodes are most strongly represented, not only in the number of different species, but also, and especially, in the number of individuals. The total thickness of this zone is not more than 1–1.5 m. It is poorly exposed.

In the section at the beach, about northeast of the bay at Djupviks Fiskläge (east of L. Hammars, Kräklingbo Parish), described by Hede (1929), this limestone, which is strongly marly there, is covered by 30 cm of dense and hard marly limestone, in which remains of *Eurypterus fischeri* Eichwald (of the extinct group of the Gigantostraca) have been found. The rock is thin bedded (1–2 cm), light grey and locally rich in ostracodes. No other exposures of similar nature and age are known.

The limestone found at the top of the above section is thinly stratified, grey to faint-brownish grey in colour, poor in fossils, almost dense to finely crystalline and marly. It was most likely deposited synchronously with the *Ilionia prisca*-containing limestone which in the environment as well as elsewhere in the Kräklingbo – Östergarn area directly overlies the very fossiliferous marly limestone, described above.

In the area of Ala Parish *Ilionia prisca* (Hisinger) was apparently present already earlier. This lamellibranch is found there stratigraphically for the first time, and in a number of exposures, in marly limestone, which directly overlies the *"Megalomus" gotlandicus* limestone. Moreover, this limestone is both lithologically and palaeontologically closely similar to the fossiliferous marly limestone of the Kräklingbo – Östergarn area. The *"Megalomus" gotlandicus* limestone in the Ala area occurs in layers of 2–10 cm thick, is mainly finely oolitic, locally, however, finely crystalline, and very fossiliferous. At one place it shows a weak anticlinal structure, most likely connected with an occurrence of reef limestone; this is also suggested by the fact that the rock there is rich in crinoids. The covering limestone, which thus is the oldest in which *Ilionia prisca* is present in the Hemse Beds, is rather marly, dense to partly very finely crystalline, bluish to brownish light grey in colour, very fossiliferous, and has layers of about 2–10 cm thick.

The *Ilionia prisca* limestone of the Kräklingbo – Östergarn area and elsewhere, which covers in the Ala area the above-described deposit, is generally thin bedded (on the average 1–2 cm). The rock is more or less marly, dominantly dense to almost dense, partly finely crystalline and sometimes finely oolitic. The colour is generally light grey or brownish light grey to greyish brown, that of a fresh surface bluish light grey. In some instances the limestone is reef-like and indistinctly stratified. Locally it contains some pyrite. The rock is very fossiliferous and in its higher parts locally bituminous. Exposures of this deposit are found, among others, along the beach between Grogarnshuvud and Herrviken. The thickness of the deposit is up to about 15 m. Upwards *Ilionia prisca* disappears from the limestone.

Apart from the Hemse Beds, *Ilionia prisca* is also known in Gotland, from parts of the Slite, Klinteberg and Hamra Beds. Thus, it is not a real index fossil. Its restricted occurrence in the profile of the Hemse Beds may, therefore, be caused by a preference for an environment fairly limited in range. Where the required conditions occurred, the lamellibranch would invade the area and spread over the sea floor during the time that the environment remained within the limits required. When at a certain moment, the conditions were no longer as ideal, the animal ceased living there and either became locally extinct or migrated to a more favourable area. The behaviour of *Ilionia prisca* is one of the appealing problems for palaeoecological study which the Silurian fauna of Gotland presents, and there are many others.

After the disappearance of *Ilionia prisca*, the lithological character of the limestone remains about the same. Among others, exposures are found in the lower part of the north of the Grogarnsberg, the lowermost parts of Gannberget, Torsburgen, Kaupungsklint and Petsarveklint, and in several places in Ardre Parish.

The upper part of the Hemse limestone succession is an alternation of more or less distinctly stratified limestones and reef limestones. This part reaches a thickness of up to 30 m. In connection with the general occurrence of reefs, the character of the stratified sediments varies greatly in both horizontal and vertical direction. There is a great number of exposures. The youngest sediment is presumably the Millklint limestone of, among other places, Millklint and Torsburgen: a finely crystalline, partly finely oolitic, stratified marly limestone (Hede, 1929).

As in the Slite Beds, the total thickness of the limestone succession is distinctly less than that of the marl succession in the same unit.

Marlstone

The Hemse marlstone, found in the south and west, is petrologically a very uniform sediment. Generally it is grey to bluish grey in colour, when weathered often brownish grey; it is soft, dense and usually more or less thinly stratified, and often foliated. Thin bands or lenses of dense to finely crystalline, grey, marly limestone are often interbedded.

Hede (1927b, p.24) has pointed out that the macrofossil content of the Hemse marlstone differs in the lower and upper part of the deposit. At the surface, the boundary between the two follows approximately the line Kvinngårde in Havdhem Parish – Stora Vasstäde in Hablingbo Parish – Mullvalds in Hemse Parish – 1.5 km southeast of Stånga Church.

In his study of the ostracodes of the family Beyrichiidae, Martinsson (1962a, pp.53–54, 60–62) found a similar faunistic boundary, which runs consistently very slightly to the east of that based on the macrofauna. He leaves the question open as to whether the boundary is a time or facies one. Martinsson also found that the picture of the Hemse marlstone is still more complicated. There is an important overlap of several faunistic elements, and the faunal successions show some compositional differences between the northern and the southern area of the northeastern facies or time unit.

The uppermost part of the Hemse marlstone is a zone of less than 10 cm thickness, known as the *Dayia* flags. The rock is a generally very fossiliferous hard, grey and dense limestone. Characteristic and very common is the brachiopod *Dayia navicula* (J. de C. Sowerby); also common is *"Strophomena" impressa* Munthe. The zone is easily recognizable and occurs over quite an extensive area, thus constituting a good marker zone.

The total thickness of the Hemse marlstone may be estimated to be about 100 m.

Reef limestones and related sediments

The Hemse Beds contain reef limestones of both the Hoburgen and the Holmhällar types. The first are by far the most common. Most exposures of the Hoburgen reef type are found inland. The ancient cliffs are formed partly by the *Ancylus* lake (e.g., Gannberg, Klinteklint, Kaupungsklint), partly by the *Littorina* sea (e.g., Grogarnsberg, Guffrideklint). An occasional outcrop is found along the present coast (Herrvik).

All exposures of Holmhällar-type reef limestones are situated around sea level along the east coast.

Hoburgen-type reef limestones

The best exposures of Hoburgen-type reef limestones and its surrounding sediments in the Hemse Beds are found in Östergarn Parish. Of these, the *Gannberg* should be mentioned first.

Directly south of the Ganne farm, in the north of the Gannberg, reef limestone of a very unorganized and marly nature is exposed, so that from the first impression one might even question whether this represents a fossil reef. However, there are reasons to consider it reef limestone. These reasons are: a rather high content of reef-building colonies, the almost entire lack of stratification, the sagging of the layers underlying it, the abutting of the lateral deposits against it and the occurrence around it of stratified limestone, very rich in crinoid remains.

Typical of the exposure is that the reef limestone occurs in very extensive, but thin patches, often one above the other (Fig.177). They may then be separated by some stratified limestone with reef debris, especially abundant in its lower part; sometimes by not much more than a single layer of such debris. The covering stratified limestone may, moreover, contain many tabular stromatoporoids of itself. The boundary between reef and overlying rock, therefore, is not always very distinct. It is notable that the surface of such a reef patch generally is flatly horizontal over long distances.

The main organic component of the reef limestone is formed by strongly recrystallized stromatoporoid colonies. Flat forms are most

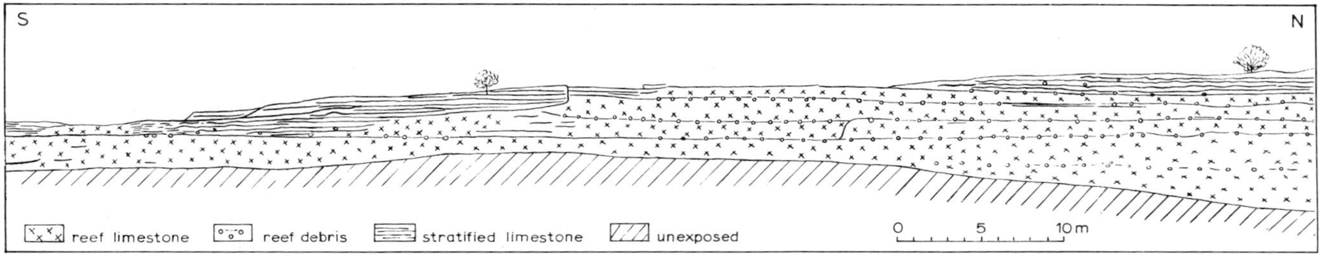

Fig.177. Gannberg, Östergarn. Thin patches of reef limestone occur on top of each other, separated by a thin zone of stratified limestone very rich in reef debris. Hemse Beds.

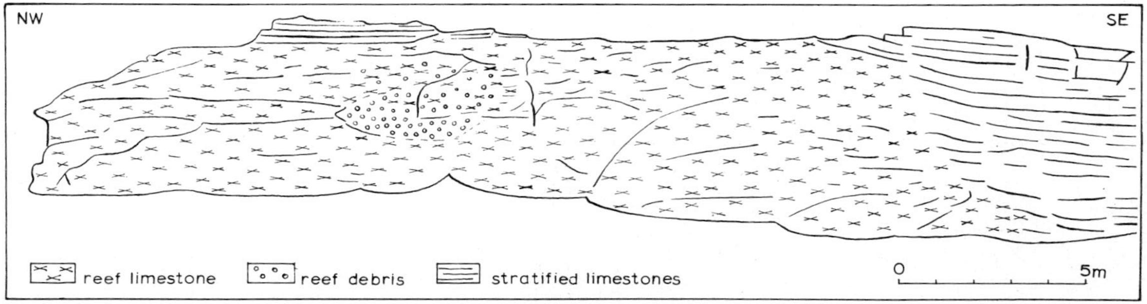

Fig.178. One of the walls of the large quarry in the Gannberg, Östergarn. Hemse Beds. A reef mainly composed of tabular stromatoporoids contains also a depression filled with, among others, the remains of many branched reef builders.

Fig.179. Gannberg, about 0.5 km southwest of the western Ganne farm; reef debris and stratified limestone. Hemse Beds.

common but rounder colonies also occur, locally up to 0.5 m or more in diameter. Such large round stromatoporoids are often found locally in several colonies together. As a general rule, the rounder colonies are least represented in the lower parts of a reef, most common in the central parts but also present in the higher parts. Tabular stromatoporoids, occurring throughout the reefs, but most common in the lower part, may show a wavy surface. Coral colonies are also common. The matrix of the reef limestone is marly. In it specially brachiopods are found; and crinoid remains often occur assembled in pockets. Portions of intercalated marl of stratified limestone occur, varying in size from only a few square centimetres to over 0.5 m The colour of the rock is grey to brownish or bluish grey, depending on the marl content. Locally it is red to reddish brown, generally because the outer few millimetres of the fossils show this colour. Further inside, these same fossils are brownish grey to grey. This red colour should not be confused

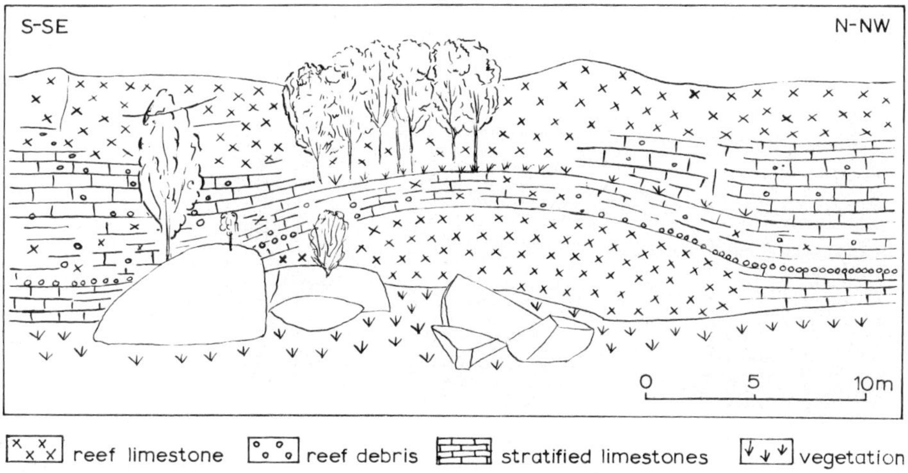

Fig.180. Part of the wall of Kyrkberget, Hemse Beds, showing reef limestone, covered by stratified limestone containing reef debris, with at the top reef limestone of a younger reef. From the older reef at both sides a zone of reef debris extends into the stratified limestone.

---

with a red weathering colour, locally found, which usually comes off and is restricted to the surface of the rock.

The lateral boundary with crinoid limestone is distinct. Where locally also larger fossils occur, the stratification in this sediment is less well developed than usual. Where the crinoid content is moderate, a marly matrix of the rock is well recognizable and bedding planes appear often to be rugged, without any directly apparent reason. The larger the amount of crinoid fragments in the rock, the stronger this is recrystallized. The crinoid remains are relatively strongly brecciated.

In the middle of the north side of the Gannberg, only a short distance east of the cross-roads, there is a large quarry. If one approaches this quarry from the east, along the wall parallel to the road, an important reef limestone exposure is found where the wall turns sharply into the quarry (Fig.178). This reef consists substantially of more or less wavy tabular stromatoporoids, and some lenticular colonies of the same, with in between these, colonies of branched and massive corals which are not so flat. The matrix is strongly marly. In it several kinds of fossils are found, such as crinoid remains (often in nests), brachiopods and bryozoan fragments.

In the quarry a filled depression in between two parts of the reef can be seen. This depression is not observable in the wall at the side of the road, although it is approximately parallel to the exposure in the quarry. This illustrates the local nature of the depression. The filling material consists, mainly, of colonies of branched corals and bryozoans, varying in size between a few centimetres and about half a metre. There are also many large and small pockets with rather unsorted crinoid remains, quite a number of fragments of tabular stromatoporoids but few intact colonies of these, solitary corals and brachiopods; the whole is embedded in a strongly marly matrix. Most of the colonies of branched reef builders, found in the depression,

Fig.181. Gannberg, Östergarn, Hemse Beds. Photograph taken about 0.5 km from the south-southeastern end of the wall. Three banks of reef limestone in vertical succession, but separated by stratified sediment.

Fig.182. Gannberg, about 0.4 km from the south-southeast end. Reef limestone with intercalated horizontal bands of stratified marly limestone. Hemse Beds.

are not in their positions of growth, but some, especially of the larger ones presumably are. This suggests that the majority of them were washed off the surrounding reef surface into the depression, where the colonies remained intact, whereas some of them apparently also continued to grow there. At the reef surface itself, water movement must have been strong to have been able to break and wash away even large colonies. Strong water movement may also explain the predominance of tabular colonies among the stromatoporoids, which could of course better resist wave action than more rounded ones. The marly matrix of the depression filling indicates deposition of terrestrial material over the reef and the local updomings in the tabular stromatoporoids can be regarded as a reaction to this; by rising somewhat over the surrounding reef surface, the mud is more easily washed off from these updomed parts which thus could keep alive.

The boundary between reef and surrounding stratified crinoid limestone is rather distinct. The flat top of the reef is overlaid by stratified limestone.

Elsewhere in the quarry crinoid limestone is exposed, locally containing several solitary corals, brachiopods and small colonies of stromatoporoids and branched and massive corals. There is occasional cross-bedding. In one place, size sorting of crinoid fragments in layers of approximately 1 cm thickness was observed. There are many well-developed stylolites.

Starting from the cross-roads (Östergarn – Katthammarsvik – Kräklingbo – Gammelgarn) southwestwards, one reaches an interesting part of the Gannberg wall some tens of metres southwest of the first farm on the southeast side of the road to Gammelgarn. Low in this wall, there is reef limestone, which locally shows some vague banking. Its matrix is very strongly marly. The top of the reef limestone is uneven, with in the middle a local continuation of reef development, at the southwest side interfingering with stratified limestone, at the northeast side with a more gradually sloping boundary. The stratified sediment overlying the main reef and enveloping the small reef outgrowth is very rich in tabular stromatoporoids, so that the difference between reef and non-reef rock is not great.

About 0.5 km southwest of the above farm very unorganized reef-like sediment is exposed over a distance of some tens of metres. It shows some stratification. Layers of limestone which are up to 10 cm thick and poor in fossils are intercalated. It is most probably debris, deposited very close to a reef (Fig.179). The transition to normal stratified limestone is gradual, through a decrease in reef material and an increase in limestone layers, both upwards and sidewards. At the south-southwest side the transition is more gradual than at the north-northeast side.

Further southwest the wall shows several reefs which need not be described in detail. It is, however, interesting to note that reef limestone of two different appearances occurs. In the first place, there is the more common Hoburgen type with stromatoporoids of varying forms, at the base generally lenticular, higher up also more semispherical and spherical colonies; after weathering, this reef limestone shows generally a brecciaceous or conglomeratic structure. The other kind of reef limestone consists almost entirely of tabular stromatoporoid colonies and, therefore, gives a more massive impression; it weathers generally as rather smooth walls. The present author calls this the Gannberg variety of the Hoburgen reef type. The normal Hoburgen type is usually exposed in the lower part of the northwest and west wall of the Gannberg, the Gannberg variety higher in these same sections.

The eastern part of the Gannberg is better known as *Kyrkberget.* After

Fig.184

Fig.183

what has been said about the Gannberg proper, it is not necessary to describe the exposed wall in detail.

In about the middle of the east wall some zones are found in the stratified limestone, which are rich in reef material (cf. Fig.180). They can be up to about 75 cm thick and originate in the reef limestone, at about the boundary between this and the overlying stratified limestone. They may extend into the stratified limestone to a few tens of metres from the exposed lateral boundary of the reef limestone, although gradually thinning out. These debris zones have a marly matrix and after weathering appear as conglomeratic interbeddings in the more massive stratified limestone.

Over a flatly truncated reef surface, another flat reef limestone occurrence may follow, which in its turn may even be again the fundament for a third reef zone, e.g., about 0.5 km before the south-southeastern end of the east wall (Fig.181). The top of the section is then usually formed by stratified limestone. Such reef limestone zones, with an almost horizontal and extensive basis and a flatly cut top surface approach in their appearance the biostromes.

About 0.4 km before the southeastern end of the east wall there is a good exposure of reef limestone, which originated through the fall of a huge block. The reef limestone is marly and brownish yellow to light grey in colour. Over almost the entire length of the exposure, three thin (2–10 cm) zones of stratified marly limestone can be observed in the reef limestone (Fig.182). These zones are relatively poor in fossils. The variety in forms of the stromatoporoids in the reef limestone is very great. Small oblique parts within the reef limestone indicate that internal displacements have taken place in the reef.

Another interesting range of exposures is found in the wall surrounding the *Grogarnsberg* in the northeast of Östergarn Parish.

The west wall exposes, in its southern part, stratified limestone with reef debris. The latter increases in abundance northwards until reef limestone of Hoburgen type crops out over a distance of about 10 m. Within the reef limestone, intercalations of stratified limestone, 2–15 cm thick and often over 1 m long are found.

Over several hundreds of metres northwards, stratified limestone is found. Reef limestone occurs about 0.3 km north of a fenced-off military area, over a height of about 6 m and a length of approximately 30 m. Slightly north of its northern boundary with the surrounding stratified limestone, the wall retreats over a few metres, thus again bringing reef limestone to the surface. This rock is an unorganized accumulation of some larger stromatoporoids, many smaller lenticular colonies of these, tabular stromatoporoids and many fragments thereof, large and small coral colonies, some solitary corals, crinoid remains, brachiopods, ostracodes; this all embedded in a very marly and often thinly parting matrix which constitutes a quite

---

Fig.183. North side of the Grogarnsberg. Hemse Beds. At the base relatively soft marlstone with lenses of marly limestone. This deposit is overlaid by reef debris. On top of this reef limestone, interrupted by another occurrence of reef debris.

Fig.184. Section in the north of the Grogarnsberg, Hemse Beds. Three stages of reef-limestone formation, interrupted by a somewhat wavy plane, eroded out as a cleft and by a zone of debris material. The such exposed reef-limestone banks are biostromal in appearance but should still be regarded as parts of reefs (bioherms).

Fig.185

Fig.186

considerable portion of the total rock volume (often 40% or more). Locally there are some small intercalations of stratified limestone. Some of the fossil fragments show indications of rounding. The exposure is likely at the northwestern (landward) periphery of a reef. Directly north of it, over some metres, irregularly stratified limestone very rich in reef builders, followed and also overlaid by more normal stratified limestone, is found.

Only a short distance northwards, reef limestone is again found. In the exposed cross-section its southern contact with the stratified limestone lateral to it is somewhat remarkable. Low in the wall, over a distance of several metres, the reef limestone is seen to retreat under the overlying stratified sediment. It may reflect a local stage of decreasing reef growth. Then the reef limestone expands in its turn at an angle of about 30° over the stratified limestone, to a height of approximately 1.5 m. Higher up, the boundary becomes obscured because it passes into a zone of vaguely stratified reef-detrital limestone.

The reef limestone can now be followed over a distance of several hundred metres, until the Grogarns farm is reached. The average height of the wall is 5 m. In the lower half of the exposed reef limestone section, rounder colonies are more common among the stromatoporoids than in the upper half. There many tabular colonies and even piles of such tabular stromatoporoids are abundant, together, particularly locally, with many flat-lenticular colonies. Some of the depressions which developed in the reef surface have been filled with stratified limestone containing some reef debris, some other depressions dominantly with such reef debris. Occasionally, the exposed reef limestone is of a talus-like nature, or is real debris exposed over the full height of the wall, coarse and very disorderly at the base, becoming finer in an upward direction and with a greater contribution of limestone.

The base of the reef limestone can be seen about 0.6 km south of the farm Grogarns. The wall there has the appearance of the underside of a staircase. Deepest eroded at the base, there is a relatively soft marly sediment with many limestone lenses, and solitary and social corals as the dominant fossils. The next step is a zone, about 1.25 m thick, of harder rock, very rich in reef debris, which upwards gets increasingly coarser; it also contains many crinoid fragments. The uppermost 20 cm of this zone is very marly and is eroded to some depth, leaving a horizontal cleft in the exposed wall. The hardest and most protruding rock is the reef limestone of the third step. At the base it is of the common Hoburgen type, upwards, the reef builders, however, get distinctly flatter.

Shortly before the northwest wall of the Grogarnsberg reaches the coast, the section of Fig.183 is found. At the base, there is marlstone, thinly parting, with lenses (5–90 cm long, on the average 6 cm thick) of very hard, somewhat marly limestone. Overlying this marlstone is 55 cm of reef debris, comprising in particular many colonies of stromatoporoids and corals, together with solitary corals, the whole being embedded in marl; in the lower

---

Fig.185. Herrviken. Hemse Beds. Reef limestone of the Hoburgen type with an intercalated 1 m-thick zone of reef limestone of its Gannberg variety. In the lower Hoburgen reef limestone a breccia-like zone, with in this and the directly overlying reef limestone several stromatoporoid colonies, including large ones, which are not in their positions of growth.

Fig.186. Herrviken. Hemse Beds. Reef talus, passing upward into stratified limestone, with on top of this again reef talus. Note in the latter several stromatoporoid colonies which are not in their growth positions; the arrow indicates one of the larger colonies.

part there are still a few limestone lenses. The fossils are sometimes covered by a network of *Aulopora*. Upwards the debris becomes distinctly coarser, with many complete colonies present. Next is 33 cm of a protruding hard rock, probably true reef limestone. On top of this, again about 50 cm of reef debris, passing upwards into reef limestone. Here, too, the higher parts of the reef are very rich in extensive tabular stromatoporoids, thus forming an intermediate step between the normal Hoburgen reef type and its Gannberg variety. The dominant flat reef builders sometimes cause a somewhat stratified appearance. The reef matrix is marly.

A comparable development in reef formation can be observed also in the north wall of the Grogarnsberg (Fig.184). The reef limestone which is exposed shows distinct signs of interruptions in its formation. At the base of the section, there is about 2.5 m of reef limestone with many round reef builders and some flatter colonies. This rock is cut off by a somewhat wavy, approximately horizontal plane and overlaid by about 3 m of reef limestone in which comparatively more flat-lenticular and tabular reef builders are present. On top of this is about 1 m stratified reef debris, with a strong decrease in the amount of debris upwards. The upper 2 m of the section show reef limestone of the Gannberg variety.

These three stages of reef development can be followed over the entire north wall; the debris zone disappear gradually eastwards.

In the northern part of the east wall of the Grogarnsberg, there is also a succession of normal Hoburgen-type reef limestone in the lower part and Gannberg-variety reef limestone higher in the section, the two being separated by a detrital zone. In both instances, the reef limestone is marly. The approximately horizontal boundary between the two can be followed for a great distance. Southwards the wall becomes lower and consequently the Gannberg variety disappears from the section. A distinct interruption in reef formation can be seen for quite a distance; locally there are also further interruptions. The plane marking the interruption may be covered by a layer of debris material.

About 0.8 km south of Grogarnshuvud, stratified limestone appears again in the wall, about 0.5 m thick, and also underlying and overlying reef limestone. Though for some distance again entirely replaced by reef limestone, it remains generally exposed southwards and from about 1.1 km south of Grogarnshuvud is the predominant sediment in the wall.

The reef limestone close to the beach about 0.7 km northwest of Herrvik is of the normal Hoburgen type, with at the base 50 cm of middle-coarse debris. At the base and top of the section is stratified limestone. The rauk-like exposures further south also show Hoburgen reef limestone.

Directly *northeast of the harbour of Herrvik*, cliff fall has brought enormous blocks down, in which reef material occurs between thick banks of stratified limestone. This reef material is very unorganized; in part it may be reef limestone belonging to the very periphery of a reef, partly it may represent reef talus. Somewhat further northeast more normal reef limestone is found exposed to a thickness of 4–5 m; it can be regarded as belonging to the Hoburgen type. Overlying it is stratified crinoid limestone. At the base of the reef limestone, very hard and splintery limestone crops out in layers of 2–15 cm thick, often separated by thinner layers of softer marly sediment. The reef limestone overlies this stratified deposit sub-horizontally. In the lower part of the reef, however, several intercalated layers of this splintery limestone still occur, up to a local thickness of 18 cm. In the reef limestone, flat colonies of reef builders are more common than rounder forms. Locally the reef consists mainly of tabular stromatoporoids. These parts show up after weathering because the wall there is smoother, in contrast to the mainly conglomeratic appearance of the reef rock. The reef limestone can be followed northeastwards for more than 100 m. Then again a great amount of debris is exposed. The following vertical succession can be observed there: At the base, again the above-named splintery limestone in layers of about 10 cm thick, but with interbedded thinly parting marly limestone in layers of up to 10 cm, very locally even 17 cm thick. Overlying these sediments is about 2 m of reef debris, with intercalated several layers of the splintery limestone, most common and thickest in its lower part. Next is about 1 m of reef limestone with some embedded debris. This is covered by a second occurrence of reef debris, about 2 m thick, in which layers of stratified limestone are increasingly common upwards. Northeast of this section, the reef limestone is replaced by debris and the thus united debris occurrence gradually thins out between stratified limestone. The debris as exposed, is presumably formed close to the north side of a reef.

Further northeast, occasionally reef limestone of Hoburgen type is exposed, with local transitions into the Gannberg variety. Fig.185 shows such an intercalation of Gannberg reef limestone in more general Hoburgen-type rock. Also here the different appearance after weathering is distinct.

Fig.186 presents another exposure of reef debris apparently deposited very close to a reef.

Along the north and northeast coast, near Kuppen, the wall exposes mainly reef limestone of about Hoburgen type, with also debris deposits or stratified limestone. The sediment overlying the reef limestone is rather strongly recrystallized, marly crinoid limestone of grey or sometimes red to reddish brown colour.

The slight accidentation of the plateau between Herrvik and Kuppen is presumably due to an alternation of reef limestone and stratified sediments.

The sediments which are exposed in Torsburgen and its direct environment (including Millklint and Herrgårdsklint) are presumably partly somewhat younger than the other rocks in the northeastern part of the Hemse Beds area, for the other part, synchronous with the higher parts of Gannberg and Grogarnsberg.

Millklint, directly south-southeast of Torsburgen, consists of light bluish grey stratified limestone, mainly finely crystalline, partly finely oolitic. The layers vary in thickness between 1–12 cm, but are generally less than 5 cm, in between the layers are films of marl. This generally fossiliferous rock is called by Hede (1929) Millklint limestone. No reef limestone was observed.

Torsburgen, in Kräklingbo Parish, is especially interesting because of its stratified limestones, extremely rich in stromatoporoids. In the western part of the north wall and in the west wall of this hill, these can be well observed. A lower and an upper stromatoporoid limestone can be distinguished. The lower stromatoporoid limestone has a biostromal character. All stromatoporoids are strongly recrystallized; they are embedded in marly limestone which is softer than the fossils are, with the result that through weathering, the rock has a strongly conglomeratic to somewhat brecciated appearance. Most of the stromatoporoids are very flat lenticular, but in the lower part of the lower stromatoporoid limestone, somewhat rounder colonies are also found locally. The largest one seen measured 60 cm long and 23 cm thick. In between the layers which are so very rich in stromatoporoids, there are some others, especially high in the lower deposit, of bluish to brownish grey marly limestone, dense to finely crystalline, relatively poor in fossils, and with bedding planes which vary from very rugged to sometimes almost smooth. The thickness of the stromatoporoid layers alternating with these limestone layers decreases upwards.

Neither the transition from lower to upper stromatoporoid limestone, nor that between the upper stromatoporoid limestone and its overlying Millklint limestone can be considered a reliable time boundary, mainly because of the occurrence of reefs.

The upper stromatoporoid limestone is generally thick bedded, locally thin bedded; there are no distinct boundaries between these two, both forms passing into each other. Usually, the rock is finely crystalline and strongly marly. Stromatoporoids, mainly in flat-lenticular and tabular colonies, are the dominant fossils, but bryozoans, crinoids and some brachiopods also occur. In an upward direction, the upper stromatoporoid limestone passes into a rock lithologically similar to the Millklint limestone.

Reefs occur in Torsburgen mainly in the upper stromatoporoid limestone. They often began growth at, or slightly above, the boundary between lower and upper stromatoporoid limestone. Their matrix is marly, most of its stromatoporoids are flat, but some are rounder.

The finely crystalline to finely oolitic Millklint limestone is remarkably poor in reefs, compared with the other Silurian limestone deposits in Gotland. Its maximum observed thickness is 7 m. Upwards, its layers are often thinner than at its base.

The above-described succession is well exposed at Tjängvide-lucka (about 0.15 km east of the topographical height 68.1 with the belvedere). At the base 3.5 m lower stromatoporoid limestone is found, partly thin, partly thicker bedded. In between the beds there is often some marl; the bedding planes are often very strongly rugged. The extremely abundant stromatoporoids of this zone are dominantly flat lenticular.

The overlying upper stromatoporoid limestone, which contains many tabular stromatoporoids, is relatively less fossiliferous than the lower one. The deposit is partly well bedded, partly vaguely bedded with beds of 1 m and more in thickness; the bedding planes are faintly rugged. The limestone is harder than the lower stromatoporoid deposit and weathers more massive.

Locally this upper deposit contains reef limestone, e.g., directly at Tjängvide-lucka itself. The exposed base of the reef limestone is slightly more than 2 m above the boundary between lower and upper stromatoporoid limestone. The reef is marly. Stromatoporoids abound; they differ from those in the stratified limestone mainly because there are so many round colonies and even quite a number with greater vertical than horizontal dimensions, such as bullet and tower-shaped forms. There are some intercalations of stratified limestone. The reef is exposed for a length of about 20 m and a height of about 2 m; it is overlaid by about 2 m Millklint limestone, poor in fossils.

West of this reef another is found, about 15 m long, followed by a still smaller one. Both of these reefs contain far less roundish fossil colonies and have a more distinct and steeper boundary with the stratified limestone at their northwest than at their southeast side. Again 20 m further west, a reef-limestone exposure of about 4 m long and 1.5 m high is found. This reef rock is rich in marly limestone and contains mainly rather round stromatoporoids. This section presumably represents the west or northwest periphery of a reef.

Another example of a reef is found in the most western part of the north wall, where at a low level in the upper stromatoporoid limestone, a reef is found of about 10 m long and in its centre 2.5 m high. It mainly consists of many flat-lenticular and faintly wavy tabular stromatoporoids in a marly matrix. There are only a few other fossils. The marl also occurs in pockets and since it is softer than the reef builders are , it is eroded away in many places, leaving holes in the exposed wall.

In the north wall, east of Tjängvide-lucka, reef limestone exposures are also found, especially in its western part. In these, flat stromatoporoids are more abundant than rounder forms, which are relatively commoner in the lower and middle parts. Also bryozoans and coral colonies contributed to reef formation, the first in colonies of up to 60 cm broad and 35 cm thick. Intercalations of stratified limestone and marl pockets are found in the reefs. The latter reach thicknesses of up to approximately 6 m. Of the exposed boundaries between reef limestone and surrounding sediments, the east boundary is almost vertical; the south one is sloping and generally sharp, although the highest reef part may contain a large amount of matrix and also reef debris. The northwest boundary is always sloping, with the reef limestone extending over the stratified sediments, but the reef limestone there is usually very unorganized and marly and layers of the surrounding limestone may penetrate one or more metres into the reef. Since in all likelihood, the exposed east and south boundaries of reef limestone represent the seaward side of the reefs, it can be concluded that the seaside of the reefs was steeper and more solidly developed than the landward side. The reef surrounding sediments are usually rich in crinoid and bryozoan fragments in the environment of the reefs.

Herrgårdsklint (Gammelgarn Parish) is subdivided into an eastern and a western part. The latter consists of Millklint limestone. The west-northwest wall of the eastern part shows stratified stromatoporoid limestone, partly detrital. It passes at the northern end of the wall into reddish brown to grey stromatoporoid reef limestone. This has a marly matrix, the stromatoporoids are mainly lenticular in shape; corals (*Heliolites, Favosites, Halysites, Aulopora*), bryozoans and brachiopods are also present, as well as some crinoid remains, gastropods, orthoceratids and trilobites (*Bumastus* sp.). In the east wall, the reef limestone can be followed for some hundreds of metres, after which it passes into stratified stromatoporoid limestone. The latter sediment is seen to overly stratified marly limestone with interbedded thin layers of marl; the limestone layers are up to 10 cm thick, the rock is dense or nearly dense and brownish grey to brown in colour and less fossiliferous than the overlying limestone.

Klinteklint, in Gammelgarn Parish, possesses a particularly interesting northeast and east wall. In the northeast of this klint, it can be seen how a reef expanded over its own debris. At the base of the section, indistinctly stratified fossiliferous limestone is exposed. This is overlaid by reef debris of an usually unorganized nature, with many fragments of fossils and many fossils in orientations other than their natural ones. The deposit is least unorganized where almost all the fossils and fragments are of a very flat form. Dips in a great many directions can be observed, but with some preference for a west-northwest dip. A large part of debris with a uniform dip of the fossils of about 20° to the south-southwest probably became detached from the reef as one single block (see also Chapter VII, p.163). Over this talus deposit, the reef expanded. Low in the reef limestone are many tabular and flat-lenticular colonies, which may give the rock a vaguely stratified appearance and may lead to more massively eroded surfaces. Upwards, rounder colonies become more common and since the matrix is marly, weathered surfaces are often of a more conglomeratic appearance. When the exposed surface in this part is smoother, this is often due to relatively recent cliff falls along joint planes. In the middle and higher parts of the reef limestone, several stromatoporoids were observed of about 70 x 30 cm, spherical colonies of a diameter up to 40 cm, also a few tower-shaped piles of stromatoporoid latilaminae of 25–30 cm high, at the base about 15 cm, at the top about 10 cm in diameter. At the top of the reef limestone, flat colonies again become more common. This is also true in the east wall, where at the top flat reef builders and a somewhat higher matrix volume locally lead to a vague stratification in the reef limestone. The maximum exposed reef thickness never exceeds 8 m. Probably the reef has not been much thicker than this, despite its great horizontal extension of presumably 250–300 m.

The zone with flat reef builders low in the reef limestone in the northeast of this klint, varies in thickness over very short distances, between less than 1 m to about the entire exposed height of reef limestone. It may also contain rounder forms, sometimes only locally or in a particular horizon. Also, intercalated layers of stratified limestone are found in it, up to 10 cm in thickness and 5 m in length. Locally, this zone with flat colonies can also be interrupted by a debris layer of 10 cm to over 1 m in thickness. Such a debris generally is less resistant to erosion and may, therefore, give lead to undercutting of wall parts or to cave formation. Flat reef builders may also predominate locally in reef parts which otherwise are characterized by rounder colonies. In several places rounder colonies not in their growth positions are rather common. The entire picture of reef formation is that of a great variation in reef vigour, with alternatingly reef expansion or retreat. All these phenomena probably took place in the more coastward part of the reef.

The east wall of Klinteklint presents a north – south section through the reef. The orientation of this reef probably was about north-northeast – south-southwest. This means that going southward the more central part of the reef is exposed, followed by the seaward side. In the wall, it can be seen how the talus floor and the lower zone with flat and tabular reef builders, described from the northeast wall, gradually disappear from the section while going south. Large reef builders of round shape become abundant. Loose blocks at the foot of the cliff indicate how the reef originally reached further east. Under an overhang low in the wall, an almost horizontal marly layer a few centimetres thick and of several metres extension crops out, with around it reef limestone showing smaller and flatter reef builders. However, this is a local phenomenon and away from this layer, the fossils again increase in size. In the last part of the exposed reef limestone, the fossils again become, on the average, distinctly smaller and the rock obtains a more debris-like appearance. At the top of the wall stratified crinoid limestone begins to occur. The rather sharp boundary between the reef limestone and the overlying sediment dips about 10°. The lateral boundary is less sharp, with limestone layers penetrating to several metres into the reef.

The last part of the east wall is taken up by the crinoid limestone, which is bluish grey, grey to brownish grey and often red-mottled.

Northwest of the section described in the first paragraph of this description of the Klinteklint, the reef debris and limestone can be followed for about 20 m, then being replaced by crinoid limestone. Some tens of metres further northwest again reef limestone of a conglomeratic nature crops out, resting over talus-like reef material with mainly much flatter colonies. Locally this reef limestone is replaced by reef rock which contains a high volume of marly matrix and predominantly flat fossils, the whole showing some stratification. It probably represents a pool in the reef surface. Via a part with many flat fossils, the reef limestone passes northwestward again into stratified limestone. This limestone further builds most of the north wall of the klint. At the place where this wall turns south into the west wall, a few small occurrences of very marly reef limestone with mainly flat stromatoporoids are found.

On the basis of the exposures described above, it seems likely that at the landward side of the main Klinteklint reef, another, but smaller reef developed, with northwest of this again a few subordinate attempts towards reef formation.

The west wall of the Klinteklint shows in the north stratified limestone with some reef limestone exposures; then a long stretch showing reef limestone, in its southern part traversed by an oblique zone of crinoid limestone, and in the south of the west wall again stratified limestone.

Petsarveklint, in Ardre Parish, shows only stratified limestone in a few small outcrops. The sediment is generally very thin bedded with irregular bedding planes, brownish light grey to greyish white in colour, finely crystalline, and very fossiliferous (crinoid remains, some flat and lenticular stromatoporoids, solitary corals, bryozoans, the large lamellibranch *"Megalomus" gotlandicus* Lindström). The rock shows much similarity to the thin-bedded crinoid limestone of Kaupungsklint. No reef limestone has been observed, though it is presumably present in the klint.

The rocks of *Kaupungsklint* can be studied in a long wall of varying height, from less than 1 m to locally about 3 m. The rocks found in this wall have been described already in Chapter VII, p.134.

It seems likely that most of the hillock of Kaupungsklint consists of one or more reefs, presumably not very thick. The rocks observed in the wall then, are either deposits from the reef surroundings or represent the periphery of the reef itself.

Aikse Bakke, the hillock south of Aikse farm, Ardre Parish, consists predominantly of stromatoporoid reef limestone of the Hoburgen type. It is exposed in a wall about 2 m high, a few hundred metres south-southeast of Aikse, in the terrain around there, in a very large quarry south of Aikse and in an old quarry directly south of the large one. Overlying crinoid limestone with reef debris is found in the two quarries. In the eastward facing slope in the northeast of the hillock, the reef limestone can be seen to overlie stratified stromatoporoid limestone, which is brownish grey, marly and finely crystalline.

Visneklint, situated east of Bofrideklint (to be described next), is best exposed in a wall of a few metres height and several tens of metres long, along the country road at the east side of the klint. In the southeast of this klint, the section has been observed which has been described in Chapter VII, pp.134–135.

Bofrideklint, in Alskog Parish, about 1.5 km north-northeast of Bofride farm, shows a variety of sediments. In the first place there is stratified limestone of a brownish white-grey colour, which is finely crystalline and sparry, and rather poor in fossils. The bedding planes are somewhat irregular, the layers generally 1–5 cm thick; locally there is some cross-bedding. Upwards this limestone may contain an increasing amount of reef debris, which in several cases shows indications of rounding. It may in its turn be overlaid by stromatoporoid reef limestone. Elsewhere at the height of the first-named limestone, there is stratified limestone, with foliaceous layers of less than 1 cm thick, which is greenish grey, dense to finely crystalline and contains some stromatoporoids and solitary corals. In some places, it is overlaid by a reef-like stromatoporoid limestone, also containing corals, bryozoans, brachiopods and some gastropods, which shows some rugged stratification, generally due to intercalated thin layers of stratified limestone or some reef debris material. The thin layered, dense limestone is also found locally in between two masses of reef limestone. At the top of the hillock, reef limestone is locally overlaid by a finely crystalline stratified limestone of light-grey to greyish-white colour, which contains *"Megalomus" gotlandicus* Lindström.

About 3.5 km north-northeast of Alskog Church, west of the road, a wall is found, up to 1.5 m high, which shows grey stromatoporoid reef limestone with a marly matrix. Further northwards, east of the road, finely crystalline stratified limestone is found, or thin, but somewhat irregularly bedded crinoid limestone, often with reef debris. Slightly southeast of the wall, in the forest, a small exposure is found, in which the crinoid limestone abuts against stromatoporoid reef limestone. The rather flat top of this reef limestone is overlaid by vaguely stratified detrital limestone with crinoids. Upwards, the reef-debris content decreases. The rather irregular distribution of stratified limestone, stratified crinoid limestone and crinoid limestone with reef debris is undoubtedly influenced by the rather irregular distribution of the reefs in this area.

Guffrideklint, about 1.1 km east-southeast of Garde Church, presents only scattered exposures, one to several square metres in size. In the majority of these, reef limestone crops out, and others show crinoid limestone. Very close to the reef limestone, the crinoid rock is very rich in reef debris. The vague and somewhat irregular bedding planes dip 12–17° down from the reef. Within a distance of some metres, the coarseness of the reef debris strongly decreases, but finer debris still remains present for quite a distance.

Directly northeast of the three-forked cross-roads to Etelhem, Garde and Lye, a quarry is found. In a 2–3 m high vertical wall there, crinoid limestone is exposed, which is locally cross-bedded and upwards shows an increasing number of stromatoporoids, and fragments of these, presumably indicating deposition close to an expanding reef.

In another quarry, mainly in Quaternary rocks, but penetrating into the Hemse Beds, the sediments underlying the crinoid limestone are visible for a thickness of 1.5 m. These are limestone and interstratified marlstone layers of on the average 1 cm thick. The lower exposed limestone layers are a few centimetres thick, but upwards their thickness increases and finally, the limestone passes into the crinoid limestone in which no marlstone layers are present. The limestone layers contain already many crinoid remains, though not in such enormous quantities and also somewhat less coarse than in the crinoid limestone. Reef debris is abundant; most of the fragments by far, have a size of less than a few millimetres, but larger remains, up to a few centimetres are commonly intermixed. Many fossil remains are enveloped by an algal crust varying in thickness from less than 1 mm to sometimes close to 1 cm. In the crinoid limestone, the reef debris is increasingly coarser upwards and also complete colonies of stromatoporoids occur there; at the top of the section, the rock even becomes reef-like.

About 0.1 km northwest and west of this exposure, some small surface exposures of stromatoporoid reef limestone are found in the field.

About 0.5 km east of Sigvalde-träsk, where the road makes a curve of about 90°, an exposure occurs which shows stromatoporoid reef limestone. The rock is strongly unorganized; larger and especially smaller intercalations of stratified limestone are common. It probably represents the outer part of a reef. South-southwestwards, stratified sediments gradually replace the reef limestone from below in an upward direction.

In the south-southwest of the section, a crinoid breccia is exposed at the base, well stratified, with layers of a few centimetres thick and with somewhat rugged bedding planes; locally there is cross-bedding. Some reef debris is enclosed. Upwards, the amount of reef debris strongly increases, the stratification becomes irregular and vague and locally the rock is almost entirely reef debris, though interfingering with crinoid limestone or having this sediment interbedded. Some of the reef fragments in the crinoid breccia are enveloped in an algal crust.

Sigvalde-träsk klint. The klint south of the eastern part of Sigvalde-träsk, shows at its base stratified marly limestone, rather rich in fossils, which partly are reef detritus. After a part with no exposures, marly reef limestone, overlying the stratified sediment, is found in a 3.5 m high wall. Higher exposures indicate that the reef limestone reaches a thickness of at least 6 m. Locally the reef limestone is vaguely bedded, elsewhere intercalations of normal limestone occur, up to 2 m long and on the average 7 cm thick. Stromatoporoids vary in size, but hardly anywhere is their thickness more than half their horizontal dimension.

The higher parts of the klint are rather badly exposed. Along the remaining southern bank of the träsk, reef limestone is also found, vaguely stratified or not at all. Present as well is some overlying stratified limestone.

Träsk klintar. In the environment of Etelhem – Lojsta, a number of klintar occurs, rising directly from or closely to a little lake (Swedish: träsk), e.g., south of Sigvalde-träsk, south of Hagby-träsk, south and west of Bro-träsk, south of Ramträsk, east of Hemträsk and southeast of Mört-träsk.

The stratified sediment in Hagby-träsk klint is a bluish to brownish grey limestone, dense to finely crystalline, somewhat marly, containing crinoid remains and reef-building organisms, the latter increasing in number towards the reef. The reef limestone varies in colour from reddish browh or rust brown with light-grey coloured fossils, through lighter colours to grey or light bluish grey. Stratified intercalations in the reef occur, especially in the lower parts.

Similar reef limestone is also found in the other klintar and may be surrounded by stratified limestone. The latter sometimes is seen dipping away from the reef, e.g., in Hagby-träsk klint (Fig.187) and Bro-träsk klint (cf. Hede, 1927b, fig.15, 16).

Tonnklint, southeast of Lojsta Church, mainly exposes crinoid limestone with a varying amount of crinoid remains, stromatoporoids and corals. Occasionally, it shows a reef-like development. Some true reef limestone also occurs, especially in the northern part of the klint.

Ausarveklint shows a great number of smaller outcrops, which present marly reef limestone, dense to finely crystalline, reddish brown crinoid limestone (which is seen in the northeast to overlie the reef limestone), and greyish marly limestone, often thinly stratified, but very rugged, due to a high fossil content of tabular stromatoporoids, compound and solitary corals, crinoids and brachiopods. The distinction with the reef limestone is not always clear.

Fig.187. Hagby-träsk klint. Hemse Beds. Stratified limestone dipping down from a reef.

Fig.188. Lindeklint. Hemse Beds. Detrital limestone in the southeast of the klint.

Lindeklint. The east wall, some hundreds of metres long and 1–7 m high, shows in the south strongly weathered stratified crinoid limestone and in the north reef limestone, also strongly weathered and generally of a brecciaceous nature. The boundary between the two is usually rather distinct. The reef limestone is very similar to that in Hoburgen, with stromatoporoids as the main reef builders and a minor contribution of corals. The majority of stromatoporoids is relatively flat. In the marly matrix, such fossils as crinoids, brachiopods, solitary corals, ostracodes and orthoceratids, occur. At the top the matrix volume increases, and the reef builders are flatter than elsewhere in this wall. In several places, these flat reef builders cause this uppermost reef limestone to show a vague pseudo-stratification. Gradual transitions from the reef limestone into stratified limestone, generally a crinoid limestone to a crinoid breccia or coquina, occur locally.

In the southeast, stratified marly limestone with a very high reef-debris content (about 35–50% of the total rock volume) is exposed (Fig.188). Flat and tabular stromatoporoids and especially fragments of these, strongly dominate. In several places, imbrication indicates that deposition of the debris took place downreef (see also Chapter IX, p.223). The thickness of the layers varies. On the average, the fossils and fossil fragments in the thicker layers are coarser. The limestone in which the debris is embedded, is light grey to brownish grey, finely to middle crystalline and marly.

In the south of the Lindeklint, similar limestone is found, but with a smaller content of reef debris. It alternates locally with more normal stratified limestone.

The wall at the north and northwest side is steep. In the northwest on the plateau of the hill, generally thin bedded, grey to brownish grey and marly crinoid limestone and hard, grey to bluish grey, marly stromatoporoid reef limestone are exposed.

The northwest wall, locally over 10 m high, consists entirely of reef limestone. Apart from the dominant stromatoporoids, coral colonies are common and both often reach large sizes. Stromatoporoids were observed with a horizontal length of up to 1 m; colonies of about 50 cm long, and in their centre 20 cm thick, are common. Colonies of *Halysites* measured up to 70 cm, with a thickness of 40 cm, and colonies of *Acervularia* 65 cm with a thickness of 35 cm. Several colonies are not in their orientation of growth. The matrix is marly, but its total volume there is somewhat below the average for Hoburgen-type reef limestone. Locally there are small intercalations of stratified limestone. The colour of the reef limestone is generally grey to light grey, locally brownish grey; as a result of weathering it is often reddish to reddish brown (elsewhere in the Lindeklint, the colour of the weathered reef limestone is usually bluish grey, grey to greyish white). Since the matrix weathers more quickly than the reef builders, weathered surfaces appear coarsely conglomeratic. In one place in this wall, a depression in the reef surface was observed, containing much debris, in between which some stromatoporoids and several corals, belonging to at least five different genera, are found, probably in their positions of growth. Upwards, the depression merges again into reef limestone of average faunal composition.

The wall at the north side of Lindeklint is closely surrounded by forest. It exposes mainly thinly bedded, marly crinoid limestone and crystalline limestone, which is hard, indistinctly bedded and bluish grey to grey in colour.

Sandarve Kulle is a hillock, about 1 km north of Fardhem Church. It is mainly covered with grass and forest; a number of small to very small exposures of mainly grey reef limestone occur here and there. In the west and southwest, there is a wall, a few metres high and about 150 m long, with thin-bedded crinoid limestone, which is light grey, marly, and comparable to that in the Lindeklint. It contains reef debris. Where this is smaller than a few centimetres, bedding is thinner than where the debris is coarser and more abundant. This sediment passes laterally into or is covered by very fossiliferous reef limestone, which is rather marly, greenish grey or reddish brown, and dense. Stromatoporoids, crinoids, bryozoans and corals are found in it, and also gastropods, cephalopods, lamellibranchs, brachiopods and trilobites. The centre of the hillock consists presumably of reef limestone.

Table XIX gives a survey of fossils found in the Hemse reef limestones and crinoid limestones. The locality given as East Sigvalde is the one described on p.369 as being situated 0.5 km east of Sigvalde Träsk, the locality Etelhem is the one close to the cross-roads south of Etelhem, mentioned on p.368.

Holmhällar-type reef limestones

The Hemse Beds are the lowest stratigraphical unit in Gotland, which also contains reef limestone of the Holmhällar type. The relevant localities where this is exposed are mentioned here from north to south. All are situated along the east coast.

Interesting exposures are found *along the coast between Herrvik and Sandvik*. They are the main source of information on the debris production of Holmhällar-type reefs and its distribution around these reefs (Chapter IX, pp.219, 223).

Around Snabben stratified limestone is exposed, which is thinly bedded, very rich in crinoid remains, and mostly grey in colour. North of Snabben, a number of raukar are found, some of which show how reef limestone is overlaid by crinoid limestone with reef debris (Fig.189). Upwards the debris becomes finer and the sediment passes into the stratified limestone, containing some reef debris, already mentioned from Snabben (see Chapter VI, p.208).

Going southwards from Snabben, two small capes are reached. The northern one is built up by reef limestone very rich in stromatoporoids, several of which are very large. Also present are corals, Algae, some crinoids and an occasional brachiopod or cephalopod. The reef has a horseshoe shape with the opening directed west-northwestwards. Locally, the reef limestone is overlaid by reef debris, which partly shows indications of rounding. The contact between reef and debris shows that the upper surface of the reef was uneven. The debris deposit shows a thickness of 30 cm or more and passes upwards into a crinoid limestone very rich in small crinoid fragments. In the second cape, erosion has caused that not much reef limestone is present any longer, but the debris and crinoid limestone are still well exposed.

Passing a smaller outcrop, in a southward direction, one reaches, about 1 km south of Snabben, a place where much reef limestone is exposed, both around sea level and in a number of raukar. The reef limestone can be classified as to be of Holmhällar type. However, the relatively high percentage of the rock volume taken up by the matrix, and the occurrence of

Fig.189. Reef limestone, overlaid by reef debris and crinoid breccia. Between Kuppen and Snabben. Hemse Beds.

Fig.190. In the middle of the southwest coast of the peninsula at Sysne remains are found of a reef which presumably had a diameter of less than 10 m. It is surrounded by reef debris, which dips away from the reef limestone. Hemse Beds.

comparatively many lenticular and tabular stromatoporoids indicate a tendency towards the Hoburgen type of reef limestone. This holds true for several exposures in this area. The reef shows a crescent shape, with the opening about west-northwestwards. Some small occurrences of stratified limestone around it indicate that this is actually the original shape of the reef. At the side of the opening, the reef limestone is different from that at the outer side. The rock at the side of the inner curve contains reef debris and the number of thin stromatoporoids is much greater. Most of these are in their attitudes of growth. On the outside of the reef, the number of large stromatoporoids is above average.

Continuing southwards along small exposures, another large outcrop of reef limestone is reached. It seems likely that three or more small reefs occur there. These are circular to oval in shape, but together are arranged in a crescent pattern with an opening towards the west-northwest.

Several other exposures follow further southwards, but they present no additional information of great importance to that described in Chapters VIII and IX. The distribution of the reef limestones and surrounding deposits has in all likelihood contributed to the present course of the coast line in this area, which shows an average direction of about N30°E.

Also along the southwest coast of the small Sysne-udd peninsula, reef limestone, reef debris and stratified limestone are exposed. Interesting to see is a small reef, which presumably measured not more than 10 m in diameter. Remains of this reef are exposed about half-way along this coast (Fig.190, see also Chapter VIII, p.190). The debris found around it is much finer than around other reefs in this area. Not many pieces are larger than 1 cm. West of it, the remains of a larger reef are found, which measured at least 50 m in chord length and which is surrounded by coarser debris. It is followed by a third reef, which is much larger again, and is exposed up to the Fiskläge.

In the northwest of this peninsula, stratified limestone is exposed mainly southeast of the country road. The limestone contains reef debris and thin marly layers are interbedded.

Going further south, the next exposures are found *along the coast northwest and west of Sandviken*. Where the road Östergarn Church – Sandviken reaches the coast, a coastal cliff of an average height of 1.7 m is found. In this cliff and directly landwards of it, reef limestone occurs, which is of Holmhällar type, with sometimes a tendency towards the Hoburgen type. In some places, crinoid limestone is found between or overlying the stromatoporoids, suggesting that it is the top of the reef limestone which is exposed. The original shape of the reef and some other reefs in this area cannot be determined with certainty, but an elongated form in about north-northeast – south-southwest orientation seems to be most likely.

Continuing along the coast southwestwards, crinoid limestones, which are rich in reef debris, are also found, as well as more normal stratified limestone, which is grey and macroscopically not very fossiliferous. The first sediment type consists of a limestone extremely rich in crinoid remains in which many - usually small - fragments of stromatoporoids and coral colonies are embedded, with solitary corals, bryozoan fragments and brachiopods. There are transitions to a more normal looking stratified limestone. In detailed inspection the latter appears to be partly a limestone rich in smaller crinoid fragments, which is various degrees have obscured during recrystallization.

At Brändåkers-udd, reef limestone is found, which represents presumably only a peripheral part of a reef of which all the rest has already been eroded. Stromatoporoids were the main reef builders, but in between these, colonies of branched and massive corals are found, varying in size from a few centimetres to a horizontal size of almost 1 m. In some approximately 1 m high vertical sections, it can be seen how the reef limestone passes rather gradually into the overlying limestone with reef debris.

Close to Grynge-udd Fiskläge, there is a large and beautiful exposure of reef limestone, overlaid locally by crinoid limestone with reef debris which upwards decreases rapidly in average size. About 20 cm above the reef limestone, the sediment passes into more normal stratified limestone.

Almost all of what should be reported about the area of *Sjausterhammar* (Fig.191) has already been said in Chapter VIII. In the north-northeast, the exposures occur which are described on pp.189–190. Slightly south of this is the next reef, which is less distinctly exposed, but a crescent shape is likely. In the north and south, stratified sediments dip away under the reef limestone. In the north this is strongly recrystallized crinoid limestone. In the south, very thinly cleaving calcareous slate is found, rich in reef debris, which is locally very coarse. A few layers of about 1 cm thick of hard, splintery limestone are interbedded; it is brownish grey and very poor in fossils. The calcareous slate is underlaid by normal, grey, fossiliferous limestone.

Next, in a southward direction, are some hundreds of metres where, around sea level, mainly limestone with reef debris is found; locally some more normal stratified limestone or some reef limestone. This is followed by the exposures described in Chapter VIII, pp.206–207. One of these exposures is shown in Fig.192.

The southernmost cape of Sjausterhammar shows mainly stromatoporoid limestone, the majority of which can be considered as being of reef-detrital origin, deposited very close to a reef. It occurs as a kind of ring, of which the west side is missing. The sediments which were present in the centre, have been eroded. The ring is overlaid by a complex of thin bedded, finely crystalline limestone, with a maximum thickness of 30 cm. The rock is brownish light grey to greyish white in colour, and remarkably rich in *"Megalomus" gotlandicus* Lindström. This sediment arches as a flat dome over the underlying stromatoporoid limestone. This structure, which is thus closely connected with reef formation, was wrongly interpreted by Munthe (1910, p.1433) as being of tectonic origin ("a somewhat folded area at Sjausterhammar").

Fågelhammar actually comprises two raukar fields, Fågelhammar North and Fågelhammar South. Of these, the southern field (Fig.193) is the largest, and raukar reach a height of up to about 6 m (Fig.194). In many aspects, it is comparable to Holmhällar. This is particularly true as far as the reef builders, their size, their growth forms and their distribution are concerned (Fig.196). The reef shows the shape of a crescent. Fissures are present, both in a radial direction and perpendicular to this. Within the reef limestone, there are in several places small slickensides, which indicate small internal displacements during consolidation of the reef limestone.

In between the two, there is a strip of about 25 m where no raukar are present and exposures only occur at beach level. The coast line shows a westward curve there.

The reef of Fågelhammar North is also of crescent shape. The opening is towards the northwest. It is smaller than the Fågelhammar South reef and the raukar are on the average also smaller. Stromatoporoid development during reef growth was less luxuriant. The average size of the stromatoporoids

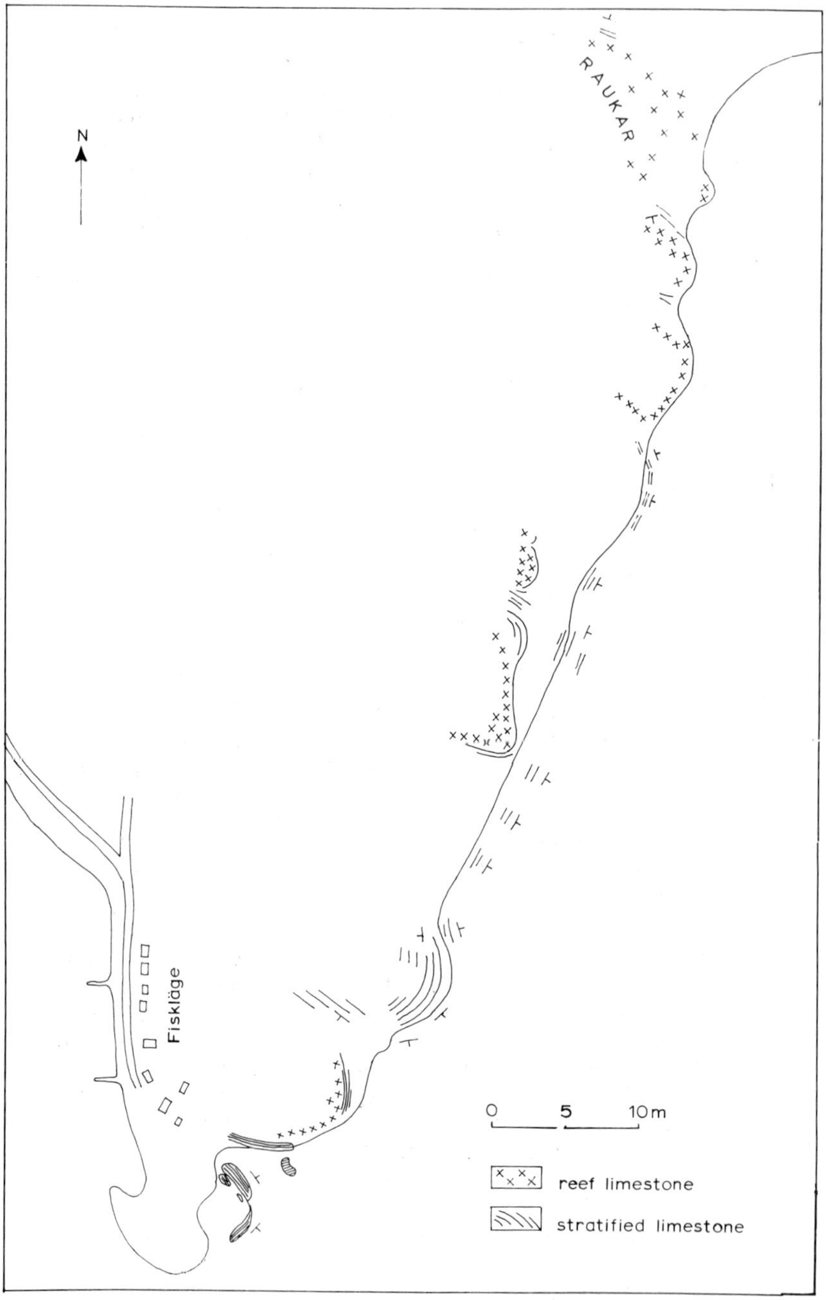

Fig.191. Sketch map, showing the distribution of reef limestone and stratified limestone (Hemse Beds) along the coast at Sjausterhammar.

Fig.192. Section at Sjausterhammar. Hemse Beds. Reef limestone occurs at the top of the succession. It overlies 0.70 m stratified marly limestone, 0.30 m very fossiliferous limestone, which probably is a reef-debris zone, and at the base of the section a calcareous slate, relatively poor in fossils.

---

is smaller. Locally some stratified limestone is intercalated (Fig.197). The percentage of the rock taken up by the matrix is higher. This causes, in some cases, a vague stratification of the reef limestone.

A few hundred metres south of Fågelhammar South, some remains are still present of an almost vanished raukar field, which is east-northeast – south-southwest orientated. The points situated farthest seawards, show, when connected, a sickle-shaped bent line. The raukar reach only little above sea level, the highest measured a height of about 1.5 m. They consist of Holmhällar-type reef limestone. The boundary of this sediment at the side of the land is not exposed.

The raukar field north of the harbour of *Ljugarn,* is a good example of a reef of Holmhällar type. The raukar are less closely together than in Holmhällar, but the field as a whole is larger. Measured along the coast, the length of the Ljugarn raukar field is approximately 750 m. The raukar are on the average also lower; most of them reach less than 2 m above sea level. The highest one measured is 2.30 m. This may explain why the number of preserved pools and debris-filled depressions is smaller than in Holmhällar. There are also fewer planes of interrupted reef growth. In view of the fact that there are only few loose blocks, which originated through undercutting of large raukar, it seems possible that the Ljugarn reef was not much higher than the top of the highest raukar. Nothing can be said about the total thickness

(Text continues on p.384)

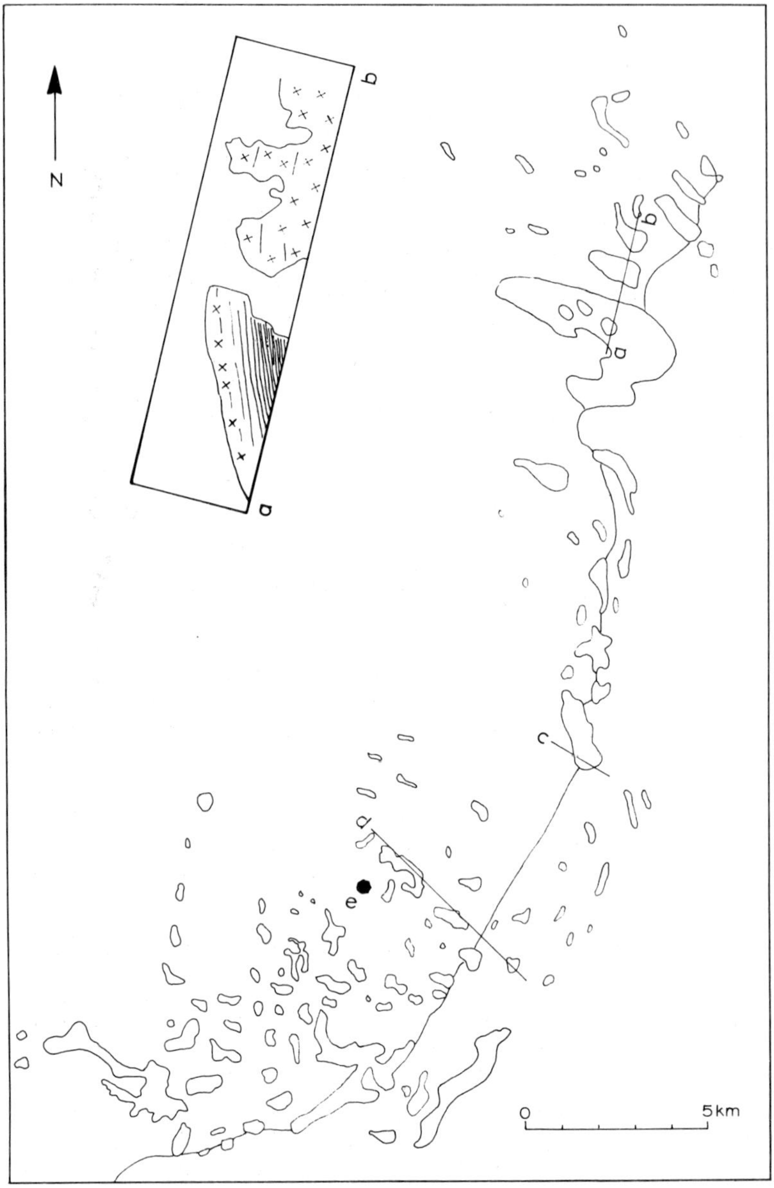

Fig.193. Map, showing the distribution of the raukar in Fågelhammar South. Almost all raukar consist of reef limestone, belonging to the Hemse Beds. The line a–b marks the position of the vertical section shown on the insertion which illustrates a local occurrence of stratified sediment; c and d indicate fissures; e is a military pill box.

Fig.194. The raukar field of Fågelhammar South. The raukar consist of reef limestone belonging to the Hemse Beds.

A B

Fig.195. Fågelhammar South. Details of the left part of section a–b in Fig.193. Stratified marlstone to very marly limestone is overlaid by reef limestone. B shows the stratified sediment in closer view.

TABLE XIX

Fossils found in the reef limestones and surrounding crinoid limestones of the Hemse Beds of Gotland

Lithology	Reef limestone											Crinoid limestone						
	Hoburgen type									Holmhällar type		Around Hoburgen-type reefs						Holmhällar type
Fossils / Localities	Gannberg	Grogarnsberg	Sigvalde Träsk	East Sigvalde	Hagby Träsk	Lindeklint	Sandarve Kulle	Tonnklint	Other localities	Fågelhammar	Ljugarn	Gannberg	Kaupungsklint	Lindeklint	Etelhem	East Sigvalde	Other localities	Snabben-Sysne
ALGAE																		
Rothpletzella sp.		+				+								+	+			
Solenopora sp.		+		+		+			+							+		
Unidentified calcareous Algae	+					+				+	+				+			+
HYDROZOA																		
Labechia conferta (Lonsdale)									+				+				+	+
Unidentified stromatoporoids	+	+	+	+	+	+	+	+	+	+	+	+	+	+	+	+	+	+
ANTHOZOA TETRACORALLA																		
Acervularia ananas (L.)						+			+	+	+			+				
Cyathophyllum bisectum						+												
Cyathophyllum sp.						+												
Diploëpora grayi (Edwards et Haime)	+																	
Entelophyllum fasciculatum Wedekind	+																	
Hedstroemophyllum sp.						+												
Rhegmaphyllum conulus (Lindström)							+											
ANTHOZOA TABULATA																		
Aulopora roemeri Foerste	+														+	+	+	
Aulopora sp.		+							+	+			+	+				+
Favosites gothlandicus Lamarck	+				+	+			+	+	+			+	+	+	+	
Favosites sp.	+					+				+			+					+
Halysites catenularius (L.)						+			+					+				
Milleporites madreporiformis Wahlenberg													+					
Roemeria kunthiana Lindström						+												
Syringopora sp.				+					+	+			+		+	+	+	+
ANTHOZOA HELIOLITIDA																		
Heliolites barrandei Penecke							+	+										
Heliolites interstinctus (L.)						+	+		+				+		+	+	+	+
Heliolites parvistella Ferd. Roemer							+											
Heliolites sp.				+		+	+			+	+		+	+		+		+
Plasmopora heliolitoides Lindström	+																	
Plasmopora rosa Lindström						+												
Plasmopora rudis Lindström					+	+	+											
Plasmopora suprema Lindström						+		+										
Unidentified corals	+	+	+	+	+	+	+	+	+	+	+	+	+	+	+	+	+	+

TABLE XIX (continued)

Lithology / Localities / Fossils	Reef limestone											Crinoid limestone						
	Hoburgen type									Holmhällar type		Around Hoburgen-type reefs						Holmhällar type
	Gannberg	Grogarnsberg	Sigvalde Träsk	East Sigvalde	Hagby Träsk	Lindeklint	Sandarve Kulle	Tonnklint	Other localities	Fågelhammar	Ljugarn	Gannberg	Kaupungsklint	Lindeklint	Etelhem	East Sigvalde	Other localities	Snabben-Sysne
ANNELIDA																		
Autodetus calyptratus (Schrenk)							+											
Conchicolites sp.															+		+	
Cornulites serpularius Schlotheim				+									+			+		
Spirorbis lewisi Sowerby															+		+	
Spirorbis sp.	+								+								+	+
Unidentified annelid remains	+																	
CRINOIDEA																		
Calceocrinus sp.																	+	
Gissocrinus sp.																	+	
Pisocrinus sp.														+				
Unidentified crinoid remains	+	+	+	+	+	+	+	+	+	+	+	+	+	+	+	+	+	+
BRYOZOA																		
Berenicea consimilis (Lonsdale)														+				
Coenites repens (Wahlenberg)							+		+					+				
Coenites sp.				+											+	+	+	
Fenestella reticulata (Hisinger)									+					+				
Fenestella sp.							+										+	
Fistulipora sp.	+					+												
Ptilodictya lanceolata (Goldfuss)														+				
Unidentified bryozoan remains	+	+	+		+	+			+	+	+	+	+	+	+	+	+	+
BRACHIOPODA																		
Atrypa reticularis (L.)	+	+	+	+	+	+	+	+	+	+		+		+	+	+	+	+
Camarotoechia diodonta (Dalman)													+					
Camarotoechia nucula (J. de C. Sowerby)													+					
Chonetes striatellus (Dalman)				+												+		+
Conchidium conchidium (L.)			+			+		+						+				
Conchidium knighti (J. Sowerby)						+	+		+					+				
Conchidium sp.														+				
Craniops implicata (J. de C. Sowerby)				+									+	+		+		+
Dayia navicula (J. de C. Sowerby)									+									
Delthyris elevata Dalman	+	+		+		+	+		+	+	+	+	+	+	+	+	+	+
Dicaelosia biloba (L.)									+									
Dinorthis rigida (Davidson)	+	+		+	+	+	+		+	+		+		+	+	+	+	+
Dolerorthis rustica (J. de C. Sowerby)														+				
Eospirifer interlineatus (Hedström, non J. de C. Sowerby)						+	+	+										
Eospirifer schmidti (Lindström)	+	+	+	+			+		+	+				+	+	+	+	
Gypidula galeata (Dalman)									+									
Leptaena rhomboidalis (Wilckens)						+	+		+	+				+				

TABLE XIX (continued)

Lithology / Localities / Fossils	Reef limestone											Crinoid limestone						
	Hoburgen type									Holmhällar type		Around Hoburgen-type reefs						Holmhällar type
	Gannberg	Grogarnsberg	Sigvalde Träsk	East Sigvalde	Hagby Träsk	Lindeklint	Sandarve Kulle	Tonnklint	Other localities	Fågelhammar	Ljugarn	Gannberg	Kaupungsklint	Lindeklint	Etelhem	East Sigvalde	Other localities	Snabben-Sysne
BRACHIOPODA (continued)																		
Leptaenoidea silurica Hedström						?												
Leptostrophia impressa (Lindström)				+					+							+		+
Levenea canaliculata (Lindström)	+	+					+		+	+		+		+		+		+
Lingula lewisi J. de C. Sowerby							+											
Lissatrypa sp.						+	+		+	+								
Nucleospira pisum (J. de C. Sowerby)															+		+	
Orbiculoidea sp.				+												+		
"Orthis" tubulata Lindström	+																	
Plectatrypa marginalis (Dalman)						+	+		+	+				+		+		+
Protoathyris didyma (Dalman)	+	+		+		+	+		+	+		+	+	+	+	+	+	+
Ptychopleurella bouchardi (Davidson)	+								+			+		+				
Rhipidomella hybrida (J. de C. Sowerby)														+	+		+	+
Rhynchospirina baylei (Davidson)	+																	
Sphaerirhynchia wilsoni (J. Sowerby)				+									+			+		+
"Spirifer" insignis Hedström					+	+	+		+									
"Strophomena" concinna Lindström in museo													+					
Trimerella sp.																	+	
Unidentified brachiopods	+	+	+	+	+	+	+	+	+	+	+	+	+	+	+	+	+	+
LAMELLIBRANCHIATA																		
Actinopterella sp.							+											
Conocardium sp.	+			+		+	+		+			+	+		+	+	+	
Cypricardinia exornata Lindström in museo							+											
Goniophora cymbaeformis (J. de C. Sowerby)									+									
Ilionia prisca (Hisinger)				+									+		+	+	+	
"Megalomus" gotlandicus Lindström													+				+	?
Pterinea nodulosa Lindström in museo						+	+		+									
Rhombopteria sp.	+								+	+								
Unidentified lamellibranchs	+	+				+						+						
GASTROPODA																		
Bellerophon gemma Lindström							+											
Bellerophon taenia Lindström	+					+												
Craspedostoma elegantulum Lindström							+											
Craspedostoma glabrum Lindström							+											
Cyclonema adstrictum Lindström							+											
Cyclonema apicatum Lindström							+											
Cyclonema cancellatum Lindström						+	+											
Cyclonema distans Lindström							+											
Cyclonema perversum Lindström						+												
Cyclonema zonatum Lindström							+		+									
Cyclonema sp.						+			+									
Euchrysalis lineolata Lindström						+	+											
Euomphalus walmstedti Lindström							+											

TABLE XIX (continued)

Lithology / Localities / Fossils	Reef limestone											Crinoid limestone						
	Hoburgen type									Holmhällar type		Around Hoburgen-type reefs						Holmhällar type
	Gannberg	Grogarnsberg	Sigvalde Träsk	East Sigvalde	Hagby Träsk	Lindeklint	Sandarve Kulle	Tonnklint	Other localities	Fågelhammar	Ljugarn	Gannberg	Kaupungsklint	Lindeklint	Etelhem	East Sigvalde	Other localities	Snabben-Sysne
GASTROPODA (continued)																		
Holopea nux Lindström						+	+											
Holopella minuta Lindström							+											
Loxonema fasciatum Lindström							+											
Loxonema strangulatum Lindström	+																	
Machrochilina bulimina Lindström						+	+											
Machrochilina cancellata Lindström						+	+											
Murchisonia attenuata (Hisinger)	+																	
Murchisonia cancellata Lindström							+											
Murchisonia cochleata Lindström		+					+											
Murchisonia compressa Lindström		+											+					
Murchisonia crispa Lindström						+	+		+									
Murchisonia deflexa Lindström						+	+											
Murchisonia imbricata Lindström							+											
Murchisonia paradoxa Lindström						+												
Onychochilus cochleatum Lindström							+											
Onychochilus reticulatum Lindström							+											
Oriostoma coronatum Lindström	+					+	+						+					+
"*Oriostoma*" *nitidissimum* Lindström						+	+											
Palaeacmaea solarium Lindström							+											
Platyceras cornutum Hisinger						+	+					+		+				
Platyceras spiratum (Sowerby)							+											
Pleurotomaria bicincta (Hall)	+																	
Pleurotomaria cirrhosa Lindström	+					+	+											
Pleurotomaria glandiformis Lindström						+	+											
Pleurotomaria gradata Lindström						+	+											
"*Pleurotomaria*" *linnarssoni* Lindström						+	+		+									
Pleurotomaria lloydi Sowerby	+																	
Pleurotomaria planorbis (Hisinger)	+			+											+	+	+	
Pleurotomaria voluta Lindström							+											
Pleurotomaria sp.	+					+			+									
Poleumita discors (J. Sowerby)	+					+							+					
Poleumita globosum (Schlotheim)						+	+											
Pycnomphalus acutus Lindström	+																	
Tremanotus compressus Lindström	+																	
Trochus cavus Lindström						+	+											
Unidentified gastropods	+	+				+	+		+				+	+	+			
CEPHALOPODA																		
Ascoceras cucumis Lindström							+											
Ascoceras decipiens Lindström						+	+											
Ascoceras manubrium Lindström						+	+											
Ascoceras pupa Lindström						+	+											
Ascoceras reticulatum Lindström							+											
Ascoceras sipho Lindström						+	+	+			+							
Ascoceras sp.						+			+	+	+							

TABLE XIX (continued)

Lithology / Localities / Fossils	Reef limestone											Crinoid limestone						
	Hoburgen type									Holmhällar type		Around Hoburgen-type reefs						Holmhällar type
	Gannberg	Grogarnsberg	Sigvalde Träsk	East Sigvalde	Hagby Träsk	Lindeklint	Sandarve Kulle	Tonnklint	Other localities	Fågelhammar	Ljugarn	Gannberg	Kaupungsklint	Lindeklint	Etelhem	East Sigvalde	Other localities	Snabben-Sysne
CEPHALOPODA (continued)																		
Glossoceras gracile Barrande						+	+											
Gomphoceras sp. (according to Hedström, 1923)									+									
Ophidioceras reticulatum Angelin		+				+	+			+								
Ophidioceras rota Lindström						+	+											
Orthoceras sp.		+				+	+		+	+	+		+				+	
Phragmoceras praecurvum Hedström							+											
Phragmoceras sp.						+			+	+	+							
Unidentified cephalopods		+				+	+			+	+							
TRILOBITA																		
Bumastus sp.	+					+	+		+					+				
Calymene spectabilis Angelin														+				
Calymene sp.	+								+	+			+					
Dalmanites obtusus (Lindström)	+								+				+					
Encrinurus punctatus (Wahlenberg)	+			+					+				+	+		+		+
Proetus conspersus (Angelin)									+			+	+					+
Proetus sp.	+			+		+	+		+	+				+		+		
Sphaerexochus laciniatus Lindström	+						+		+	+				+				
Sphaerexochus sp.						+			+							+	+	
Unidentified trilobites	+	+											+	+				
OSTRACODA																		
Hemsiella maccoyana (Jones)									+					+				
Leperditia gigantea Roemer																	+	
Leperditia gregaria Kiesow				+								+	+			+		
Leperditia phaseolus (Hisinger)												+						
Neobeyrichia nodulosa (Boll)									+					+				
Neobeyrichia sp.				+									+			+		
Unidentified ostracodes				+								+	+	+				

of the reef limestone since it is not known how deep this limestone reaches below sea level.

Stromatoporoids are the dominant element in the reef limestone. Their development seems to have been even more luxuriant than in the type area, especially in the centre of the reef. In between them are Algae, some colonies of branched and massive corals, cephalopods, and some solitary corals, brachiopods and lamellibranchs. Also present are crinoids, which

Fig.196. Fågelhammar South. Reef limestone rich in large stromatoporoid colonies. Hemse Beds.

Fig.197. Detail of a rauk in Fågelhammar North, with stratified sediments intercalated in the unstratified reef limestone. Hemse Beds.

contributed to the matrix, especially at the edges of the reef, as can be seen particularly in the north and south. The shape of the reef is that of a crescent, with the opening about westwards.

A survey of fossils found in the Holmhällar-type reef limestones of the Hemse Beds, in Fågelhammar and Ljugarn, and in the reef-detrital limestones between Snabben and Sysne is presented in Table XIX.

Discussion

It is clear that the deposition of the Klinteberg Beds must have been followed by an increase in water depth. This is particularly demonstrated by the presence of the wide strip of Hemse marlstone (which is comparable to that of the Slite marlstone) but to some extent also by changes in the nature of the limestones. Perhaps the lowermost limestones reflect a deepening of the water that may have taken place at the time of their deposition. In the remaining part of Early Hemse time the depth of the water may have remained more or less the same.

In the second half of Hemse time, the water became shallower again. The best developed reef of Holmhällar type, that of Ljugarn, is found around present sea level close to the boundary between the limestone and marlstone areas. Reefs found close to this boundary, inland, occur at a higher topographical and stratigraphical level, and are of the Hoburgen type (Lindeklint, Guffrideklint). The decrease in water depth is further reflected by the nature of part of the Hoburgen-type reef limestone exposures, particularly in the Östergarn area. At the very end of Hemse time, the rate of this drop of water depth, became more rapid (deposition of the *Dayia* flags).

Whereas during Klinteberg time, the depth contours in the area of central Gotland might have shown a tendency to return to a more northeast – southwest orientation, in the course of Hemse time, their direction became again more east – west, most presumably due to the same epeirogenetic movements which caused the variations in water depth. This, then, will explain the direction of the boundary between the Hemse limestone and marlstone areas.

The occurrence of the Holmhällar-type reef belt in a direction of about 55° may indicate that this was approximately the direction of the depth contours at the time when these reefs began their development. But whereas the reefs at Ljugarn and Fågelhammar could develop into large, typical reefs, that of Ljugarn being the largest, the reefs further northeast along the coast between Snabben and Sysne especially remained smaller and less typical of Holmhällar type, even showing some signs of transition to the Hoburgen type. This suggests that the change in the direction of the depth contours took place roughly at the time of formation of the Holmhällar-type reefs. Also, similarly the somewhat younger reef limestone at Herrvik is not of the characteristic Hoburgen type.

Again somewhat younger are the reefs of Grogarnsberg and Gannberg, in Östergarn Parish. They are the most northern of the several Hoburgen-type reefs in the Upper Hemse Beds. It is just these which show the most distinct signs of formation in shallow water. This indicates that by that time, the direction of coast line and depth contours was about east – west.

EKE BEDS

The parish of Eke, after which the Eke Beds have been named, is in the middle of a strip, with a maximum width of about 5 km, in which the beds are either exposed or form the solid rock directly underneath a cover of Quaternary material. The strip extends from Näsudden in the southwest, over parts of the parishes of Näs, Havdhem, Grötlingbo, Eke, Alva and Rone, to Hummelbosholmen in Burs Parish in the northeast. Some further exposures are found northeast of this strip, in the parishes of När, Lau and Burs. Only in the latter area do reef limestones occur.

The thickness of the Eke Beds is about 14 m in the west of southern Gotland and about 10 m in the east.

Stratified sediments

In most of the area where Eke Beds occur, they are built up by bluish grey marlstone, which after weathering is often somewhat brownish. The rock is often somewhat micaceous and sandy and rather hard; stratification is not always very apparent and the fossil content is high. Northeastwards the content of calcareous matter increases.

Marly limestone is only found in the northeast, e.g., in the higher parts of Lau Backar, along the coast between Nyudden and Nabbens Fiskläge in the northeast of När Parish, north of Österviken in the east of När Parish, and in a few small islands off the coast of Lauviken (Lau Holmar). In general, this rock is distinctly stratified, bluish or brownish grey in colour, rather hard; the marl content varies.

Included in the stratified marly limestone are some reefs.

Reef limestones and related sediments

One of the best exposures of Eke reef limestone is found in an old quarry, west of the road Lau – Gunnor, *about 0.7 km south-southwest of Lau Church*. Best exposed is the northwest wall (Fig.198, 198), which shows the basal part of a reef, which in all likelihood, was originally quite large, to judge from the rather strong dips in the stratified limestone close to the reef (generally about 10°, locally up to 25° towards the reef). This stratified limestone is so rich in large and small remains of crinoids that it should rather be called a crinoid coquina; also present in it are some brachiopods, bryozoans and corals; the matrix is marly. The stratification is rather distinct; in a few places there is cross-bedding. The boundary with the reef limestone is comparatively distinct.

The reef limestone is strongly marly and contains many flat-lenticular colonies of reef builders, which give the reef a vaguely stratified appearance. Corals and stromatoporoids are the most common; the latter do not dominate as strongly as in other reefs of comparable (Hoburgen) type, and the majority of their colonies are relatively small. Bryozoans are also very common. Algae are present. In the matrix, crinoid remains and brachiopods dominate, but the former are not nearly as abundant there as they are in the reef-surrounding rock. Marl pockets are common but generally small; usually they show a very thin stratification. The reef as a whole gives a less

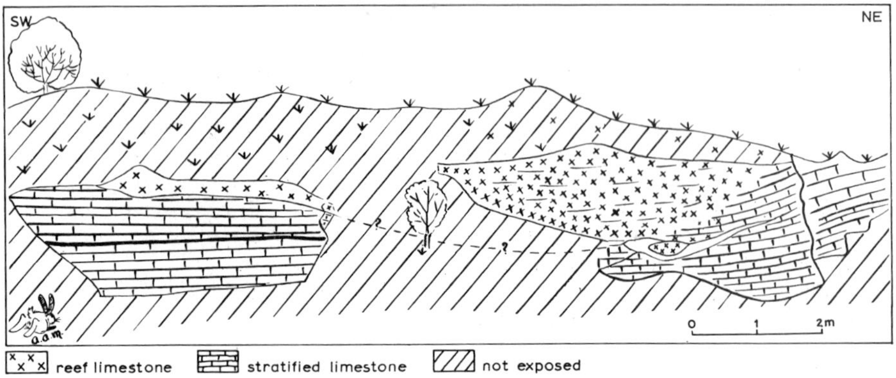

Fig.198. Drawing of the exposures in the northwest wall of an old quarry about 0.7 km south-southwest of Lau Church, at the west side of the country-road Lau – Gunnor. A photograph of the right two-thirds of this wall is shown as Fig.199.

---

unorganized impression than the majority of Hoburgen-type reefs, even though reef builders out of their growth orientations are rather occasionally found.

In the southwest wall of the same quarry, only stratified crinoid limestone is exposed, dipping slightly towards the reef. The northeast wall also shows such crinoid limestone, locally passing upwards into reef limestone of the same nature as in the northwest wall (Fig.200).

On the east side of *Lau Backar*, a section of a few metres high is found west of Hallsarve, 1.25 km east of Lau Church. The lowest part is hidden behind scree material. The higher parts show stromatoporoid reef limestone with a very marly matrix, brownish in colour through weathering. Also present in the reef limestone are several corals, and bryozoans, crinoids and brachiopods. Locally, the reefs show a vague and irregular stratification; in these parts crinoid fragments are distinctly more abundant. The exposed thickness of the reef limestone is 1–3 m. Laterally the rock passes into stratified crinoid limestone with a varying content of crinoid remains and reef debris, generally rather rich in brachiopods; the colour of the sediment is brownish grey to light violet grey. Underlying the reef limestone is 3–4 dm of finely to middle crystalline, light grey to brownish light grey limestone in layers of an average of 1 cm thick. In its upper part, that limestone is poor in fossils; in the lower part, brachiopods, bryozoans and crinoids are represented. The sediment slightly sags under the reef limestone. Underneath this stratified limestone, locally the *Dayia* flags, the uppermost Hemse Beds, crop out, with on top of these, a thin, dark layer of phosphorite with glauconite

Northeast of the previous locality, reef limestone can be seen in a few other exposures. About half-way between Hallsarve and Botvide, crinoid limestone crops out to a thickness of about 4 m. It is a greyish white rock, extremely rich in crinoid fragments, and with a varying content of marl, which locally, through weathering, gives the rock a brownish colour. In some parts, the stratification is very distinct, but in others, very vague. Locally

Fig.199. Reef limestone overlying stratified limestones, as exposed in an old quarry about 0.7 km south-southwest of Lau Church. This picture shows part of the northwest wall of the quarry (cf. Fig.198).

Fig.200. Reef limestone and stratified limestones of the Eke Beds, as exposed in the northeast wall of an old quarry, about 0.7 km south-southwest of Lau Church, along the countryroad Lau – Gunnor.

it can be seen that this crinoid limestone occurs in close connection with reef limestone.

At Botvide, in the northeast of Lau Backar, the exposed wall is divided into two parts. The upper part is some tens of metres west of the road. In a number of outcrops, 1–7 m high, reef limestone is predominantly found, brownish to bluish grey in colour and with a rather strongly marly matrix; it brecciates under the influence of weathering. In addition to stromatoporoids, which are generally thin, corals and bryozoans also occur as reef builders. The rock is rather unorganized, with several colonies not in their positions of growth. Marly crinoid limestone is intercalated as pockets in the reef limestone.

Locally the reef limestone is covered by light grey to light brownish grey, thin and very irregularly stratified limestone, extremely rich in crinoid fragments. The lower 30–40 cm are often also very rich in reef debris, which is partly coarse. The maximum observed thickness of this crinoid limestone is about 2 m.

The lower part of the wall is found along the road. At the base, lenses of marly limestone in a bluish grey marl are exposed. This deposit belongs to the top of the Hemse marlstone. It is overlaid by a 5 cm thick layer of hard, dense, splintery limestone, very rich in *Dayia navicula* (J. de C. Sowerby); the colour of the rock is blue to bluish grey; after weathering it is often more brownish. Locally the layer is divided into two layers, with an interbedded marl film. The overlying Eke Beds are represented by marly reef limestone, exposed to a thickness of up to about 1.5 m, alternating with irregularly stratified marly crinoid limestone. The *Dayia* layer sags underneath reef limestone occurrences.

Along the coast about 0.5 km *west-southwest of Nyudden*, stromatoporoid reef limestone is found, in an exposed thickness of only 1–2 dm. It overlies stratified limestone rich in reef debris (remains of stromatoporoids and bryozoans) and crinoid fragments.

A similar thin remnant of reef limestone has been described by Munthe (1902, p.263) from the beach about 1 km west-southwest of Nyudden. Fig.201, taken from the detailed map of that area given by Munthe, shows the distribution of reef and stratified limestone at that locality. Presumably all reef-limestone exposures formed part of the same reef, which then was at least 12.5 m long and 4 m wide. The stratified limestone is marly and partly splintery. It sags in the shape of a basin under the reef limestone, with dips of up to as much as about 35°.

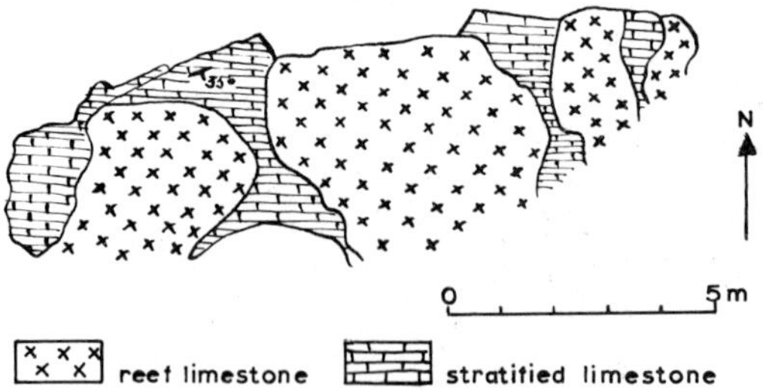

Fig.201. Map, showing the distribution of reef and stratified limestone (Eke Beds) on the beach approx. 1 km west-southwest of Nyudden (När Parish). Presumably all reef limestone formed part of one reef with a somewhat undulating basis.

Along the coast of *Hammaren,* in the east of När Parish, and close to sea level, distinctly stratified grey, splintery limestone is found, which is generally very rich in fossils. Within short distances (order of magnitude of metres) it shows strong variations in the direction of dip, presumably caused by an earlier coverage by reef limestone. Roundstones on the beach, which are in places large, and consist of remains of coral colonies, may represent the last remnants of these reefs. The exposed stratified limestone is locally rich in crinoid fragments, bryozoans, corals and Algae; these parts are less well stratified and may have been formed close to the reefs. Large exposures, also showing reef limestone, are not found in this area.

Sediments comparable to those of Hammaren are also found in the *environs of the Maldes farms*, about 1–4 km southeast of När Church. At the base the *Dayia* flags are exposed, representing the top of the Hemse Beds. Next is a very thin layer of phosphorite. The lowermost Eke Beds consist of a zone of stratified crystalline limestone, about 10 cm thick, very rich in fossils which are partly worn. This rock is very similar to the limestone cropping out at Hammaren. Here, however, this rock is overlaid by reef limestone, locally up to just over 1 m thick. Originally the reef limestone must have been thicker; much has been eroded.

A list of fossils found in the Eke reef limestones and directly surrounding sediments is included in Table IX (pp.60–67). All fossils which have been identified, came from Lau Backar and the old quarry south-southwest of Lau Church.

Discussion

The Eke reef limestone, which in east Gotland, follows almost directly over the Hemse marlstone, seems to have been formed in shallower water. In the Lau – När district (e.g., in the environment of the Maldes farms), the *Dayia* flags, concluding the Hemse Beds, are covered by a thin layer of phosphorite with glauconite, which is believed to represent a stratigraphical hiatus (Spjeldnaes, 1950). Since both phosphorite and glauconite are assumed to form slowly on the sea bottom, the occurrence of a thin layer of these in the Lau – När area may identify a rather long non-sedimentary interlude not related to emergence or erosion. On the basis of graptolite distribution, the top of the Hemse Beds is placed somewhere in the zone of *Monograptus scanicus*; the Burgsvik Beds are certainly of Upper Ludlowian age. Thus, in between the Hemse and Burgsvik Beds, sediments should occur representing a long time interval, comprising a major part of the upper Lower Ludlowian (part of the zone of *Monograptus scanicus* and the entire zone of *Monograptus tumescens*), the entire Middle Ludlowian and the beginning of the Upper Ludlowian. Since the Eke Beds do not themselves indicate a slower rate of deposition than the other sedimentary complexes in Gotland, the presence of a long break in the stratigraphical sequence is indeed likely. It would be illogical then to assume that the break might only be a local phenomenon in the Lau – När district, as was suggested by Spjeldnaes (1950). In the west the thickness of the Eke Beds is only slightly more than in the east. No indications of a hiatus have so far been found in the west, however, and the problem of the geographical and time extension of the hiatus noted in the Lau area has at this stage to be left open.

That one or more breaks should be present *within* the Eke Beds is most unlikely.

Reefs are restricted to the Lower Eke Beds. These and the marly limestones equivalent to them were presumably formed in deeper water than the somewhat arenaceous and micaceous marlstone occurring in the west and in the Upper Eke Beds in the east. This might suggest that the direction of the depth contours at the beginning of Eke time was probably north-northeast – south-southwest; at the end of Eke time, more northeast – southwest. In the east, this resulted in a decrease in water depth during the time of formation of the Eke Beds.

BURGSVIK BEDS

Burgsvik, after which the Burgsvik Beds have been named, is a harbour and adjoining settlement in the northwest of the southern peninsula of Gotland, belonging to Öja Parish.

The Burgsvik Beds are exposed on the surface, or are overlaid by only a thin cover of Quaternary material, in large parts of Grötlingbo Parish. They also occur in a usually rather broad zone, going from there southwards, east of the eastern beach of Burgsviken (the bay on which the harbour Burgsvik is situated), down to Fide. There it shows an eastward extension towards Tubode. From Burgsvik this zone runs westwards, to Valar and from there again, in a generally narrow belt along the coast, southwards, down to slightly south of Hoburgen.

From a petrographical point of view, the Burgsvik Beds present a rather heterogenous picture. The main component is an even- and very fine-grained calcareous sandstone, rich in mica. Often, particularly in the Lower Burgsvik Beds, the sandstone is shaly. Locally the sandstone contains lenses of harder, more strongly calcareous material. In addition, the Burgsvik Beds also comprise clayish marlstone and claystone, while in the uppermost part of the unit a pure limestone occurs as well. The latter is normally developed as an oolite, and in part also contains conglomerates.

Between the sandstone, the clayish marlstone and the claystone, several transitions occur, the same being true for between the sandstone and the oolite.

In the west, the Burgsvik Beds reach a thickness of about 47 m. Towards the east they thin out rather rapidly. The maximum thickness cannot be established by any direct observation. It is known from two deep borings, the first being the well-known boring at Burgsvik (Table XX), the other a lesser-known boring at Vamlingbo. The latter was a boring for water, carried out early in this century near the shop, about 0.5 km north-northeast of Vamlingbo Church. There, first about 40 m of the so-called "kopphäll" (Hamra limestone) were bored through, after which "sandstone with clay" was found to a thickness of at least 40 m, whereupon the boring was stopped. At a depth of about 70 m below the surface (or about 55 m below present sea level) in the sandstone an unusually hard layer, 0.3 m thick, was found, which presumably was a lense of "flinta" (very hard, strongly calcareous sandstone). From the above information, it seems probable that the Burgsvik Beds also beneath Vamlingbo reach a thickness of more than 40 m and have their boundary with the Hamra-Sundre Beds at a depth of about 25 m below present sea level.

A few more borings have actually been carried out in the Burgsvik Beds, but not to great depths; nowhere did these borings reach the base of the Burgsvik Beds.

For the sake of convenience, the present author will divide the Burgsvik Beds into three subunits, based mainly on the results of the boring at Burgsvik (Table XXI).

TABLE XX

The Burgsvik Beds in the core drilling at Burgsvik

Stratigraphical unit	Lithology	Thickness (m)
Upper Burgsvik Beds	typical Burgsvik sandstone	0.33
	somewhat oolitical sandstone	0.24
	typical Burgsvik sandstone	0.48
	argillaceous, shaly sandstone	2.07
	finely-oolitical limestone	0.38
	claystone and clayey shales	0.60
	gap in the core	1.01
	claystone	0.05
	oolite	1.96
Middle Burgsvik Beds	typical Burgsvik sandstone	0.35
	gap in the core	0.80
	argillaceous, shaly sandstone	0.79
	typical Burgsvik sandstone	8.02
	argillaceous, shaly sandstone	0.04
	claystone	0.12
	argillaceous, shaly sandstone	0.07
	typical Burgsvik sandstone	12.51
Lower Burgsvik Beds	argillaceous, shaly sandstone	1.20
	claystone and clayey shales	0.15
	argillaceous, shaly sandstone	2.90
	claystone	0.12
	argillaceous, shaly sandstone	7.00
	claystone and clayey shales	0.07
	argillaceous, shaly sandstone	0.76
	"flinta" (hard, very calcareous sandstone)	0.20
	argillaceous, shaly sandstone	0.30
	gap in the core	0.70
	claystone and clayey shales	4.00
		47.22

TABLE XXI

Summarized lithological composition of the Lower, Middle and Upper Burgsvik Beds[1]

Subunit	Claystone	Argillaceous shaly sandstone	Sandstone	Oolite	Total thickness
Upper Burgsvik Beds	0.65	2.07	1.05	2.34	7.11
Middle Burgsvik Beds	0.12	0.91	20.88	–	22.71
Lower Burgsvik Beds	4.34	12.16	0.20	–	17.40

[1]All thicknesses are given in metres.

Stratified sediments

The succession of strata within the Burgsvik Beds in the west of southernmost Gotland is rather constant in main lines, but varies strongly in detail. The shaly sandstone is characteristic of the Lower Burgsvik Beds, the sandstone for the Middle Burgsvik Beds and the alternating occurrence of shaly sandstone, sandstone and oolite for the Upper Burgsvik Beds. The latter also contain some claystone. The Upper Burgsvik Beds are much more commonly exposed than the two other subunits. Good exposures are found particularly in the west of the southern peninsula. Within these Upper Burgsvik Beds, the strata, with the exception only of the uppermost oolite horizon, thin out laterally and are replaced by other sediments. Sandstone often plays a more important part in the Upper Burgsvik Beds than it does in the succession found in the Burgsvik boring, whereas claystone is only scarcely represented or even completely lacking in some localities.

Some students of the geology of Gotland have correlated the Burgsvik Beds with the Upper Whitcliffe Flags in the English type succession, which at that time were considered to represent the uppermost Silurian (Upper Ludlowian) (Hede, 1921; Säve-Söderbergh, 1941). Murchison (1846, p.27) considered the sandy and calcareous sediments, now collected under the name of Burgsvik Beds, to represent a passage from the Silurian into the Devonian. When White (1950) redefined the Silurian – Devonian boundary in England and placed this at the base of the Ludlow Bone-Bed, Spjeldnaes (1950), drew the consequences of this for Gotland and rejuvenated Murchison's view by drawing the boundary Silurian – Devonian between the Middle and Upper Burgsvik Beds. The international Devonian symposium, held in Calgary in Canada in 1967, moved the Silurian – Devonian boundary up to the basis of the *Monograptus uniformis* Zone. The consequences of this was that the Burgsvik Beds, and also the Hamra-Sundre Beds, of Gotland now have to be considered again as belonging to the Silurian.

Sandstone

The typical Burgsvik sandstone, particularly as found in the Upper Burgsvik Beds, is a fine-grained, calcareous quartz sandstone. On a fresh surface the rock has a bluish-grey colour; a weathered surface often is more or less greyish brown.

Quartz constitutes over 90% of the typical Burgsvik sandstone, the grains being usually angular or only slightly rounded. The size of the grains is generally between 0.05 and 0.1 mm; grains exceeding 0.2 mm are rare. The calcareous cement generally constitutes about 7–8% of the Burgsvik sandstone (Munthe, 1921b). Dispersed through the sandstone are grains of feldspars, which are of the same size as the quartz grains. Flakes of mica, both muscovite and biotite, are common, particularly on bedding planes. Very locally rather large crystals of pyrite were observed.

Locally, the Burgsvik sandstone contains lenses of hard and more calcareous material. On a weathered surface, they appear as lumps, e.g., at the foot of Hoburgen. The people of Gotland call these "flinta". According to Munthe (1921b), these concretions contain about 40% $CaCO_3$.

The thickness of the sandstone layers varies greatly. At one place some layers of 1.5 m thick may be found, without lamination; in other places such layers may show lamination; and elsewhere the sandstone has layers of only 1–4 cm thick. A great variation in the thickness of the layers, particularly in a vertical direction, may even be found within one cross-section.

The typical Burgsvik sandstone may, in the Upper Burgsvik Beds, pass into deposits of a more mixed nature. Thus, layers are found with such a high calcium-carbonate content that the rock is rather an arenaceous limestone. In general, the content in clay is very small. Certain layers, however, are so rich in clay that transitions to arenaceous marlstone and claystone occur. The more-argillaceous layers often show a shaly character.

Oolite

The oolites of the Upper Burgsvik Beds are rather hard limestones, light white grey or greyish white in colour. The rocks are distinctly stratified, with layers varying in thickness from some decimetres up to a few metres. The oolites are exposed at several places, e.g., over a rather large area south and north of Öja Church and in a number of larger and smaller vertical sections which occur along the western shore of the southern peninsula. Locally, however, the oolites may also be absent, e.g., in the western part of Grötlingbo-udd and near Rommunds (northeast of Fide Church). The $CaCO_3$ content of the oolites varies between about 90 and 96%.

The uppermost oolite horizon, forming the top of the Burgsvik Beds, is found in the great majority of the outcrops, which expose rock of this age. For this reason, it is mentioned in the older literature (e.g., Munthe, 1910) as a separate stratigraphical unit. The other oolite layers are not continuous, but wedge out laterally, as is the case with the claystone.

The ooids of the Burgsvik oolites show a great diversity in size and form. In general, they have a diameter of 1–4 mm, but ooids of only microscopic size also occur. Often the ooids are more or less rounded in shape, with a concentric or a radial structure, or they form rosette-like balls, "twins", "triplets" and "multiplets", which are together enveloped by a common oolitic mantle. Moreover, also other, larger ooids of a more elongated or more irregular shape are common. Some of these possess only a very small nucleus of foreign material, such as a quartz grain, but some may have relatively large fragments of shells, corals or bryozoans as a nucleus. The latter often show more the character of incrustations than that of ooids. If the size of the enclosed fossil or fossil fragment is much greater than the thickness of the enveloping oolitic crust and, moreover, the nucleus has a form diverging from that of a sphere, this form may be recognizable also, through the oolite mantle, in the final product. These larger incrustations of fossils are clearly distinguishable from the true, more or less round, egg-shaped or even somewhat elongated ooids. Thus, at first sight, one may tend to consider them as quite different formations. In reality, however, there are many transitions which make it impossible to draw a boundary between both types. This is all the more difficult because both show the same structure.

The ooids are generally sorted in layers according to size. Thus, e.g., the more typical oolites and those incrustating larger objects are often not found together in one layer, but may well occur immediately above each other. The dimensions of the individual ooids may be a standard for the mobility of the water (Cayeux, 1935).

Although the ooids generally have only one nucleus, specimens with two or more are also found. In most of these cases, ooid formation had started around one grain or fossil fragment. When other grains came into contact with the first one, they may have become surrounded as well. In a similar way, two developing ooids may have united laterally to become one final ooid. In both cases, such a growing together had its influence on the form of the final product. In other instances, with single ooids, the thickness of the successive laminae is not the same everywhere. This has also led to divergences in shape. The deviations from a spherical shape, however, are not so great and not so common as in the quiet-water oolites described by Freeman (1962) and Davis (1966). The latter are characterized by a not more than low to moderate sphericity, by eccentric nuclei and by laminae which abut against the nuclei. A further difference between the agitated-water oolites, such as found in Gotland, and the quiet-water oolites is that the latter are not or at best moderately sorted, whereas, as stated, the former can clearly show a certain sorting to size.

In addition to the more or less pronounced concentric lamination, the ooids often also show a radial structure, which presumably is not of primary but of secondary character. It may have been caused by recrystallization (Cayeux, 1935, p.224). The radial structure reveals itself in the arrangement of the calcite crystals. In general, however, this structure is not as clear as the spheroidal lamination, and it is often only to be seen with difficulty. The radial structure may be found both in the normal ooids and in the larger incrustations.

In between the ooids free grains and fragments of fossils are also found. These may be either rounded or angular, and do not differ in their shape in any way from the surrounded grains. Compared to those which served as nuclei, however, the free grains are, on the average, of smaller size. The same was found by Bersier and Vernet (1956) of oolites in the molasse of the Alps. They measured an average of 0.27 mm for the diameter of 100 ooid nuclei against an average diameter of only 0.14 mm in 100 free grains. This is reminiscent of the formation of crystals. There, too, larger ones grow at the cost of the smaller ones, as a result of surface tension. The greater average size of the ooid nuclei is, therefore, a strong argument in favour of oolite formation through chemical precipitation.

The Burgsvik oolites partly occur in close connection with algal balls. The latter, however, reached their maximum development in the algal limestone which overlies the uppermost oolite horizon and which represents the lowermost part of the next stratigraphical unit, the Hamra-Sundre Beds. The similarity between ooids and algal balls may be so great as to be hardly discernible in the field. In microscopic slides, however, the coarser organogenic structure of the algal balls makes the latter easily distinguishable from the ooids. Like the ooids, the algal balls often show a nucleus of foreign material and a tendency to develop a spheroidal shape. With them, too, the final form was greatly influenced by the shape and size of the nucleus and the thickness of the mantle. In his limestone classification, Wolf (1960) united ooids and Algae-encrusted grains with pisolites and Foraminifera-encrusted grains under the collective name of "coated grains" (cf. Bissell and Chilingar, 1967, p.96).

The occurrence of ooids and algal balls close together suggests a preference for a similar environment of formation. Under suitable conditions Algae may even have promoted oolite formation. Goldberg (1957) and Holland et al. (1963) found that oolites may contain a higher Sr/Ca ratio than normal sea water. Higher Sr/Ca ratios are also found in calcareous green Algae (Odum, 1957). Associated with their metabolism are large diurnal pH changes which cause solution and precipitation. This may cause a higher Sr content. If rapid precipitation is required for oolite genesis, the very great vital activity of Algae (photosynthesis) may also in this respect be partly responsible (Odum, 1957). High Sr/Ca ratios may be a palaeoecological indication of algal photosynthesis (Wolf et al., 1967, p.89).

Cayeux (1935) and Newell et al. (1960) list, as favourable environmental conditions for oolite formation: a warm sea, supersaturated with calcium salts, and with water that is pure, very shallow and strongly agitated. Further studies showed that the causes of oolite formation are still more complex and that chemical factors play a more important part with respect to physical factors than was originally thought. The above-mentioned possible influence of pH changes caused by algal photosynthesis is an example of this. Further, Usdowski (1963) demonstrated that also the Mg/Ca ratio and the salinity are of influence on oolite formation: the Mg/Ca ratio should be between 2/1 and 8/1; the lowest limit of salt content lies somewhere between 3.6 and 0.5% (cf. also Wolf et al., 1967, pp.97–98).

On the other hand it can be questioned whether all conditions which are generally listed in literature as being required for oolite formation, are really essential. Although the present author is convinced that the Burgsvik oolites were formed in agitated water, the studies by Freeman (1962) on oolites presently being generated in Laguna Madre, Texas, and by Davis (1966) on Ordovician oolites of Minnesota, have shown that at least some kind of oolite can form in a low-energy environment. With respect to Gotland, another traditionally listed condition is questionable, viz. whether the water in which the Burgsvik oolites were formed, was really pure. It is true that the oolites present themselves as rather pure limestones. But this may also have been caused by the fact that during the time of their formation small clay particles, which were present in the water, did not have the opportunity to settle because of the strong agitation of the water. The presence of claystone as local lenses in the Upper Burgsvik Beds shows that there was a supply of such fine terrigenous debris, which, however, only settled where the water was more quiet. Moreover, if oolite samples are dissolved, a small fraction, generally 1–2%, of insoluble material may remain.

The $CaCO_3$ which forms the matrix of the oolites, has perhaps been chemically precipitated directly from the water to a hard and compact limestone. That is to say, not as a detrital lime mud. An indication of this is found in the fact that higher in the oolite layers, rounded pebbles are found which consist of oolite limestone, both ooids and matrix. Apparently these pebbles were worked loose from lower parts of the layer in question. The time between the formation of the first oolitic limestone and the inclusion of pebbles thereof higher in the same layer must have been relatively short and completely inadequate for the compaction and hardening of a mud under normal subsea conditions. Similar oolite pebbles have also been described as "des morceaux de calcaires oolithiques remaniés" by Cayeux (1935, p.225), who classified these among the so-called pseudo-oolites, and by Bersier and Vernet (1956, fig.10), as "fragments polyoolithiques remaniés". Purdy and Imbrie (1964), working on Recent sediments of the Great Bahama Bank, consider it likely that subaerial exposure also played an important part when lithification of a carbonate deposit took place within a very short interval of time. Newell et al. (1960) state that waters less than 6 ft. deep are optimum for oolite formation. With such slight depth small environmental changes may indeed have lead to temporary subaerial exposure. However, no other evidence for subaerial exposure has been found in the Upper Burgsvik oolites; in contrast to the uppermost Burgsvik sandstone.

As said before, almost all oolite occurrences in the Upper Burgsvik Beds are local and thin out laterally. Probably they were formed in those parts of the sea where the water was most strongly agitated. Only the uppermost oolite deposit is found over most of the area where the Upper Burgsvik Beds occur at, or closely to, the present surface of Gotland. It seems that at the time of its formation, the water was strongly agitated over a large area. Unevennesses which occurred in the sea floor, such as offshore bars (see later) were buried by this deposit, which forms a good index horizon.

Other stratified sediments

As was said while discussing the sandstone, in the Upper Burgsvik Beds the latter sediment may pass laterally into arenaceous marlstone and claystone. In the Lower, and to a much lesser extent, also in the Middle Burgsvik Beds, claystone may be found as a more independent element in the stratigraphical succession. It then forms continuous layers. In the Upper Burgsvik Beds, the claystone is found rather in local lenses of greatly varying size, and in local thin layers.

Generally, the argillaceous marlstone is a rather loose rock, with distinctly lower average grain size than the sandstone. It contains mica. The calcium-carbonate content is about the same as that of the sandstone (Munthe, 1921b). The claystone is usually a somewhat harder rock, very fine-grained and very poor in sandy or calcareous material.

It should be noted, that, with a great variety of mixed sediments occurring, nowhere are transitions between claystone and oolite found. Even oolite underlying or overlying claystone is rare compared to oolite resting upon or being overlaid by sandstone. Marlstone takes an intermediate position in this respect. When the mobility of the water decreased, coarse terrigenous detritus settled first.

The absence of a mixed form of claystone and oolite, thus, suggests that deposition of both sediments took place under different circumstances. The claystone is believed to be a quiet water deposit, while the oolite is a sediment from agitated water.

The present author assumes that the claystones found in the Burgsvik Beds represent deposits from two different environments:

(1) Claystones laid down in water deep enough for their normal sedimentation. To this belong the claystones and argillaceous sandstones of the Lower and Middle Burgsvik Beds. They form a normal succession to the Eke marlstones.

(2) Claystones laid down on flats, in the littoral zone, either in local sheltered depressions within these flats, or on higher places which were only inundated periodically (e.g., at flood tides). Most of the claystone found in the Upper Burgsvik Beds may have been deposited under such conditions.

Sedimentary characteristics

It has already been mentioned that the claystone of the Upper Burgsvik Beds generally occurs in the form of local lenses of greatly varying size or in local thin layers, and was probably formed in very shallow water. Also the formation of oolite gives some indications about the environment in which the Upper Burgsvik Beds are laid down. In this section, some further characteristics of the environment of deposition of these beds will be presented.

Cross-bedding, discordant bedding. In some localities, cross-bedding can be seen in the Upper Burgsvik sandstone. Cross-lamination within sandstone layers (Fig.202) is more common.

Occasionally the boundary between a sandstone and an oolite layer shows discordant bedding. There is no evidence to suggest that such an unconformity may result from by-passing or lack of sedimentary deposition for some time. Therefore, it is most likely that the discordant bedding is produced by the

Fig.202. Transverse section through Upper Burgsvik sandstone, quarry Hansén and Co., Valar. Cross-bedding in an alternation of light and dark grey lamellae. The cut-of indicates that the upper part probably belonged to the filling sediment of a small depression in a beach. At the top an erosion level. (After Manten, 1966a, fig.3.)

erosive action of moving water in a very shallow environment. As will be seen from some of the further characteristics to be described, there are also other indications that strong erosion alternated with sedimentation.

Ripple marks. Fossil ripple marks were observed in the Upper Burgsvik sandstone, very close to the beach of a small bay just slightly north of the peninsula Killingholmen, and about 1.9 km southwest of Valar. The marks were regular, parallel ridges, having a symmetrical shape and with sharp depressions between them. The distance between successive crests was about 4 cm. The orientation of the crests was north-northeast – south-southwest. The marks may be considered as oscillation ripples.

Similar ripple marks have been found by Hadding (1929) on a sandstone layer at Burgsvik. The crests there had an east – west orientation.

Offshore bars. In a chapter "Some remarks as to the tectonics", Munthe (1910) mentioned a "folding of the solid rock" from the environment of Burgsvik. This "folding" is still exposed fragmentarily in the surroundings of the pond Kroksteats Brye. It appears in the field as a small pseudo-anticline, formed by a single oolite layer. Munthe assumed a post-Silurian tectonic origin of this structure. Since, however, in Gotland, no other distinct indications of folding were observed, this structure near Burgsvik has been examined more closely. It has been found than, that the oolite layer in this area wedges out from a thickness of 1.5 m at the lower flank to at the most 0.5 m at the top. This shows that the oolite has been sedimented over an uneven surface. Therefore, the structure cannot be of a tectonic origin (Fig.203). The axis of the structure has an east – west direction.

A second "fold" has been reported by Munthe from a locality some kilometres northward, west of Fide; in this case with a north – south directed axis. The occurrence in one, restricted area of two similar structures with axes that are perpendicular to each other is further proof that these structures are non-tectonic.

A third example has been found by the present author at Uddvide, 3.3 km north of Fide Church. During the summer of 1956, in a quarry in this locality, a claystone layer was observed, which wedged out from a thickness of at least 35 cm to a minimum of about 5 cm on top of a "fold-like" structure. The axis of this structure quite probably had an approximately north-northwest south-southeast direction.

There can, thus, be little doubt that in all these three cases, deposition took place on an uneven surface, and that the unevennesses in these surfaces were of sedimentary origin. The most likely explanation is that these were offshore bars. Seawards of most sandy coasts of the present day, offshore bars are found, in an orientation parallel to the coast. All conditions required for the formation of the offshore bars of the present day, were present in southern Gotland too: a coast with shallow water, wave action, sufficient sandy material and a flat sea bottom. The dips of modern offshore bars (about 6–10°) agree perfectly with those found in the Burgsvik Beds of Gotland.

Distribution of fossils. The great majority of the Burgsvik Beds is very poor in fossils. This certainly holds for the Middle Burgsvik Beds and the lower Upper Burgsvik Beds, which only locally contain one or a few fossils.

In sharp contrast to these beds, the uppermost sandstone, the oolite horizon at the top, and the transitions between these sandstone and oolite, are generally very fossiliferous. Altogether remains of 85 different species were found in these deposits. Among these fossils, there are 23 species of brachiopods, 19 species of lamellibranchs, 14 gastropode species, 5 species of ostracodes and 5 species of trilobites. Particularly the lamellibranchs

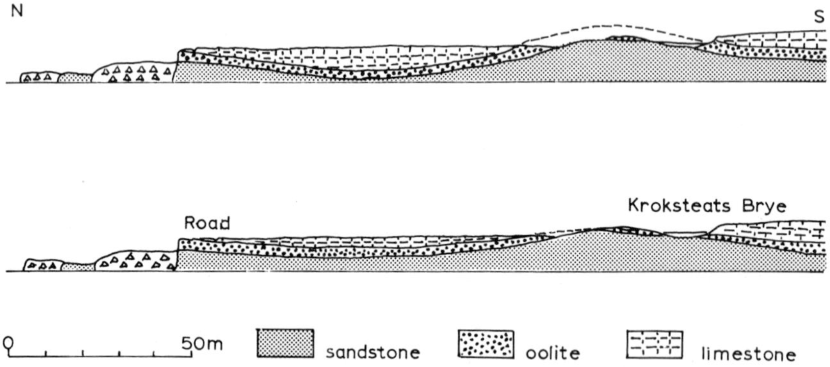

Fig.203. Two views on the "folding" near Burgsvik. Above the interpretation of Munthe (1910), as a tectonic structure. Below the interpretation of the present author, showing how an oolite layer thins over an uneven surface, presumably representing a fossil offshore bar.

and some brachiopods are present in a great number of specimens. Many of the lamellibranchs have thick shells.

The distribution of fossils in the Burgsvik Beds may be compared to that in very shallow parts of the seas of the present day. The deposits of these very shallow marine facies are the terminal products not of one uninterrupted phase of sedimentation, but of a long-lasting succession of alternate sedimentation and erosion. Every time a certain amount of sediment is reworked, it goes together with a separation of the material according to grain size. The shells tend thereby to become concentrated on the beach, whereas they are very scarce elsewhere.

If applied to the Burgsvik Beds, this suggests deposition in very shallow water, of decreasing depth, until the uppermost deposits were laid down very close to the shore line.

Rounded pebbles. In addition to grains of oolite and quartz, the transitions between sandstone and oolite also often contain rounded fossils. The latter are in many cases surrounded by a calcareous film. As already mentioned when the Burgsvik oolite was described, rounded pebbles of oolite may also be found. These are sometimes enveloped by a thin cover of algal tissue. The rounded fossils and oolite pebbles together may give the oolitic-sandstone layers a somewhat conglomeratic character.

Oolite pebbles may also be found in normal oolite layers. The presence of such pebbles suggests that the sea must have been very shallow during the deposition of the Upper Burgsvik Beds. Wave action was so strong that material already deposited could be worked loose again, to be embedded in slightly younger layers. Rounded material is found up to and in the uppermost oolite horizon.

At the level where the fossiliferous, uppermost sandstone passes into the oolite horizon, locally a layer of a few centimetres thick is found in which fossils are sorted according to size.

Burrows. The author has observed only one trace of a burrowing organism in the Upper Burgsvik Beds of Gotland.

Most sea bottoms seem to have supported a variety of burrowing and scavenging animals. Evidence of burrowing and organic disturbance of sediment is found in deposits of many ages. They occur particularly in marine shales and fine-grained sandstones. In deposits which were laid down on an open shore, they are, however, very rare.

Erosion channels. In several cross-sections through Burgsvik sandstone, channels have been seen (Fig.204). They may be up to several metres wide and have a depth of several decimetres, and may cut entirely through from two to several layers of sandstone. As the channels are all exposed in vertical sections, it could not be ascertained whether their course was straight or sinuous, neither could their length be determined. The main direction of the channels is west – east and their sides are usually smooth. The channel filling is generally structureless or horizontally stratified.

Pothole-like excavations. In a few places, the Upper Burgsvik sandstone was found to contain shallow filled-in excavations (Fig.205). These excavations may be circular or irregular in outline and generally have diameters of 7–30 cm. They are filled with thin layers of sandstone which

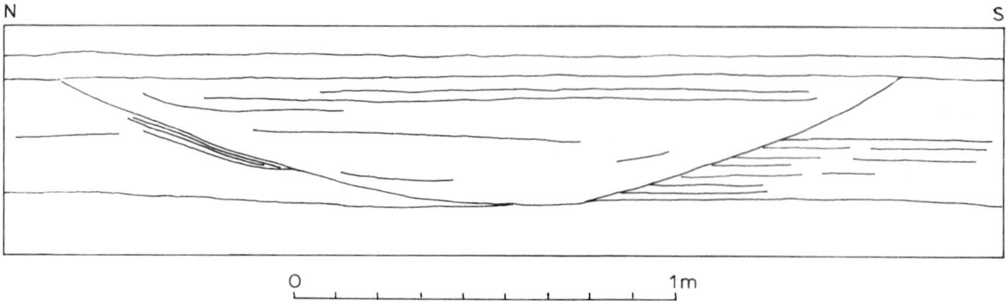

Fig.204. Channel in a cross-section through the Upper Burgsvik Beds at the foot of the second hillock of Hoburgen, near the south of the west wall of this hillock.

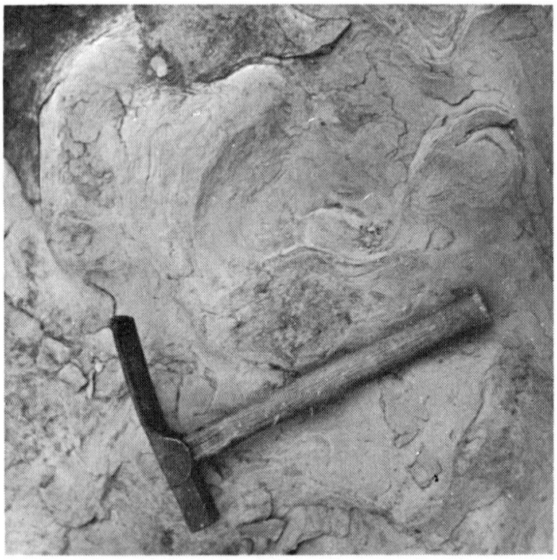

Fig.205. Part of the bottom of a layer of Upper Burgsvik sandstone, showing the filling sediment of pothole-like excavations which were present in the surface of the underlying layer. Quarry Hansén and Co., Valar. (After Manten, 1966a, fig.4.)

usually follow the contours of the excavation. Very occasionally the upper bedding plane of such a layer is vaguely rippled.

The structures probably originated on an open beach, under the influence of interference of water currents (A.H. Bouma, personal communication, 1965

Flute marks. Further erosion phenomena are the flute marks, which were observed in a quarry near Burgsvik and at the foot of Hoburgen. These a discontinuous bulbosities of an elongated form. They may have been produced by eddies resulting from currents, flowing over the sea floor, where these met a small original irregularity in the floor or by a chance scour of a softer patch of sediment.

Mud cracks. At the foot of Hoburgen, mud-cracked surfaces were occasionally seen in layers of Upper Burgsvik sandstone. Generally such layers are a few centimetres thick. The cracks are up to some millimetres wide at the top and narrow downwards. They are usually not visible on the bottom side of the layers. The roughly polygonal pattern of the cracks shows no discernible preferred orientation. This indicates that a stress pattern larger than the local stresses acting centripetally within the polygonal cells, was absent.

Mud cracks are generally the result of shrinkage due to subaerial dehydration. The Burgsvik sandstone layers which show these cracks were probably exposed and dried between tides, or during longer periods.

Rill marks. Rill marks were twice found in the Upper Burgsvik Beds, again at the foot of Hoburgen (Fig.206). They form a dendritic pattern of marks, 1–16 mm wide. The wider marks have a more or less median groove, which is 1–2 mm deep compared to its direct surroundings. This groove excluded, the rill marks are 1–4 mm deep, and their main direction is from west to east.

The marks were presumably formed by currents on a beach which periodically fell dry.

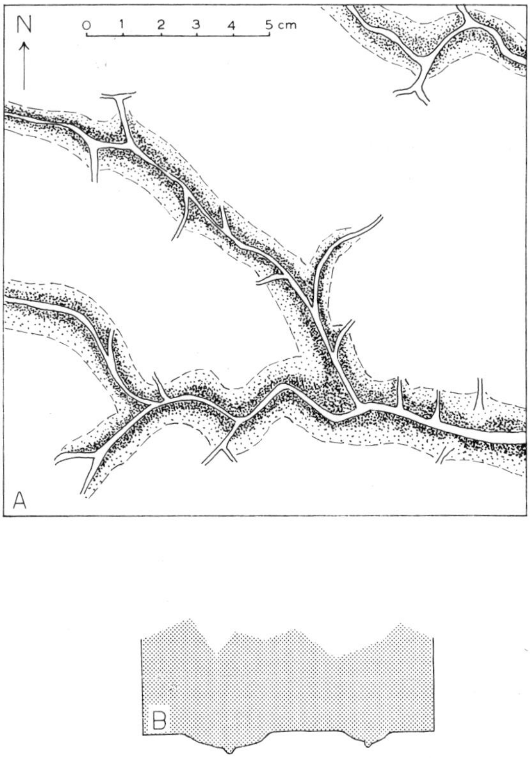

Fig.206. Part of a dendritic pattern of rill marks in the bottom plane of a layer of Burgsvik sandstone, south of the second hillock of Hoburgen, about 1 m below the top of the Burgsvik Beds. A = surface of bottom plane; B = cross section. (After Manten, 1966a, fig.5.)

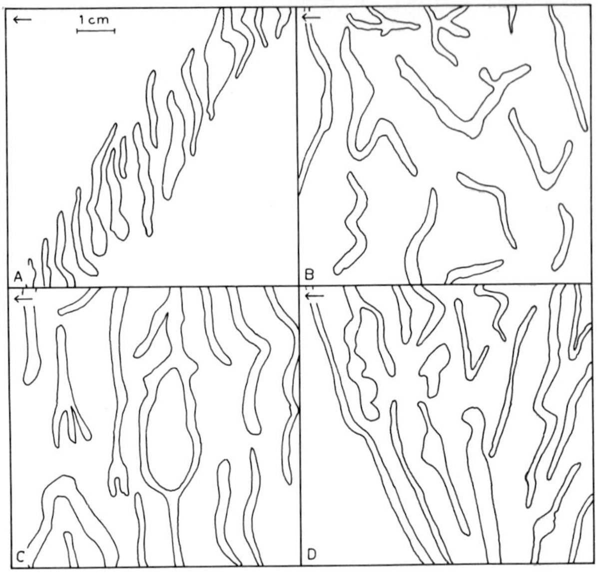

Fig.207. Structure found in Upper Burgsvik sandstone, south of the second hillock of Hoburgen, about 1.30 m below the top of the Burgsvik Beds. The structure consists of collections of small and shallow (1–2 mm) grooves which may occur en-echelon in a belt (A), irregularly distributed (B), roughly parallel (C) or in a fan-like pattern (D). The grooves are found in the top surface of sandstone layers, in parts of one to a few square decimetres large, and are not rare. In cross-section the grooves are, on the average, symmetrical. The structure represents presumably not true rill marks (cf. Fig.206), which show a somewhat different pattern and have a median subgroove, which occurs nowhere in the above marks. Perhaps, however, these grooves are the earliest stages in rill-mark formation. (After Manten, 1966a, fig.6.)

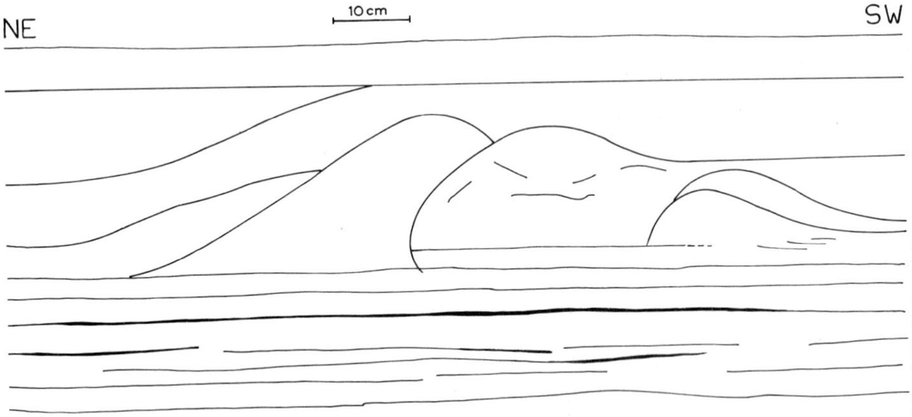

Fig.208. Remarkable lumps, found in a layer of Burgsvik sandstone, 6 cm thick, south of the second hillock of Hoburgen, about 1.50 m below the top of the Burgsvik Beds. The boundaries of the lumps are of similar nature as the stratification planes northeast and southwest thereof. There is no lithological difference between the rock of the lumps and of other parts of the same layer. No other structures of this kind have been seen. The mode of origin of the structure is unknown. (After Manten, 1966a, fig.7.)

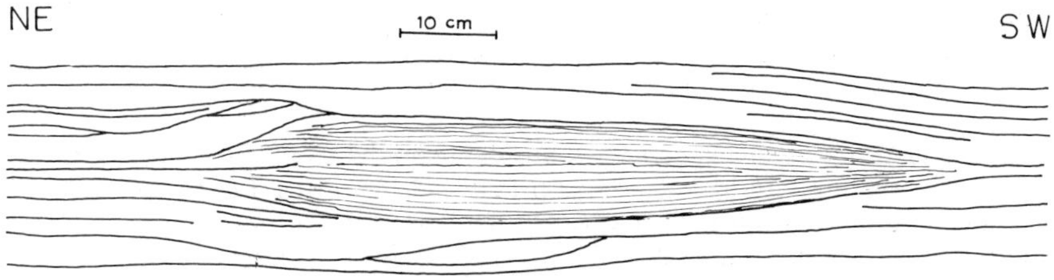

Fig.209. Lens-like intercalation within thinly parting shaly sandstone, Upper Burgsvik Beds, south of the second hillock of Hoburgen, about 1.60 m below the top of the Burgsvik Beds. The lens shows a very thin lamination. In composition the lens is slightly more clayish than the surrounding shaly sandstone, but the difference is not great. No indications were found for deposition in a somewhat sheltered environment; the origin of the structure is still problematic. (After Manten, 1966a, fig.8.)

Some problematic structures. In addition to the structures described above, three structures were also observed, which the present author is unable to explain (Fig.207, 208, 209). However, if one studies sedimentary features in present-day littoral environments, one is struck by the great variety of patterns which can form. Great variability of the bottom is found there even in areas where no differences could be ascertained in wave regimes, currents, bottom topography or sediment types. Consequently, no speculations are made with regard to offering specific explanations for structures as drawn in Fig.207–209.

Reef limestones and related sediments

Only a few subordinate outcrops of reef limestones are found within the Burgsvik Beds. These occur in Burgen and in Närsholm, and belong to the Upper Burgsvik Beds.

In Burgen, the reef limestone shows a bluish to brownish-grey colour and a vague bedding. It has a somewhat unorganized character and is strongly marly. Among the remains of reef builders, fragments of bryozoans play an important and conspicuous part. The compound stromatoporoids and corals are also present and presumably were more important as frame builders. Most stromatoporoids are rather flat. The exposures are usually small and do not give much information about the original dimensions of the reefs.

Between the reefs, there is a sediment extremely rich in crinoid remains. Part of it even forms a crinoid coquina. This limestone is usually brownish, the colour being caused by weathering in combination with the high marl percentage. It is clearly stratified; the individual layers generally being between 5 and 30 cm thick. The bedding planes are somewhat wavy and rugged. In the upper parts a coarse cross-bedding occurs, with dips in varying directions, but the majority pointing towards the northwest (Fig.210). Within these layers, a finer cross-bedding sometimes also occurs. Locally a certain sorting as to size of the crinoid material can be seen. The sediment is very rich in reef debris; mainly fragments of bryozoans, corals and stromatoporoids. This debris is especially abundant in the higher parts; in part it is clearly rounded. Solitary corals are also common.

This reef growth continued into the lower parts of the Hamra Beds.

In Närsholm, a grey reef limestone is exposed in a few small outcrops on the beach, around sea level, and in a few raukar, situated southeast of the lighthouse. The reef limestone shows a marly matrix. As reef builders, stromatoporoids, corals and bryozoans occur. The stromatoporoids are found locally in remarkably large colonies, always wider than they are thick. Bryozoans are usually represented by masses of small fragments. Intact colonies are rare; the largest one observed was 0.80 m thick with a maximum width of 1.42 m. Crinoid stem remains are also extremely common, both in larger and smaller pockets, in between the reef-forming organisms, and scattered in the matrix.

Enormous amounts of Recent debris are found along the present shore. Among this there is a wealth of pebbles, cobbles and boulders derived from crinoid limestone and reef builders. This makes it likely that formerly much more reef limestone was present in that area. Perhaps the faint vaulting in the topography, upon which the lighthouse is situated, hides an unexposed reef.

Compared to the reefs in Burgen, those in Närsholm show characteristics of formation in slightly deeper water. The matrix contains less

Fig.210. Cross-bedding in crinoid limestone, Burgsvik Beds, Burgen.

terrigenous debris and the stromatoporoid colonies are larger, as are also the rare intact colonies of branched bryozoans. The reefs in Närsholm may have been larger too. Finally the cross-bedding and sorting as to grain size, as found in the reef-surrounding sediments in Burgen, are also indications of deposition in shallower and more agitated water.

The Närsholm reefs seem to have formed entirely in Burgsvik time, while the Burgen reefs are enclosed by Upper Burgsvik and by Lower Hamra-Sundre Beds (Hamra limestone). This suggests that reef growth started somewhat later in a westward direction. Still further west, in the area of the present west coast of the southern peninsula, reef formation began still later, viz., in the algal limestone, which represents the lowermost Hamra-Sundre Beds.

Discussion

On the basis of the information presented in the previous pages, it seems most likely that during deposition of the Burgsvik Beds, the water became somewhat shallower. This reached its culmination in Late Burgsvik time. The Upper Burgsvik Beds, as found in the west of the southern peninsula, evidently were laid down in extremely shallow water, very close to the shore line, in a littoral zone which faintly sloped towards an open sea. The occurrence of dessication (mud) cracks and rill marks suggests that at least parts of the sea floor fell dry from time to time, between tides or for longer periods.

The Upper Burgsvik Beds in the east (Närsholm, Burgen) were presumably laid down in slightly deeper water. From that direction, reefs began to develop again at the end of Burgsvik time, forming the onset of the reef formation as this took place in Hamra-Sundre time. A real deepening of the water only started again with the uppermost oolite horizon.

It is interesting to note that the few oscillation ripples and the few offshore bars, found in the Upper Burgsvik Beds, show similar orientations, east – west in some places and north – south in others.

Symmetrical transverse ripples define a line of transport or movement, perpendicular to the strike of their crests. In very shallow water close to the shore, water movement is generally perpendicular to the coast line. This makes it likely that the strike of the crests of the ripples roughly indicates the direction of the coast line during Late Burgsvik time.

As said before, offshore bars are generally parallel to the coast as well.

A combination of the directions found in the oscillation ripples and offshore bars might thus suggest that the Late Burgsvik coast line ran about south-southeast – north-northwest from Hoburgen to north of Valar, then bent eastwards, but after a short distance returned to an about south – north direction (Fig.211).

This direction of the Late Burgsvik coast line also helps to understand the direction of the present west coast of the southern peninsula of Gotland, including the bay Burgsviken. All along this coast the Upper Burgsvik Beds are exposed around sea level. The beds have a primary strike in a north – south direction, parallel to the coast line of Late Burgsvik time, and a slight primary dip perpendicular to the coast line of deposition, that is also perpendicular to the present coast line. The waves of the present Baltic Sea are broken on the protruding ends of the Burgsvik sandstone layers.

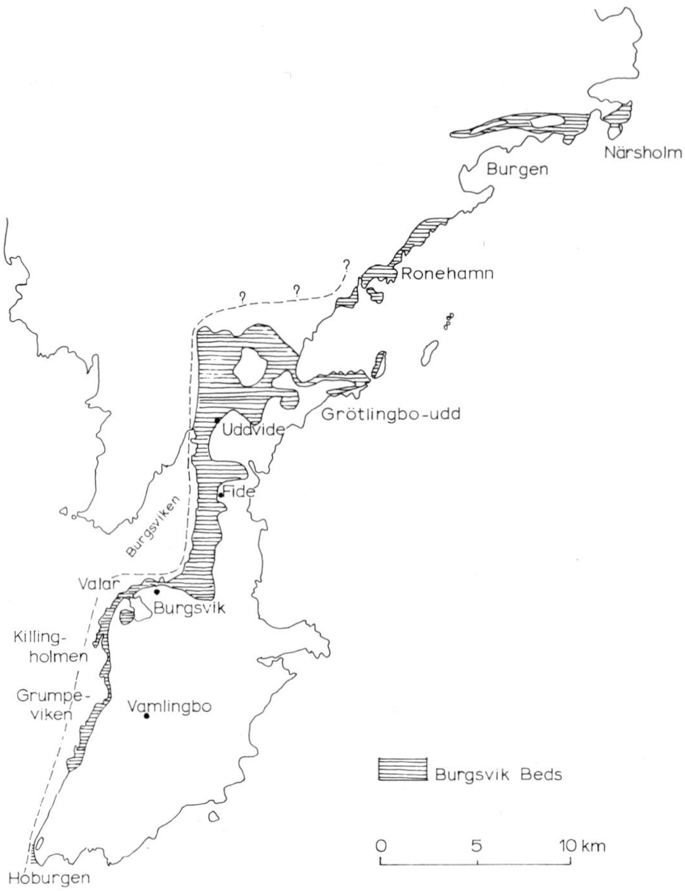

Fig.211. Map of the southern peninsula of Gotland, showing the distribution of the Burgsvik Beds and the supposed direction of the coast line during Late Burgsvik time.

HAMRA-SUNDRE BEDS

As already mentioned in Chapter III, the present author does not accept the boundary which Hede drew between his Hamra and Sundre limestones. Especially at Hoburgen it is obvious that the red crinoid limestone, which Hede considered to be the basal Sundre limestone, is a facies directly related to the Hamra reefs. It is nevertheless probably true that part of what Hede reckoned to be the Sundre limestone is younger than the Hamra limestone, but the present author did not manage to find an acceptable stratigraphical boundary between the two which could replace that proposed by Hede and, therefore, rather presents the two together as the Hamra-Sundre Beds.

Stratified sediments

Hamra limestone

The Hamra limestone is the next to youngest unit in Hede's stratigraphy. The name is derived from Hamra Parish, in the east of the southern peninsula of Gotland. The unit consists of a number of different sediments, such as rather pure to faintly-marly limestone (the typical Hamra limestone), marly limestone, sometimes alternating with layers of marlstone or argillaceous marlstone, marly reef limestone and crystalline limestone. All these sediments can occur at the same stratigraphical height and pass laterally into each other.

Also taken with the Hamra Beds is an algal limestone at the base of the unit. This deposit has been described in the literature as *Girvanella* or *Sphaerocodium* limestone. Hede considers this algal limestone to be petrographically nearer to the Hamra limestone than to the Burgsvik oolite, even though he admits that this oolite can also show transitions to algal limestone. The present author would favour drawing the stratigraphical boundary still somewhat lower, viz., between the Burgsvik sandstone and the upper oolite level. He admits that neither is this boundary sharp, since the two are rather closely connected and present transitions and alternations. But it was shown in previous pages that the Upper Burgsvik sandstone shows a number of characteristics indicative of deposition during a period of shallowing water, which shallowing presumably ended at the time of formation of the above petrographical boundary. The upper oolite is found over nearly the whole area where the Upper Burgsvik crops out and probably was deposited when the shallow water gradually began to deepen again. This deepening of the water continued during formation of the overlying Hamra limestone. Genetically, therefore, the drawing of the boundary below the upper oolite horizon would be more appropriate.

The Hamra Beds occur on the surface over a rather large area of southernmost Gotland, east of the line from Ytterholmen in the northeast, via the southern part of Grötlingbo-udd to Öja Church and from there farther over Valar to south of Hoburgen. In the southeast of the southern peninsula it is overlaid by sediments, which Hede ascribed to the Sundre limestone. It should be noted, however, that even southeast and south of the Sundre limestone, outcrops of Hamra Beds can be found, e.g., along the beach northeast and southwest of Klehammars-udd.

Only in the southwest can the total thickness of the Hamra Beds be assessed by direct observation. This is in the steep cliffs of Marbardshue and Hoburgen, where it reaches a height of 20–25 m. A boring near the shop north of Vamlingbo Church, mentioned by Munthe (1921b), resulted, however, in a much greater thickness. There about 40 m of "kopphäll" was bored through before sandstone was struck. In Gotland "kopphäll" is a somewhat nodular limestone. This must have been Hamra limestone, presumably rich in stromatoporoids. Since no Sundre limestone is known from that locality, the Hamra Beds must have a thickness there of about 40–45 m.

The thickness of the algal limestone varies somewhat, ranging on the average between a few decimetres and about 1.25 m. As a rule it overlies the Burgsvik oolite and is covered by the Hamra limestone. Between the former and the latter, it can also be developed as an intermediate form between oolite and clayish limestone, which joins in rather closely with the Hamra limestone proper. Finally it occurs locally in a reef-like development.

The algal limestone crops out mainly in the northern part of the area in which the Hamra Beds are found at the surface. It is usually found there as a more or less narrow belt between the oolite and the Hamra limestone, or, where the former is missing, between the Burgsvik sandstone and Hamra limestone. In the environment of Grötlingbo Church, this belt broadens somewhat, while from there an eastern offshoot extends to Grötlingbo-udd and Grötlingbo-holm. South of this belt the algal limestone can mainly be observed in some places where more or less vertical sections are exposed.

The problem of the true nature of the algal remains has been touched on in Chapter V. The balls include *Rothpletzella*, with some *Girvanella*, alternating layers of *Spongiostroma*-like material and even some encrusting bryozoan material (cf. pp.73–74).

The algal balls are nearly always developed around one or other nucleus which is surrounded on all sides by algal tissue. Anything could act as a substratum. Knolls of widely different shapes are formed, according to the nature of this nucleus. More or less spherical balls are most common. If, on the other hand, the nucleus was more elongated, as, for instance, part of a colony of the bryozoan *Ptilodictya lanceolata*, then more longdrawn concretions were formed. As a rule, the surrounding algal tissue is not very thick, the maximum being about 1.5 cm. Their external form is often reminiscent of certain knolls of *Lithothamnium*. On a fracture surface an irregularly concentric structure is often recognizable even with the naked eye

The Hamra algal limestone is usually very fossiliferous. In certain parts, the algal balls, together with a number of other fossils, occur so profusely that the rock acquires the character of a biogenic conglomerate.

Locally a more or less clayish, reef-like limestone is found as an equivalent of the Hamra algal limestone. The character of this rock is defined by the presence, together with algal balls, of great numbers of stromatoporoids and corals.

An example of such a reef-like development is found near Kettelviken, in the southwest of Vamlingbo Parish. There, the normal algal limestone overlies the Burgsvik oolite and passes upwards into the mentioned reef-like variety. Both forms of the algal limestone together reach a thickness of 1–2 m Farther northeast such a reef-like development has been observed at Grumpevik. It rests there on crinoid limestone, which shows an oolitic character close to the Burgsvik sandstone.

Munthe (1921b, pp.46–47) described clayish and reef-like algal limestone, in a thin biostromal development, from a quarry near the country road south of Uddvide (Grötlingbo Parish). There, it is more or less clearly inserted between the oolite and algal limestone which is still rather rich in ooids. Munthe gave the following section:

0.10 m	dense to finely-oolitic limestone
0.08 m	algal limestone, oolitic
0.05 m	stromatoporoid reef limestone, argillaceous
0.09 m	oolite
1.30 m+	argillaceous marlstone, alternating with sandstone layers

This section thus shows a close connection between the various types of sediment. The reef-like variety of the algal limestone is even able to replace oolitic limestone, in the same way as elsewhere oolite, sandstone and argillaceous marlstone can replace each other.

Finally, still another variety of the algal limestone should be mentioned. This is reported by Hede (1921, p.76) from, among others, the southern beach of Grötlingbo-udd. There algal balls occur only scarcely, and are nowhere characteristic of the deposit, whereas other fossils play a very prominent part.

The typical Hamra limestone is more or less clearly bedded and, on a fresh fracture, dark grey to bluish grey in colour. In a weathered condition, it is grey to sometimes brownish grey. Usually it is only slightly contaminated with marl. The calcium-carbonate content ranges generally from 84 to 94%. The rock is often slightly bituminous. Fossils, especially shells of ostracodes, are common. The typical Hamra limestone is mainly found in the east of the southern peninsula.

Different developments of the Hamra limestone are found in the west of the peninsula of south Gotland, where facies differences within the Hamra Beds are most strongly developed. In contrast to the typical Hamra limestone, in which the author observed neither cross-bedding nor rounded fossils, these were found in a number of places in the west. There, the limestone is generally also more marly.

In the western and southwestern parts of the parishes of Vamlingbo and Sundre, sediments are found which are no doubt synchronous with the typical Hamra limestone. In some places, they appear as beds of normal Hamra limestone, a few centimetres thick, alternating with layers of argillaceous marlstone. This is the case, e.g., in extensive parts of the area between Västlands and Kvarna (Vamlingbo Parish). In other parts, it is developed as marlstone or marl shales with limestone bands in between. As such it is mainly found in the 20–30 m high cliffs between Kettelviken and Hoburgen, as well as in nearly the whole of the southern wall of Hoburgen, underneath the "Hoburg marble". The calcium-carbonate content of the marlstone is generally about 50–55%. The sediment is usually very fossiliferous, and a certain amount of these fossils may presumably be considered as reef wash.

The other varieties of the Hamra limestone will not be discussed here individually. The reader is referred to the description of the geological map Burgsvik by Munthe (1921b).

Sundre limestone

The Sundre limestone is the youngest unit in Hede's stratigraphy of the Palaeozoic of Gotland. The name is derived from Sundre, the southernmost parish of the island, in which the sediments of this unit have their largest distribution.

In some places the Sundre limestone is seen to overlie Hamra limestone, e.g., in Storburg (Hoburgen), where the Sundre limestone is developed as a red crinoid limestone, the so-called "Hoburg marble". The contact is concordant and indistinct.

Since no deposits other than those of the Quaternary overlie the Sundre limestone, the original thickness of the unit can not longer be determined. Where Hede (1921) speaks of a thickness of a few metres up to about 10 m, this is an indication only of the thickness remaining after erosion.

Exposures of Sundre limestone are found mainly in the southeast and south of the southern peninsula of Gotland, southeast of the line between Faludden in the northeast and Sundre in the southwest. Locally occurrences are also found northwest of this line, e.g., west of Hamra Church and in Vamlingbo Parish.

Munthe (1921b) distinguished four subunits in the Sundre limestone, viz. (a) a red crinoid limestone ("Hoburgsmarmor", "Hoburg marble"); (b) a grey limestone with some crinoid fragments ("grå krinoidkalk"); (c) a grey limestone, which contains crinoid remains, but also colonies of corals and stromatoporoids ("grå revkalk"; grey reef limestone; it will be shown that this is no true reef limestone); (d) red-brown reef limestone ("rödbrun revkalk"; this is a true reef limestone, described by the present author as reef limestone of Holmhällar type).

"Hoburg marble" is known mainly from Hoburgen, where, however, much of it has been quarried, especially in the first hillock. It is still present mainly on top of the high, vertical cliffs. Another well-known occurrence is at Hallbjäns, where it is still periodically quarried.

The rock is a sparry limestone, built up by numerous crinoid remains in a limestone matrix. Other fossils are poorly represented. The crinoid fragments are generally strongly recrystallized; in both transversal and radial sections, they show a fine net-like porosity. The colour of the "Hoburg marble" is red, caused by a thin film of Fe_2O_3 covering the walls of the numerous pore spaces in the crinoid material. The matrix is finer and not porous and is grey in colour.

The "Hoburg marble" is usually very thickly bedded. It is often traversed by broad vertical cracks, which give rise to the formation of large blocks. At the east and southeast side of Storburg, several of these have loosened and have fallen down.

The *grey crinoid limestone* is generally a grey and middle coarse limestone, distinctly poorer in crinoid remains than the "Hoburg marble". Other fossils are rather poorly represented. Occasionally small crystals of pyrite are found, or yellowish-brown spots where such crystals were weathered. Exposures are found, e.g., in the barren plains west of Sundre, near Enviken and south of Stockviken.

Generally the rock is more or less distinctly stratified. Locally there is some cross-bedding, exposed, e.g., in weathered walls such as those bounding cracks in the plains near Sundre. There, occasionally stromatoporoid or coral colonies were found which were somewhat ground and rounded. Also small insertions of a clayish character are found there. Eastward, the sediment is purer. According to Munthe (1921b), the calcium-carbonate content of the grey crinoid limestone in the east is on the average 95–98%.

The stratification is generally horizontal with local dips of a few degrees occurring, however. Munthe (1921b) measured the following examples: south and southeast of Skoge (Sundre Parish) dips of 4–5° to the south-southwest; southeast of this, locally a similar dip in a west-southwest direction; south of Hamra Church, dips of a few degrees to the southeast or south-southeast; south of the southern Sällmung farm (Hamra Parish), 5–6° eastward; southeast of Faludden Lighthouse 10–15° west-southwestward; west of the western Sibbenarve farm on Faludden 10° in about the same direction. On the southern beach, south of Bringes farm, a slight arching of the layers was observed. It may be possible that these slight dips are connected with accumulations of reef builders underlying the exposed grey crinoid limestone

In the opinion of the present author, the *grey reef limestone* of Munthe (1921b) is a variety of the grey crinoid limestone described in the preceding paragraph. The difference between the two is that the "reef" limestone also contains such fossils as stromatoporoids, corals and bryozoans. They are generally not abundant and certainly are not characteristic of the rock. Only in a few restricted localities is the number of stromatoporoids and other potential reef builders much larger; there, they created a rock of real unstratified reef-limestone character. An example is found northeast of Marbardshue, where the reef limestone overlies the "Hoburg marble". In general such reef developments are an exception. The great majority of Munthe's grey reef limestone is thickly bedded. Bedding planes usually appear horizontally, but in a few instances slight and local deviations are found, e.g., in the plains east and west of Sundre. They may be caused by small reef-like developments, which are not yet exposed.

Reef limestones and related sediments

Two kinds of reef limestone are found in the Hamra-Sundre Beds, viz. of the Hoburgen and of the Holmhällar type.

Hoburgen-type reef limestones

Exposures of Hamra reefs are found mainly in the west of the southern peninsula of Gotland, in a number of steep cliffs close to the shore. The most important of these are the cliffs of Hoburgen. The reef limestone which these show has already been discussed in Chapter VII; only little needs to be added to that in this chapter.

Other exposures are situated near Kettelviken, and a little more landward in the cliff near the farm Juves (Sundre Parish) which is known as Klev. This cliff forms part of the highest erosion level of the *Littorina* sea. In Burgen, farther to the north, reefs occur in the lowermost part of the Hamra-Sundre Beds.

The great weight of the reef limestone masses led, in most cases, to the formation of faintly basin-shaped depressions in the sediments underneath the reefs. This can be observed very clearly at Hoburgen, where these downward bucklings occur not only underneath the present hillocks, but also in parts of the Burgsvik sandstone exposed more seawards (Fig.212). This indicates that in former times, reefs must have been present there as well. Similar faint and local depressions in the Burgsvik sandstone are found, among others, on the western beach of Killingholmen (in the north of Vamlingbo Parish) and in the northeastern part of Grötlingbo-udd, where nothing of the reefs themselves is present any more.

The dimensions of the reefs in Hoburgen are generally moderate. The thickness of the exposed reef limestone is generally less than 20 m, very often even less than 10 m. The thickest exposures are found in the west wall of the first hillock (Storburg), where the wall reaches locally up to 25 m in height (Fig.212). Usually, however, not the whole section consists of reef limestone or of the material of only one reef. It is unlikely that any of the reefs, exposed in that hillock, has ever been much thicker than the reef limestone which is present. The surface of the Storburg is faintly accidentated, the higher parts being probably determined by the occurrence of reefs.

Fig.212. Hoburgen, Storburg, western cliff. Reef limestone and stratified limestone, belonging to the Hamra-Sundre Beds. At sea level Burgsvik sandstone is exposed. Note the dip in this sandstone, caused by differences in the weight of the younger sediments, particularly by the heavier reef limestones.

Where reef limestone is exposed, it can be seen in several places how the stratified sediments from all sides advanced over the reef. Locally the uppermost reef limestone shows a vaguely stratified character caused by a predominance of flat-lenticular reef builders and a comparatively high volume of matrix; many crinoid are present there, both scattered in the matrix and in intercalations of crinoid limestone. Crinoid limestone directly on top of the reefs shows much reef debris, but about 30–50 cm higher only little such debris is found any more in the crinoid limestone. All these data suggest that the surface of the hillock represents about the end of the reefs. Similar situations can also be found in a number of places on the other hillocks.

The fact that in Hoburgen hardly any indications of interruptions in reef growth affecting the entire surface or a major part of a reef are found (Fig.217), indicates that the water in which the reefs developed was always deep enough to prevent erosion of the reef surface. More local interruptions in reef growth, represented by intercalated parts of stratified limestone do, however, occur (Fig.213).

It was already demonstrated in Chapter VII that in several places in the Hoburgen area a number of reefs may have started growth close to each other, leading to severe competition during the later stages of their development and frequently to the end of some of the reefs.

In the south of the west cliff of the second hillock of Hoburgen, a reef-limestone part is found almost entirely built up by stromatoporoids (Fig.214). Not far from the northern end of the Storburg (Fig.77), a part is seen which is quite a massive coral section. Generally the composition of the reef limestone is that of the varied and rather unorganized rock described in Chapter VII. In crinoid limestone underneath the reefs exposed in the Storburg, several roundstones and rounded fossils were observed.

The second burg of Hoburgen (Fig.215) contains a peculiar and rather complex example of reef development, showing how different various parts of one reef can be.

The reef limestone in the short southwest-facing cliff near the southern end of this second hillock is characterized by the presence, at the bottom of that cliff, of many distinctly dipping stromatoporoids (see p.164; Fig.69). A little of the reef limestone in which these stromatoporoids occur, is also exposed in the main westward facing cliff, around the corner, north of the small cliff. There it is found to be overlaid by stratified marly limestone, containing a number of flat compound corals and stromatoporoids, particularly in a few of the layers. This stratified sediment is again overlaid by reef limestone. The latter is very marly and unorganized in its lower parts, with many fossil fragments and fossils not in their growth orientations, and also with a 3 m long marlstone lens which is about 10 cm thick in its centre. Higher upwards, reef growth seems to have met with more favourable conditions. The reef surface then split into two or perhaps even more centres of growth. One of these was almost completely built by stromatoporoids, as illustrated by Fig.214; locally in between the stromatoporoids there are a few small corals or bryozoans or a little marl with solitary corals, brachiopods and an occasional crinoid fragment may also occur. Southwards the rock passes into gradually less massive reef limestone, with smaller reef builders, some of which dip or lie upside down.

Between this growth centre and a second, exposed in the short northwest – southeast wall, there is a depression which contains red-mottled, sparry crinoid limestone ("Hoburg marble"). The second growth centre is represented by very unorganized reef limestone, which partly approaches the character of reef talus, with many fossils out of their growth positions, many fossil fragments and much marl. Nevertheless it is reef limestone, although it remains difficult to believe that it was formed synchronously with the stromatoporoid rock described above and even formed part of the same reef. In the course of its development, this centre of reef growth

Fig.213. Intercalations of stratified limestone within a reef. Photograph taken to the south of the western cliff, Storburg, Hoburgen. Hamra Beds.

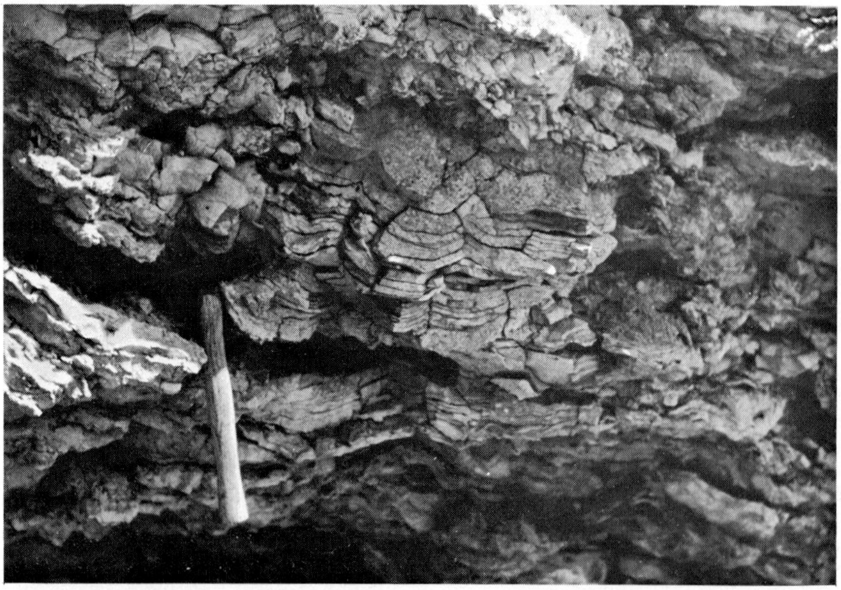

Fig.214. Reef part almost entirely built by large stromatoporoids. Hoburgen, second hillock. The photograph was taken at a locality about 9 m from the reef part drawn at the right margin of Fig.215. Hamra-Sundre Beds.

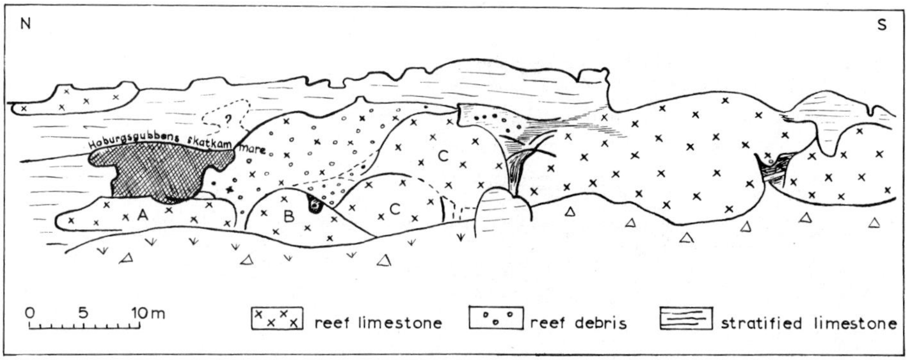

Fig.215. The western cliff of the second hillock, Hoburgen. Hamra-Sundre Beds. At the left a cave, known as "Hoburgsgubbens Skatkammare" (the treasure chamber of grandfather Hoburgen). The reefs A, B and C presumably grew more or less synchronous, probably together with a fourth reef (D), which was more eastward, behind the three which are drawn. It seems likely that D started growth somewhat earlier, and supplied the debris found in between A and B. The latter two reefs rather soon died, perhaps under the influence of D. Reef C kept on growing and filled a small depression in the surface of B with its debris. This debris consists of colonies and fragments of stromatoporoids, in random positions, inbedded in a matrix of marly limestone with a few crinoid remains. Also part of the surface of B itself became buried under debris of reef C. However, much more debris was supplied by the west-northwestward expanding reef D. Little true reef limestone of this reef is exposed. The top of the cliff consists of stratified crinoid limestone.

The southern part of this western section is also mainly built by reef limestone. Details are shown in Fig.53 and 69.

---

moved its area of maximum growth somewhat to the southeast; initially it was situated in the left of this small wall, but at the end of reef growth about in the middle of that wall. The thickness of the reef limestone is at its maximum 6 m. In the southeast where this growth centre directly overlies the older reef with the dipping stromatoporoids (Fig.69) this younger reef limestone interfingers somewhat with stratified limestone.

The reef limestone at Kettelsård is comparable to that in Hoburgen. The best exposure is at the west side, where a lenticular reef about 7 m long and, in its centre, 3 m thick is found (Fig.218). Almost all reef builders are comparatively thin and the matrix is marly. Most other reefs are thicker. Much crinoid limestone is exposed in this site.

Reefs belonging to the oldest part of the Hamra limestone are found in Burgen. They already started growth during Late Burgsvik time. The character of the reefs is even more unorganized than in Hoburgen and the matrix is very marly. Among the reef builders, bryozoans played an important part, but stromatoporoids and compound corals were the main frame builders. The reef limestone often shows a vague stratification and there are many intercalations of marly limestone rich in crinoid fragments. The exposures

Fig.216. Hoburgen, west side of the third hillock. Hamra Beds.

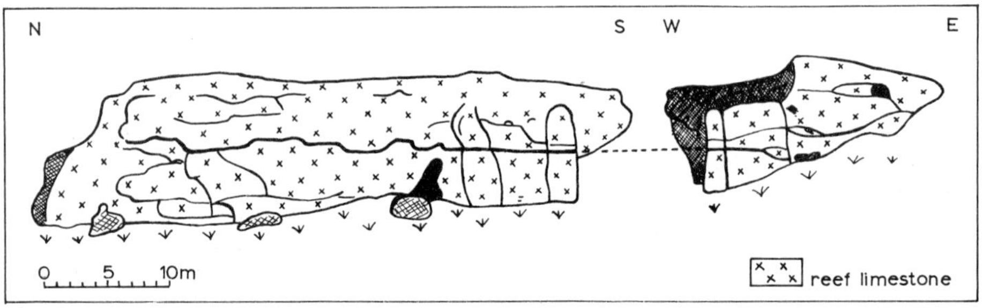

Fig.217. Hoburgen, the west and south side of the fourth hillock, exposing reef limestone which over most of its exposed surface shows intersection by a more or less horizontal plane, which may represent a level at which reef growth over most of the reef surface was interrupted (cf. Chapter VII, p.131). This is the only presumed interruption in reef growth of this extent known from Hoburgen.

---

are generally small and no complete sections through reefs have been seen. It seems most likely, however, that the dimensions of the reefs were small, less than of those in Hoburgen. They presumably formed in shallower water than the reefs at Hoburgen. Surrounding the bluish to brownish-grey reef limestone in Burgen is an irregularly-stratified crinoid limestone, which is often rich in reef debris. Several crinoid fragments and parts of reef debris are clearly rounded and the sediment shows coarse cross-bedding.

Holmhällar-type reef limestones

The reef limestone found in the raukar field of Holmhällar, about 6.8 km east of Sundre Church, at the east coast has been discussed in detail in Chapter VIII. Nothing need be added here. A few further exposures of comparable nature should, however, be described briefly here. All are found in the east of the southern peninsula of Gotland.

Roughly 0.75 km northeast of Holmhällar, in the territory of Hamra Parish, is the raukar field of Hammarshagahällar. The similarity to

Fig.218. Reef, surrounded by stratified limestones. Hamra Beds. Kettelsård.

Fig.219. Raukar of Holmhällar-type reef limestone in the forest near Austre. Sundre Beds.

Holmhällar is striking. The red-brown variety of the reef limestone dominates; in several places red and greyish green reef limestone occur in great confusion. Calcite-filled veins are common, generally of restricted length and of varying thickness, though rarely thicker than 1 cm. As in Holmhällar, there are debris-filled depressions, pools with a different fossil content, fissures, lines indicating interruptions in reef growth, etc. The number of crinoid remains seems to be slightly higher than in Holmhällar. The highest raukar reach to about 8 m above present sea level, suggesting that the thickness of the reef in the centre must have been at least 10 m. Reconstruction of the reef leads to a crescent shape also there.

The small island of Heliholm (Vamlingbo Parish), about 0.6 km south-southeast of Holmhällar presents raukar in a great variety of forms and contains niches, rooms, caves, gates, etc. (Fig.78, 85, 87, 88). With the exception of a part at the north and northeast side, the entire island is surrounded by raukar, together constituting a zone of 15–50 m, on the average 30 m wide and about 2 km long. The highest raukar reach about 6 m above present sea level; the largest raukar are found in the south. All features found in the Holmhällar reef are also found in Heliholm.

Southeast of Rems, near Austre, 1.2–1.8 km west of Holmhällar, a smaller raukar field is found in the forest, partly hidden by wind-driven sand (Fig.219). The raukar are smaller and more scattered than in the three previously described raukar fields; they have been excavated by the *Littorina* sea. Detailed study of the reef limestone in the field is hindered by the brecciation which has taken place and by the patina which covers the weathered rock. There seems to be much similarity with the reef limestone of Holmhällar.

The reef limestone in the small island of Skäret, east of the bay Enviken, has not been seen by the author.

In the southeast of Faludden (Öja Parish), reef limestone is exposed around sea level. It contains large stromatoporoids and coral colonies. The military authorities who operate a radar observation station there, did not allow detailed studies of these outcrops.

A moving trough?

In 1956, F.P. Agterberg paid the author a visit of some weeks in Gotland While familiarizing himself with the geology of southern Gotland, Agterberg became much impressed with the differences in thickness which some of the stratigraphical units there seem to exhibit. He attempted to explain these by assuming a north – south orientated trough, which gradually moved eastwards during the time that the youngest Palaeozoic beds of Gotland were deposited. This moving trough could be the reflection, at the surface, of viscous matter, presumably magma, flowing at great depth below Gotland, on account of a difference in load between the area around the basin and the basin centre (Agterberg, 1958).

The present author is unable to share the conclusions reached by Agterberg.

In the first place, the observed differences in thickness of the stratigraphical units in southern Gotland, as far as these are realistic, can be explained in a simpler way.

The thickness of the Eke Beds which is about 10 m in the east and about 14 m in the west, does not show abnormal differences.

The Burgsvik Beds, about 50 m thick in the west of the southern peninsula, thin out rather rapidly towards the east, and in the Burgen area this thickness does not amount to more than a few metres. It should be noted, however, that the Burgsvik Beds were laid down under rather special conditions, partly very close to the coast. Moreover, the conditions at the time of deposition, were not identical in Burgen and, e.g., Hoburgen. The boundary between the Eke and Burgsvik Beds strikes about N 50° E and it dips about 0°.25'. During Burgsvik time, epeirogenetic movements seem to have taken place, as discussed in Chapter IV. These were similar to those which are assumed to have influenced the depositional pattern also of part of the older stratigraphical units in Gotland, as indicated in a few places earlier in this chapter. The boundary between the Burgsvik and Hamra Beds strikes about N 30° E and has an average dip of about 0°.30'. Whereas Burgen remained at about the same distance from the coast during the whole of Burgsvik time, Burgsvik and Hoburgen came to lie closer to the beach; the latter alteration of environment no doubt also influenced the sedimentation of the Burgsvik rocks in that area.

The differences in the thickness of the Hamra limestone, as described in earlier literature, are disputable, because the upper boundary of the Hamra Beds, as defined by Hede, is not a time boundary. It was mentioned that in Hoburgen, this boundary is drawn between the grey and red crinoid limestone, both of which, no doubt, are facies related to the reefs. In other places, such red crinoid limestone was also assumed to represent the basal Sundre limestone. Consequently it may very well be that there are no real differences in the thickness of the Hamra Beds, but that these were only suggested by the absence in certain areas of red crinoid limestone.

Although the data which led Agterberg to his theory are less puzzling than he supposed and certainly do not need such a far-fetched explanation, the question may also be put whether a flow of material in the depth is at all plausible under the given conditions.

Agterberg referred to the squeezing out of salt towards a "salt pillow", as Trusheim (1957) assumed to have taken place in northwestern Germany. The situation there, however, was not directly similar: salt is a material of low specific weight and it flowed at a depth of only a few kilometres. The magma flow, as supposed by Agterberg, must, as he himself stated, have taken place at greater depth. Moreover, the specific gravity of the material that probably flowed below Gotland must have been higher and its viscosity lower.

The Silurian basin, as is evident from several data presented earlier in this book, must have been a large and shallow-bottomed sea bordered by low-elevation continents. The load differences must thus have been very small and at great depth this can hardly have played a part of any importance, as the following reasoning illustrates.

Let us assume that in the centre of the Silurian basin, water depth was 200 m and the height of the borderland at some distance from the coast was 50 m. If we take the specific gravity of the continental rock to be as high as 2.9, at a depth of 200 m below sea level there will have been pressures of 72.5 kg/cm^2 below the land and 20 kg/cm^2 at the floor of the basin centre. Let us further assume that on the continent, heavy crystalline rocks (sp. gr. 2.9) constituted the entire succession, whereas underneath the basin centre, 1 km of sediment (sp. gr. 2.4) and further only lighter crystalline rocks (sp. gr. 2.7) are present. At a level of 10 km below sea level, which is the least which will have to be assumed for a flow of material such as assumed by Agterberg, the pressures will then have been 2,914.5 and 2,636 kg/cm^2, respectively. With a horizontal distance of 100 km, the difference in load could not have been more than 2.785 kg/cm^2/km, or only about 0.1%. If, as is more likely, the crystalline rock that constituted the continent and the basement under the basin centre, had a similar specific gravity, the difference in load would have been only about 0.074%.

In summarizing, therefore, the present author sees neither a need nor a sound basis for assuming a flow of viscous material in the depth below Gotland during the time that the Eke, Burgsvik and Hamra-Sundre Beds were deposited.

Discussion

The stratified Hamra limestone and the varieties thereof suggest that the deposits found in the west were laid down in somewhat shallower and more agitated water than the rocks farther to the east. The stratigraphical succession in Hoburgen, from the uppermost Burgsvik Beds, via the algal limestone, to the Hamra limestone with the reefs and the surrounding and covering crinoid limestones makes it most likely that during most of the time that the majority of the sediments found at Hoburgen were deposited, the water became gradually deeper. In the sediment at the foot of the Storburg, rounded fossils are much more common than in the younger crinoid limestone and the reef debris embedded therein is often angular. The reefs also show the likelihood that the water depth slowly increased; the thickness of the reefs in proportion to the horizontal extension suggests this, but also the fact that growth interruptions of some extent are hardly found. At any rate, a decrease of the water depth during the time of their formation, as supposed by Hadding (1933, pp.49–50), seems very unlikely. This even more, when we compare the reefs of Hoburgen with those in, e.g., the Hemse Beds, which the present author assumes to have developed in shallowing water and which bear several characteristics of such an environment. It is more difficult to draw any conclusions from the youngest deposits, such as the Sundre limestone in the southeast, as to alterations in water depth, but it is not unlikely that the depth remained roughly the same during the time it was laid down.

The distribution of both the Hoburgen and Holmhällar-type reef limestones suggest a roughly south-southwest – north-northeast course of the depth contours at the time of their formation, in the north probably turning slightly to a more northeastward direction.

Chapter XII

PALAEOECOLOGICAL OBSERVATIONS ON SOME FOSSILS AND FOSSIL GROUPS

FOSSILS OF GOTLAND IN THE LITERATURE

Fossils from Gotland have been actively collected and distributed all over the world ever since naturalists discovered this fascinating island. A great many publications have resulted from the study of a part of these fossils. Much of this literature is useful for scientists who are more than superficially interested in the Palaeozoic history of Gotland. It may, therefore, be of value to open this chapter with a short index to these publications.

Plants
Algae: Andersson (1895), Hadding (1939, 1950, 1956, 1959), Rothpletz (1908, 1913), Stolley (1896, 1897).
Psilophytales: Halle (1920).

Hystrichosphaeridae
Eisenack (1954a, 1958).

Protozoa
Foraminifera: Eisenack (1954b), Smith (1915).
Chitinozoa: Eisenack (1955, 1962, 1964a), Taugourdeau and De Jekhowsky (1964).

Porifera – Spongiae
Dames (1874), Rauff (1893–1894), Schlüter (1884).

Coelenterata
Anthozoa: Dybowski (1873–1874), Lindström (1865, 1866, 1868, 1870a,b,c, 1873, 1882b, 1896, 1899), Manten (1960c, 1961a), Minato (1961), Tripp (1933), Wedekind (1927).

Annelida and other worm phyla
Bergenhayn (1955), Hinde (1882), Martinsson (1960b).

Arthropoda
Crustacea: Aurivillius (1892), Boll (1862), Botke (1916), Chapman (1901), Hedström (1923a), Jaanusson and Martinsson (1956), T.R. Jones (1887, 1888), Jones and Woodward (1888), Kolmodin (1869, 1879), Krause (1877, 1889, 1891, 1892), Kuiper (1916), Kummerow (1924), Lindström (1885a), Martinsson (1955, 1956, 1960a, 1962a, 1966a,b), Reuter (1885), Spjeldnaes (1951), Von Kiesow (1888).

Arachnoidea: Thorell and Lindström (1885).
Hexapoda: Holm (1892).

Mollusca
Amphineura: Bergenhayn (1943, 1955).
Gastropoda: Lindström (1884).
Pteropoda: Lindström (1884).
Lamellibranchiata: Soot-Ryen (1964), Walmsley (1962).
Cephalopoda: Hedström (1917a), Lindström (1890), Troedsson (1931, 1932).

Tentaculata
Bryozoa: Bassler (1911), Borg (1964), Eisenack (1964b), Hennig (1905–1908), Martinsson (1964).

Brachiopoda
Böger (1968), Boucot (1957, 1962), Davidson and King (1874), De Verneuil (1848), Hedström (1917b, 1923e), Lindström (1860), Wright (1965).

Echinodermata
Crinoidea: Angelin (1878), Bather (1893), Manten (1970), Springer (1920), Ubaghs (1956a,b, 1958).
Pelmatozoa non Crinoidea: Regnéll (1945, 1956).
Asteroidea: Rasmussen (1952).
Echinoidea: Regnéll (1956).

Hemichordata
Graptolithina: Hede (1919a, 1942), Holm (1890), Linnarsson (1879), Wiman (1897b).

Chordata
Pisces: Gross (1968), Lindström (1895), Martinsson (1966a), Säve-Söderbergh (1941), Spjeldnaes (1950).

PERSISTENT FOSSILS AND GUIDE FOSSILS

When examining in some detail the distribution of the Palaeozoic fossils of Gotland, two main groups can be distinguished.

One group is formed by those taxa whose distribution seems to be restricted to one particular type of sediment or to closely related sediments. Several of the fossils are real facies fossils; these will be mentioned in the next section of this chapter. Many other taxa may seem to fall in this first group only because their distribution in the series of strata of Gotland is not yet sufficiently studied.

The second group consists of taxa which are found in several lithofacies. The best guide fossils are to be sought in this group, which comprises primarily the remains of organisms that ranged widely when living, so that their remains became naturally entombed in the various lithofacies of a set of contemporary sediments. This is particularly important in Gotland, because of the great variety in environmental conditions which generally existed while the Middle Palaeozoic sediments were being laid down.

Therefore, attention will first be given in this chapter to fossils with a wide palaeoecological range. It will be seen that this group is not very large. Moreover, several of these have a vertical range which is too long to be useful for a close determination of the relative age of the strata in Gotland.

Persistent fossils

Persistent fossils with a wide palaeoecological range are the tabulate corals *Favosites gothlandicus* Lamarck and *Aulopora* sp. (*Aulopora* cf. *roemeri* Foerste), the heliolitid coral *Heliolites interstinctus* (Linnaeus), the bryozoans *Fenestella reticulata* (Hisinger) and *Ptilodictya lanceolata* (Goldfuss), and the annelid *Cornulites serpularius* Schlotheim. Among the brachiopods, fossils with a wide horizontal and vertical range are species like *Atrypa reticularis* (Linnaeus), *Camarotoechia diodonta* (Dalman), *Camarotoechia nucula* (J. de C. Sowerby), *Delthyris elevata* Dalman, *Howellella elegans* (Muir-Wood), *Leptaena rhomboidalis* (Wilckens), *Craniops implicata* (J. de C. Sowerby), *Rhipidomella hybrida* (J. de C. Sowerby) and *Sphaerirhynchia wilsoni* (J. Sowerby). The second, third and fourth named species are not known from the Visby marlstones. The lamellibranchs *Conocardium* and *Cypricardinia* also have a wide distribution in both directions. Persistent fossils among the gastropods are *Euomphalopterus alatus* (Wahlenberg), *Platyceras cornutum* Hisinger and *Tremanotus longitudinalis* Lindström. Also the trilobites *Calymene tuberculata* (Brünn) and *Encrinurus punctatus* (Wahlenberg) may be mentioned here, even though these two species are not known to the author to occur in the Lower Visby marlstones and in the Burgsvik and Hamra-Sundre Beds.

Guide fossils

The number of fossils with a wide palaeoecological range and a restricted vertical range is small in Gotland.

Among the brachiopods which fulfil the above requirements, there are two which are only known from the Slite Beds. These are *Atrypa lamellosa* (Lovén) and *Conchidium tenuistriatum* (Walmstedt). The stratigraphical range of *Camarotoechia borealis* (Buch) in Gotland extends from the Lower Visby until the Klinteberg Beds. In the English Silurian, this species does not seem to occur in deposits younger than lowermost Ludlowian (Squirrel and Tucker, 1960, p.174). "*Atrypa*" *pusilla* (Hisinger) is only present from the Eke Beds onwards.

Ostracodes which occur in various lithofacies and which have a vertical range of probable stratigraphical importance, include *Leperditia hisingeri* Schmidt, of the Lower Visby Beds, *Leperditia baltica* (Hisinger), of the Slite Beds, *Leperditia phaseolus* (Hisinger) and *Primitia mundula* (Jones), both of which occur from the Halla-Mulde Beds onwards, and four species which are found from the Hemse Beds upwards, viz. *Beyrichia buchiana* (Jones), *B. maccoyiana* (Jones), *B. nodulosa* Boll, and *Cytherellina siliqua* (Jones). Several of the ostracodes which are more or less restricted to the marlstone facies may also have a vertical distribution of stratigraphical value (cf. Martinsson, 1962a). They cannot, however, be used to correlate the various lithofacies within a supposed time-stratigraphical unit, which is such an important aspect of the geology of Gotland.

Fossils with a vertical distribution of stratigraphical importance - though being facies fossils - also occur in other groups. Thus, *Porpites porpita* (Linnaeus) is a common coral in the Lower Visby marlstones, which is rare in the Upper Visby marlstones and unknown from the marlstones in the Slite, Halla-Mulde, Hemse and Eke Beds.

Further fossils which are facies fossils but which nevertheless are of particular value, especially for the correlation with deposits elsewhere in the world, are the graptolites (Hede, 1919a, 1942). *Monograptus spiralis* (Geinitz) is found in the Upper Visby Beds, *M. flemingi* (Salt.) occurs in the Slite Beds, *M. dubius* (Suess) is known from the Slite and Mulde marlstones. Also present in the Mulde marlstone is *Gothograptus nassa* (Holm). *Dictyonema* sp. is found in the Mulde and Hemse marlstones. Furthermore, also *Monograptus bohemicus* (Barrande), *M. chimaera* (Barrande), *M. nilssoni* (Barrande) and *M. varians* Wood are known from the Hemse Beds.

FACIES FOSSILS

Many of the Palaeozoic fossils in Gotland are facies fossils. If fossil finds are sorted out according to the lithofacies in which they were made, it appears that several have exclusively or predominantly been collected from one kind of sediment.

The marly facies

The marlstones of Gotland particularly contain a great number of fossils which are not or only occasionally found in other sediments, even if these were deposited more or less contemporarily with the marlstones. All graptolites belong to this group of facies fossils, but also many corals, bryozoans, brachiopods and ostracodes; other groups of organisms also had representatives with a distinct preference for a muddy environment.

A great many tetracorals are characteristic of a marly environment. Some of these have only been found in the marlstones of Gotland, others also occur in the marly matrix of some reefs, more especially those of the Upper Visby and Lower Högklint Beds (cf. Table IX) (see also Manten, 1961a). The following species are examples of tetracorals that had a preference for a muddy environment: *Aeropoma prismaticum* (Lindström), *Aulacophyllum angelini* Wedekind, *Au. linnarssoni* Wedekind, *Calostylis denticulata* (Kjerulf), *Clisiophyllum involutum* Edwards et Haime, *Cystiphyllum cylindricum* Lonsdale, *C. siluriense* Lonsdale, *C. tenue* Wedekind, *C. visbyense* Wedekind, *Dinophyllum hisingeri* (Edwards et Haime), *Goniophyllum pyramidale* (Hisinger), *Hedstroemophyllum articulatum* Wedekind, *Holophragma calceoloides* (Lindström), *Ketophyllum annulatum* (Wedekind), *K. elegantulum* Wedekind, *K. hoegbomi* (Wedekind), *K. subturbinatum* (Edwards et Haime), *Kyphophyllum conicum* Wedekind, *Phaulactis angusta* (Lonsdale), *Ph. irregulare* (Wedekind), *Ph. tabulatus* (Wedekind), *Polyorophe glabra* Lindström, *Porpites (Palaeocyclus) porpita* (Linnaeus), *Rhegmaphyllum slitense* Wedekind, *Rhizophyllum elongatum* Lindström, *R. gotlandicum* (Roeme

Schlotheimophyllum patellatum (Schlotheim), *Syringaxon dalmani* (Edwards et Haime), *Zaphrentis vortex* Lindström.

Several tabulate corals from Gotland, of which *Favosites* spp. and *Halysites* spp. are the most well-known also seem to have had a preference for an environment with muddy water. Especially typical for the marlstones are *Catenipora escharoides* Lamarck, *Roemeria kunthiana* Lindström and *Thamnopora lamellicornis* (Lindström).

Eight species of heliolitid corals are characteristic for the marlstones, viz. *Cosmiolithus halysitoides* Lindström, *C. ornatus* Lindström, *Heliolites repletus* Lindström, *H. spongodes* Lindström, *Plasmopora calyculata* Lindström, *P. petalliformis* (Lonsdale), *Propora conferta* Edwards et Haime and *P. tubulata* (Lonsdale). Most common in marlstone, but occurring also in marly limestones are: *Heliolites barrandei* Penecke, *H. interstinctus* (Linnaeus) (occasionally even found in rather pure limestones), *H. parvistella* Ferd. Roemer, *Plasmopora scita* Edwards et Haime, and *Propora eurycantha* Lindström.

As could be expected, annelid remains have been reported mostly from the marlstones although marly limestones and the Burgsvik sandstone and oolite have also yielded some. They have been classified in the genera *Autodetus*, *Conchicolites*, *Cornulites*, and *Spirorbis*.

Of the bryozoan genera that are represented in the Middle Palaeozoic of Gotland, only three have a distribution of some importance also outside the marlstones, although even they showed a preference for an environment in which marl could be deposited. *Coenites* sp. has been found also in several reefs; *Fenestella* spp. are common in marlstones, reefs and stratified marly limestones, as is *Ptilodictya lanceolata*, which in addition is also common in the Burgsvik sandstone and oolite. Ten other genera are almost exclusively known from marlstones. They include the following species: *Berenicea consimilis* (Lonsdale), *Ceramopora lindströmi* Hisinger, *Corynotrypa* cf. *dissimilis* (Vine), *Crepipora lunariata* Hisinger, *Cyclotrypa inflata* (Hisinger), *Fistulipora membranacea* Hisinger, *F. mutabilis* Hisinger, *Helopora lindströmi* Ulrich, *Mesotrypa suprasilurica* Hisinger, *Phaenopora lindströmi* Ulrich, and *Stomatopora minor* Hisinger. It will appear from Table IX that the above-named species of *Berenicea*, *Helopora* and *Phaenopora* have also been observed in a few reefs and related deposits. When in their original position, they are found there in marl nests between the actual reef builders. This occurrence, therefore, does not exclude them from the group of fossils typical of the marlstones.

Among the brachiopods, several species are typical for marl deposits. To these belong *Antirhynchonella linguifera* (J. de C. Sowerby), *Chilidiopsis pecten* (Linnaeus), *Dayia navicula* (J. de C. Sowerby), *Dicaelosia biloba* (Linnaeus) and *D. verneuilana* (Lindström), *Eospirifer plicatellus* (Linnaeus) var. *interlineatus* (Lindström) and *E. radiatus* (J. de C. Sowerby), *Glassia compressa* (J. de C. Sowerby) and *G. obovata* (J. de C. Sowerby), *Isorthis lovéni* (Lindström), *Leptaena laevigata* (J. de C. Sowerby), *L. lovéni* De Verneuil, *Lingula lewisi* J. de C. Sowerby and *L. striata* J. de C. Sowerby, *Nucleospira pisum* (J. de C. Sowerby), *Orbiculoidea rugata* (J. de C. Sowerby), *Pentamerus gotlandicus* Lebedev and *P. sphaera* Lindström, ?*Plectambonites inconstans* (Haupt) and *P. segmentum* (Angelin), *Plectatrypa imbricata* (J. de C. Sowerby) and *P. marginalis* (Dalman), *Plectodonta transversalis* (Dalman) and *P. transversalis* var. *lata* (Jones), *Skenidioides acuta* (Lindström),

Scenidium lewisi (Davidson), *Protozeuga bicarinata* (Angelin). Also *Cyrtia* is a genus from marlstones and marly limestones. *C. exporrecta* (Wahlenberg) is reported from the Upper Visby marlstone, marly limestones of the Lower Högklint Beds and the Slite marlstone; *C. trapezoidalis* (Hisinger) from the Mulde marlstone and the Klinteberg Beds, while it is not impossible that it also occurs in the Hemse marlstone. *Gypidula galeata* (Dalman) is mainly a species from the marlstones (Slite, Mulde, Hemse), but also occurs in marly limestones from the Klinteberg and Hemse Beds and in the Burgsvik sandstone. The same applies to *Strophonella euglypha* (Hisinger). *Amphistrophia funiculata* (McCoy) is known from the Slite, Mulde and Hemse marlstones, but also from marly Slite and Klinteberg limestones. *Ptychopleurella bouchardi* (Davidson) is noted from the Slite, Mulde, Hemse and Eke marlstones and from marly limestones in the Slite and Klinteberg Beds. *Chonetes* spp. also seem to have had a certain preference for a marly environment. *C. cingulatus* Lindström is known from the Slite sandy limestone in the south (see p.275) and the Mulde marlstone; *C. gotlandicus* Hede occurs in the Slite and Mulde marlstones and in marly limestones of the Klinteberg Beds. Less restricted to the marlstones is *C. striatellus* (Dalman) which, although known from the Hemse and Eke marlstones, also has been found in marly limestones and limestones of the Lower Högklint, Hemse and Hamra Beds, and in the Burgsvik sandstone and oolite. Whether *Chonetes* sp. which is found in the Slite limestones and marlstone may also be attributed to *C. striatellus* can not be stated with certainty. *Resserella* too, is a genus that is most common in marlstones, even though *R. elegantula* (Dalman) may also be found in Högklint and Slite limestones. Some of the persistent brachiopods, though occurring in several lithofacies, also showed a preference for a marly environment. Examples are *Atrypa reticularis* (Linnaeus), *Craniops implicata* (J. de C. Sowerby), *Leptaena rhomboidalis* (Wilckens) and *Rhipidomella hybrida* (J. de C. Sowerby).

It is not surprising that many ostracode species of the Middle Palaeozoic of Gotland are mainly or completely restricted to the marlstones. Only some will be mentioned here. *Aechmina bovina* Jones and *Beyrichia spinigera* Boll occur in the marlstone and marly limestones in the Slite Beds, the Mulde marl and the Lower Klinteberg Beds. *Primitia valida* Jones et Holl is known from marlstones in the Slite, Mulde, Lower Klinteberg and Hemse Beds. *Thlipsurella discreta* (Jones) occurs in both the Slite and Mulde marlstones. *Leperditia baltica* is not uncommon in both the marlstones and limestones of the Slite Beds; both species may be of stratigraphical importance. *Colpos insignis* Moberg, *Entomis migrans* Barrande and *Leperditia grandis* Schrenck are found in the Hemse marlstone, *Beyrichia steusloffi* Krause occurs in both marlstones and marly limestones of this stratigraphical unit. *Craspedobolbina clavata* (Kolmodin) is found in marlstones, but also marly limestones, from the Högklint to the Hemse Beds, *Beyrichia lauensis* Kiesow in the Hemse and Eke marlstones. Many other ostracode species could easily be added to this list. Thus Martinsson (1962a), in his monograph on the Beyrichiidae of Gotland, reports that already of this family alone not less than 114 different species occur in marly sediments in Gotland.

As far as can be judged from the observations that have been recorded, the following species of lamellibranchs seem to have a distribution that is mainly or exclusively limited to the marlstones: *Ambonychia punctata* Lindström, *Cornellites damesi* (Philippi), *C. sowerbyi* (McCoy),

Ctenodonta sulcata (Hisinger), *Folmanella duplicata* (Lindström in museo), *Nucula anglica* (d'Orbigny) (known also from the Burgsvik sandstone and oolite), *Posidonomya glabra* Münster, *Rhombopteria* cf. *mira* Barrande (also known from the Burgsvik oolite) and *Salweyia* cf. *striata* (J. de C. Sowerby).

About 175 different species of gastropods from Gotland have been described, especially by Lindström. Unfortunately after he had published his fundamental monograph (Lindström, 1884), little attention was paid to the distribution and palaeoecology of the many genera and species. No doubt several of them have their distribution mainly in the marlstones, more especially species of the genera *Bellerophon*, *Cyclonema*, *Loxonema*, *Pleurotomaria* and *Subulites*.

Pellet-like bodies and traces of borings, as were observed in some places in the reef limestone, were only very occasionally seen in the marlstones. This may be simply because the author has paid comparatively less detailed attention to these deposits. The usually high organic content of the muds provided a rich food supply also in that environment. But because of the fine texture of the mud and presumably often high colloidal content, there was little circulation of interstitial water. This lack of circulation, together with the organic content, may have led to local and post-depositional anaerobic conditions (as also suggested by the local presence of pyrite) with consequent limitations on the fauna able to inhabit these deposits.

The limestone facies

Compared with the marlstone facies, the limestone facies in Gotland presents only few fossils which are restricted to this lithofacies. It is not unexpected to find that the Algae, found in the Middle Palaeozoic of the island, are predominantly restricted to the limestones. Among the corals the tetracoral *Acervularia*, the tabulate coral *Striatopora* and the heliolitid *Thecia swinderniana* (Goldfuss) seem to have thrived much more abundantly in clear than in muddy water. The same holds for the brachiopods *Dinobulus davidsoni* (Salt.), *Linoporella punctata* (De Verneuil) and *Platystrophia*. In the group of the lamellibranchs *Ilionia prisca* (Hisinger) and "*Megalomus*" are typical for the limestone facies.

The reef facies

There is no evidence of fossils occurring only in the reef limestones and their directly-surrounding sediments. The occasional specimens of the brachiopod *Cliftonia lindströmi* Ulrich et Cooper and the trilobite *Bumastus sulcatus* Lindström that were observed, were found in reef limestone, but it is likely that additional observations will reveal their presence also in other kinds of deposits. Of the more common fossils, none was found to be restricted to the reef environment.

PALAEOECOLOGY OF CORALS

Several kinds of studies on the palaeoecology of corals can be made on the basis of the rich fossil fauna of Gotland. Although the author has made a number of observations on this subject, he has not gone into it in great detail, however tempting this might have been. A number of observations will be reported in the next few pages. Thereafter, opening the section on stromatoporoid palaeoecology, a short discussion will also be presented on the competition between corals and stromatoporoids. Palaeoecological data which are relevant to both the corals of Gotland and the reefs in Gotland in general, will be summarized in Chapter XIV.

Some observations on solitary corals

It is somewhat amazing that Woodford (1965, p.42) wrote that most corals are outstanding examples of facies fossils, being restricted to clear and shallow seas. Particularly the preference for clear water is untrue. In fact modern solitary corals are remarkable for their powers of removal of surface sediment. Fossil solitary corals are commonly found associated with such rocks as marlstones and shales. As mentioned earlier in this chapter, in Gotland, too, several solitary corals were found to be characteristic for such lithofacies. It will be discussed later in this chapter that also the fossil compound corals present evidence that they thrived well in muddy water.

It is significant to mention here as well that the reefs in which solitary corals were found to be most abundant, in both number of species and number of individuals, are those with the highest volume and highest degree of impurity of the matrix; specifically the Upper Visby reefs and some reefs in the Lower Högklint Beds.

A particularly interesting coral in this connection is *Schlotheimophyllum patellatum* (Schlotheim). This tetracoral is characterized by a narrow and deep calyx, with axial twisting of the septa, surrounded by a broad reflected rim, which is formed by the deposition of a large stereozone. Sections through this rim show successive layers, partly separated by films of marl. As the fossil is always found in a marly lithofacies, the impression is gained that the coral was repeatedly covered with mud. Only by retreating into its deep calyx could it escape complete suffocation. From there it thereupon expanded again over its own rim and the sediment cover laid down upon it.

Solitary corals which occur in the stratified marlstone deposits often show a horn-like bent shape, whereas those in the limestone layers and also the majority of those found in reefs possess subcylindrical forms. The curved forms generally broaden rapidly, the basic angle is rather large. The plane of the edge of the calyx is oblique to the axis of the polyps. The growth form can be explained by disturbance of equilibrium of corals which were attached to small objects lying on a soft sea floor. They became tilted, together with their substratum, when they grew heavier, until the resistance from the sea floor led again to a restoration in equilibrium. Corals growing on a larger substratum enlarged and strengthened their attachment and thus developed their characteristic subcylindrical forms.

Influences of mud sedimentation and growth rate on the growth form of colonies

Vaughan (1915a,b, 1919), Marshall and Orr (1931) and Yonge (1935) have shown, both through observations in nature and through experiments, that round coral colonies are better adapted to an environment with mud sedimentation than flat forms. Flat colonies, especially those with small polyps, have to rely on the movement of the water to remove sediment from their upper surfaces. Similar observations were made by Motoda (1940a).

He found that young specimens of *Goniastrea aspera* Verrill started with the formation of an irregular, flat colony on the top surface of their substratum, after which they increased their growth upwards until a globular or semi-globular shape was reached.

These data help to explain growth forms found in heliolitid corals in Gotland (Manten, 1960c, 1961a). In rather strongly marly facies the *Heliolites* colonies are frequently either small and relatively flat, or larger and rounded. In the less marly facies, also the larger colonies often have a distinctly stronger horizontal than vertical extension. In strongly marly deposits and in reef debris, some extreme forms were also observed that approached the shape of a mushroom.

Coral colonies which started growth in a period of comparatively little mud sedimentation and which later had to cope with increased sedimentation were often faced with serious difficulties. In many cases, they had developed flat colonies and then were confronted with changed conditions under which a round form would be more advantageous. But owing to their age, they were unable to completely change the form of the colony.

Stephenson and Stephenson (1933) showed that the growth rate of some corals decreased in inverse proportion to their size and age. Goreau (1961) found that colonies of *Manicina areolata* weighing approximately 0.05 g deposited calcium nearly one hundred times faster, per unit of tissue nitrogen, than did colonies weighing 150 g.

Several *Heliolites* colonies in Gotland apparently found a way out of coping with increased sedimentation during a later stage of their existence. Being unable to change the entire form of the colony to a rounder one, they developed semi-globular knobs and finger-like extensions on their surface (Fig.220), thus reaching at least partly the more suitable form (cf. Manten, 1960c, p.159, 1961a, fig.3, 1962, fig.7). Similar observations were made on the

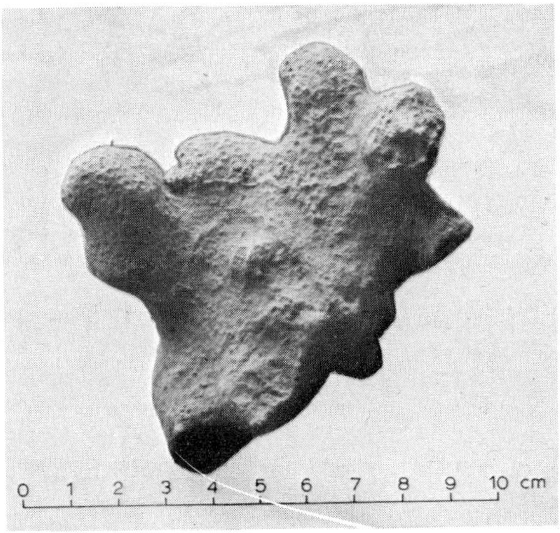

Fig.220. *Heliolites* sp. from an Upper Visby reef at Kneippbyn, Västerhejde Parish. Note the knobby outgrowths in the upper part of the colony which have presumably been caused by an increase in mud sedimentation during the development of the coral colony. (After Manten, 1962, fig.7.)

tabulate coral *Planalveolites fougti* (Edwards et Haime). The transformation of the form of the colony will presumably have been comparatively easiest when the rate of sedimentation increased gradually. It is known that many recent organisms can tolerate a greater range of variation in a given environmental factor if the change takes place slowly than if the change is rapid.

Worth mentioning at this point is also the influence which sedimentation may have had on the form of branched corals. In a classic paper, F.W. Jones (1907) already reported that in parts of such corals which were exposed to sedimentation, the mouth fields were often small, the theca extended above the surface of the colony and the coenenchym was somewhat sculpted. Similar observations have been made on fossil branched corals of the Gosau Formation in Austria (Kühn, 1925), and from Gotland (Manten, 1960c). The conditions under which the branched corals developed were also susceptible to changes. In this case an increase or decrease in sedimentation was best recorded by corals which had been damaged. The regenerated part could then adapt itself to the new conditions and in this way be different from the older part of the colony.

Some differences between reef edge and core

In part of the reefs, particularly in north and middle Gotland, some faunal differences can be observed between the edge of the reef and the reef core.

One of the examples is provided by the coral *Heliolites parvistella* Ferd. Roemer. In the centres of several reefs, a branching form thereof can be found, which is known as the variety *caespitosa*. At the same time, the massive form of this coral species is observed to be more common around the margins of some reefs. Laminar colonies with a few knobs on their surfaces suggest that the branched and laminar forms are no genotypical varieties.

Another example is presented by the heliolitid coral, *Thecia swinderniana* (Goldfuss). In the great majority of cases, this coral occurs in a massive, laminar growth form in the marginal regions. Thin laminar expansions of this coral may also locally be found in the stratified rocks immediately around the reefs. Occasionally, however, small branching forms or otherwise vertically attenuated forms may be observed, but then generally in the more interior parts of the reefs. Similar behaviour is displayed by the stromatoporoid *Labechia conferta* (Lonsdale).

It is more than likely that environmental factors have been responsible for the development of massive growth forms at the reef margins. The effect of waves, in particular, in inhibiting branching has been described by several authors. They have also shown that even slight differences in position might produce large effects in the growth form. The classic example is that of a log, chained beneath the surface of the water, where the coral colonies on its upper surface were flattened bosses, and on its lower side showed delicate branching forms of the same coral (F.W. Jones, 1907, p.536). Thus, even the distance of part of the circumference of a tree trunk is sufficient to lead to a significant variation in growth form.

The question which of the growth forms is to be regarded as "normal", if indeed the word can be used in this connection, and which is the

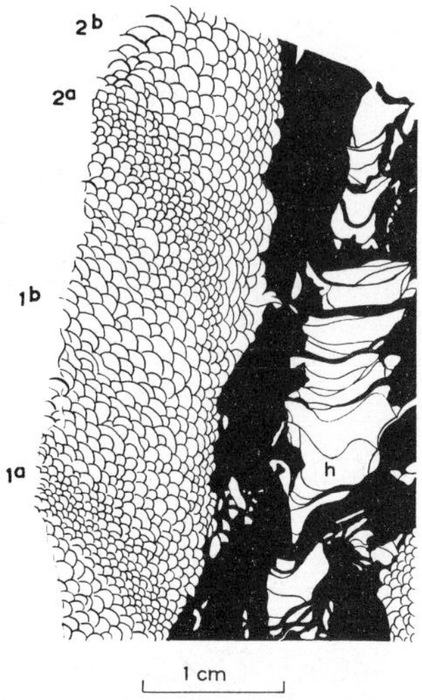

Fig.221. Two bands (1 and 2) of a rhythmic growth pattern in an unidentified tetracoral from the Upper Visby marlstone, Högklint. Each band consists of a part with minute, densely crowded and thickened dissepiments or vesicles (a), followed by a layer which possesses larger, less crowded and not thickened dissepiments or vesicles (b); in general the alternating layers show rather imperceptible transition from one to another, but sometimes the upward gradation from a layer with minute elements to another with larger ones is more or less abrupt; h = polyp cavity. (After Manten, 1961a, fig.5.)

modification brought about by environmental factors is, therefore, academic. This is clear in the case of *Heliolites*. In the examples of *Thecia* and *Labechia*, one might be inclined to consider the massive form as "normal" because it occurs so much more abundantly. But how far is this due to the generally small size of the reefs of Gotland and how would the ratio between massive and branched forms be under another set of environmental conditions?

Rhythmic growth patterns

In the nineteen-thirties, Ma (1933, 1934, 1937) found that many invertebrate skeletons exhibit rhythmic growth patterns. These are particularly common in corals, both recent and ancient. Also corals in Gotland show such rhythmicity (Fig.221). The rhythmic banding strongly recalls the rhythmic growth of tree rings. Ma, therefore, drew the conclusion that the invertebrate fossils which show such banding also developed in seasonal climates. However, Fischer (1963) argued that at least in the case of the corals, this climatological explanation is not satisfactory. In the discussion of Fischer's paper, Durham pointed out that rhythmic growth patterns in other organisms do not always seem to be related to climate either, and Lowenstam said that, of the modern corals growing off Bermuda, those growing offshore seem to show no rhythmic growth, whereas those from the restricted inshore environment do show such a rhythmic growth banding. In contrast to these views is the observation which Wells made in that same year. He counted in Devonian corals on the average 400 fine bands within a supposed annual band. This figure agrees with the number of days in the Devonian year as obtained from calculations on the effect of tidal friction (Wells, 1963). Independently of Wells, Scrutton (1965) made the observation that some corals also have monthly bands. With Middle Devonian corals he found evidence for 13 months per year, with an average of 30.6 days per month. Also Runcorn and co-workers are of the opinion that the banding in certain corals can be regarded as palaeontological clocks (Runcorn, 1966).

Probably related to the above-named rhythmic growth patterns is the phenomenon of "layered" coral colonies. Several colonies, particularly of *Heliolites* and *Favosites* show growth forms which can best be described as a pile of discs separated by constrictions (Fig.222). The thickness of the discs within one colony is rather constant. Between the various colonies which show this pattern the thickness of the discs varies between 8 and 20 mm. The rhythmic growth patterns in coral skeletons, mentioned before, show band thicknesses of 12–25 mm in the corals of Gotland. Both phenomena thus are in the same order of thickness. The author considers it possible that (seasonal?) fluctuations in the supply of continental debris may have been a major reason causing the observed growth forms of these coral colonies.

Worth mentioning in this respect is a *Favosites* colony found in the reef limestone at Spinnbjersbacke (6 km south of Boge Church, Slite Beds). The reef limestone around the coral colony contained a local layer, about 1 cm thick, of marlstone. Around the upper surface of this layer, the *Favosites* colony showed a constriction, about 1.5 cm thick and some centimetres deep.

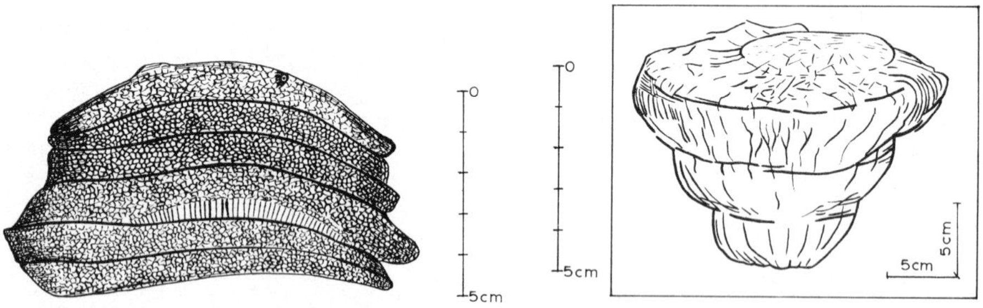

Fig.222. Layered coral colonies from marlstone deposits. Left: *Favosites gothlandicus* Lamarck from Hemse marlstone, Petesviken, Hablingbo Parish; right: *Halysites* sp. from Upper Visby marlstone, Luseklint, Lummelunda Parish. (After Manten, 1961a, fig.4.)

Percentage of coral colonies in life orientation

In almost every reef, coral colonies in position (orientation and possibly also place) of growth are found together with colonies which distinctly are not in their original position. If the percentage of the coral colonies which are found in life orientation is calculated, it appears that there are notable differences. In Table XXII the observations within individual stratigraphical units are taken together and the percentages from the various units are then arranged in four groups, according to the supposed water depth in which the reefs, in which the corals were found, have developed. Coral colonies, of which it could not be determined with any certainty whether they were in growth orientation or not, have been left out of consideration. This does not significantly influence the percentages obtained. The highest number of doubtful cases was met in the four localities in the Slite IV Beds (32 out of a total of 159). If these are counted as having been either all in life orientation or not, respectively, the percentage of coral colonies in original position would have been 60 or 40. In other words, even the highest extreme value does not yet reach the average value found for the next group. Similarly, for the coral colonies in one of the reefs of Lilla Karlsö (23 uncertain cases of a total of 223), extreme percentages of 83 and 73, respectively, are obtained.

Another restriction that was made is that only the massive coral colonies were considered. There are no indications that for branched colonies of corals and bryozoans significantly different percentages would be found, but with these colonies it is often less easy to determine whether they are in orientation of growth. In two localities comparatively good observations, however, could be made on branched corals. One was Galgberg extension (Middle Högklint Beds), with 24 colonies of which 10 were in growth orientation, 10 were not and 4 were doubtful cases; percentage of colonies in life orientation was 50% of the certain cases. In the raukar in the northwest of Lilla Karlsö 40 branched colonies were observed, of which 31 were in growth orientation, 4 were not and 5 were doubtful cases; percentage of colonies in original orientation was 89% of the certain cases.

In order to eliminate the influence of the position of the colonies within the reefs from the final result (e.g., possible differences between the marginal and more interior parts of the reefs), the observations were spread as much as possible over the total reef bodies.

Considering other factors which could have influenced the orientations of the observed colonies, three deserve special attention: (1) pressures within the reefs; (2) the shape of the colonies; (3) destructive action by moving water.

In at least part of the cases which were considered as of uncertain orientation, as well as in a number of colonies which were found in an oblique position, it is likely that they still were in about their place of growth, but that the corals had been tilted after their death. This tilt may have been caused by differential pressures which originated within the reefs when these grew higher. That such pressures occurred and led to small displacements is proved by phenomena like small slickensided surfaces (generally 100–250 cm^2) and bucklings in local stratified intercalations within the reefs.

It is worth mentioning in this respect that in some reefs an increase in the percentage of colonies in original orientation can be found while going

TABLE XXII

Percentage of massive coral colonies found in growth orientation in reef limestones

Stratigraphical unit	Kind of reefs	Localities	Total number of colonies	Total number in orientation of growth	Percentage in orientation of growth
Upper Visby Beds	Upper Visby reef type	North of Kneippbyn	50	40	80
? Slite IV Beds	?Stäurnasar reef type (lower part)	Raukar of Norderslätt, in the northwest of Lilla Karlsö	200	162	81
Hemse Beds	Holmhallär reef type	Fågelhammar	37	30	81
Lowermost Högklint Beds	Hoburgen reef type	About 0.7 km north of Stenkyrkahuks Fyr	56	42	75
Upper Slite III Beds	Hoburgen reef type	Hejnum Hällar	25	17	68
Slite IV Beds	Hoburgen reef type	Lännaberg, Solklint, Patvalds, Stora Vede	127	64	50
Middle Högklint Beds	Hoburgen reef type	Lickershamn, Galgberg extension	100	46	46
Hemse Beds	Hoburgen reef type	Gannberg	36	21	58

upwards in the reef. This can be the fact even in a rather small reef. An example is provided by a Lower Högklint reef, about 4 m long and 2.5 m thick, which is found approximately 0.7 km north of Stenkyrkahuks Fyr. In the lowermost metre of this reef, 14 out of 21 specimens (67%) of the massive coral colonies were found in growth orientation against 28 out of 35 colonies (80%) in the uppermost 0.5 m. Many of the colonies which were not in their original orientation, particularly those in the lower part of the reef, were found to lie obliquely, with the axis of the colony generally making an angle of 30–50° with the vertical.

The shape of the coral colonies can certainly have played a part in determining the position in which the colonies were found fossilized. Particularly in the fourth group (last three values in Table XXII) it could be noted that the flat colonies were more frequently found in original orientation than the rounder colonies. This was particularly true in Galgberg extension. For the branched colonies in that locality this was less evident, probably because the latter grew in more sheltered places. In the reefs of the Upper Visby Beds, the shape of the colonies may also have been of influence, but not to a significantly stronger degree than in the reefs of the fourth group. Neither can the influence of the shape of the colonies explain the high percentage of colonies that were found in their proper position in Fågelhammar and in Lilla Karlsö, where the great majority of colonies displayed great vertical dimensions.

Thus the conclusion remains that the major factor which determined the final position of the coral colonies must be something other than those which have been thus far discussed. This major factor must have been the movements of the water.

In the stratified limestone very close to the reef which is found about 0.7 km north of Stenkyrkahuks Fyr, the percentage of corals which occurred in a position that could have been their growth position was 48% (22 of 46 specimens). This percentage is low compared with both the figure found for the corals in the reef (67%) and with that for corals found in the stratified deposits of the Lower Högklint Beds at some distance from the reefs (around 85%). This observation makes it likely that a large number of the colonies present in the reef-surrounding sediment came from the reef. They may have been washed off the reef after their death, but to some this may have happened during their life. In some colonies in reef-surrounding deposits the corallites were oblique in the lower part and roughly vertical in the upper part.

Coral colonies in various orientations may also be found in filled depressions within the reefs. The majority of these were apparently washed in from elsewhere on the reef surface. Such depressions are in sharp contrast to pools in the reef surface where clearly many colonies lived in a sheltered environment and where the percentage of colonies in growth orientation generally is equal to or higher than elsewhere in the reef (cf. the figures obtained for branched colonies).

If, thus, water action could cause the detachment and displacement of coral colonies, it is also logical to assume that loosened colonies could remain lying somewhere on the reef surface.

A reef surface growing at a water depth of only one or a few metres will have been more strongly subjected to the destructive action of agitated water than one developing at a depth of some tens of metres. This may be the main explanation of why decreasing percentages of corals in growth orientation are found if the observations are arranged in an order of assumed decreasing water depth (Table XXII).

PALAEOECOLOGY OF STROMATOPOROIDS

With so many stromatoporoids preserved in the Middle Palaeozoic rocks of Gotland, it is inevitable that something should also be said here about their palaeoecology. This is even more desirable since, compared with other groups of organisms, not much has so far been written on the subject, and what has been written is, moreover, somewhat confusing. The best paper on stromatoporoid palaeoecology, known to the present author, is by Flügel (1959, pp.233–246).

Competition between corals and stromatoporoids

There has been some discussion in literature on the environmental factors which influenced the ratio between stromatoporoids and corals in a reef part. In a well-known study of Frasnian (lower Upper Devonian) reefs in the Dinant basin (Belgium), Lecompte (1958) found a notable faunal succession in the larger reefs (1–4 km in diameter, 200–250 m thick). At the base there is a zone of lamellar corals. This is followed by a thin intermediate zone of lamellar stromatoporoids. A thick upper zone, forming the main part of the reef, is built essentially of massive stromatoporoids and, in an accessory role, large subglobular corals. Lecompte's interpretation was that progressive subsidence was the controlling factor of the faunal changes, the most notable of which is the absence of stromatoporoids beneath the zone of turbulence. He assumed (Lecompte, 1956a, 1958) that corals and stromatoporoids could coexist in the zone of turbulence. Corals could attain great size there, but they suffered greatly in competition with the stromatoporoids, which dominate in the reef parts formed in this zone. Beneath the zone of turbulence, according to Lecompte, only corals were capable of building reefs, not because they thrived best there, but on the contrary because there they did not have to compete with the stromatoporoids.

Lecompte's opinion that the waning of importance of the stromatoporoids is caused by an increase in depth is, however, in the opinion of the present author, open to serious criticism. An earlier publication by Lecompte (1951, p.53) already indicates that another factor was of great influence on the development of the stromatoporoids, viz. mud tolerance. In that paper, he described how stromatoporoids are abundant in limestones, seldom found in shales, and are apparently more affected by mud than the corals.

In the Niagaran (Middle Silurian) reefs of Illinois, stromatoporoids also decrease in importance as reef builders in the direction of increase of the muddy impurities (Lowenstam, 1950, p.483, 1957), that is from the clastic-free belt, across the low-clastic belt to only a subordinate contribution at the southern border of this belt (see also the section on the North American reefs in Chapter XIII, pp.458–463).

Therefore, it is much more likely that the degree of muddiness, and not primarily water depth, has been the fundamental determining factor in the occurrence of stromatoporoids. Increase in muddiness, leading to a waning in dominance of the stromatoporoids, may be but is not necessarily correlated with an increase in depth. The English Wenlockian presents mud but should, nevertheless, be considered a shallow-water deposit.

Gotland produces several facts which further support the conclusion reached about stromatoporoid palaeoecology.

First, there is the stratigraphical succession of the Upper Visby Beds with its small reefs, dominated by corals, and the Högklint Beds with larger reefs of Hoburgen type, in which stromatoporoids are the predominant reef builders. Thus, a gradual decrease in water depth, leading to a succession of marlstones by marly limestones and limestones, is accompanied by a replacement of corals by stromatoporoids as the main reef-forming organisms. Together with the stromatoporoids, the Algae become more common. As is well-known, the distribution of Algae is strongly governed by the illumination of the water, and the latter is again strongly influenced by the degree of muddiness.

There is also a correlation between the average size of the stromatoporoids and the lithofacies in which they are found. Those which are found in very marly limestone are generally small, and those in rather pure limestone large. This rule also holds for lithofacies found in close vicinity to each other, where there has not been any question of significant differences in water depth.

The Hamra algal limestone shows intercalations of more or less clayish, reef-like limestone, containing fossilized stromatoporoids, corals and algal balls as major organogenic constituents. However, both the number and average size of the stromatoporoids are significantly smaller in these reef-like intercalations than in the Holmhällar-type reefs of the Hamra-Sundre Beds, the matrix of which is notably purer. No doubt these Holmhällar-type reefs formed in deeper water than the algal limestone. Thus, a similar effect in stromatoporoid development is reflected here as in the Upper Visby – Högklint succession, but with an opposite trend in water depth.

Next, there are several examples of corals which settled on a living stromatoporoid, but which were overgrown by the latter. In stromatoporoids collected from rather pure limestone many examples can be found of such overgrown corals, but generally these corals remained very small and are only visible in polished sections. Thus a few stromatoporoids from Holmhällar-type reefs in Ljugarn (Hemse Beds) and Holmhällar (Hamra-Sundre Beds) were found to contain on the average 65 corals per 100 cm^2 of vertical section through stromatoporoid colonies. The average size of the sagittal sections through the corals was not more than 10 mm^2. In stromatoporoids collected in more marly limestone, the number of corals fully enclosed by the stromatoporoid is much smaller. A stromatoporoid, about 75 cm long, and in its centre around 30 cm thick, collected in the Spillingsklint (Othem Parish, Slite Beds), showed in a full cross-section only three overgrown favositid corals, which were several square centimetres large. From strongly marly limestones, no examples of corals which became completely enveloped by a stromatoporoid are known to the author. In the marly reefs of Upper Visby type, one example was found. This was in a rather exceptional reef, with some large stromatoporoids, that occurs north of Kneippbyn (cf. Chapter VI, p.108). One of the stromatoporoids enclosed a *Favosites* colony a few centimetres in size. It should be pointed out that the above-mentioned

Fig.223. Rauk of about 1 m high, in the raukar field of Fågelhammar South, Hemse Beds, reef of Holmhällar type. The entire rauk consists of the remains of one stromatoporoid colony.

---

envelopments of corals by stromatoporoids should not be confused with symbiotic occurrences of corals and stromatoporoids, such as are known under the name "*Caunopora*" (cf. Flügel, 1959, p.240).

Also of great palaeoecological value is the fact that stromatoporoids are most abundant and largest (Fig.223) just in those reefs in which also Algae reached their richest development, the reefs of Holmhällar type.

Many further observations of value for the understanding of stromatoporoid palaeoecology have been made in the reefs of Holmhällar type. These are of special importance because variations in stromatoporoid abundance or in the size and shape of the colonies within one reef can not be ascribed to any significant difference in water depth. Other differences in environmental conditions must than have played a part; variations in the purity of the water seem again to have been the most prominent ones.

Stromatoporoids in Holmhällar-type reefs

In the reefs of Holmhällar type the stromatoporoids are more dominant reef builders than in the other types of reefs (cf. Chapter VIII). The colonies usually also reached large dimensions there. Generally they are separated by thin layers of calcareous mud or Algae. The volume of reef rock consisting of stromatoporoids is more than inversely proportionate to the amount of reef matrix. This is the case even though the matrix in the Holmhällar-type reefs is purer than in the average reef of Hoburgen type.

Near the distal ends of the reefs the contribution made by the stromatoporoids to the total amount of reef limestone is distinctly smaller than in the more central reef parts.

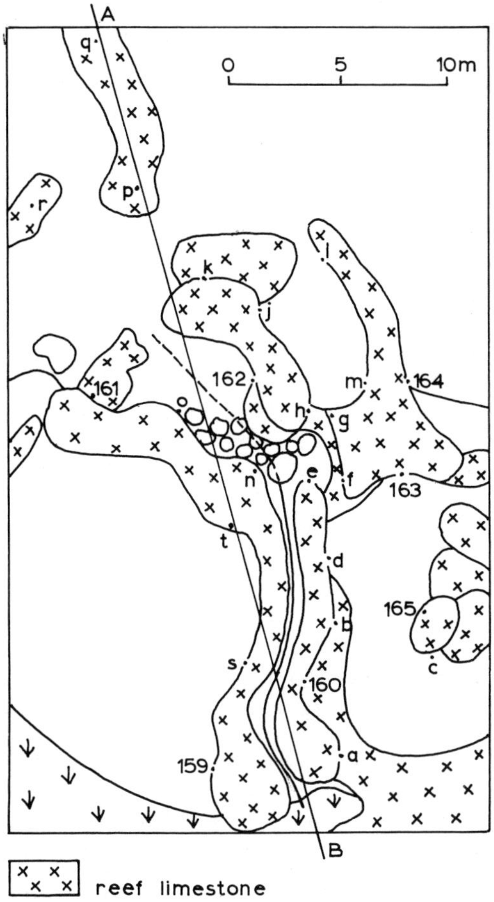

Fig.224. Detail of the map of the Holmhällar raukar field with the observation points 159–165 and the supplementary observation points a–t. Data from these points together form the basis of the curves in Fig.225.

In sheltered pools within the growing reef surface, where somewhat more mud and also fine reef debris were deposited, apparently several stromatoporoids died and were replaced by corals and bryozoans (Fig.225).

At the observation points 161 and 162 in Holmhällar (see the map of this raukar field) a higher matrix volume was noted, which was also reflected by the fact that the stromatoporoids do not occur in the usual large and massive colonies. A similar situation was met at observation point 171, where locally the stromatoporoids are smaller and flatter; more matrix is present than in the surroundings, where stromatoporoids had built so compact a reef limestone that the individual colonies could hardly be recognized.

At observation point 182, in the bottom part of the rauk, a zone was found which contained comparatively more corals and less stromatoporoids

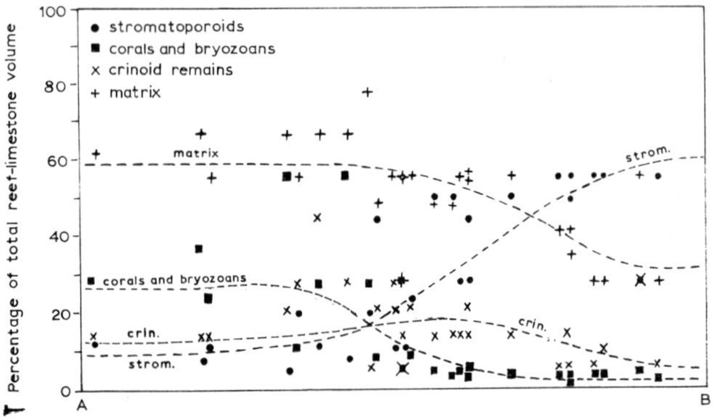

Fig.225. Composition of the reef limestone of Holmhällar along the line AB in Fig.224. The right part of the graph gives the composition of the "normal" reef limestone, the left part that of the rock formed in a large pool within the reef. The crinoid percentage is exaggerated by a factor of two. Even though the amount of terrigenous mud deposited in the pool was small, the fact that it was laid down in rather stagnant water was apparently sufficient to make many stromatoporoids give place to corals and bryozoans.

---

than the remaining parts of the rauk. This zone could also be recognized in point 181, although it is less distinct at that place. In point 180, the zone was no longer developed in the same way; the rock there is again very rich in stromatoporoids, but among these there are notably many which are "tower shaped" (latilaminae strongly bulging upwards in places); some others are tabular in shape. This zone may reflect less favourable conditions for stromatoporoids, leading to a decrease in their number or to aberrant growth forms. Where stromatoporoids retreat corals replace them.

Around planes, which seem to represent interruptions in reef growth (cf. Chapter VIII), an inverse relation between stromatoporoid volume and amount of reef matrix could be established also in several cases.

Any theory on the subject of palaeoecology must admit a certain flexibility. Exceptions to the general rule may be found. Thus, in observation point 121, a high contribution by corals to the reef limestone was observed, even though the matrix volume was low. Also the reverse, many stromatoporoids in places comparatively rich in matrix, was observed in a few instances. It may have been possible, however, that the surfaces of stromatoporoid colonies were kept free of sediment by moving water in a similar way as the surfaces of stones on a recent sand beach are continuously washed clean.

In almost all cases where the matrix sedimentation was very great, however, the stromatoporoids distinctly gave way. In the observation points 161 and 162, where the matrix volume was estimated to comprise about 50% of the total rock volume, against about half that percentage in the close surroundings, the stromatoporoids constituted only about 10% of the reef-limestone volume, against about 40% in the direct surroundings and about 60% at some greater distance.

Different growth forms

In reef limestones both of Holmhällar and of Hoburgen type, different growth forms are exhibited by the stromatoporoids. The main types have been described in Chapter V. On a soft sea floor the nature of the weak substratum often forced a development in the form of covers or thin lenses. In the reefs, however, where a solid substratum was generally present, upward growth was easier. There one often gets the impression that sedimentation stimulated such upward growth, particularly in the interior parts of the reefs, where moving water contributed less to keeping the colonies clear of sediment than at the reef edges. In a Hamra-Sundre reef in the south of the second hillock of Hoburgen at three different places ten specimens of stromatoporoids were measured. The reef section was northeast – southwest. At the northeast side, close to the reef margin, the average length of the stromatoporoids was 41 cm and the average thickness 9 cm. At the same level, about 3 m from the reef margin, these measurements were 34 cm and 11 cm, respectively. In the reef core, which in this case was about 6 m from the margin, the measurements were 27 cm and 13 cm, respectively. From reef margin to reef core, the thickness/length ratio thus was found to increase from 0.22 to 0.48. In the small reefs in the Klinteberg Beds the difference in size between stromatoporoids in the reef margins and core is much less. In a few Upper Visby reefs where measurements of this kind were made, no significant differences were found at all.

A few examples of different stromatoporoid growth forms are shown in Fig.226.

A stromatoporoid limestone of reef-detrital character, occurring at Sjausterhammar, has been mentioned in the discussion of the debris floor and talus mantle of Holmhällar-type reefs, in Chapter VIII. Some of the stromatoporoids found there, may have grown at the place where they are found. In a few thin and relatively more-marly layers low in sections through the stromatoporoid limestone, laminar stromatoporoids are present. Higher up, in a section at the south side of the reef, some stromatoporoids are apparently also in their original positions. Part of these show a high and narrow growth form, somewhat comparable to torpedos (cf. Fig.226 J,M), which is probably caused by the high sedimentation rate, due to reef-debris deposition. Still higher in the section, true reef limestone is exposed, indicating complete colonization of the area by the stromatoporoids. In the basal part of the reef most stromatoporoids have flat shapes; upwards the colonies become rounder. This example shows how different growth forms may be found even in one restricted locality.

The author has insufficient data to be able to say in how far the different growth forms are genotypical differences and in how far they are to be attributed to environmental factors. However, it seems probable that both factors have been at work. The example of *Labechia* is recalled here (cf. p.432). This genus occurs typically in a laminar growth form in the marginal regions of the reefs, but may be found occasionally in a branched form in more sheltered places. Environmental factors may have modified the growth form of some taxa and may also have made a certain environment more favourable for other taxa which normally developed the growth form most suitable for life in that part of the reef or reef environment.

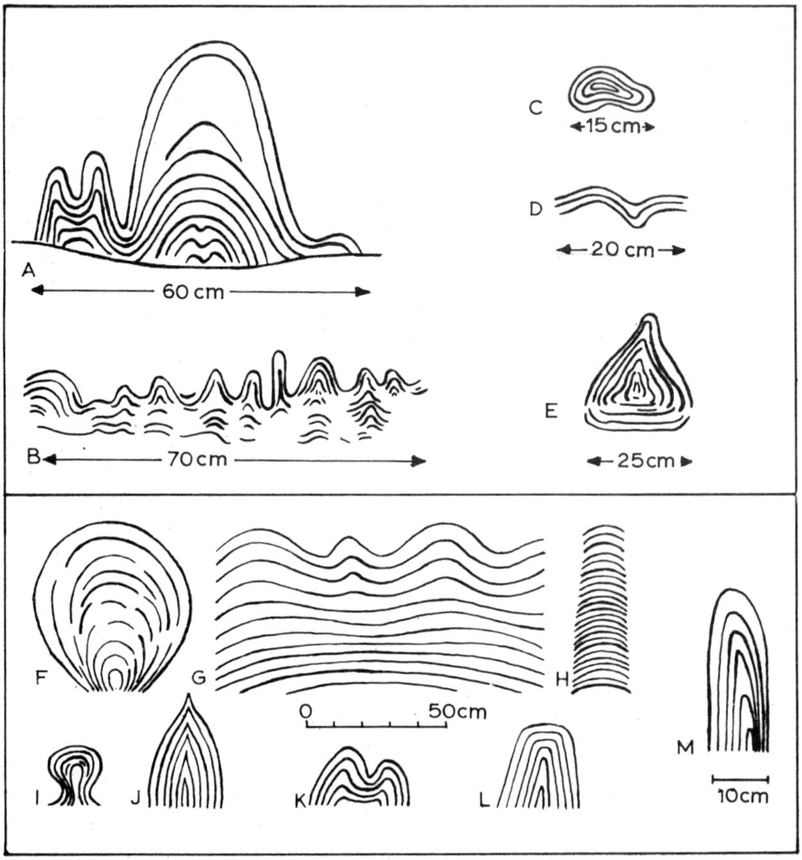

Fig.226. Some growth forms exhibited by stromatoporoids in reefs of Holmhällar type. A. Heliholm, Hamra-Sundre Beds; B–E. Holmhällar, Hamra-Sundre Beds; F–L. Fågelhammar South, Hemse Beds; M. Herrvik, Hemse Beds.

Possible explanation of stromatoporoid palaeoecology

There is no direct apparent explanation for the intolerance of mud on the part of the stromatoporoids. However, Colter (1957, p.260) presented two hypotheses which are worth mentioning here.

One reason may be similar to that postulated for the predominance of solitary corals in extremely muddy conditions, namely the size of the polyp. The smaller a coral polyp, the less able it is to cope with mud. Whatever the form of the stromatoporoid individuals has been, they have obviously been extremely small. Consequently it is likely that they will have been unable to remove sediment which has fallen onto the colony.

Another possible reason for the intolerance of mud displayed by the stromatoporoids may have been connected with feeding habits. Yonge (1940, pp.362–363) has shown that those corals with small polyps which have adapted their ciliary systems for feeding are less able to cope with sediment than those whose cilia can be used for the removal of sediment. The feeding habits of the stromatoporoids are not known, but if they were ciliary feeders, then they will have met the same difficulties as the ciliary-feeding small coral polyps.

Finally, in spite of all the emphasis laid in the preceding pages on the great susceptibility of the stromatoporoids to mud sedimentation, caution is still advised. Considering the importance of this aspect of stromatoporoid palaeoecology, the last word on that subject as a whole has certainly not been said. Recent ecological work has demonstrated that only rarely is a particular variation in environmental conditions the single cause or probable cause of animal behaviour. Also, that more than one factor may, under different conditions, produce the same effect. Much more work on the palaeoecology of stromatoporoids needs to be done, particularly much detailed work. Wherever possible, this should go together with taxonomic studies.

Latilaminae

As has been mentioned in Chapter V, several stromatoporoids show latilaminae. These are layers, generally 1–20 mm thick (Fig.227), which are composed of many laminae or cysts. The latilaminae are best seen in weathered specimens, and in rather pure and massive reef limestone may

Fig.227. Stromatoporoid colony, falling apart along latilaminar planes. Gannberg, Ostergarn Parish, Hemse Beds.

make the stromatoporoids distinguishable in the field from the matrix or the algal growths. Even within one and the same species, specimens may or may not show latilaminae and consequently they are of no taxonomic importance. Their development seems to be connected with pauses in growth, reproductive or perhaps seasonal pauses. In this connection it is noteworthy that some corals also show a kind of coarsely-laminated structure; this phenomenon is more common and more-distinctly developed in localities where, comparatively, the stromatoporoid latilaminae are also best developed.

PALAEOECOLOGY OF CRINOIDS

After the discussions on the palaeoecology of the two main groups of reef builders, some space needs also be devoted to the most important group of associated organisms, the crinoids. As mentioned in Chapter V, there is hardly a reef in Gotland in which crinoid fragments are not found. But far more abundant than in the reef matrix are crinoid remains occurring in the deposits which surround many of the reefs. In many instances these deposits deserve the name of crinoid limestone or even crinoid breccia.

Occurrence of crinoids in reef-surrounding sediments

The crinoid limestones are generally built up, for the main part, of small and large, disarticulated crinoid-skeletal remains and a calcareous mud which fills the interspaces and cements the whole. Embedded in the deposit, there is generally a varying amount of reef debris. The larger crinoid fragments, generally strongly recrystallized, have retained their original forms. In both transverse and radial sections the remains may show a fine porosity, a net-like structure. Crinoid sand is also a very common constituent of the crinoid limestones. It is generally most abundant at the original seaward side of the reefs, i.e., usually the southeastward side.

Around the reefs of Hoburgen type, crinoids developed on all sides. The crinoid limestone which they built is found directly around the reefs or, where these are surrounded by a talus mantle, directly against this mantle. The most characteristic of the crinoid limestones is a real crinoid breccia. This is usually best developed around the higher parts of a reef. Around the lower parts and also around the crinoid breccia higher up, the more usual crinoid limestone with reef debris or otherwise a limestone with crinoids and reef debris is generally present. Around several reefs, the crinoid breccia is missing. Only the two latter sediments envelop the reef limestone there.

In a crinoid limestone with reef debris in the Bogeklint or Klinteklint (Boge Parish, Slite Beds), the vertical distributions of the larger crinoid stem fragments have been studied (Fig.228). It appeared that there is a rough correlation with the number of pieces of reef debris in the same rock. At the base of the section, neither is abundant and at the top, both decline. Crinoid material of smaller size remains abundant there, but the larger stem fragments decrease in number.

In the Solklint (Slite, Slite Beds) the distribution of the larger crinoid remains along a horizontal line was studied (Fig.229). It was found that directly adjacent to the reef their number was highest and it rapidly decreased with increasing distance from the reef.

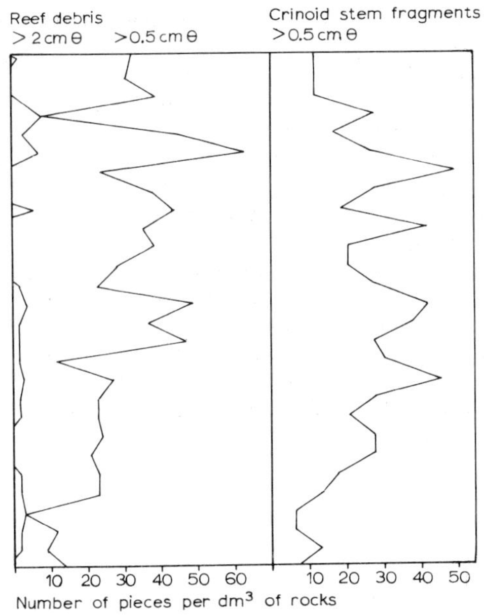

Fig.228. Vertical distribution of reef debris and crinoid stem fragments in a section through crinoid limestone with reef debris in the Bogeklint, Boge Parish. Only the material with a size, resp. diameter, of more than 5 mm has been counted. An additional curve shows the number of pieces of reef debris larger than 2 cm. The section is about 4 m high. The rock has been deposited at the southwest side (lateral) of a reef, the material exposed at the base presumably at 4–5 m distance from the reef, the material at the top at 6–7 m from the reef. The highest part of the reef extended about 1.5 m above the top of the crinoid-limestone section. (After Manten, 1970, fig.2.)

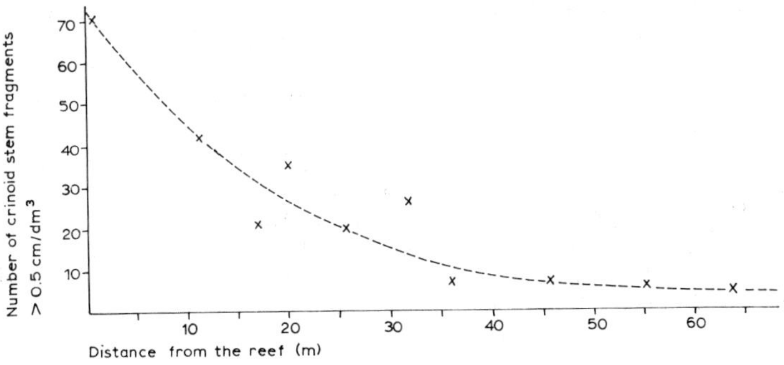

Fig.229. Number of crinoid-stem fragments with a diameter of more than 5 mm, per cubic decimetre of sediment deposited in the vicinity of a reef. The curve is based on the reef-surrounding deposits found in the Solklint (Slite, Slite Beds). Note that the number of larger crinoid remains rapidly decreases with increasing distance from the reef.

Both the crinoid breccia and the crinoid limestone with reef debris are generally thick-bedded. In some instances they are cross-bedded and in a few cases they show evidence of wave sorting.

In the Upper Visby Beds no crinoid limestones are found to envelop the reefs. Around the reefs of Holmhällar type they were also present, but in Recent time they were eroded from most of the localities where these reefs are exposed.

In the stratified limestones and marly limestones which were deposited at greater distances away from the reefs, crinoid remains are found scattered at random through the sediment, together with other marine invertebrates, such as brachiopods, bryozoans, corals and occasional molluscs.

In conclusion, the distribution of crinoid remains in the limestones of Gotland shows that these organisms could grow almost everywhere on the sea floor at the time that these rocks were laid down but that they were only abundant in the direct vicinity of reefs. When reef growth ended in a particular locality, apparently the conditions for crinoid development also became less suitable.

Occurrence of crinoids in reef limestone

The sediments surrounding the reefs of Hoburgen type always contain many more crinoid fragments than the reef limestones themselves, which may even be poor in such fossils. Comparatively speaking, the most common are crinoids represented in the reef limestones within the Klinteberg and Eke Beds.

Generally the reefs of Holmhällar type contain many more crinoid remains than those of Hoburgen type (cf. Chapter VIII). In the central parts of the reefs they may constitute 1–6% of the total volume of the reef limestone. Towards the periphery this percentage increases, and close to the margin crinoid remains may form 10–25% of the rock volume. The highest percentages are found towards the distal ends of the larger, crescent-shaped reefs. In the surface of the developing reef depressions occurred which were either filled with debris or became pools in which a different fauna developed, particularly corals and bryozoans. In some of the debris-filled depressions crinoid stem fragments are extremely abundant; in others they are much less numerous. In general, there seems to be a certain relationship between the amount of crinoid remains in a depression and the distance of the depression from the reef margin.

In the reefs of Hoburgen type debris-filled depressions can also be found, but there they are generally less characteristically developed and nowhere do they contain large amounts of crinoid remains.

As will be discussed in more detail on p.452, the above-outlined distribution of crinoid remains in the reef limestones suggests that crinoids lived on the reef flanks rather than on the reef top.

Average diameter of crinoid stem fragments

There are notable differences in the size of the crinoid stem remains found in the various localities. In an attempt to investigate whether some general lines could be detected in crinoid development, average diameters

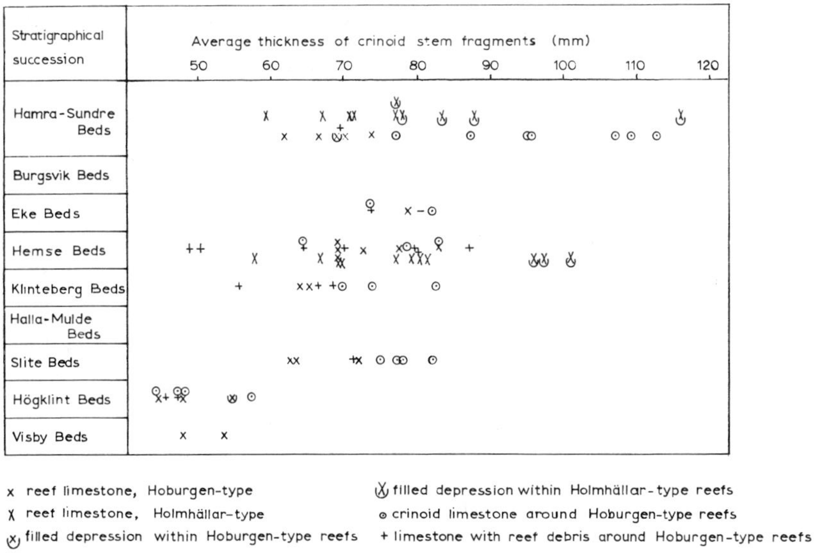

Fig.230. Graphic representation of variations in the average diameter of crinoid-stem fragments over the stratigraphical units in Gotland. (After Manten, 1970, fig.3.)

of stem fragments were determined. A total of 84 samples was studied and per sample 100 diameters were measured. The average values which were found are shown graphically in Fig.230 and are summarized in Table XXIII. The size-frequency distribution per sample was found to be rather normal.

The main conclusions are:

(1) The average diameter is larger in crinoid stem fragments found in depressions within the reefs than in those found scattered in the reef limestone proper, and presumably also than in those found in the crinoid limestones around the reefs.

(2) The average diameter of crinoid stem fragments is distinctly larger in crinoid limestones than in the reef limestone which they enclose or in limestone with reef debris which often occurs somewhat farther away from the reefs than the real crinoid limestones.

(3) There is no significant difference in the diameters of crinoid stem fragments found in reef limestone and in limestone with reef debris.

(4) In reef limestone of Holmhällar type the crinoid stem fragments are, on the average, thicker than in reef limestone of Hoburgen type.

(5) From the Upper Visby Beds up to the Hamra-Sundre Beds the crinoid stems tend to attain greater average thicknesses. From the Hemse Beds upwards, this trend is more pronounced in the figures from the reef-surrounding sediments than in those from the reef limestones.

(6) The averages found for the thickness of crinoid stem fragments in the Klinteberg Beds are below the values that would be expected on the basis of the increase noted under (5). (As argued in Chapter XI, the Klinteberg Beds were, however, deposited in shallower water than the majority of Slite and Hemse Beds.)

TABLE XXIII

Variations in stem diameter in crinoids of Gotland

Stratigraphical unit	Kind of reefs	Average diameter of crinoid stem fragments (mm) and number of samples (in brackets)			
		Reef limestone	Depression within reef	Crinoid limestone	Limestone with reef debris
Hamra-Sundre Beds	Holmhallär reef type	7.14 (7)	9.58 (3)	—	—
Hamra-Sundre Beds	Hoburgen reef type[1]	6.84 (5)	—	10.25 (5)	8.07 (3)
Eke Beds	Hoburgen reef type	7.86 (1)	—	7.79 (2)	7.72 (2)
Hemse Beds	Holmhallär reef type	7.27 (8)	9.80 (3)	—	—
Hemse Beds	Hoburgen reef type	7.15 (5)	8.32 (1)	7.55 (3)	6.87 (7)
Klinteberg Beds	Hoburgen reef type	6.46 (2)	—	7.55 (3)	6.35 (3)
Slite Beds	Hoburgen reef type	6.61 (3)	—	7.80 (4)	7.14 (1)
Högklint Beds	Hoburgen reef type	4.60 (2)	5.46 (1)	4.91 (4)	4.62 (2)
Upper Visby Beds	Upper Visby reef type	5.03 (2)	—	—	—

[1]Also two samples of stratified algal limestone with crinoid remains, found directly underneath reef limestone with an average diameter of the crinoid stem fragments of 7.16 mm.

(7) The difference in average crinoid diameter in reef limestone and in surrounding deposits is greatest in the Hoburgen-type reefs in the Hamra-Sundre Beds. (This may be caused by the origin of the samples; the reef-limestone samples were collected low in the reef-sections of Hoburgen, except for one with an average diameter of 7.38; the samples from the surrounding deposits are of younger age; as shown in Chapter XI, during the formation of these reefs, the water depth, initially very shallow, increased.)

(8) The average diameter of crinoid stem fragments in the reef limestone is closest to that in the surrounding sediments with the small reefs in the Eke Beds.

(9) The difference between the sample with the highest and that with the lowest value of average crinoid stem diameter, as obtained from the reef limestones, is greater for the reefs of Holmhällar type than for the nearest comparable reefs of Hoburgen type (in horizontal plane the latter generally are smaller). In the Hemse Beds these differences amount to 2.25 and 0.85, respectively, in the Hamra-Sundre Beds to 1.84 and 1.26.

Additional observations on crinoid-stem diameters are:

(10) In the deposits which envelop the reef limestone, the average diameter of crinoid stem fragments is, as a rule, somewhat smaller at the original seaward side of the reefs than at the original landward side.

(11) The average crinoid diameter in the reef-surrounding deposits generally decreases with increasing distance from the reef.

(12) Crinoid remains with notably large diameters (over 20 mm) occur in scattered fashion from the Slite Beds upwards. They are most common in the crinoid limestone of the Hamra-Sundre Beds ("Hoburg marble") and in the filled depressions within the reefs of Holmhällar type.

Factors which influenced crinoid distribution and size

An attempt will now be made to determine the various factors which have influenced the development of the crinoids found in the Palaeozoic deposits of Gotland. Only those factors which are of specific value in understanding the observed differences in crinoid distribution and size will be discussed here. Environmental conditions which influenced more generally both reef growth and crinoid development will be dealt with in Chapter XIV.

Different crinoid taxa

A major problem in the study of crinoids is the loosely articulated skeletal structure of the living animal. Minor agitation of the water over the sea floor has presumably already resulted in major disarticulation of the skeletal parts of defunct individuals. Well-preserved crowns are rarely found in the reefs or on their flanks. Some are found locally in limestone with crinoid remains and reef debris, which was laid down as inter-reef deposits. The taxonomic information which the author has obtained from these crowns is rather scattered and one-sided. Therefore, he is unable to prove whether some or all of the variations found in crinoid development should be attributed to the occurrence of different taxa. That different taxa have been present, however, is certain (Table IX) and that the ecological requirements and average sizes of each of these were identical is unlikely.

Taxonomic identifications of crinoids can not be made on the basis of stem fragments alone. There are some morphological differences between the stem remains, even within one sample, but it is known that these may occur even within one species. Thus the generally cylindrical columnals of *Crotalocrinus* species passed into somewhat more pentagonal forms down the stem. (In earlier literature, these pentagonal *Crotalocrinus* stem fragments are incorrectly described as *Cyathocrinites pentagonalis* Goldf.).

For the Niagaran (Middle Silurian) reefs in the North American Great Lakes region, Lowenstam (1948, 1950, 1957) found that in the development from the quiet to the semi-rough and rough water stages of reef development, more and more camerate crinoids, with their massive box-like calices, began to occur in addition to the more fragile inadunate crinoids (see also the section on the North American reefs in Chapter XIII).

Linkage to the reef environment

In Middle and Upper Palaeozoic times crinoids were generally associated with reef structures (Laudon, 1957, p.961). In Gotland it is only in very close connection with reefs that crinoids appear to have lived in tightly-knit gregarious coenoses.

The question can be put here as to whether the crinoids lived predominantly on the reef surface, on the reef flanks or in the immediate reef surroundings.

The author believes that it is the second environment which has been the most important. If the majority of crinoids lived on the reef surface, one would expect to find many more crinoid remains in the matrix of the Hoburgen type reefs. The different hydrodynamic properties of the crinoid remains, due to the very porous nature of the crinoid skeletons, compared to other fossil material of comparable size, should, of course, be taken into account. But where other, smaller fossils could be embedded in between the reef builders, together with terrigenous debris, why should not many more crinoid fragments have been preserved in reef interstices, if many crinoids had grown and become disarticulated on the reef surface?

The fact that the average diameter of crinoid stem fragments is larger in the crinoid limestones around the Hoburgen-type reefs than in the reef limestones also suggests that the crinoid limestones are not formed by the washing off of crinoid material from the reef surface.

The increase in crinoids from the central parts of the Holmhällar-type reefs towards the margins suggests that also these crinoids were by far the most abundant on the reef flanks, even though several may also have been growing on the surface of these larger reefs.

On the other hand, the presence of the most characteristic crinoid limestones directly against several reefs of Hoburgen type, or in some cases against their talus mantles, shows that the densest communities of crinoids must have been located very close to the reefs. Much reef debris even moved down from the reef over the crinoid thanatocoenoses to a final position somewhat further away from the reef. In the further reef surroundings also many crinoids presumably grew, but nowhere do they seem to have been so particularly abundant as on the reef flanks, irrespective of whether these flanks consisted of the reef frame proper or of a mantle of coarse talus material deposited around the actual reef.

The crinoid curve in Fig.228 suggests that the conditions for crinoid development became less favourable with the death of the reef around which they grew.

The typical situation of a reef with forests of crinoids on its flanks, developed particularly in the case of the larger reefs. The smaller the reefs were (Klinteberg Beds, Eke Beds), the less distinction there apparently was between the fauna of reef surface and reef flanks.

Water depth

As has been shown in the Chapters VI, VII, VIII and XI, all fossil reefs of Gotland developed in shallow water. In contrast to most crinoids of the present day, which are deep-water forms, Palaeozoic crinoids thus flourished abundantly in shallow water. Even in very shallow water deposits crinoid remains are found in large numbers.

One of the most characteristic crinoid limestones of Gotland is the "Hoburg marble", which is linked to the Hoburgen-type reefs of the Hamra-Sundre Beds in southwestern-most Gotland, particularly to the younger parts of these reefs. It was there that the highest average crinoid stem diameters were found. There are indications that these younger reef parts were formed in slightly deeper water than the older parts of these reefs and than most of the other Hoburgen-type reefs.

The next-largest average stem diameters are found in the reefs of Holmhällar type. Crinoids are also much more abundant in these reefs than in the other reefs of Gotland. It is likely that at least some of the most characteristic reefs of the Holmhällar type developed in somewhat deeper water than the majority of Hoburgen-type reefs.

There are, thus, some indications that in comparatively deeper shallow water (probably deeper than 10–15 m), the crinoids became larger and presumably also more abundant than in very shallow water. It may be assumed that the fragile crinoids must have been extremely sensitive to water agitation. At relatively greater depth, water mobility on the average may have been somewhat less. That may have been a greater influence on this facet of the crinoid distribution pattern than water depth itself.

Mobility of the water

There are several further indications that the crinoids grew larger in less agitated water than in strongly agitated water. The larger average diameters of crinoid stems at the original landward side of the reefs as compared to the original seaward side, and in filled-in depressions within the reefs as compared to the reef matrix, no doubt are functions of water mobility. This also holds for the greater abundance of crinoids at the distal ends of the crescent-shaped reef of Holmhällar compared to the seaward side (cf. Chapter VIII). The same may be true for the greater variation found in crinoid diameters within reef limestone of Holmhällar type. In larger reefs there is more variation in local environmental conditions between the various parts of the developing reef.

The presence of unusually large amounts of crinoid columns in the reef environments and the marked scarcity of skeletal remains of the crowns of crinoids in these places may perhaps also be attributed to the destructive action of agitated water. Another possibility is that predators fed on the crowns of the crinoids, allowing only the other skeletal parts to accumulate on the sea floor.

Sediment content of the water

Against the widespread belief that crinoids could only live in a clear sea, Ager (1963, p.132) demonstrated that at least some crinoid taxa could also thrive very well in muddy seas. He observed that in the Mississippian of Indiana (U.S.A.) autochthonous crinoid remains (long stems, calices, holdfasts) are commonly found embedded in shales.

The fossil crinoids of Gotland, however, belong to taxa which preferred relatively clear water. Some crinoid remains are found in stratified marlstone deposits but are only randomly scattered. In stratified limestones in reefless areas crinoid remains are more commonly found than in the marlstones, though they are never abundant there either. The stratified deposit in which they are most common is the algal limestone at the base of the Hamra-Sundre Beds. The marly to somewhat clayish matrix of some reef-like intercalations in this algal limestone indicates that there was a certain supply of terrigenous debris. On the other hand the general abundance of Algae in the algal limestone suggests that this deposit has been laid down while the water was not truly muddy. At least not to such a degree that a silt or clay deposit sedimentated from it, such as was the case in the Mississippian of Indiana.

As far as the reefs of Gotland are concerned, the matrix of the reefs of Hoburgen type, too, generally shows that the water in which the reefs grew contained terrigenous debris, but the mobility of the water prevented a substantial deposition of this material. Crinoids are distinctly least common in and around the reefs with the strongest marly matrix, those of the Upper Visby type. The reefs with the purest matrix, those of the Holmhällar type, are the richest in crinoids.

Better adaptation to the environment in the course of time

It is difficult to give a satisfactory explanation for the increase in average diameter of the crinoid stems in the course of geological time, as found in Gotland (Manten, 1970). As this trend does not directly continue in Devonian and younger formations in other areas, it is apparently a regional phenomenon. The reefs of the Upper Visby Beds are the first which formed in the west part of the Baltic Sea of Palaeozoic times. In the following younger geochronological unit, the Högklint Period, crinoids began to occur in crowded communities around reefs. From Högklint time onwards, the crinoids appear to have become increasingly better adapted to the reef environment, as appears from both their increase in size and in abundance. It would be interesting to know whether this development took place within the same taxa, or whether in the course of time new taxa evolved in the area which were more specialized towards this environment, or whether perhaps other taxa migrated into the area from elsewhere, to find in the reef environment their optimum living conditions. The author is unable to answer this intriguing question at his present stage of knowledge.

Chapter XIII

COMPARISON OF THE REEFS OF GOTLAND WITH REEFS IN SOME OTHER AREAS

GREAT BRITAIN

Whereas conditions suitable for the establishment of reefs occurred several times during the deposition of the Silurian rocks of the Baltic area, reef growth in the British Wenlockian was generally confined to the time of deposition of one sequence of bedded limestones of approximately 30 m thickness, alternating in part with thin marly layers. This sequence is underlaid by marl shales with nodules or lenses of marly limestone, and is generally overlaid by a sequence of crinoid beds, about 30–45 m thick. Within the limestone sequence reef building could apparently start and finish at any time, depending upon local conditions. Some reefs started quite near the bottom of the sequence and continued until deposition of the overlying crinoid limestones took place. Murchison (1839, p.211) recorded a reef limestone 80 ft. (approx. 25 m) thick, and Crosfield and Johnston (1914, p.200) listed various quarries which might, at one time, have contained reef masses with a vertical extent of 50–60 ft. (approx. 15–18 m). At the time of these publications, however, these reefs were nowhere higher than 30–40 ft. (9–12 m), with smaller reef masses of approximately 4 m high occurring commonly. According to Colter (1957, p.7), the lateral extent of any single reef exposed at that time did not exceed 100 ft. (30 m), and many were only a few tens of feet across.

It appears from these brief descriptions that the Wenlockian reefs of Great Britain compare the best with the Upper Visby reefs of Gotland. The stratified Wenlockian sediments may sag beneath and arch over the reefs, and laterally they may dove-tail irregularly into them. Laminar organisms may also spread out over the surrounding sediments, and this fact, together with the often gradual transition from reef to stratified limestone, shows that the reefs generally rose gently from the sea floor. Exceptions, where steeply-dipping laminar stromatoporoids at the reef edge indicate steeper reef flanks, are also found in the British Wenlockian.

The most important reef builders in the Wenlockian reefs are the tabulate corals, and among these, probably the most important single species is *Heliolites parvistella* Ferd. Roemer, which occurs in both branched and massive forms (Colter, 1957). In the Upper Visby reefs of Gotland, corals play an even more important part than stromatoporoids and other fossils, in contrast to the larger reefs of Hoburgen type in which stromatoporoids prevail. Branched corals, however, do not belong to the prominent reef components in the Upper Visby Beds, as they do in the Wenlockian, even though their colonies are known in the reefs of Gotland. Algae are present in both the Upper Visby and the Wenlockian reefs, but they are not as common as in the reefs of Hoburgen and Holmhällar type. In general, masses of *Solenopora* sp. and incrustations of *Rothpletzella* spp. are common in the Wenlockian, where they seem to be better developed than in the Upper Visby reefs, but much less so than in the other reefs of Gotland.

According to Crosfield and Johnston (1914), 87.5–97.5% of the favositids, stromatoporoids and cyathophyllid corals in the Wenlockian reefs are in positions of growth, against only 7.5–30% in the reef-surrounding sediments. Since these figures are obtained in a somewhat different way from that of the percentages in Chapter XII, they are not directly comparable, but the general tendency seems to be the same as that in the Upper Visby reefs (Table XXII, pp.435, 437). From the great difference in percentages, Twenhofel (1950) drew the conclusion that the Wenlockian reefs seem to have grown under conditions not favourable for lateral expansion.

A tentative conclusion which may be reached from the comparison between the British Wenlockian sediments and those of Gotland, is that the facies of the Wenlockian falls between that of the Upper Visby Beds and the limestone deposits of Gotland, for both its stratified and unstratified rocks. Its stratified sediments are impure limestones, but not as rich in marl as in the Upper Visby Beds. In the reefs, corals are usually dominant, but stromatoporoids are also important.

The stratified stromatoporoid Tofta limestones may be compared to stromatoporoid limestones found at Dudley (Colter, 1957, p.261). Both, no doubt, represent shallow water conditions, but the occurrence of branching forms at Dudley suggests still water. The large *Spongiostroma* balls, which are associated with the stratified stromatoporoid rocks of the Tofta facies, seem to have no directly comparable equivalents in the Wenlockian limestone.

ESTHONIA

As is apparent from the review of the Esthonian Palaeozoic in Chapter II, reefs occur in Esthonia at seven different stratigraphical levels. These levels are the Vasalemma Stage in the uppermost Middle Ordovician, the Pirgu Stage in the Upper Ordovician, the Porkuni Stage in the lowermost Llandoverian, the Hilliste member of the Tamsalu Stage in the Middle Llandoverian, the Raikküla Stage in the upper Middle Llandoverian (locally), the Lower Jaagarahu Stage in the Middle Wenlockian, and the Adla member of the Paadla Stage in the Lower Ludlowian.

Although reef development thus already occurred during the Ordovician, this took place much more commonly during the Silurian. At the end of the Ordovician, the Taconic phase of the Caledonian orogeny occurred simultaneously with a general regression of the sea. The transgression, which began during the middle of Porkuni time, did not bring the sea as far as before. In the Esthonian area, the sea remained shallow during the whole of the Silurian, thus contributing to favourable conditions for reef growth.

Reefs are especially abundant in Esthonia in the Jaagarahu strata, where they form lentoid and irregularly shaped bodies which are clearly distinguishable from the surrounding rock. They are the best exposed along the northern shore of the islands of Saaremaa (Ösel) and Muhu (Moon). The reef dolomites are found in the upper parts of the eminences, whereas the stratified dolomites and marlstones lie bare in the lower parts. Due to the hardness of the biohermal rock, the reefs frequently stand out from the surrounding rock as a consequence of the action of the land ice, and erosion and abrasion by the post-Glacial Baltic Sea. They form isolated eminences or cliffs, which are often steeply terraced and irregular in shape.

The Jaagarahu Stage reefs occur in both the Kesselaid dolomite and the Jaagarahu limestone. The main reef builders in the former are calcareous Algae (Stromatolithi and others) and bryozoans (*Coenites repens* (Wahlenberg)); corals are less common and are mainly represented by Rugosae, to a lesser degree by *Favosites* sp. and locally by *Halysites* sp. In the Jaagarahu limestone, dominant reef builders are stromatoporoids and rugose corals, each mainly represented by only one species, viz., *Stromatopora typica* Ros. and *Favosites mirandus* Sok.

It is likely that the deposit, described by Hoppe (1932, p.49) as a stromatoporoid reef built up by unluted, angular stromatoporoid fragments of various dimensions, originates from the fore-reef facies rather than from the reef proper.

The diameter of some reefs attains several kilometres and they may be up to 16 m thick. However, most of them are smaller, with a diameter between some metres and a few hundred metres. The larger reefs are mainly found in the islands.

There is an apparent connection between the increase in reef size in the western direction and the organic composition of the reefs. The dolomites with algal and bryozoan reefs occur in the east of the Jaagarahu Stage (cf. Chapter II). Towards the west, limestones are found and corals and stromatoporoids increase in importance as reef builders; Algae remain present but are less abundant in the west. It is supposed that the basin increased in depth westward, although the Algae are an indication that the depth was not more than some tens of metres at any one place.

The Jaagarahu reefs occur in a belt parallel to the coast line of that time, which was north of the reefs. There is great similarity between these and the reefs of the Högklint and Slite Beds of Gotland, with one of which the Jaagarahu reefs may correlate and probably even form one belt. Martinsson's (1958) study on the submarine morphology of the Baltic (see also Fig.3) suggests a correlation with the Högklint klint complex, although that author had to admit (p.23) that in Saaremaa the klints lie higher in the sequence than the coast klint of Gotland.

The Jaani Stage in Esthonia consists in its lower part of marlstone and clayey marlstone, passing upwards into fossiliferous marly limestone with marlstone. In view of this lithological sequence and also of the relatively great thickness (up to 56.5 m), it may well represent the equivalent of the Visby and Högklint Beds of Gotland, as has already been supposed by, e.g., Luha (1930).

It is interesting to note that Aaloe (1956, p.94) assumed that since the Jaani – Jaagarahu boundary in the Esthonian area the water depth in the Silurian basin gradually and slightly increased, thus leading to favourable conditions for reef growth. In the second half of Jaagarahu time, water depth decreased again, reef development came to an end, an terrigenous components became more common, as well as calcareous Algae. The latter locally formed algal biostromes, which can be considered the very-shallow water equivalent of a reef facies. In Kaarma time in the Esthonian mainland, deposits of a lagoon-like facies were laid down. This sequence of events appears to be similar to that which happened in Gotland in Slite time and the Halla element of Halla-Mulde time.

The stromatoporoid-coral reefs of the Paadla Stage in Esthonia probably correlate with the Hemse reef belt of Gotland. These Esthonian reefs are found in the southwest of Saaremaa. Northeastwards, the reef-containing Adla limestone passes into a somewhat marly, coarse detrital limestone without reefs. Similar sediments also overlie the reef member. The whole is suggestive of water that gradually became shallower, and a moving down of the coast line in an approximately southwesterly direction. Deposits from the last two Silurian stages in Esthonia, the Kaugatuma and Ohesaar Stages, are found still farther southwest and, thereafter, a hiatus in the stratigraphical column of Esthonia occurs until in the Middle Devonian.

NORTH AMERICA

It has already been mentioned in Chapter V, that Silurian reefs are extensively distributed in the eastern half of North America. They have been especially well studied in the Great Lakes area of the United States (Chamberlin, 1877; Carozzi and Zadnik, 1959; Cumings, 1930a,b, 1932; Cumings and Shrock, 1927, 1928a,b; Fenton, 1931; Ingels, 1963; Lowenstam, 1948, 1949, 1950, 1952, 1957; Pinsak and Shaver, 1964; Shrock, 1929, 1939; Textoris and Carozzi, 1964; Textoris, 1966). Silurian reefs in the adjacent Canadian area, and further north and east of this, have been described by Bell (1886), Savage and Van Tuyl (1919), Thorsteinsson and Fortier (1953) and Twenhofel (1927).

Great Lakes area

In the Great Lakes area, reefs occur in the Niagaran, which is roughly equivalent to the Upper Llandoverian and the Wenlockian of the European stratigraphical column. In that area, the Niagaran comprises three broad sedimentary belts, in the trend southwest – northeast. From northwest to southeast, these are the clastic-free belt, the low-clastic belt and the high-clastic belt (Lowenstam, 1949, 1950). The first two contain reefs, the last one does not.

The reefs vary in form across the strike of the sedimentary belts. According to Lowenstam (1950), the water in the northwest, in the clastic-free belt, was very shallow. Reefs formed there are lenticular in vertical section and lack steeply-dipping flank beds. They are presumed to have risen only a little above the surrounding sea floor (Shrock, 1939, p.555).

Towards the southeast the water became deeper. Reefs developing there could rise to considerable heights above their surroundings, until they entered the turbulent surface waters.

Distribution of these two types of reefs in the two sedimentary belts is a good general rule, and not quite without the exceptions that are needed to prove a good rule (Lowenstam, 1949, p.28).

The reefs of the first type, those from the clastic-free belt, are small in area, being no more than 30 m or so in diameter (Lowenstam, 1949, p.29, 1950, p.478). Vertical expansion was limited by the shallowness of the water over most of the area, hence the development of low-lying, lenticular to tabular reefs. They stood only a metre or so - definitely less than 3 m - above the surrounding sea floor (Lowenstam, 1957, p.226).

The reefs in the Great Lakes area are dolomitized, but corals and stromatoporoids can still generally be recognized. The fauna of the reefs is comparable in composition to that in the surrounding stratified sediments, differences being more in degree than in kind. As there was generally little change in environment between the initial and final stages of reef growth, there are little, if any, differences in reef builders and reef-dwelling populations between the lower and higher parts (Lowenstam, 1957, p.226).

In the clastic-free belt, the substrate of the reefs, as a rule, consisted of sand-sized calcareous bioclastics with negligible amounts of terrigenous material.

The surrounding stratified sediments of the reefs are said to show signs of deposition in shallow water and of the reworking of material swept off the reef. The absence of talus flanks around the reefs of this type is explained by the debris being swept away from the reef across the surrounding sea bed, which was only a little deeper than the reef surface itself.

In the southernmost part of the clastic-free belt, some reefs with detrital flank beds are present. These reefs are comparatively small, ranging approximately from 10–300 m in diameter and their water depth is presumed to have been generally greater than that farther inside the belt. Their faunal characteristics differ little from the inter-reef bottoms, except for the presence of reef builders. A large admixture of reef-derived, skeletal material tends to obscure further any minor differences than may have been present initially.

The reefs of the low-clastic belt could reach a size of up to several kilometres in diameter and almost 300 m in thickness. Spacial configuration of the reefs is known only in a few instances and indicates circular, ellipsoidal, hemispheroidal, cuboidal, mound or ridge shapes. The Marine reef in the subsurface of Madison County, Illinois, is the largest one known in this region, and has an area of more than 15 km^2. Its surface outline in its final growth stage is horseshoe-shaped, with the convex side facing south, which illustrates the moulding effect of the prevailing southerly winds (Lowenstam, 1957, p.223). No other reefs of this shape have been recorded from the American Silurian.

In certain instances, closely spaced growth centres probably coalesced to form composite reef bodies, but in the main sites, the reefs have maintained their individuality. Subsidiary growth centres developed on the talus slopes of some reefs, but they were generally short-lived owing to burial by debris from the main reef.

Reef building in the low-clastic belt also, always occurred on unconsolidated bottoms. In many instances the substrate consisted of sand-sized, calcareous bioclastics; others have been found rooted in argillaceous, carbonate muds and calcareous siltstones (Lowenstam, 1957, p.219). A number of reefs have settled into the argillaceous bottom deposits.

Stages of reef development

Following a study of the reefs of northeastern Illinois and northern Indiana, Lowenstam (1957) presented a survey of the vertical differentiation of environment and the composition of the reef fauna. He distinguished a *quiet water stage*, a *semi-rough water stage* and a *rough water or wave-resistant stage*.

The quiet water stage comprised the time between initial colonization and the building of a solid reef framework upward to the effective wave base. According to Lowenstam (1957) the first organisms to settle were almost exclusively *Syringopora* colonies. These built the initial reef surface, together with colonies of *Favosites*, which were the first to join them. Unfortunately, dolomitization is extensive, making recognition of reef builders rather difficult. In a reef at Wabash, Indiana, and in others elsewhere, Lecompte (1938) noticed the occurrence of structures resembling the problematical *Stromatactis* of the Belgian Devonian. These structures present themselves as cavities, with a relatively flat bottom and digitated top, filled with sparry calcite (Textoris, 1966). Although Lowenstam is not certain whether the structures represent stromatoporoids or Algae, he considers these as the builders of a loose network, together with favositids. According to Lowenstam, this network should have acted as the framework of the developing reef. Textoris and Carozzi (1964) indicated a fistuliporoid bryozoan control of the *Stromatactis* cavities. These authors were able to discover some non-dolomitized to partly-dolomitized outcrops of reefs in Indiana. On the basis of these they were able to establish six ideal stages of reef development, as opposed to the three stages of Lowenstam. The first three stages compare with the quiet water stage. They consisted of the development of a mound of calcisiltite containing crinoid, bryozoan, and ostracode bioclastics and *Stromatactis*, but notably enough, essentially no stromatoporoids or corals. These three stages of development include the Wabash reef. Living locally on and within the reef frame, in addition to crinoids, bryozoans, and ostracodes, were trilobites, brachiopods and sponges. The fauna of the quiet water stage was characterized by a low population density and a small number of species of both the reef builders and the reef dwellers. As a result of this, the reef framework was mainly filled with inorganic, terrigenous sediment. The reef flank may have consisted of inclined-bedded core-type rock with only a subordinate amount of bioclastic material (Lowenstam, 1950). The contrast between the reef fauna and that of the surrounding stratified sediments was not as marked in the quiet water stage as it is in the next two stages, as distinguished by Lowenstam.

During the semi-rough water stage (stage 4 of Textoris and Carozzi, 1964), the reefs were subjected to the action of storm waves, remaining in this stage until the surf base was just about reached. Lowenstam believed that the reef frame in this phase was built of *Stromatactis*-like structures and stromatoporoids, the latter having considerably increased in proportion over the former. However, Textoris (1966) found that in the best exposure thusfar known of reef limestone from this stage, the organisms which capped the quiet water mounds, thus allowing them to continue to evolve into more agitated environments, were not encrusting stromatoporoids but spongiostromid Algae. Corals were represented in the semi-rough water stage by such genera as *Favosites*, *Halysites*, *Heliolites* and *Syringopora*, but were not of great importance. Reef dwellers, in order of their relative abundance, were crinoids, brachiopods, bryozoans, trilobites, sponges, cephalopods, gastropods and lamellibranchs. In comparison to the first stage, the population density and the number of species in the reefs had increased. The reef frame was now filled mainly with detritus.

In the rough water stage (stage 5 of Textoris and Carozzi, 1964), the number of species and the population density were considerably higher, which was true for both the reef builders and the reef dwellers. *Stromatactis*-like forms were still present, but stromatoporoids now irrefutably participated in reef formation and dominated among the reef builders. Corals too, were very abundant; they were now represented by other genera, such as *Arachnophyllum*, *Thecia*, *Alveolites*, *Fletcheria* and *Coenites*, and actively participated in frame building. Among the reef dwellers some new groups had appeared, viz. inarticulate brachiopods, cystoids, blastoids and conularids. Material swept from the reef by water turbulence rolled down the slope into the deeper water, where it was not subjected to wave action. As a result, steeply-dipping talus layers were formed, possessing a synsedimentary inclination, in contrast to the result of differential compaction (Shrock, 1939).

The final phase in reef building is represented by the subsurface Marine reef, which bears on its upper side, an accumulation of bioclastic debris, interpreted by Lowenstam (1950) to be material deposited on the reef surface by the winds prevailing at the time of the emergence of the reef. Growth was then restricted to a narrow zone on the windward side.

Crinoids

Crinoids were always the most important contributors to the debris, as this is found in both the reef limestone itself and in the flank deposits. In the quiet water stage, reef dwellers including crinoids were sparingly represented; in the semi-rough water stage, they were much more abundant, making up the bulk of the debris. Locally, however, the brachiopods increased in reef-flank deposits to proportions close to those of the crinoids. In the rough water stage, crinoids were by far the most common element among the reef dwellers.

A marked pattern of change has been noted in the crinoid group (Lowenstam, 1957, p.239). At the initial stage of reef development, the Inadunata were the sole crinoid representatives. They were first accompanied by Camerata at the intermediate upgrowth stage; and by the terminal upgrowth stage, the Inadunata had become only a subordinate faunal component, as were the Flexibilia. More than 75% of the crinoids at that stage were camerates. Morphologically, they were particularly conditioned to occupy the rough water niches successfully, not only those on the reefs but also those of the inter-reef and open shelf habitats.

Part of the crinoid remains were stabilized on the reef surface, but by far the most were on the flank of the reefs. In the flank-free reefs of the clastic-free belt, much crinoid debris was carried away from the reefs and scattered over the inter-reef bottoms (Lowenstam, 1957, p.241).

Comparison with Gotland; the criterion of wave resistance

In conclusion, it can be stated that the reefs in the Niagaran of eastern North America show a greater variety of composition and of size than those in Gotland. Of the two main classes, distinguishable in the Niagaran, those of the clastic-free belt are the most comparable to the reefs of Gotland, in that they developed in a shallow sea, with no chance to grow up to a great height above the surrounding sea floor. They correspond in size and form, and in their relation to the stratified sediments around them. There are generally no well-developed, more or less steeply-dipping flank beds, but the material swept from the reefs was distributed over the surrounding sea floor.

Lowenstam (1950) also compared the flankless reefs of the clastic-free belt with those of Gotland, the reef nature of which he had surprisingly denied earlier in the same paper. Following Ladd (1944), Lowenstam considered wave resistance as the most important criterion by which reefs should be judged. Even though he admitted the existence of a "grading spectrum of allied habitats", he insisted on the maintenance of a rigid separation of wave-resisting structures from those that are not wave-resisting (Lowenstam, 1950, p.435). Lowenstam admitted that the structures of Gotland show frame building and sediment binding, and that they developed above wave base, as is shown by the presence of bioclastic debris, swept off them, in the stratified sediments. However, it is Hadding's belief that the structures of Gotland developed not immediately below the ebb level, but at a depth estimated to be greater than 5 m (Hadding, 1941, p.75) which is taken by Lowenstam (1950, p.438) as indicating that they were not actually wave resistant.

By a strange coincidence, in the same issue of the journal in which Lowenstam advanced his opinion, Hadding (1950, p.405) showed that the water around the reefs was restricted in some places and had free play in others, resulting in the deposition of muddy sediments alongside pure limestones. This is evidence of "reef-induced turbulence", which Lowenstam considers as an indication of wave resistance.

It is with approval, that the present author quotes Colter (1957, p.233), who opposed the view that wave resistance is some universal constant by means of which a structure may be measured, in order that it might be accepted or rejected as a reef. Colter then points out that "the depths of wave action and the breaker zone vary considerably with size and form of local waves, and a structure which resisted the action of breakers in one environment might well crumble before the onslaught of larger waves. Similarly, a structure might well develop in quite shallow water, but be beneath the local breaker zone". Even if it were true that the reefs of Gotland did not grow in water shallower than 5 m - and the present author indeed believes that it is not true - this is not an adequate criterion by means of which the reef character of these structures may be denied. The application of such a standard takes no account of the variation in composition and potential which is possible in organic structures. As Colter (1957, p.234) states, there "is probably no gap or clean-cut distinction in nature between structures which are wave-resisting and those which are not, more especially as the very properties necessary for wave resistance are variable, and the standard applied inconstant". Acceptance of a structure as a reef should also take into account such factors as the associations of organisms responsible for it and the relationships of the structure to the environment.

CONCLUSIONS

The observations and views quoted on Silurian reefs of areas outside Gotland put reef development in Gotland into rather better perspective, and also confirm a number of conclusions which were drawn in earlier chapters for the reefs of Gotland:

(1) Reefs begin to occur much lower in the stratigraphical column of the east Baltic area than in the west.

(2) The succession of strata in the British Wenlockian shows that sedimentation of argillaceous matter should be below a certain limit to permit reef growth. This is in agreement with the picture presented by the Upper Visby Beds of Gotland.

(3) Reef building generally took place on unconsolidated sea bottoms.

(4) Not every one of the reef belts of Gotland corresponds with a similar development in the east Baltic area. Only during two of the five main periods of reef formation in Gotland (presumably Slite and Hemse) did reefs develop also in the east of the basin.

(5) The distribution of reefs in eastern North America shows that water depth is the more important factor in determining the dimensions of the reefs, provided that the water is not very muddy, which would cause a premature destruction of the reefs (cf. the reefs in the middle Upper Visby Beds of Gotland).

(6) In shallow water reefs generally rose not more than a few metres above the surrounding sea floor. Under such circumstances, differences between the fauna of the reefs and that of their surroundings everywhere, are more in degree than in kind. Within the reefs there is no apparent vertical zonation in faunistic composition. However, a deviating fauna and vertical zonation are found in reefs which grew in deeper water. In Gotland, the reefs of the main island all correlate with shallow-water reefs elsewhere, but the large reefs of Karlsöarna are rather of the deep-water type.

(7) The facies of the reef-bearing sequence in the British Wenlockian is intermediate between that of the Upper Visby Beds and the Högklint Beds. The contribution of stromatoporoids to the Wenlockian reefs is also intermediate between that of the reefs in the two stratigraphical units of Gotland. This makes a correlation of stromatoporoid development with decreasing deposition of inorganic detritus more likely than a correlation with shallower water.

(8) In the Niagaran of North America, stromatoporoids increase in number with a decrease in water depth, but the same is true for the corals. Lowenstam (1950, p.483, 1957, p.240) has also found that the stromatoporoids decrease in importance as reef builders from the clastic-free belt across the low clastic belt, and are subordinate at the southern border of this region. This decline coincides with an increase in the muddy impurities. These observations also suggest that a decrease in sedimentation was of greater influence on their expansion than preponderance in their competition with the reef-building corals in shallower water.

(9) Water turbulence was of great influence on the type of sediment surrounding the reefs. In deeper water, material swept from the higher parts of the reefs rolled down and formed steeply-dipping talus layers. In shallow water, where the bottom was also under the influence of wave action, reef-derived material was to a large extent swept across the surrounding sea bed. The latter situation is found in Gotland, with the possible exception of the large reefs of Karlsöarna.

(10) Reef growth everywhere is generally accompanied by a strongly-increased development of crinoids. The marked pattern of change observed in the large reefs of the American low clastic belt, indicates that the

taxonomic composition of the crinoid community there was linked with the depth of the growing reef surface, which suggests that the crinoids grew on the reef. Whether they were more common high up on the reef flanks, or on the upper side of the reef, is difficult to judge on the basis of the available literature. Where small reefs developed (clastic-free belt in North America; British Wenlockian) crinoid remains are found also most abundantly in the immediate environment of the reefs.

Chapter XIV

GENERAL CONDITIONS UNDER WHICH THE REEFS OF GOTLAND FORMED

Together with the discussions of the various reef types of Gotland, including Karlsöarna, of the stratigraphy of the island, and of the palaeoecology of some groups of fossils, much information has already been given on the conditions under which the reefs of Gotland formed. A few additional remarks on this subject, of a general rather than of a specific nature, will be added in this chapter.

WATER TEMPERATURE

In the assessment of ancient climatic conditions, reliance may be placed on some general comparisons of ancient life with its modern equivalents.

Both ancient and recent reefs show a great variety of organisms with calcareous parts. Over half a century ago, Vaughan (1911) argued that the physiology of organic calcium-carbonate precipitation is governed by physico-chemical processes which have probably remained similar over the years.

Although the details of the carbonate system in sea water are not yet fully known, in general calcium-carbonate precipitation occurs more readily under conditions of high temperature, high salinity, and low carbon-dioxide concentration. Such conditions are typical of tropical shallow waters, away from the influence of fresh water, where evaporation raises the salinity and active plant growth reduces the CO_2 content.

Most papers on modern reefs stress the fact that reef building takes place in warm and shallow water. The high rate of calcium-carbonate precipitation in stromatoporoids, corals, crinoids, calcareous Algae, and other contributors to the reefs of Gotland, suggests that also these reefs developed under shallow and warm-water conditions, similar to those which occur in tropical regions of the present day.

Oolites are present in various stratigraphical units of Gotland. They also indicate a warm climate and a shallow sea during their deposition.

Probably even a second comparison between present and ancient marine life may be made. Most marine organisms display absolute-temperature limits of distribution. Generally these are not lethal temperatures, but mechanisms such as feeding, reproduction, and general activity are often modified already at temperatures less extreme than those at which immediate or rapid death occurs. In the warmer seas, it is generally the summer maximum which determines the possibility of survival, whereas the winter minimum determines the possibility of breeding (Moore, 1958, p.19). Temperature-tolerance ranges are usually smallest for developing eggs and larvae, and tend to increase as the organism grows older. Early in this century, Mayer (1914) already showed that forms inhabiting warm waters predominantly live closer to their upper limiting temperature than to their lower limit. Several later studies have confirmed this. Particularly for reef building, which requires the regular addition of new organisms, a temperature minimum well above the lower limit for breeding is extremely important.

It may, thus, be assumed that the reefs of Gotland developed in water with temperatures similar to those in at least the subtropical, if not tropical, seas of the present day.

WATER DEPTH

In several of the preceding chapters evidence has been produced suggesting that the reefs of Gotland developed in shallow water (e.g., pp.176–177, 210–212, 311, 336, 462–463). This evidence came both from the organic and general features of the reefs themselves and from the nature of the surrounding sediments. Some additional remarks will be made here about the role played by Algae.

Most papers on modern reefs stress the association between Zooxanthellae and reef corals. This association is vital to the former, but not to the individuals of the latter. Nevertheless, it may be an essential factor in the attainment of the high rates of metabolism which are necessary for the establishment and maintenance of the coral-reef community. Moreover, photosynthesis assists in increasing the calcium content of sea water. The photosynthesis connected with the association is one of the main reasons why present-day reef building is confined to shallow water.

Unfortunately, there is no evidence of an association between Palaeozoic Algae and corals. However, if there were not such an association, the economy of the reef must have been different from that of their modern counterparts.

Yonge (1940, p.368) suggested that present-day reef corals also rely very strongly on the Algae for the removal of certain waste products; particularly phosphoric and probably also nitrogenous excretory products (cf. also Moore, 1958, p.328). If the Palaeozoic corals and stromatoporoids were not associated with symbiotic Algae, it must be assumed that these ancient reef builders were able to take care of this waste removal alone.

In conclusion, if the Palaeozoic reef builders contained no Zooxanthellae, they must have been more efficient reef builders than the modern reef corals. In itself this is not impossible, but a symbiosis of Algae and reef builders in Palaeozoic times is more likely.

In addition to the possible presence of Zooxanthellae, the reefs of Gotland also contain other Algae. Calcareous Algae of the genus *Solenopora* are fairly well represented in several reefs. The genus may have belonged to the red Algae. *Rothpletzella*, which precipitated lime from calcium-rich water, was a genus of blue-green Algae.

Cloud (1952a) believed that a depth of not more than 30 m is likely for the formation of any rock, deposited under marine conditions, in which blue-green Algae are common. The red Algae are unable to use blue light. Although light of this wave length also decreases exponentionally with depth, it is able to penetrate to greater depths than visible light of other wave lengths. Nevertheless, Cloud stated that branching and nodular coralline red Algae are also rarely abundant below 30 m.

Calcareous Algae are common in the reefs of Hoburgen and Holmhällar type in Gotland. In several of the Upper Visby reefs, they are not found at all and occur only rarely in some others. Perhaps conditions for their development were marginal in these reefs. However, the fact that some are present suggests that the Upper Visby reefs cannot have developed below the lower depth limit of algal growth. This is even truer if the corals in these reefs should indeed have lived in symbiosis with Zooxanthellae. The presence of marlstones in the Visby Beds suggests that the water in which the Upper Visby reefs developed must have been frequently rather muddy. This would have cut down the penetration of sunlight over the reefs considerably.

In conclusion, it is likely that all the reefs of Gotland developed in water less than 50 m deep (if not at even shallower maximum depth). For the reefs of Hoburgen type a water depth of less than 30 m may then generally be assumed. The characteristic Holmhällar-type reefs grew in rather clear water and may have had a slightly deeper limit, but probably not more than 40 m.

As discussed, e.g., while describing the Hoburgen-type reefs, several reefs grew during periods in which the water gradually became shallower or deeper. This had its effect on the lateral and vertical extension of the reefs. The growing parts will have always remained within the depth zones which were suggested in the previous paragraphs.

WATER AERATION

The general characters of the reefs of Gotland and their associated sediments, combined with the high density of organisms in both, show that the water in which the rocks were formed and in which the organisms lived was well aerated. The water must have usually been in movement. It is true that most attached-living animals produce some water currents by themselves. Generally, however, they benefit from a more general movement of water past them, even though they are liable to mechanical damage from such water movement. The latter ensures the supply not only of oxygen, but also of their food. It also assists in the removal of waste products. With the abundance of benthos in the reefs of Gotland, and in their surroundings, it may be assumed that there has been a general movement of the water during at least a major part of the day.

Occasional heavy movements of the water were generally not beneficial to the reef builders. As several of the reefs show, these movements damaged the more-fragile reef builders and detached more-massive colonies; or, in severe cases, even damaged larger parts of the reef.

As discussed in Chapter VII, when treating the interrelations between Hoburgen-type reefs developing in a particular reef zone, the reefs growing furthest toward the side of the open sea may have met with more favourable conditions for a rapid and healthy development than those with a position closer to the coast.

On the sea floor also, oxidizing conditions will have prevailed. The presence of pyrite in some of the stratified deposits does certainly not indicate general reducing conditions at the sediment – water interface. The pyrite was presumably formed diagenetically under reducing conditions developing only after the burial of the sediment in which the pyrite is found.

RATE OF REEF GROWTH

It is very difficult to make a reliable estimate of the rate of upward growth of the reefs of Gotland, but nevertheless, an attempt will be made at this point.

The growth rate of coral colonies in recent reefs varies strongly in different colonies of the same species and within the same habitat. One of the main reasons for this may be a decrease in growth rate with increasing age of the colony. Furthermore, it appears that growth of corals takes place intermittently. An average annual upward growth of about 1 cm may be assumed (cf. Moore, 1958, p.332).

The same figure may be a reasonable approximation for the upward growth of the coral and stromatoporoid colonies in the reefs of Gotland. For the total reef surface, however, the rate will have been lower. Living and dead colonies will have alternated. On the various parts of the reef surface, conditions for reef growth were different, very favourable in one place and submarginal in others.

Along with addition to the reef by growth, there has been loss by various causes. Water turbulence damaged the reef builders and washed both fragments and complete colonies off the reef. In reefs growing in very shallow water, even the entire reef surface or major parts thereof, may temporarily have been subjected to erosion (Upper Högklint Beds, Upper Hemse Beds). Pressure exerted by the younger reef parts on the underlying reef parts caused compaction. Locally, the mechanical strength of the reef frame may also have been further weakened by boring organisms (cf. p.76).

Taking all these factors into consideration, the average annual growth of the reefs in upward direction may have been not more than about 1 mm. For the development of a reef of 10 m thick, a period of 10,000 years may consequently have been required.

Field observations make it likely that a living reef of that thickness did not extend more than about 2–4 m above its surroundings. The total weight of the reef mass caused a sagging of the underlying stratified sediments. If this is assumed to have caused a subsidence of the central part of the reef of about 0.5 m, then in the 10,000 years which reef development took, about 6 m of stratified sediments must have been laid down in the area around the reef. This is an average of 60 cm per 1000 years.

In the present Moluccan area, where there is both a high productivity of lime-secreting organisms and a fair amount of detrital-lime supply by rivers, about 20 cm of calcium carbonate is deposited per 1000 years (Kuenen, 1950, p.379). This seems to be one of the highest figures for lime sedimentation that has been established.

The continent bordering the epicontinental sea in which the reefs of Gotland formed, probably consisted for a major part of crystalline rocks. As the shift of the successive reef zones of Gotland towards the southeast shows, the Palaeozoic Baltic basin was constricted in the course of time. The continental area closer to the coast may, therefore, have consisted of elevated Cambrian and Ordovician sedimentary deposits, consisting in part of limestone. Nevertheless, the supply of lime from the continent will presumably have been distinctly lower than in the case of the Moluccan area. This the more so as the relief on the continent bordering the Palaeozoic Baltic basin presumably was much lower. There is no reason to assume that the productivity of lime-secreting organisms in reefless areas in Gotland has exceeded the high productivity of the Moluccan area; it has rather been lower. Therefore, with regard to areas on the Silurian sea floor which were devoid of reefs, a total lime accumulation (organogenic-detrital and terrigenous-detrital) of 10–11 cm per 1000 years may perhaps be a realistic assumption. Adding to this the contribution of the terrigenous mud to the marly limestone, the total figure for normal sedimentation may have been in the order of magnitude of 12 cm per 1000 years. Considering further that the stratified sediments in the reef environment consist for about 80% of crinoid and reef debris, a total figure for these parts of the sea floor of 60 cm per 1000 years may have been possible.

In conclusion, average upward growth of the reefs of Gotland will presumably not have exceeded 1 m per 1000 years in absolute sense or 0.40 m per 1000 years with respect to the surrounding sea floor. In the initial stage of reef development, this figure was presumably higher, whereas in a late stage of reef growth it was distinctly less. This conclusion is in agreement with Moore (1958, p.333), who stated that many modern reefs maintain more or less a balance between growth and loss.

Chapter XV

CONCLUDING REMARKS AND SUMMARY

This chapter is to end the monograph on the Silurian of Gotland, but the author realizes that it is not a satisfactory end. Several subjects treated in the preceding pages still need a more thorough and detailed study. Many other subjects which have hardly been touched upon or not at all in the present work are equally intriguing. However, one has to make a break somewhere, just to create order in the collection of information which has been brought together and to define one's position. The end of this book is only such a break, certainly as far as the study, in a general way, of the Silurian deposits of Gotland is concerned. The author hopes that for him personally it is also only a break. There are very few places in the world which enchant him more than this Baltic island. He would be glad to be enabled to return there from time to time to learn more about its geological and historical past, and about its present development.

In the next sections, a short summary will be given of the main conclusions reached in the work that has been presented here.

STRATIGRAPHY OF THE MIDDLE PALAEOZOIC OF GOTLAND

The stratigraphical subdivision arrived at by the present author is shown in Table XXIV. Compared to the stratigraphy drawn up by Hede (1921, 1925a), there are four modifications: (1) The Lower and Upper Visby marl-stones are united in one main unit; the subdivisions proposed for other main units by the present author are often thicker than the Lower or Upper Visby

TABLE XXIV

Stratigraphical units in the Middle Palaeozoic of Gotland

Stratigraphical unit	Main lithology	Maximum thickness (m)
Hamra-Sundre Beds	limestones, reef limestones	50
Burgsvik Beds	sandstone, claystone, oolite	50
Eke Beds	marlstone, marly limestone, reef limestone	15
Hemse Beds	limestones, reef limestones, marlstone	100
Klinteberg Beds	limestones, reef limestones	65
Halla-Mulde Beds	limestones, reef limestones, marlstone	25
Slite Beds	limestones, reef limestones, marlstone	100
Högklint Beds	limestones, reef limestones	35
Visby Beds	alternating marlstone and marly limestone	20
Total maximum thickness, about		460

Beds, but they remain subunits. Moreover, as the names Lower and Upper Visby Beds already suggest subunit rank, and there is great similarity also in lithological composition, it is only logical to bring them together in one main unit. (2) The Tofta limestone is included as a facies in the (Upper) Högklint Beds. (3) The Halla limestone and Mulde marlstone are considered to be synchronous deposits, and are, therefore, united in the Halla-Mulde Beds. (4) The Hamra limestone and Sundre limestone are united in the Hamra-Sundre Beds, because the boundary between the two, as drawn by Hede, is distinctly a facies boundary, and no satisfactory time boundary could be fixed.

The present author has not done any work himself on the correlation of the succession of strata of Gotland with the English Silurian succession. A correlation based on literature data is given in Table VI (p.43).

OCCURRENCE OF TRUE FOSSIL REEFS

In each of these stratigraphical units reef limestones are found, sometimes as a conspicuous component and sometimes only very subordinate.

Although a universal definition of "organic reefs" is very difficult or even impossible to give, there can be little doubt that the word "reef" may be used for the structures constituting the reef limestones in the Silurian of Gotland. They have formed in a shallow epicontinental basin, through the joint activities of various kinds or organisms, among which there were colonial frame builders. The structures grew up above the surrounding sea floor and induced water turbulence. They also provided material swept off their surface which contributed to the formation of surrounding detrital deposits.

GENERAL DEVELOPMENT OF THE REEFS OF GOTLAND

The Visby Beds were laid down in a sea gradually decreasing in water depth. This was presumably caused by small epeirogenetic movements of the basin floor. The shallowing of the sea led to a decrease in marlstone deposition and an increase in the sedimentation of marly limestone towards the top of the Visby Beds. As a result, the situation on the sea floor was also modified, but not uniformly over the entire area. During middle Late Visby time, the alteration in environmental conditions ran ahead of the general trend in some restricted or very restricted parts. More organisms populated these areas. Their remains contributed to extra limestone deposition and provided suitable places for the settling of sedentary organisms. Among these were many potential reef builders, which, in several cases, started the formation of small reefs. These are the reefs of the Upper Visby type. Corals, in a rather great variety of species and forms, strongly dominate the organic element in these reefs. The reef matrix is voluminous and strongly marly. The deposition of this marl apparently made the conditions for the stromatoporoids rather unfavourable and this applied even more strongly to the calcareous Algae. Mantles of stratified limestone are present around many of the reefs, but there are no crinoid limestones.

With a continued decrease in water depth and consequent gradual alteration of other environmental conditions also towards the end of Visby time, reef development was increasingly favoured. The foundations were thus laid for much richer reef growth in the next, the Högklint Period. A number of Högklint reefs have their roots in the Upper Visby Beds. Moreover, several new reefs began to grow early in the Högklint Period. Together they formed a broad reef zone.

The reefs of the Högklint Beds are the first, in Gotland, of the Hoburgen reef type. These reefs are larger than those of the Upper Visby time. Stromatoporoids were the dominant reef builders, with corals taking second place and Algae also being common. The average size of the reef builders is larger. On the reef flanks and directly around the reefs crinoids usually grew abundantly. Their remains contributed significantly to the crinoid limestones with reef debris which were often formed around the reefs.

During Early Högklint time, the gradual decrease in water depth presumably continued. While the Upper Högklint Beds were being laid down the sea was probably very shallow. These Högklint Beds include the Tofta limestone as a facies. Small alterations in sea level occurred, and a few times the sea floor probably even temporarily fell dry. Almost no reefs began to develop, but in comparatively somewhat deeper parts of the Late Högklint sea some reefs already in existence continued growth.

The limestones of the Slite I Beds were deposited in very shallow water; these beds are missing in the north of the present Slite limestone area. During Slite II time, water depth increased, probably with fluctuations. Slite II Beds are found over the entire present Slite limestone area. A further increase in water depth took place during Slite III time, whereas Slite IV time was again characterized by a gradual shallowing of the water. During the entire Slite time marl was laid down in a zone at the seaward side of the marlstone zone.

During Slite III time, and, to a much stronger degree, during Slite IV time, there was a second main period of extensive reef formation. Reefs of the Hoburgen type developed.

Halla-Mulde time is characterized by a change in the direction of the hinge line of epeirogenetic movement of the basin floor, from about southwest – northeast to about west – east. This caused a decrease in water depth in the northeast and an increase in depth in the southwest. Towards the southwest, limestone was deposited in water of increasing depth, until limestone sedimentation was followed by marlstone deposition. In the area of limestone formation, reef growth continued, with only small reefs in the northeast, and reefs of larger size towards the southwest.

The fluctuations in water depth during Slite III, IV and Halla-Mulde time were most pronounced in the area of Karlsöarna. In Slite III and IV time large reefs grew in that area. They are called reefs of the Stäurnasar type. The lower parts of these reefs were presumably mainly built by huge coral colonies, while the upper parts contain stromatoporoids and corals together. In Lilla Karlsö one of these large reefs continued growth in Halla-Mulde time. On the flanks of the large reefs smaller ones developed. They have been named reefs of Fanterna type. These smaller reefs were mainly built by bryozoans and corals. Some of the younger reefs of Fanterna type seem to have been able to survive the main reefs on whose flanks they began to grow, and became strongly extended.

During Late Halla-Mulde time the water began to shallow again. The Klinteberg Beds were laid down in a sea of slight depth. Presumably water depth was less than that in which most of the other limestone complexes of Gotland were deposited.

Small reefs, roughly of Hoburgen type, occur throughout most of the Klinteberg Beds, and represent a third period of reef development in the area of Gotland.

During Late Klinteberg time and very Early Hemse time the water depth again increased. In the remaining part of Early Hemse time it may have remained more or less the same. In Upper Hemse time a renewed gradual decrease of the water depth occurred in the east. The epeirogenetic movements which caused the alterations in water depth during Klinteberg and Hemse times probably went together with alterations in the direction of the hinge line. This is reflected by the assumed directions of the depth contours. During Klinteberg time these had a tendency to return to a more northeast – southwest direction. In the course of Hemse time their direction again became more east – west.

Reefs are particularly common in the Upper Hemse limestones. These were formed during a fourth period of reef development. Probably as a result of an alteration of the direction of the depth contours, Hemse reefs only developed in the east, where the water became shallower. In the west the water remained deeper and mainly marlstone formation took place. Reefs of the Hoburgen type are the most common, but in the east reefs of another, the Holmhällar type, are also found, as well as transitional forms between both reef types. The typical reefs of Holmhällar type are rather large, often crescent-shaped in ground plan and consist very predominantly of very large stromatoporoids. There are many Algae and the matrix consists of rather pure limestone.

In the east, the Hemse and Eke Beds are separated by a stratigraphical hiatus. Such a hiatus may probably also have been present in the west, but this could not, so far, be proved.

During Eke time also, small epeirogenetic movements of the sea floor took place, resulting in variations in water depth, together with alterations in the direction of the depth contours. At the beginning of Eke time, the latter probably ran north-northeast – south-southwest and at the end of Eke time, more northeast – southwest. This resulted particularly in a decrease in the depth of the water in which the Eke Beds in the east were laid down. Some comparatively small reefs are present in the eastern Lower Eke Beds. In the area of Gotland, however, Eke time was not a period of extensive reef formation.

During deposition of the Burgsvik Beds a shallowing of the water took place, which reached its culmination in Late Burgsvik time. The Upper Burgsvik Beds in the west were laid down in extremely shallow water, very close to the shore line, in a littoral zone which gently sloped towards the basin centre.

An increase in water depth set in with the formation of the uppermost oolite horizon of the Burgsvik Beds and continued well into Hamra-Sundre time, whereafter there was presumably a more constant sea depth. Formation of Hoburgen-type reefs began in Late Burgsvik time in the east and extended from there westward in Hamra-Sundre time. Later in the Hamra-Sundre Period and further east than the Hoburgen-type reefs, characteristic reefs of Holmhällar type developed.

MAIN PERIODS OF REEF FORMATION

Summarizing, Gotland contains the remains of five main periods of extensive reef formation.

I. Upper Visby Beds – Lower Högklint Beds.
II. Slite III and IV Beds – Halla-Mulde Beds, including Karlsöarna.
III. Klinteberg Beds.
IV. Upper Hemse Beds.
V. Upper Burgsvik Beds–Hamra-Sundre Beds.

The reefs often occurred randomly distributed in a zone of varying width (Fig.74), rather than in a single row (Fig.75).

The alternation of periods with and without reef formation was mainly determined by generally small epeirogenetic movements of the basin floor, causing variations in water depth. Alterations in the direction of the hinge line of these movements led to alterations in the direction of the depth contours, and therewith to variations in orientation of the various reef zones, and to variations in the length of the reef zones.

REFERENCES

Aaloe, A.O., 1956. The reef facies in the Jaagarahu Horizon of the Silurian of Esthonia. Esti N.S.V. Teaduste Akad., Geol. Inst. Uurimused, 1: 89–94 (in Russian).
Aaloe, A.O., 1958a. Jaagarahu lademe pangad. Eesti Loodus, 1958: 14–18.
Aaloe, A.O., 1958b. Stratigraphy of the Juuru (GI) and Tamsalu (GII) Stages. Eesti N.S.V. Teaduste Akad., Geol. Inst. Uurimused, 3: 81–98 (in Russian; Esthonian and English summaries).
Aaloe, A.O., Mark, E.J., Männil, R.M., Müürisepp, K.K. and Orviku, K.K., 1958. Review of the stratigraphy of the Palaeozoic and Quaternary deposits of the Esthonian S.S.R. Izv. Akad. Nauk Est. S.S.R., Inst. Geol., 1958: 3–42 (in Russian).
Adamczak, F., 1966a. On kloedenellids and cytherellids (Ostracoda Platycopa) from the Silurian of Gotland. Stockholm Contrib. Geol., 15(2): 7–21.
Adamczak, F., 1966b. Morphology of two Silurian metacope ostracodes from Gotland. Geol. Fören. Stockholm Förh., 88: 462–475.
Ager, D.V., 1963. Principles of Paleoecology, McGraw-Hill, New York, N.Y., 371 pp.
Ager, D.V., 1965. The adaptation of Mesozoic brachiopods to different environments. Palaeogeography, Palaeoclimatol., Palaeoecol., 1(2): 143–172.
Agterberg, F.P., 1958. An undulation of the rate of sedimentation in southern Gotland. Geol. Mijnbouw, N. Ser., 20: 253–260, 449.
Andersson, G., 1895. Om några växtfossil från Gotland. Geol. Fören. Stockholm Förh., Sci., Holmiae, 64 pp.
Angelin, N.P. et Lindström, G., 1880. Fragmenta Silurica e dono Caroli Henrici (Opus postumum). Edited by G. Lindström and S. Lovén. Roy. Swedish Acad. Sci. Holmiae, 64 pp.
Angelin, N.P. et Lindström, G., 1880. Fragmenta Silurica e dono Caroli Henrici Wegelini. Stockholm, 60 pp.
Arkell, W.J., 1933. The Jurassic System in Great Britain. Clarendon Press, Oxford, 681 pp.
Arkell, W.J., 1935. On the nature, origin, and climatic significance of the coral reefs in the vicinity of Oxford, England. Quart. J. Geol. Soc. London, 91: 77–110.
Aurivillius, C.W.S., 1892. Über einige obersilurische Cirripeden aus Gotland. Bihang Kgl. Svenska Vetenskapsakad. Handl., 18 (IV, 3): 1–24.

Badgley, P.C., 1965. Structural and Tectonic Principles. Harper and Row, New York, N.Y., 521 pp.
Bassler, R.S., 1911. The Early Paleozoic Bryozoa of the Baltic Provinces. U.S. Natl. Museum Bull., 77.
Bather, F.A., 1893. The Crinoidea of Gotland, Part I. The Crinoidea Inadunata. Kgl. Svenska Vetenskapsakad. Handl., 25(2): 1–200.
Bather, F.A., 1914. Note on some reef-structures in Gotland. In: M.C. Crosfield and M.S. Johnston, 1914. A study of ballstone and the associated beds in the Wenlock limestone of Shropshire. Proc. Geologists Assoc. London, 25(3): 193–228.
Bekker, H., 1922. A new brachiopod (*Leptestia*) from the Kuckers Stage in Estonia. Geol. Mag., 59: 361–365.
Bekker, H., 1924. Stratigraphical and paleontological supplements on the Kukruse Stage of the Ordovician rocks of Eesti (Estonia). Publ. Geol. Inst. Tartu, 1: 3–19.

Bell, A., 1886. Reef structures on the Attawapiskat River. Geol. Surv. Can., Ann. Rept., 2: 27G–28G.
Bergenhayn, J.R.M., 1943. Preliminary notes on fossil Polyplacophoras from Sweden. Geol. Fören. Stockholm Förh., 65: 297–303.
Bergenhayn, J.R.M., 1955. Die fossilen schwedischen Loricaten nebst einer vorläufigen Revision des Systems der ganzen Klasse Loricata. Lunds Univ. Arsskr., N.F., Avd.2,51 (8): 3–47. Also: Skrifter Mineral. Paleontol.-Geol. Inst. Lund, 22: 3–47.
Bersier, A. et Vernet, J.P., 1956. Un cas d'oolithisation en faciès détritique. Bull. Lab. Géol. Minéral. Geophys. Musée Géol. Univ. Lausanne, 116.
Bissell, H.J. and Chilingar, G.V., 1967. Classification of sedimentary carbonate rocks. In: G.V. Chilingar, H.J. Bissell and R.W. Fairbridge (Editors), Carbonate Rocks, A. Origin, Occurrence and Classification. Elsevier, Amsterdam, pp.87–168.
Böger, H., 1968. Paläoökologie Silurischer Chonetoidea auf Gotland. Lethaia, 1: 122–136.
Bohlin, B., 1949. The *Asaphus* Limestone in northernmost Öland. Bull. Geol. Inst. Univ. Uppsala, 33: 529–670.
Bohlin, B., 1955. The Lower Ordovician limestone between the *Ceratopyge* shale and the Platyurus limestone of Böda hamn. With a description of the microlithology of the limestones by Valdar Jaanusson. Bull. Geol. Inst. Univ. Uppsala, 35: 111–173.
Boll, E., 1862. Die Beyrichien der norddeutschen silurischen Gerölle. Arch. Verein. Freunde Naturgesch. Mecklenburg, 16: 114–115.
Borg, F., 1964. A comparative and phyletic study on fossil and recent Bryozoa of the suborders Cyclostomata and Trepostomata. Arkiv Zool. Kgl. Svenska Vetenskapsakad., 17(1): 1–91 (on cover dated 1965).
Botke, J., 1916. Het geslacht *Aechmina* Jones et Holl. Verhandel. Geol. Mijnbouwk. Genoot. Ned. Koloniën, Geol. Ser., 3: 21–30.
Boucot, A.J., 1957. Revision of some Silurian and Early Devonian spiriferid genera and erection of *Kozlowskiellinae*, new subfamily. Senckenbergiana Lethaea, 38: 311–334.
Boucot, A.J., 1962. The Eospiriferidae. Palaeontology, 5(4): 682–711.

Carozzi, A.V. and Zadnik, V.E., 1959. Microfacies of Wabash reef, Wabash, Indiana. J. Sediment. Petrol., 29: 164–171.
Cayeux, L., 1935. Les Roches Sédimentaires de France. Roches Carbonatées (Calcaires et Dolomies). Masson, Paris, 463 pp.
Chamberlin, T.C., 1877. Geology of Wisconsin. Wisconsin Geol. Surv., 2: 360–371.
Chapman, T., 1901. On some fossils of Wenlock age from Mulde, near Klinteberg, Gotland. With notes by Prof. T. Rupert Jones and Dr. F.A. Bather. Ann. Mag. Nat. Hist., Ser. 7, 7: 141–160.
Chilingar, G.V., Bissell, H.J. and Fairbridge, R.W. (Editors), 1967a. Carbonate Rocks, A. Origin, Occurrence and Classification. Elsevier, Amsterdam, 471 pp.
Chilingar, G.V., Bissell, H.J. and Fairbridge, R.W. (Editors), 1967b. Carbonate Rocks, B. Physical and Chemical Aspects. Elsevier, Amsterdam, 413 pp.
Cloud, P.E., Jr., 1952a. Facies relationships of organic reefs. Bull. Am. Assoc. Petrol. Geologists, 36: 2125–2149.
Cloud, P.E., Jr., 1952b. Notes on stromatolites. Am. J. Sci., 240: 363–379.
Colter, V.S., 1957. The Palaeoecology of the Wenlock Limestone. Thesis, Univ. Cambridge, 315 pp. (unpubl.).
Crosfield, M.C. and Johnston, M.S., 1914. A study of ballstone and the associated beds in the Wenlock limestone of Shropshire. With note on some reef-structures in Gotland, by F.A. Bather. Proc. Geologists Assoc. London, 25(3): 193–228.
Cumings, E.R., 1930a. List of species from New Corydon, Kokomo, and Kenneth Formations of Indiana, and from reefs in the Mississinewa and Liston Creek Formations. Proc. Indiana Acad. Sci., 39 (1929): 204–211.
Cumings, E.R., 1930b. Silurian reefs near Tiffin, Carey and Marseilles, Ohio. Proc. Indiana Acad. Sci., 39 (1929):199–204.

Cumings, E.R., 1932. Reefs or bioherms? Bull. Geol. Soc. Am., 43: 331–352.
Cumings, E.R. and Shrock, R.R., 1927. The Silurian coral reefs of northern Indiana and their associated strata. Proc. Indiana Acad. Sci., 36: 71–85.
Cumings, E.R. and Shrock, R.R., 1928a. Niagaran coral reefs of Indiana and adjacent states and their stratigraphic relations. Bull. Geol. Soc. Am., 39: 579–620.
Cumings, E.R. and Shrock, R.R., 1928b. The geology of the Silurian rocks of nort' Indiana. Indiana Dept. Conservation, Div. Geol., Publ., 73: 226 pp.
Cuvillier, J. et Sacal, V., 1963. Présence de tintinnoïdiens dans le Dévonien supérieur du Sahara septentrional. Rev. Micropaléontol., 6(2): 73.

Dames, W., 1874. Über Spongien von Gotland. Z. Deut. Geol. Ges., 26: 613–614.
Dames, W., 1890. Über die Schichtenfolge der Silurbildungen Gotlands und ihre Beziehungen zu obersilurischen Geschieben Norddeutschlands. Sitz. Ber. Kgl. Preuss. Akad. Wiss., Phys. Math. Kl., 42: 1111–1129.
Davidson, T. and King, W., 1874. On the Trimerellidae, a Palaeozoic family of the Palliobranchs or Brachiopoda. Quart. J. Geol. Soc. London, 30: 124–173.
Davis, R.A., Jr., 1966. Quiet-water oolites from the Ordovician of Minnesota. J. Sediment. Petrol., 36(3): 813–818.
De Sitter, L.U., 1956. Structural Geology. McGraw-Hill, New York, N.Y., 551 pp.
De Verneuil, E., 1848. Note sur quelques brachiopodes de l'île de Gothland. Bull. Soc. Géol. France, 2 Sér., 5: 339–347.
Dybowski, W., 1873–1874. Monographie der Zoantharia sclerodermata rugosa aus der Silurformation Estlands, Nord-Livlands und der Insel Gotland. Arch. Naturk. Liv-, Esth- und Kurlands, Ser.1, 5(3): 257–414, 5(4): 415–531.

Eisenack, A., 1954a. Hystrichosphären aus dem baltischen Gotlandium. Senckenbergiana Lethaea, 34: 205–211.
Eisenack, A., 1954b. Foraminiferen aus dem baltischen Silur. Senckenbergiana Lethaea, 35: 51–72.
Eisenack, A., 1955. Neue Chitinozoen aus dem Silur des Baltikums und dem Devon der Eifel. Senckenbergiana Lethaea, 36: 311–319.
Eisenack, A., 1958. *Tasmanites* Newton 1875 und *Leiosphaeridia* n.g. als Gattungen der Hystrichosphaeridea. Palaeontographica, A, 110: 1–19.
Eisenack, A., 1962. Chitinozoen aus Sedimenten Gotlands (Vorläufige Mitteilung). Neues Jahrb. Geol. Paläontol., Monatsh., 1962(4): 218–219.
Eisenack, A., 1964a. Mikrofossilien aus dem Silur Gotlands. Chitinozoen. Neues Jahrb. Geol. Paläontol., Abhandl., 120(3): 308–342.
Eisenack, A., 1964b. Mikrofossilien aus dem Silur Gotlands. Phosphatische Resten. Paläontol. Z., 38: 170–179.
Elles, G.L. and Wood, E.M.R., 1914. A monograph of British graptolites, 10. Monographs Palaeontogr. Soc. (London), 1913: 487–526.

Fåhræus, L.E., 1967. Upper Ludlovian deposits of Gotland defined by means of conodonts. Geol. Fören. Stockholm Förh., 89: 218–220.
Fenton, C.L., 1931. Niagaran stromatoporoid reefs in the Chicago Region. Am. Midland Naturalist, 12: 203–212.
Fischer, A.G., 1963. Growth patterns of Silurian Tabulata as palaeoclimatologic and palaeogeographic tools. In: A.E.M. Nairn (Editor). Problems in Palaeoclimatology. Interscience (Wiley), London, pp.608–617, 641.
Flügel, E., 1959. Die Gattung *Actinostroma* Nicholson und ihre Arten (Stromatoporoidea). Ann. Naturhist. Museums Wien, 63: 90–273.
Frebold, H., 1926. Zur Gliederung des Obersilurs in Gotland und im Ostbaltikum nebst Beschreibung eines neuen ostpreussischen Obersilurgeschiebe. Centr. Mineral. Geol., B, 9: 297–313.
Freeman, T., 1962. Quiet-water oolites from Laguna Madra, Texas. J. Sediment. Petrol., 32: 475–483.
Fromm, E., 1943. Havsbottnens morfologi utanför Stockholms södra skärgård. Geograf. Ann., 1943 (3/4): 137–169.

Goldberg, E.D., 1957. Biogeochemistry of trace metals. In: J.W. Hedgpeth (Editor), Treatise on Marine Ecology and Palaeoecology, 1. Ecology - Geol. Soc. Am., Mem., 67: 345–358.

Gorbatschev, R., 1962. Deep boring on Gotska Sandön, 2. The pre-Cambrian sandstone of the Gotska Sandön boring core. Bull. Geol. Inst. Univ. Uppsala, 39(11): 1–30.

Goreau, T., 1961. Problems of growth and calcium deposition in reef corals. Endeavour, 20: 32–39.

Gross, W., 1968. Fragliche Actinopterygier-Schuppen aus dem Silur Gotlands. Lethaia, 1: 184–218.

Hadding, A., 1927a. The pre-Quaternary sedimentary rocks of Sweden, 1. A survey of the pre-Quaternary sedimentary rocks of Sweden. Lund Univ. Årsskr., N.F., Avd.2, 23(5): 3–39. Also: Kgl. Fysiograf. Sällskap. Lund Handl., 38(5): 3–39. Also: Medd. Lunds Geol. Mineral. Inst., 32: 3–39.

Hadding, A., 1927b. The pre-Quaternary sedimentary rocks of Sweden, 2. The Paleozoic and Mesozoic conglomerates of Sweden. Lund Univ. Årsskr., N.F., Avd.2, 23(5): 42–171. Also: Kgl. Fysiograf. Sällskap. Lund Handl., N.F., 38(5): 42–171. Also: Medd. Lunds Geol. Mineral. Inst., 32: 42–171.

Hadding, A., 1929. The pre-Quaternary sedimentary rocks of Sweden. 3. The Paleozoic and Mesozoic sandstones of Sweden. Lunds Univ. Årsskr., N.F., Avd.2, 25(3): 3–287. Also: Kgl. Fysiograf. Sällskap. Lund Handl., N.F., 40(3): 3–287. Also: Medd. Lunds Geol. Mineral. Inst., 41: 3–287.

Hadding, A., 1932. The pre-Quaternary sedimentary rocks of Sweden. 4. Glauconite and glauconitic rocks. Lunds Univ. Årsskr., N.F., Avd.2, 28(2): 3–175. Also: Kgl. Fysiograf. Sällskap. Lund Handl., N.F., 43(2): 3–175. Also: Medd. Lunds Geol. Mineral. Inst., 51: 3–175.

Hadding, A., 1933. The pre-Quaternary sedimentary rocks of Sweden, 5. On the organic remains of the limestones. Lunds Univ. Årsskr., N.F., Avd.2, 29(4): 3–93. Also: Kgl. Fysiograf. Sällskap. Lund Handl., N.F., 44(4): 3–93. Also: Medd. Lunds Geol. Mineral. Inst., 55: 3–93.

Hadding, A., 1939. Observations on Teicholite, an algal limestone with stromatolitic structure. Kgl. Fysiograf. Sällskap. Lund Förh., 9(4): 1–9. Also: Medd. Lunds Geol. Mineral. Inst., 75: 1–9.

Hadding, A., 1941. The pre-Quaternary sedimentary rocks of Sweden, 6. Reef limestones. Lunds Univ. Årsskr., N.F., Avd.2, 37(10): 1–137. Also: Kgl. Fysiograf. Sällskap. Lund Handl., N.F., 52(10): 1–137. Also: Medd. Lunds Geol. Mineral. Inst., 90: 1–137.

Hadding, A., 1950. Silurian reefs of Gotland. J. Geol., 58: 402–409.

Hadding, A., 1956. The lithological character of marine shallow water limestones. Kgl. Fysiograf. Sällskap. Lund Förh., 26(10): 1–18. Also: Skrifter Mineral. Paleontol. Geol. Inst. Lund, 33: 1–18.

Hadding, A., 1958a. Origin of the lithographic limestones. Kgl. Fysiograf. Sällskap. Lund Förh., 28(4): 21–32. Also: Publ. Inst. Mineral. Paleontol. Quatern. Geol. Univ. Lund, 41: 21–32.

Hadding, A., 1958b. Hidden hiatuses and related phenomena. Some lithological problems. Kgl. Fysiograf. Sällskap. Lund Förh., 28(17): 159–171. Also: Publ. Inst. Mineral. Paleontol. Quatern. Geol. Univ. Lund, 60: 159–171.

Hadding, A., 1958c. The pre-Quaternary sedimentary rocks of Sweden, 7. Cambrian and Ordovician limestones. Lunds Univ. Årsskr., N.F., Avd.2, 54(5): 1–262. Also: Kgl. Fysiograf. Sällskap. Lund Handl., N F., 69(5): 1–262. Also: Publ. Inst. Mineral. Paleontol. Quatern. Geol. Univ. Lund, 50: 1–262.

Hadding, A., 1959. Silurian algal limestones of Gotland. Lunds Univ. Årsskr., N.F., Avd.2, 56(7): 1–26. Also: Publ. Inst. Mineral. Paleontol. Quatern. Geol. Univ. Lund, 70: 1–26.

Halle, T.G., 1920. *Psilophyton ? hedei* n. sp., probably a land-plant from the Silurian of Gothland. Svensk Botan. Tidskr., 14: 258–260.

Hede, J.E., 1917. Faunan i kalksandstenens märgliga bottenlager söder om Klintehamn på Gotland. Sveriges Geol. Undersökn., Ser.C, 281: 1–32. Also: Sveriges Geol. Undersökn., Årsbok, 11(2): 1–32.
Hede, J.E., 1919a. Om några nya fynd av graptoliter inom Gottlands Silur och deras betydelse för stratigrafien. Sveriges Geol. Undersökn., Ser.C, 291: 1–31. Also: Sveriger Geol. Undersökn., Årsbok, 12 (1918) (8): 1–31.
Hede, J.E., 1919b. Djupborrningen vid Burgsvik på Gottland, 1915. Paleontologisk-stratigrafiska resultat. Sveriges Geol. Undersökn., Ser.C. 298: 1–59. Also: Sveriger Geol. Undersökn., Årsbok, 13(7): 1–59.
Hede, J.E., 1921. Gottlands Silurstratigrafi. Sveriges Geol. Undersökn., Ser.C, 305: 1–100. Also: Sveriger Geol. Undersökn., Årsbok, 14 (1920) (7): 1–100.
Hede, J.E., 1923a. Svar på H. Hedströms inlägg i frågan om Gottlands Silurstratigrafi. Geol. Fören, Stockholm Förh., 45: 198–208.
Hede, J.E., 1923b. Genmäle till H. Hedström med anledning av hans inlägg "Ytterligare om Gottlands Silurstratigrafi". Geol. Fören. Stockholm Förh., 45: 349–353.
Hede, J.E., 1925a. Beskrivning av Gotlands Silurlager. In: H. Munthe, J.E. Hede och L. von Post, Gotlands Geologi, en Översikt. Sveriges Geol. Undersökn., Ser.C, 331: 13–30. Also: Sveriges Geol. Undersökn. Arsbok, 18 (1924) (3): 13–30.
Hede, J.E., 1925b. Berggrunden (Silursystemet). In: H. Munthe, J.E. Hede och L. von Post, Beskrivning till kartbladet Ronehamn. Sveriges Geol. Undersökn., Ser.Aa, 156: 14–51.
Hede, J.E., 1927a. Berggrunden (Silursystemet). In: H. Munthe, J.E. Hede och G. Lundqvist, Beskrivning till kartbladet Klintehamn. Sveriges Geol. Undersökn., Ser.Aa, 160: 12–54.
Hede, J.E., 1927b. Berggrunden (Silursystemet). In: H. Munthe, J.E. Hede och L. von Post, Beskrivning till kartbladet Hemse. Sveriges Geol. Undersökn., Ser.Aa, 164: 15–56.
Hede, J.E., 1928. Berggrunden (Silursystemet). In: H. Munthe, J.E. Hede och G. Lundqvist, Beskrivning till kartbladet Slite. Sveriges Geol. Undersökn., Ser.Aa, 169: 13–65.
Hede, J.E., 1929. Berggrunden (Silursystemet). In: H. Munthe, J.E. Hede och G. Lundqvist, Beskrivning till kartbladet Katthammarsvik. Sveriges Geol. Undersökn., Ser.Aa, 170: 14–57.
Hede, J.E., 1933. Berggrunden (Silursystemet). In: H. Munthe, J.E. Hede och G. Lundqvist. Beskrivning till kartbladet Kappelshamn. Sveriges Geol. Undersökn., Ser.Aa, 171: 10–59.
Hede, J.E., 1936. Berggrunden. In: H. Munthe, J.E. Hede och G. Lundqvist, Beskrivning till kartbladet Farö. Sveriges Geol. Undersökn., Ser.Aa, 180: 11–42.
Hede, J.E., 1940. Berggrunden. In: G. Lundqvist, J.E. Hede och N. Sundius, Beskrivning till kartbladen Visby och Lummelunda. Sveriges Geol. Undersökn., Ser.Aa, 183: 9–68.
Hede, J.E., 1942. On the correlation of the Silurian of Gotland. In: G. Regnéll et 10 al., Lunds Geologiska Fältklubb 1892–1942. Carl Bloms Boktryckeri, Lund, pp.205–229. Also: Medd. Lunds Geol. Mineral. Inst., 101: 205–209.
Hede, J.E., 1958. [The Silurian articles]. In: Lexique stratigraphique international, 1(2c) Suède-Sweden-Sverige. Publ. sous l'égide du Congr. Géol. Internat. par les soins du Comm. Français de Stratigraphie. Centre Natl. Rech. Sci., Paris, 498 pp.
Hede, J.E., 1960. The Silurian of Gotland. In: G. Regnéll and J.E. Hede, The Lower Palaeozoic of Scania; the Silurian of Gotland. Excursion Guide 21st Intern. Geol. Congr., Copenhagen, 1960, pp.44–89.
Hedström, H., 1904. Detaljprofil fran skorpionfyndorten - *Pterygotus* lagret - i Siluren strax söder om Visby. Geol. Fören. Stockholm Förh., 26: 93–96.
Hedström, H., 1910. The stratigraphy of the Silurian strata of the Visby district. Geol. Fören. Stockholm Förh., 32 (Congr. guide): 1455–1484.
Hedström, H., 1912. Om grundvattens förhållandena i trakten af Visby. Sveriges Geol. Undersökn., Ser.C, 239: 1–70.

Hedström, H., 1917a. Über die Gattung *Phragmoceras* in der Obersilurformation Gotlands. Sveriges Geol. Undersökn., Ser.Ca, 15: 1–35.
Hedström, H., 1917b. Über einige mit der Schale befestigte Strophomenidae aus dem Obersilur Gotlands. Sveriges Geol. Undersökn., Ser.C, 276: 1–14. Also: Sveriges Geol. Undersökn., Årsbok, 10 (1916) (2): 1–14.
Hedström, H., 1923a. Om *Leperditia phaseolus* (His.) från Visbytrakten. Geol. Fören. Stockholm Förh., 45: 335–336.
Hedström, H., 1923b. Till frågan om Gottlands Silurstratigrafi. Geol. Fören. Stockholm Förh., 45: 167–168.
Hedström, H., 1923c. Ytterligare om Gottlands Silurstratigrafi. Geol. Fören. Stockholm Förh., 45: 341–348.
Hedström, H., 1923d. Remarks on some fossils from the diamond boring at the Visby Cement Factory. Preliminary report. Sveriges Geol. Undersökn., Ser.C., 314: 1–26. Also: Sveriges Geol. Undersökn., Årsbok, 16 (1922) (4): 1–26.
Hedström, H., 1923e. Contributions to the fossil faunas of Gotland, 1. Sveriges Geol. Undersökn., Ser.C, 316: 1–24. Also: Sveriges Geol. Undersökn., Årsbok, 16 (1922) (6): 1–24.
Hedström, H., 1928a. Om asfaltit och några andra mineral från Gottlands Silurlager. Geol. Fören. Stockholm Förh., 50: 443–448.
Hedström, H., 1928b. Om ett fosforitkomglomerat på Gotland. Geol. Föran. Stockholm Förh., 50: 429–435.
Hennig, A., 1905–1908. Gotlands Silur-Bryozoer, 1–3. Ark. Zool., 1905, 2(10): 37 pp.; 1906, 3(10): 62 pp.; 1908, 4(21): 64 pp.
Henson, F.R.S., 1950. Cretaceous and Tertiary reef formation and associated sediments in Middle East. Bull. Am. Assoc. Petrol. Geologists, 34(2): 215–238.
Hermes, J.J., 1966. Tintinnids from the Silurian of the Betic Cordilleras, Spain. Rev. Micropaléontol., 8(4): 221–214.
Hessland, I., 1953. Studies in the lithogenesis of the Cambrian and basal Ordovician of the Böda Hamn sequence of strata. Bull. Geol. Inst. Univ. Uppsala, 35: 35–109.
Hinde, G.J., 1882. On the annelid remains from the Silurian strata of the isle of Gotland. Bihang Kgl. Svenska Vetenskapsakad. Handl., 7(5): 28 pp.
Hisinger, W., 1798. Minerographiske anmärkningar öfver Gotland. Kgl. Svenska Vetenskapsakad. Handl., 1798: 276–287.
Hisinger, W., 1826. Gotland geognostik beskrifvit. Kgl. Svenska Vetenskapsakad. Handl. Reprinted in a hardly altered form in: Hisinger, W., 1828. Anteckningar i Physik och Geognosie under resor uti Sverige och Norrige, Vol.4, Stockholm.
Hodgson, R.A., 1961. Regional study of jointing in Comb Ridge-Navajo Mountain area, Arizona and Utah. Bull. Am. Assoc. Petrol. Geologists, 45: 1–38.
Holland, C.H., Lawson, J.D. and Walmsley, V.G., 1959. A revised classification of the Ludlovian succession at Ludlow. Nature, 184: 1037–1039.
Holland, H.D., Bodine, M.W., Borcsik, M., Collins, P., Kirsipu, T.V., Rosenberg, P.E., Sawkins, F.J. and Tsusue, A., 1963. The coprecipitation of metallic ions with calcium carbonate. Ann. Rept. Princeton Univ., Princeton, N.J., Contract AT (30)-1-2266, 57 pp.
Holm, G., 1890. Gotlands graptoliter, Bihang Kgl. Svenska Vetenskapsakad. Handl., 16 (IV,7): 32 pp.
Holm, G., 1892. (Förevisar i Geol. För. i Stockholm en i *Palaeophonus*-lagret vid Visby af v. Schmalensee funnen insekt.) Geol. Fören. Stockholm Förh., 147(14): 559.
Holm, G., 1893. Sveriges Kambrisk-Siluriska Hyolithidae och Conulariidae. Sveriges Geol. Undersökn., Ser.C, 112: III–IX, 1–173.
Holm, G., 1901. Uttalande om Gotlands lagerföljd. In: Upplysningar till geologiska öfversiktskarta öfver Sveriges berggrund, upprättad och utgifven af Sveriges Geologiska Undersökning. Sveriges Geol. Undersökn., Stockholm.
Holmsen, P., 1958. Alkalisyenitter av Oslo-feltets eruptivstamme i Danmarks dypere undergrunn. Medd. Dansk Geol. Foren., 14: 61–62.

Hoppe, K.H., 1932. Die Coelolepiden und Acanthodier des Obersilurs der Insel Oesel. Palaeontographica, 76: 35–94.

Ingels, J.J.C., 1963. Geometry, paleontology, and petrography of Thornton Reef complex, Silurian of northeastern Illinois. Bull. Am. Assoc. Petrol. Geologists, 47: 405–440.

Jaanusson, V., 1944. Übersicht der Stratigraphie der Lyckholm-Komplexstufe. Bull. Comm. Géol. Finlande, 132: 92–100.

Jaanusson, V., 1945. Über die Stratigraphie der Viru-Serie in Estland. Geol. Fören. Stockholm Förh., 67: 212–224.

Jaanusson, V., 1955. Description of the microlithology of the Lower Ordovician limestones between the *Ceratopyge* shale and the *Platyurus* limestone (Böda hamn). Bull. Geol. Inst. Univ. Uppsala, 35: 153–168.

Jaanusson, V., 1957. Unterordovizische Illaeniden aus Skandinavien. Mit Bemerkungen über die Korrelation des Unterordoviziums. Bull. Geol. Inst. Univ. Uppsala, 37: 79–165.

Jaanusson, V., 1960. On the series of the Ordovician System. Intern. Geol. Congr., 21st., Copenhagen, 1960, Rept. Session, Norden, 7: 70–81.

Jaanusson, V. and Martinsson, A., 1956. Two hollinid ostracodes from the Silurian Mulde marl of Gotland. Bull. Geol. Inst. Univ. Uppsala, 36: 401–410. Also: Publ. Palaeontol. Inst. Univ. Uppsala, 13: 401–410.

Jaanusson, V. and Strachan, I., 1954. Correlation of the Scandinavian Middle Ordovician with the graptolite succession. Geol. Fören. Stockholm Förh., 76: 684–696.

Jones, F.W., 1907. On the growth forms and supposed species in corals. Proc. Zool. Soc. London, 1907: 518–566.

Jones, T.R., 1887. Notes on some Silurian Ostracoda from Gothland (Including a table of stratigraphical distribution by G. Lindström). Stockholm, 8 pp.

Jones, T.R., 1888. Note on the Palaeozoic bivalved Entomostraca. On some Silurian Ostracoda from Gothland. Ann. Mag. Nat. Hist., Ser.6, 1: 395–411.

Jones, T.R. and Woodward, H., 1888. On some Scandinavian Phyllocarida. Geol. Mag., 5: 96–100, 146–147.

Jonker, H.G., 1905–1906a. Bijdragen tot de kennis der sedimentaire zwerfsteenen in Nederland, 1. De Hondsrug in de provincie Groningen. 2. Bovensilurische zwerfsteenen - Eerste mededeeling. Zwerfsteenen van den ouderdom der oostbaltische zone G. Kon. Akad. Wetensch. Amsterdam, Verslag Gewone Vergaderingen Wis- en Natuurk. Afdeel., 13: 548–565.

Jonker, H.G., 1905–1906b. Bijdragen tot de kennis der sedimentaire zwerfsteenen in Nederland. 1. De Hondsrug in de provincie Groningen. 2. Bovensilurische zwerfsteenen - Tweede mededeeling: Zwerfsteenen van den ouderdom der oostbaltische zonen H en I. Kon. Akad. Wetensch. Amsterdam, Verslag Gewone Vergaderingen Wis- en Natuurk. Afdeel., 13: 758–770.

Jux, U., 1957. Die Riffe Gotlands und ihre angrenzenden Sedimentationsräume. Stockholm Contrib. Geol., 1: 41–90.

Kaljo, D. and Sarv, L., 1966. On the correlation of the Baltic Upper Silurian. Eesti N.S.V. Teaduste Akad. Toimetised, 15 (Füusika- Matem. Teknik. Ser.2). (in Russian).

Kaufmann, R., 1931. Die Klufttektonik des Kambrosilurs von Gotland, Öland und dem Kalmargebiet. Geol. Rundschau, 22: 292–306.

Kiaer, J., 1908. Das Obersilur im Kristiania-gebiete. Videnskapsselskapets Skrifter, I. Mat. Naturv. Kl., Kristiania, 1908.

Kiesow, J., 1888. Uber Gotländische Beyrichiën. Z. Deut. Geol. Ges., 40(1): 1–16.

Kolmodin, L., 1869. Bidrag till kännedomen om Sveriges Siluriska ostracoder. Inaug. Dissertation, Univ. of Upsala, Upsala, 22 pp.

Kolmodin, L., 1879. Ostracoda Silurica Gotlandiae. Öfvers. Kgl. Svenska Vetenskapsakad. Förh., 1879(9): 133–139.

Krause, A., 1877. Die Fauna der sogen. Beyrichien- oder Choneten-Kalke des norddeutschen Diluviums. Z. Deut. Geol. Ges., 29: 1–49.

Krause, A., 1889. Beyrichien und verwandte Schalenkrebse in märkischen Silurgeschieben. Sitz. Ber. Ges. Naturf. Freunde Berlin, 1889: 12–14.

Krause, A., 1891. Beitrag zur Kenntniss der Ostrakoden-Fauna in silurischen Diluvialgeschieben. Z. Deut. Geol. Ges., 43: 488–521.

Krause, A., 1892. Neue Ostrakoden aus märkischen Silurgeschieben. Z. Deut. Geol. Ges., 44: 383–400.

Kuenen, Ph.H., 1950. Marine Geology. Wiley, New York, N.Y., 568 pp.

Kühn, O., 1925. Variationsuntersuchungen an rezenten Korallen und ihre Anwendbarkeit auf die Fossilen. Verh. Zool. Bot. Ges. Wien, 1925: 129–134.

Kuiper, W.N., 1916. Eene nieuwe ostracode uit de bovensilurische mergel van Mulde op Gotland. Verhandel. Geol. Mijnbouwk. Genoot. Ned. Koloniën, Geol. Ser., 3: 119–121.

Kummerow, E., 1924. Beiträge zur Kenntnis der Ostracoden und Phyllocariden aus nordischen Diluvialgeschieben. Jahrb. Preuss. Geol. Landesanstalt (Berlin), 44 (1923): 405–448.

Ladd, H.S., 1944. Reefs and other bioherms. Rept. Comm. Marine Ecology as related to Paleontology, 1943–1944, Natl. Res. Council, Washington, D.C., 4: 26–29.

Laudon, L.R., 1957. Crinoids. In: H.S. Ladd (Editor), Treatise on Marine Ecology and Paleoecology, 2. Paleoecology - Geol. Soc. Am. Mem., 67(2): 961–972.

Lawson, J.D., 1960. The succession of shelly faunas in the British Ludlovian. Intern. Geol. Congr., 21st., Copenhagen, 1960, Rept. Session, Norden, 7: 114–125.

Lecompte, M., 1938. Quelques types de "récifs" siluriens et dévoniens de l'Amerique du Nord. Essai de comparaison avec les récifs coralliens actuels. Bull. Museum Hist. Nat. Belgique, 14(39): 1–51.

Lecompte, M., 1951. Les stromatoporoides du Dévonien moyen et supérieur du Bassin de Dinant. Première partie. Mém. Inst. Sci. Nat. Belgique. 116/117: 1–215.

Lecompte. M., 1952. Les stromatoporoides du Dévonien moyen et supérieur du Bassin de Dinant. Deuxième partie. Mém. Inst. Sci. Nat. Belgique, 116/117: 216–369.

Lecompte, M., 1954a. Quelques données relatives à la genèse et aux caractères écologiques des "récifs" du Frasnien de l'Ardenne. Volume jubilaire Victor Van Straelen, Inst. Roy. Sci. Nat. Belg., Bruxelles, 1: 153–181.

Lecompte, M., 1954b. Données relatives à la genèse et aux caractères écologiques des récifs du Frasnien de l'Ardenne. Compt. Rend. Congr. Géol. Intern., Alger, 13: 315.

Lecompte, M., 1956a. Quelques précisions sur le phénomène récifal dans le Dévonien de l'Ardenne et sur le rythme sédimentaire dans lequel il s'intègre. Bull. Inst. Sci. Nat. Belgique, 32(21): 1–39.

Lecompte, M., 1956b. Nouvelles données sur la genèse des récifs du Dévonien de l'Ardenne belge. Resumenes de los Trabajos presentados, 20 Congr. Geol. Intern., Mexico, p.234.

Lecompte, M., 1956c. Le rythme sédimentaire du Dévonien moyen et supérieur de l'Ardenne belge. Resumenes de los Trabajos presentados, 20 Congr. Geol. Intern., Mexico, p.265.

Lecompte, M., 1956d. Stromatoporoidea. In: R.C. Moore (Editor), Treatise on Invertebrate Paleontology, part F, Coelenterata. Kansas Univ. Press, Lawrence, Kans., pp.107–144.

Lecompte, M., 1957. Les récifs dévoniens de la Belgique. Bull. Soc. Géol. France, 70: 1045–1068.

Lecompte, M., 1958. Les récifs paléozoiques en Belgique. Geol. Rundschau, 47: 384–401.

Lecompte, M., 1959. Certain data on the genesis and ecologic character of Frasnian reefs of the Ardennes. Intern. Geology Rev., 1(7): 1–23.

Lindblom, A. och Svahnström, G., 1959. Gotländska Stenmästare. Allhem, Mälmo, 167 pp.

Lindström, G., 1857. Till Gotlands geologi. Öfvers. Kgl. Svenska Vetenskapsakad. Förh., 1857: 33–34. Also: Gotl. Läns. Tidning, 1851 (33, 34, 40).
Lindström, G., 1860. Bidrag till kännedomen om Gotlands brachiopoder. Öfvers. Kgl. Svenska Vetenskapsakad. Förh., 1860: 337–382.
Lindström, G., 1865. Nagra iakttagelser öfver Zoantharia rugosa. Kgl. Svenska Vetenskapsakad. Förh., 22: 271–294.
Lindström, G., 1866. Some observations on the Zoantharia rugosa. Geol. Mag., 3: 356–362, 406–414.
Lindström, G., 1867. Nomina Fossilium siluriensium Gotlandiae. Stockholm.
Lindström, G., 1868. Om tvenna nya öfversiluriska koraller från Gotland. Öfvers. Kgl. Svenska Vetenskapsakad. Förh., 1868: 419–428.
Lindström, G., 1870a. A description of the Anthozoa perforata of Gotland. Kgl. Svenska Vetenskapsakad. Handl., 9(6): 1–12.
Lindström, G., 1870b. Om operkularbildningen hos några nytidiga och Siluriska koraller. Kgl. Svenska Vetenskapsakad. Förh., 1870(9): 921–926.
Lindström, G., 1870c. On some operculated corals, Silurian and Recent. Geol. Mag., 8(1871): 122–126.
Lindström, G., 1873. Några anteckningar om Anthozoa tabulata. Öfvers. Kgl. Svenska Vetenskapsakad. Förh., 1873(4): 3–20.
Lindström, G., 1882a. Anteckningar om Silurlagren pa Carlsöarnw. Ofvers. Kgl. Vetenskapsakad. Förh., 30: 5–30.
Lindström, G., 1882b. Om de Palaeozoiska formationernas operkelbärande koraller. Bihang Kgl. Vetenskapsakad. Handl., 7(4): 1–92.
Lindström, G., 1884. On the Silurian Gastropoda and Pteropoda of Gotland. Kgl. Svenska Vetenskapsakad. Handl., 19(6): 1–250.
Lindström, G., 1885a. Förteckning på Gotlands Siluriska crustacéer. Öfvers. Kgl. Svenska Vetenskapsakad. Förh., 42(6): 37–99.
Lindström, G., 1885b. List of the fossils of the Upper Silurian Formation of Gotland. Palaeontol. Dept. Swedish State Museum (Nat. Hist.), Stockholm, 20 pp.
Lindström, G., 1888a. Uber die Schichtenfolge des Silur auf der Insel Gotland. Neues Jahrb. Mineral. Geol. Petr.-Kunde, 1: 147–164.
Lindström, G., 1888b. List of the fossil faunas of Sweden, 2. Upper Silurian. Palaeontol. Dept. Swedish State Museum (Nat. Hist.), Stockholm, 29 pp.
Lindström, G., 1890. The Ascoceratidae and the Lituitidae of the Upper Silurian Formation of Gotland. Kgl. Svenska Vetenskapsakad. Handl., 23(12): 1888–1889.
Lindström, G., 1895. On remains of a *Cyathaspis* from the Silurian strata of Gotland. Kgl. Svenska Vetenskapsakad. Handl., 21(3): 1–15.
Lindström, G., 1896. Beschreibung einiger obersilurischen Korallen aus der Insel Gotland. Bihang Kgl. Svenska Vetenskapsakad. Handl., 21 (Afd.4, no.7): 1–50.
Lindström, G., 1899. Remarks on the Heliolithidae. Kgl. Svenska Vetenskapsakad. Handl., 32(1): 1–140.
Lindström, M., 1963. Sedimentary folds and the development of limestone in an Early Ordovician Sea. Sedimentology, 2: 243–292.
Linnaeus, C. (Carl Linnaei), 1745b. Ölandska och Gotländska Resa på Riksens Högloflige Ständers befallning förrättad ahr 1741. Partly reprinted as: Gotländska Resa förrättad 1741. Natur och Kultur, Stockholm, 1957, 187 pp.
Linnaeus, C. (C. von Linné), 1745a. Corallia baltica. Dissertatio, Univ. Upsala, 40 pp.
Linnarsson, G., 1879. Om Gotlands graptoliter. Öfvers. Kgl. Svenska Vetenskapsakad. Förh., 1879(5): 3–11.
Lowenstam, H.A., 1948. Biostratigraphic studies of the Niagaran interreef formations of north-eastern Illinois. Illinois State Museum Sci. Papers, 4: 1–146.
Lowenstam, H.A., 1949. Niagaran reefs in Illinois and their relation to oil accumulation. Rept. Illinois Geol. Surv., Investigation 145, 36 pp.
Lowenstam, H.A., 1950. Niagaran reefs of the Great Lakes area. J. Geol., 58: 430–487.
Lowenstam, H.A., 1952. Some new observations on Niagaran reefs in Illinois. Illinois Acad. Sci. Tr., 45: 100–107.

Lowenstam, H.A., 1957. Niagaran reefs in the Great Lakes area. In: H.S. Ladd (Editor), Treatise on Marine Ecology and Paleoecology, 2. Paleoecology - Mem. Geol. Soc. Am., 67(2): 215–248.

Luha, A., 1930. Über Ergebnisse stratigraphischer Untersuchungen in Gebiete der Saaremaa-(Osel-)Schichten in Eesti. Acta Commentationes Univ. Tartuensis (Dorpatensis), A, 18(6): 1–18.

Luha, A., 1938. Muhu pangad. The cliffs of Muhu. Looduskaitse, 1: 1–8.

Luha, A., 1940. Kesselaid. Looduskaitse, 2: 1–8.

Lundquist, G. (Redaktör), 1940. Gotland. Svenska Turistfören. Årsskrift 1940, 394 pp.

Ma, T.Y.H., 1933. On the seasonal change of growth in some Palaeozoic corals. Proc. Imp. Acad. Japan, 9(8): 407–409.

Ma, T.Y.H., 1934. On the seasonal change of growth in a reef coral, *Favia speciosa* (Dana), and the water temperatures of the Japanese seas during the latest geological times. Proc. Imp. Acad. Japan, 10: 353–356.

Ma, T.Y.H., 1937. On the seasonal growth in Palaeozoic Tetracoralla and the climate during the Devonian Period. Palaeontol. Sinica, Ser.B, 2: 1–106.

Magnusson, N.H., Lundqvist, G. och Granlund, E., 1957. Sveriges Geologi. Scandinavian University Books, Svenska Bokförlaget/Norstedts, Stockholm, 557 pp.

Maillieux, E., 1933. Les récifs siluriens de l'île de Gotland. Bull. Soc. Belge Géol., 43: 206–208.

Manten, A.A., 1958. Een Gotlandse zwerfsteen. Grondboor Hamer, 1958(6/7): 143–147.

Manten, A.A., 1959. Heiligen en Duivels. Op den Uitkijk, 1959–60(3): 118–121.

Manten, A.A., 1960a. Het Cambrium van het Zweedse Oostzeegebied. Grondboor Hamer, 1960(2): 29–37.

Manten, A.A., 1960b. De Ordovicische afzettingen in het Zweedse Oostzeegebied. Grondboor Hamer, 1960(2): 41–56.

Manten, A.A., 1960c. De groeivormen van koralen als milieu-indicatoren. Grondboor Hamer, 1960(5): 145–165, 203.

Manten, A.A., 1960d. Een getuigenis van vroomheid uit het verleden. De Spiegel, 1960(29): 34–37.

Manten, A.A., 1961a. Korallengestalten als Kennzeichen des Milieus. Geol. Rundschau, 51: 663–671.

Manten, A.A., 1961b. Gotland, de schone slaapster. De Kern, 31(2): 64–71.

Manten, A.A., 1962. Some Middle Silurian reefs of Gotland. Sedimentology, 1: 211–234.

Manten, A.A., 1965. Book review of *Principles of Paleoecology* by D.V. Ager. Palaeogeography, Palaeoclimatol., Palaeoecol., 1(2): 173–176.

Manten, A.A., 1966a. Some problematic shallow-marine structures. Marine Geol., 4(3): 227–232.

Manten, A.A., 1966b. Note on the formation of stylolites. Geol. Mijnbouw, 45: 269–274.

Manten, A.A., 1968. Pre- or post-induration formation of stylolite seams: a reply. Geol. Mijnbouw, 47(2): 114–115.

Manten, A.A., 1970. Palaeoecology of Silurian crinoids of Gotland (Sweden). Palaeogeography, Palaeoclimatol., Palaeoecol., 7: 171–184.

Marr, J.E., 1882. On the Cambrian (Sedgw.) and Silurian rocks of Scandinavia. Quart. J. Geol. Soc. London, 38: 313–327.

Marshall, S.M. and Orr, A.P., 1931. Sedimentation on Low Isles and its relation to coral growth. Great Barrier Reef Exp., Sci. Repts., Brit. Museum Nat. Hist., 1(5): 93–133.

Martinsson, A., 1955. Studies on the ostracode family Primitiopsidae. Bull. Geol. Inst. Univ. Uppsala, 36: 1–33. Also: Publ. Palaeontol. Inst. Univ. Uppsala, 4: 1–33.

Martinsson, A., 1956. Ontogeny and development of dimorphism in some Silurian ostracodes. A study on the Mulde marl fauna of Gotland. Bull. Geol. Inst. Univ. Uppsala, 37: 1–42. Also: Publ. Palaeontol. Inst. Univ. Uppsala, 14: 1–42.

Martinsson, A., 1958. The submarine morphology of the Baltic Cambro-Silurian area. Bull. Geol. Inst. Univ. Uppsala, 28: 11–35. Also: Publ. Palaeontol. Inst. Univ. Uppsala, 21: 11–35.

Martinsson, A., 1960a. The origin of the crumina in beyrichiid ostracodes. Bull. Geol. Inst. Univ. Uppsala, 39: 12 pp. Also: Publ. Palaeontol. Inst. Univ. Uppsala, 25: 1–12.
Martinsson, A., 1960b. Two assemblages of polychaete jaws from the Silurian of Gotland. Bull. Geol. Inst. Univ. Uppsala, 39: 8 pp. Also: Publ. Palaeontol. Inst. Univ. Uppsala, 26: 1–8.
Martinsson, A., 1962a. Ostracodes of the family Beyrichiidae from the Silurian of Gotland. Bull. Geol. Inst. Univ. Uppsala, 41: 1–369. Also: Publ. Palaeontol. Inst. Univ. Uppsala, 41: 1–369.
Martinsson, A., 1962b. The concealed Silurian of the Baltic area. Geol. Fören. Stockholm Förh., 84: 539–541.
Martinsson, A., 1964. Phosphatic linings in bryozoan zooecia. Geol. Fören. Stockholm Förh., 86: 404–408. Also: Publ. Palaeontol. Inst. Univ. Uppsala, 68: 404–408.
Martinsson, A., 1965a. The Siluro-Devonian ostracode genus *Nodibeyrichia*, and faunally associated klorfrniinrd. Geol. Fören. Stockholm Förh., 87: 109–138.
Martinsson, A., 1965b. Remarks on the Silurian ostracode genus *Craspedobolbina* from the Baltic area and Britain. Geol. Fören. Stockholm Förh., 87: 314–325.
Martinsson, A., 1965c. Nielsens Grund och Hoburgs Bank i Östersjön. Geol. Fören. Stockholm Förh., 87: 326–336.
Martinsson, A., 1966a. Beyrichiacean ostracodes associated with the earliest Silurian vertebrates from Gotland. Geol. Fören. Stockholm Förh., 88: 327–339. Also: Publ. Palaeontol. Inst. Univ. Uppsala. 68: 327–339.
Martinsson, A., 1966b. *Ampirulum*, a new genus of beyrichiacean ostracodes. Geol. Fören. Stockholm Förh., 88: 68–74.
Martinsson, A., 1967. The succession and correlation of ostracode faunas in the Silurian of Gotland. Geol. Fören. Stockholm Förh., 89: 350–386.
Martinsson, A., 1968. The Ordovician – Silurian hiatus below Gotland. Geol. Fören. Stockholm Förh., 90: 561–563.
Martinsson, A., 1969. The series of the redefined Silurian System. Lethaia, 2: 153–161.
Martna, J., 1955. Studies on the *Macrourus* and Slandrom formations, 1. Shell fragment frequencies of the *Macrourus* formation and adjacent strata at Fjäcka, Gräsgård, and File Haidar. Geol. Fören. Stockholm Förh., 77: 229–256.
Maxwell, W.G.H., 1968. Atlas of the Great Barrier Reef. Elsevier, Amsterdam, 242 pp.
Mayer, A.G., 1914. The effects of temperature upon tropical marine animals. Papers Tortugas Lab., 6(1): 1–24.
Milliman, J.D., 1967. Carbonate sedimentation on Hogsty Reef, a Bahamian atoll. J. Sediment. Petrol., 37(2): 658–676.
Minato, M., 1961. Ontogenic study of some Silurian corals of Gotland. Stockholm Contrib. Geol., 8: 37–100.
Moberg, J.C., 1911. Historical-stratigraphical review of the Silurian of Sweden. Gotland. Sveriges Geol. Undersökn., Ser.C, 229: 40–57.
Moore, H.B., 1958. Marine Ecology. Wiley, New York, N.Y., 493 pp.
Motoda, S., 1940a. A study of growth rate in the massive reef coral *Goniastrea aspera* Verrill. Palao Trop. Biol. Station Studies, 2: 1–6.
Motoda, S., 1940b. The environment and the life of massive reef coral, *Goniastrea aspera* Verrill, inhabiting the reef flat in Palao. Palao Trop. Biol. Station Studies, 2: 61–104.
Munthe, H., 1902. Stratigrafiska studier öfver Gotlands Silurlager. Geol. Fören. Stockholm Förh., 24: 221–273. Also: Sveriges Geol. Undersökn., Ser.C, 192: 221–273.
Munthe, H., 1907. Om stratigrafien inom Syd-Gotlands Silur. Föredrag i Geol. Fören. i Stockholm jemte yttranden med anledning däraf. Geol. Fören. Stockholm Förh., 29: 130–140.
Munthe, H., 1910. On the sequence of strata within southern Gotland. Geol. Fören. Stockholm Förh., 32(5) (Congr. guide): 1397–1453.
Munthe, H., 1912. Om kalkalger och hydrozoer i Gotlands Silurlager. Geol. Fören. Stockholm Förh., 34(5): 472–473 (summary of lecture).

Munthe, H., 1915a. Om en nyligen slutförd diamantborrning i Siluren vid Burgsvik på Gotland. Geol. Fören. Stockholm Förh., 37: 346–349.
Munthe, H., 1915b. Oolit med kraftiga böljslagsmärken vid Klintebys på Gotland. Geol. Fören. Stockholm Förh., 37: 430–434.
Munthe, H., 1915c. Stora Karlsö topografiska och geologiska hufvuddrag. In: Stora Karlsö (Minnesskrift). Karlsö Jagt- och Djurskyddsförenings A.B., Stockholm, p.5.
Munthe, H., 1916. Meddelande om en detalj i Silurstratigrafien inom SÖ:a Gotland. Geol. Fören. Stockholm Förh., 38: 229–231.
Munthe, H., 1920. Strandgrottor och närstående geologiska fenomen i Sverige. Sveriges Geol. Undersökn., Ser.C, 303: 1–67. Also: Sveriges Geol. Undersökn., Årslok, 14(4): 1–67.
Munthe, H., 1921a. Sveriges raukar jämte exempel på pseudoraukar. Sveriges Geol. Undersökn., Ser.C, 303: 1–39. Also: Sveriges Geol. Undersökn., Årsbok, 14(5): 1–39.
Munthe, H., 1921b. Berggrunden (Silursystemet). With contributions by J.E. Hede and J.C. Moberg. In: H. Munthe, Beskrivning till kartbladet Burgvik jämte Hoburgen och Ytterholmen. Sveriges Geol. Undersökn., Ser.Aa, 152: 16–79.
Munthe, H., 1923. Till diskussionen om Gottlands Silurstratigrafi. Geol. Fören. Stockholm Förh., 45: 337–341.
Munthe, H., Hede, J.E. och Von Post, L., 1925. Gotlands geologi. Sveriges Geol. Undersökn., Ser.C, 331: 1–130. Also: Sveriges Geol. Undersökn., Årsbok, 18(1924)(3): 1–130.
Munthe, H. och Lohmander, H., 1946. Ur Gotlands geologiska utvecklingshistoria. In: B. Pettersson och K. Curry-Lindahl (Redaktörer), Natur på Gotland. Bokförlaget Svensk Natur, Stockholm, pp.14–30.
Murchison, R.I., 1839. The Silurian System founded on Geological Researches. London.
Murchison, R.I., 1846. On the Silurian and associated rocks in Dalecarlia, and on the succession from Lower to Upper Silurian in Smoland, Öland and Gothland, and in Scania. Quart. J. Geol. Soc. London, 3: 1–46.

Nathorst, A.G., 1884. Upplysningar till geologisk öfversigtskarta öfver Sverige. Sveriges Geol. Undersökn., Ser. Ba, 4: I–IV, 3–39 (Gotland, pp.23–24).
Nathorst, A.G., 1886. Anmärkningar med anledning af A. Hellands uppsats: "Svenska geologer om indsjöerne". Geol. Fören. Stockholm Förh., 8: 322–345.
Newell, N.D., 1955. Depositional fabric in Permian reef limestones. J. Geol., 63: 301–309.
Newell, N.D., Rigby, J.K., Fischer, A.G., Whiteman, A.J., Hickox, J.E. and Bradley, J.S., 1953. The Permian Reef Complex of the Guadelupe Mountains Region, Texas and New Mexico - A Study in Paleoecology. Freeman, San Francisco, Calif., 236 pp.
Newell, N.D., Purdy, E.G. and Imbrie, J., 1960. Bahamian oolitic sands. J. Geol., 68: 481–497.
Nilsson, B., 1913. Gotland. Chart accompanying G. de Geer, 1913. Oversiktskarta över södra Sveriges landformer. Sveriges Geol. Undersökn., Ser.Ba, 9.
Noreen, S.E., Holmström, R. och Svensson, S.A. (Redaktörer), 1959. Gotland. Allhem, Mälmo, 502 pp.

Odum, H.T., 1957. Biochemical deposition of strontium. Texas Univ., Inst. Marine Sci., 4: 39–114.
Öpik, A.A., 1934a. Über Klitamboniten. Acta Commentationes Univ. Tartuensis (Dorpatensis), A, 26(5): 1–239. Also: Publ. Geol. Inst. Tartu, 39: 1–239.
Öpik, A.A., 1934b. *Ristnacrinus*, a new Ordovician crinoid from Estonia. Acta Commentationes Univ. Tartuensis (Dorpatensis), 27(8): 3–7. Also: Publ. Geol. Inst. Tartu, 40: 3–7.

Peres, J.M., 1965. Réflexions sur les rapports entre l'écologie et la paléoécologie marines. Palaeogeography, Palaeoclimatol., Palaeoecol., 1(1): 51–68.
Pia, J., 1926. Pflanzen als Gesteinsbildner, Bornträger, Berlin, 355 pp.

Pilkey, O.H., Schnitker, D. and Pevear, D.R., 1966. Oolites on the Georgia continental shelf edge. J. Sediment. Petrol., 36(2): 462–467.
Pinsak, A.P. and Shaver, R.H., 1964. The Silurian formations of northern Indiana. Indiana Dept. Conserv., Geol. Surv. Bull., 32: 87 pp.
Pokorný, V., 1958. Grundzüge der zoologischen Mikropaläontologie, 1. Deutscher Verlag Wiss., Berlin, 582 pp.
Prokopovich, N., 1952. The origin of stylolites. J. Sediment. Petrol., 32: 212–220.
Purdy, E.G. and Imbrie, J., 1964. Carbonate Sediments, Great Bahama Bank. A Guidebook for Field Trip 2, Geol. Soc. Am. Conv., Miami, Fla., 58 pp.

Rasmussen, H.W., 1952. A new Silurian asteroid from Gotland, Sweden. Geol. Fören. Stockholm Förh., 74: 17–24.
Rauff, H., 1893–1894. Palaeospongiologie. Palaeontographica, 40–41.
Regnéll, G., 1940. Om faunan i *Planilimbata* kalkstenen vid Köping på Öland. Kgl. Fysiograf. Sällskap. Lund Förh., 10(1): 1–16. Also: Medd. Lunds Geol. Mineral. Inst., 80: 1–16.
Regnéll, G., 1942. Stratigraphical and paleontological remarks on the Lower Ordovician of central and northern Oland. With special reference to Köpings Klint. In: G. Regnéll et 10 al., Lunds Geologiska Fältklubb 1892–1942. Carl Bloms Boktryckeri, Lund, pp.168–184.
Regnéll, G., 1945. Non-crinoid Pelmatozoa from the Paleozoic of Sweden. A taxonomic study. Thesis, Univ. of Lund, 255 pp. Also: Medd. Lunds Geol. Mineral. Inst., 108: 1–255.
Regnéll, G., 1948. Ölands geologi. Öland, 1: 17–57.
Regnéll, G., 1956. Silurian echinoids from Gotland. Arkiv Mineral. Geol. Kgl. Svenska Vetenskapsakad., 2: 155–178.
Regnéll, G., 1960. The Lower Palaeozoic of Scania. In: G. Regnéll and J.E. Hede, The Lower Paleozoic of Scania. The Silurian of Gotland. Excursion guide 21st. Intern. Geol. Congr., Copenhagen, 1960, pp.3–43.
Reuter, G., 1885. Die Beyrichien der obersilurischen Diluvialgeschiebe Ostpreussens. Z. Deut. Geol. Ges., 37: 621–679.
Rice, C.M., 1954. Dictionary of Geological Terms. Rice, Princeton, N.J., 465 pp.
Roemer, F., 1856. Bericht von einer geologisch-paläontologischen Reise nach Schweden. Neues Jahrb. Mineral. Geol., 1856: 794–815.
Roemer, F., 1858. Die Versteinerungen der silurischen Diluvial-Geschiebe von Groningen in Holland. Neues Jahrb. Mineral., 1858: 257–272.
Rõõmusoks, A., 1960. Stratigraphy and paleogeography of the Ordovician in Esthonia. Intern. Geol. Congr., 21st., Copenhagen, 1960, Rept. Session, Norden, 7: 58–69.
Roosval, J., 1950. Den Gotländske Ciceronen. Svenska Turistföreningen, Stockholm, 193 pp.
Roosval, J., 1952. Gotlands Kyrkokonst. Generalstabens Litografiska Anstalts Förlag, Stockholm, 115 pp.
Rosvall, S. och Pettersson, B., 1951. Gotlands Orkidéer. Alb. Bonniers Förlag, Stockholm, 108 pp.
Rothpletz, A., 1891. Fossile Kalkalgen aus den Familien der Codiaceen und der Corallinieen. Z. Deut. Geol. Ges., 43: 295–322.
Rothpletz, A., 1908. Über Algen und Hydrozoen im Silur von Gotland und Oesel. Kgl. Svenska Vetenskapsakad. Handl., N.F., 43(5): 1–25.
Rothpletz, A., 1913. Uber die Kalkalgen, Spongiostromen und einige andere Fossilien aus dem Obersilur Gotlands. Sveriges Geol. Undersökn., Ser.Ca, 10: 1–56.
Runcorn, S.K., 1966. Corals as paleontological clocks. Sci. Am., 215(4): 26–33.
Rutten, M.G., 1956. Devonian reefs from Belgium: relations between geosynclinal subsidence and hinterland erosion. Am. J. Sci., 254: 685–692.
Rutten, M.G., 1958. Detailuntersuchungen an Gotländischen Riffen. Geol. Rundschau, 47: 359–384.

Sanders, J.E. and Friedman, G.M., 1967. Origin and occurrence of limestones. In:

G.V. Chilingar, H.J. Bissell and R.W. Fairbridge (Editors), Carbonate Rocks, A. Origin, Occurrence and Classification. Elsevier, Amsterdam, pp.169–265.

Savage, T.E. and Van Tuyl, F.M., 1919. Geology and stratigraphy of the area of Paleozoic rocks in the vicinity of Hudson and James Bays. Bull. Geol. Soc. Am., 30: 339–378.

Säve-Söderbergh, G., 1941. Remarks on "Downtonian" and related vertebrate faunas. Geol. Fören. Stockholm Forh., 63: 229–244.

Schieferdecker, A.A.G. (Editor), 1959. Geological Nomenclature. Royal Geol. Mining Soc. The Netherlands/Noorduyn, Gorinchem, 521 pp.

Schlüter, C., 1884. Uber *Astylospongia gothlandica* n. sp. Correspondenzbl. Naturh. Verein Rheinlande, 41: 79–80.

Schmidt, F., 1858. Untersuchungen über die Silurische Formation von Esthland, Nord-Livland und Oesel. Arch. Naturk. Liv-, Ehst- u. Kurlands, 1 Ser., 2: 1–248.

Schmidt, F., 1859. Beitrag zur Geologie der Insel Gotland, nebst einigen Bemerkungen über die untersilurische Formation des Festlandes von Schweden und die Heimat der norddeutschen silurischen Geschiebe. Arch. Naturk. Liv-, Ehst- u. Kurlands, 1 Ser., 2(6): 403–464.

Schmidt, F., 1881. Revision der ostbaltischen silurischen Trilobiten nebst geognostischer Übersicht des ostbaltischen Silurgebiets, 1. Mém. Acad. Imp. Sci. St. Pétersbourg, 7 Sér., 30(1).

Schmidt, F., 1882. On the Silurian (and Cambrian) strata of the Baltic provinces of Russia, as compared with those of Scandinavia and the Baltic Isles. Quart. J. Geol. Soc. London, 38: 514–536.

Schmidt, F., 1890. Bemerkungen über die Schichtenfolge des Silur auf Gotland. Neues Jahrb. Mineral. Geol., 1890(2): 249–266.

Schmidt, F., 1891. Einige Bemerkungen über das baltische Obersilur in Veranlassung der Arbeit des Prof. W. Dames über die Schichtenfolge der Silurbildungen Gotlands. Bull. Acad. Imp. Sci. St. Pétersbourg, Nouv., Sér., 34(2): 119–138.

Schot, E.H. and Park, W.C., 1968. Note on the formation of stylolites. Discussion on a paper by A.A. Manten. Geol. Mijnbouw, 47(2): 112–113.

Schwarzbach, M., 1949. Fossile Korallenriffe und Wegeners Drifthypothese. Naturwissenschaften, 36: 229–233.

Scrutton, C.T., 1965. Periodicity in Devonian coral growth. Palaeontology, 7(4): 552–558.

Seward, A.C., 1898. Fossil Plants: for Students of Botany and Geology. Cambridge Univ. Press, Cambridge, 452 pp.

Shaub, B.M., 1939. The origin of stylolites. J. Sediment. Petrol., 9: 47–61.

Shaub, B.M., 1949. Do stylolites develop before or after the hardening of the enclosing rock? J. Sediment. Petrol., 19: 26–36.

Shaub, B.M., 1955. Origin of stylolites: a discussion. J. Sediment. Petrol., 25: 60–66.

Shinn, E.A., 1966. Coral growth rate, an environmental indicator. J. Paleontol., 40(2): 233–240.

Shrock, R.R., 1929. The klintar of the Upper Wabash Valley in northern Indiana. J. Geol., 37: 17–29.

Shrock, R.R., 1939. Wisconsin Silurian bioherms (organic reefs). Bull. Geol. Soc. Am., 50: 529–562.

Silvester, R., 1965. Coral reefs, atolls and guyots. Nature, 207(4998): 681–688.

Smith, J., 1915. Upper Silurian Foraminifera of Gotland. Ann. Mag. Nat. Hist. London, Ser.8, 15: 301–309.

Söderberg, B.G., 1948. Gotland. Wahlström och Widstrand, Stockholm, 238 pp.

Soot-Ryen, H., 1964. Nuculoid pelecypods from the Silurian of Gotland. Arkiv Mineral. Geol. Kgl. Svenska Vetenskapsakad., 3(28): 489–519.

Spjeldnaes, N., 1950. On some vertebrates fossils from Gotland, with some comments on the stratigraphy. Arkiv Mineral. Geol. Kgl. Svenska Vetenskapsakad., 1(8): 211–218.

Spjeldnæs, N., 1951. Ontogeny of *Beyrichia jonesi* Boll. J. Paleontol., 25: 745–755.

Spjeldnæs, N., 1959. Silurian bentonites from Gotland, Sweden. Geol. Fören. Stockholm Förh., 81: 582–587.

Spjeldnæs, N., 1966. N.P. Angelin's work on fossil ostracodes. Geol. Fören. Stockholm Förh., 88: 407–409.
Springer, F., 1920. The Crinoidea Flexibilia. Smithsonian Inst. Publ., 2501: 1–486.
Squirrell, H.C. and Tucker, E.V., 1960. The geology of the Woolhope inlier, Herefordshire. Quart. J. Geol. Soc. London, 116: 139–185.
Stephenson, T.A. and Stephenson, A., 1933. Growth and asexual reproduction in corals. Great Barrier Reef Exp., Sci. Repts., Brit. Museum Nat. Hist., 3(7): 167–205.
Stetson, H.C., 1936. Geology and paleontology of the Georges Bank Canyons. Bull. Geol. Soc. Am., 47: 339–440.
Stockdale, P.B., 1922. Stylolites: their nature and origin. Indiana Univ. Studies, 9(55): 1–97.
Stockdale, P.B., 1943. Stylolites: primary or secondary. J. Sediment. Petrol., 13: 3–12.
Stolley, E., 1896. Über gesteinsbildende Algen und die Mitwirkung solcher bei der Bildung der skandinavisch-baltischen Silurablagerungen. Naturwiss. Wochenschr., Berlin, 1896: 173–178.
Stolley, E., 1897. Die silurische Algenfacies und ihre Verbreitung im skandinavisch-baltischen Silurgebiet. Schr. Naturwiss. Verein. Schleswig-Holstein, 11: 109–148.
Stolley, E., 1900. Geologische Mitteilungen von der Insel Sylt, 2. Archiv Anthropol. Geol. Scleswig-Holsteins, 4(1): 3–49.

Taft, W.H., 1967. Modern carbonate sediments. In: G.V. Chilingar, H.J. Bissell and R.W. Fairbridge (Editors), Carbonate Rocks, A. Origin, Occurrence and Classification. Elsevier, Amsterdam, pp.29–50.
Talent, J.A. and Philip, G.M., 1956. Bioherms and biostromes in the Middle Palaeozoic of Victoria. Resumenes de los Trabajos presentados, 20 Congr. Geol. Intern., Mexico, p.237.
Taugourdeau, P. et De Jekhowsky, B., 1964. Chitinozoaires siluriens de Gotland; comparaison avec les formes sahariennes. Rev. Inst. Franç. Pétrole Ann. Combust. Liquides, 19(7/8): 845–871.
Teichert, C., 1928. Stratigraphische und paläontologische Untersuchungen im unteren Gotlandium (Tamsal-Stufe) des westlichen Estland und der Insel Dagö. Neues Jahrb. Mineral. Geol., Beil.B, Geol. Paläontol., 60: 1–112.
Teichert, C., 1952. Fossile Riffe als Klimazeugen in Australien. Geol. Rundschau, 40: 33–38.
Textoris, D.A., 1966. Algal cap for a Niagaran (Silurian) carbonate mud mound of Indiana. J. Sediment. Petrol., 36(2): 455–461.
Textoris, D.A. and Carozzi, A.V., 1964. Petrography and evolution of Niagaran (Silurian) reefs, Indiana. Bull. Am. Assoc. Petrol. Geologists, 48: 397–426.
Thorell, T. and Lindström, G., 1885. On a Silurian scorpion from Gotland. Kgl. Svenska Vetenskapsakad. Handl., 21(II,9): 33 pp.
Thorslund, P., 1948. Silurisk bentonit från Gotland. Geol. Fören. Stockholm Förh., 70: 245–246.
Thorslund, P., 1958. Djupborrningen på Gotska Sandön. Geol. Fören. Stockholm Förh., 80: 190–197.
Thorslund, P., 1960. The Cambro-Silurian. In: N.H. Magnusson, P. Thorslund, F. Brotzen, B. Asklund and O. Kulling, Description to Accompany the Map of the Pre-Quaternary Rocks of Sweden. Sveriges Geol. Undersökn., Ser.Ba, 16: 69–110 (Gotland, pp.99–100).
Thorslund, P., 1968. The Ordovician – Silurian boundary below Gotland. Geol. Fören. Stockholm Förh., 90: 443–451.
Thorslund, P. and Westergård, A.H., 1938. Deep boring through the Cambro-Silurian at File Haidar, Gotland. Sveriges Geol. Undersökn., Ser.C, 415: 1–48. Also: Sveriges Geol. Undersökn., Arsbok, 32(5): 1–48.
Thorsteinsson, R. and Fortier, Y.O., 1953. Report of progress on the geology of Cornwallis Island, Arctic Archipelago, Northwest Territories. Geol. Surv. Can. Paper 53–24: 1–25.
Tripp, K., 1933. Die Favositen Gotlands. Palaeontographica, A, 79: 75–142.

Troedsson, G.T., 1931. Studies on Baltic fossil cephalopods, 1. On the nautiloid genus *Orthoceras*. Lunds Univ. Arsskr., N.F., Avd.2, 27(16): 5–36. Also: Kgl. Fysiograf. Sällskap. Handl., N.F., 42(16): 5–36.

Troedsson, G.T., 1932. Studies on Baltic fossil cephalopods, 2. Vertically striated or fluted orthoceracones in the Orthoceras Limestone. Lunds Univ. Årssko, N.F., Avd.2, 28(6): 5–38. Also: Kgl. Fysiograf. Sällskap. Handl., N.F., 43(6): 5–38.

Troedsson, G.T., 1936. The Ordovician–Silurian boundary in Europe, mainly in the Scandinavian–Baltic region. Intern. Geol. Congr., 16th., Washington, 1933, Repts., pp.495–504.

Trusheim, F., 1957. Über Halokinese und ihre Bedeutung für die strukturelle Entwicklung Norddeutschlands. Z. Deut. Geol. Ges., 109: 111–158.

Tullberg, S.A., 1883. Über die Schichtenfolge des Silurs in Schonen, nebst einem Vergleiche mit anderen gleichaltrigen Bildungen. Z. Deut. Geol. Ges., 35: 223–269.

Twenhofel, W.H., 1916. Expedition to the Baltic provinces of Russia and Scandinavia, 1914. Part 3. An interpretation of the Silurian section of Gotland. Bull. Mus. Comp. Zool. Harvard Coll., 56: 341–354.

Twenhofel, W.H., 1927. The geology of Anticosti Island. Can. Geol. Surv. Mem., 154: 1–107.

Twenhofel, W.H., 1950. Coral and other organic reefs in geologic column. Bull. Am. Assoc. Petrol. Geologists, 34: 182–202.

Ubaghs, G., 1956a. Recherches sur les crinoides camerata du Silurien de Gotland (Suède). Introduction générale et partie I: Morphologie et paléobiologie de *Barrandeocrinus sceptrum* Angelin. Arkiv Zool. Kgl. Svenska Vetenskaps-akad., Ser.2, 9(26): 515–550.

Ubaghs, G., 1956b. Recherches sur les crinoides camerata du Silurien de Gotland (Suède). Partie II: Morphologie et position systématique de *Polypeltes granulatus* Angelin. Arkiv Zool. Kgl. Svenska Vetenskapsakad., Ser. 2, 9(27): 551–572.

Ubaghs, G., 1958. Recherches sur les crinoides camerata du Silurien de Gotland (Suède). Partie III: Melocrinicae. Avec des remarques sur l'évolution des Melocrinidae. Arkiv Zool. Kgl. Svenska Vetenskapsakad., Ser.2, 11(16): 259–306.

Usdowski, H.E., 1963. Der Rogenstein des norddeutschen Unteren Buntsandsteins, ein Kalkoölith des marinen Faziesbereichs. Fortschr. Geol. Rheinland Westfalen, 10: 337–342.

Van Bemmelen, R.W., 1954. Mountain Building. Martinus Nijhoff, The Hague, 177 pp.

Van Hoepen, E.C.N., 1910. De Bouw van het Siluur van Gotland. Thesis, Technische Hoogeschool, Delft, 161 pp.

Vaughan, T.W., 1911. The physical conditions under which Paleozoic coral reefs were Bull. Geol. Soc. Am., 22: 238–252.

Vaughan, T.W., 1915a. The geologic significance of the growth rate of the Flridian and Bahaman shoal-water corals. Washington Acad. Sci. J., 5: 591–600.

Vaughan, T.W., 1915b. Growth rate of the Floridian and Bahaman shoal-water corals. Carnegie Inst. Washington, Year Book, 14: 221–231.

Vaughan, T.W., 1919. Corals and the formation of coral reefs. Annual Rept. to the Board of Regents of the Smithsonian Institution, 1917: 189–276.

Vaughan, T.W. and Wells, J.W., 1943. Revision of the suborders, families and genera of the Scleractinia. Geol. Soc. Am., Spec. Papers, 44: I–XV, 1–363.

Von Bromell, M., 1727. Lithographiae Suecanae. Acta literaria Sueciae, 2–3. Upsaliae et Stockholmiae.

Von Bubnoff, S., 1931. Zur Deutung des südbaltischen Kluftnetzes. Geol. Rundschau, 22: 306–308.

Von Bubnoff, S., 1943. Der Südrand Skandinaviens. Geol. Rundschau, 34: 197–208.

Von Bubnoff, S., 1952. Fennosarmatia. Geologische Analyse des europäischen Kerngebietes. Akademie Verlag, Berlin, 450 pp.

Von Kiesow, J., 1888. Uber Gotländische Beyrichien. Z. Deut. Geol. Ges., 40(1): 1–16.

Waern, B., 1948. The Silurian strata of the Kullatorp core. Bull. Geol. Inst. Univ. Upsala, 32: 433–474.
Waern, B., 1949. Palaeontology and stratigraphy of the Cambrian and lowermost Ordovician of the Bödahamn core. Bull. Geol. Inst. Univ. Uppsala, 34: 223–250.
Waern, B., 1960. Bentonit och fjärr konnektion. Anmärkningar med anledning av N. Spjeldnæs' uppsats "Silurian bentonites from Gotland" (Bentonite and long-distance correlation. Comments on N. Spjeldnæs' paper "Silurian bentonites from Gotland".). Geol. Fören. Stockholm Förh., 82: 342–344.
Waern, B., Thorslund, P. and Henningsmoen, G., 1948. Deep boring through Ordovician and Silurian strata at Kinnekulle, Västergötland. Bull. Geol. Inst. Univ. Upsala, 32: 337–474.
Walmsley, V.G., 1962. The identity and a new description of *Pteronitella retroflexa* (Wahlenberg) from the Upper Silurian of Gotland and the Welsh borders. Geol. Fören. Stockholm Förh., 84: 351–362.
Wedekind, R., 1927. Die Zoantharia Rugosa von Gotland (bes. Nordgotland). Nebst Bemerkungen zur Biostratigraphie des Gotlandium. Sveriges Geol. Undersökn., Ser.Ca, 19: 1–94.
Wedekind, R., 1932. Die biostratigraphische Gliederung des Gotlandiums von Gotland. Neues Jahrb. Mineral. Geol. Paläontol., Beil. B, Geol. Paläontol., 68: 391–402.
Wedekind, R. und Tripp, K., 1930. Die Korallenriffe Gotlands. Ein Beitrag zur Lösung des Problems von der Entstehung der Barrierriffe. Centr. Mineral. Geol., B, 1(8): 295–304.
Wells, J.W., 1963. Coral growth and geochronometry. Nature, 197(4871): 948–950.
Westerberg, A., 1895. En dolomitisk öfversilurisk kalksten på Gotland. Geol. Fören. Stockholm Förh., 17: 415–423.
Westergård, A.H., 1922. Sveriges olenidskiffer. Sveriges Geol. Undersökn., Ser.Ca, 18: 1–205.
Westergård, A.H., 1929. A deep boring through Middle and Lower Cambrian strata at Borgholm, Isle of Öland. Sver. Geol. Undersökn., Ser.C, 355: 1–19.
Westergård, A.H., 1936. *Paradoxides oelandicus* beds of Öland, with the account of a diamond boring through the Cambrian at Mossberga. Sveriges Geol. Undersökn., Ser.C, 394: 1–66. Also: Sveriges Geol. Undersökn., Årsbok, 30(1): 1–66.
Westergård, A.H., 1944. Borrningar genom alunskifferlagret pa Öland och i Östergötland 1943. Sveriges Geol. Undersökn., Ser.C, 463: 1–22. Also: Sveriges Geol. Undersökn., Årsbok, 38(5): 1–22.
Westergård, A.H., 1946. Agnostidea of the Middle Cambrian of Sweden. Sveriges Geol. Undersökn., Ser.C, 477: 1–140. Also: Sveriges Geol. Undersökn., Årsbok, 40: 1–140.
Westergård, A.H., 1947. Nya data rörande alunskifferlagret på Öland. Sveriges Geol. Undersökn., Ser.C, 483: 3–12. Also: Sveriges Geol. Undersökn., Årsbok, 41(2): 3–12.
White, E.I., 1950. The vertebrate faunas of the Lower Old Red Sandstone of the Welsh Borders. Bull. Brit. Museum, Geol., 1(3): 49.
Wiman, C., 1897a. Über Silurische Korallenriffe in Gotland. Bull. Geol. Inst. Univ. Uppsala, 3: 311–326.
Wiman, C., 1897b. Über den Bau einiger gotländischen Graptoliten. Bull. Geol. Inst. Univ. Uppsala, 3: 352–368.
Wiman, C., 1901. Über die Borkholmer Schicht im Mittelbaltischen Silurgebiet. Bull. Geol. Inst. Univ. Uppsala, 5(2): 151–222.
Wiman, C., 1933. Ein neuer Typus gotländischer Korallenriffe. Bull. Geol. Inst. Univ. Uppsala, 24: 279–286.
Winterer, E.L. and Murphy, M.A., 1960. Silurian reef complex and associated facies, central Nevada. J. Geol., 68: 117–139.
Wolf, K.H., 1960. Simplified limestone classification. Bull. Am. Assoc. Petrol. Geologists, 44: 1414–1416.
Wolf, K.H., Chilingar, G.V. and Beales, F.W., 1967. Elemental composition of carbonate skeletons, minerals, and sediments. In: G.V. Chilingar, H.J. Bissell and R.W. Fairbridge (Editors), Carbonate Rocks, B. Physical and Chemical Aspects. Elsevier, Amsterdam, pp.23–149.

Wood, A., 1948. "*Sphaerocodium*", a misinterpreted fossil from the Wenlock Limestone. Proc. Geol. Assoc. London, 59: 9–22.
Woodford, A.O., 1965. Historical Geology. Freeman, San Francisco, Calif., 512 pp.
Wright, A.D., 1965. The shell punctation of *Dicoelosia biloba* (Linnaeus). Geol. Fören. Stockholm Förh., 87: 548–556.

Yonge, C.M., 1935. Studies on the biology of Tortugas corals. Carnegie Inst. Washington Publ., 452: 185–208.
Yonge, C.M., 1940. The biology of reef-building corals. Great Barrier Reef Exp., Sci. Repts., Brit Museum Nat. Hist., 1(13): 353–391.

Ziegler, A.M., 1965. Silurian marine communities and their environmental significance. Nature, 207(4994): 270–272.

INDEX[1]

[1] English alphabetization is used throughout this index. Consequently, the Swedish Å and Ä are treated as A's, and the Ö is treated as O.

ENCLOSURES

Geological map of Gotland

Note: For reasons of readability, not every one of the large number of localities mentioned in the text is shown on this map. The more important sites and the names of all parishes from which localities have been mentioned are indicated. In the index, the parish is stated in which each locality is to be found. This still enables readers to establish the approximate position of each locality.

Map of the Holmhällar raukar field